Soil and Fertilizers

Advances in Soil Science

Series Editors
Rattan Lal
B. A. Stewart

For more information about this series, please visit: https://www.crcpress.com/
Advances-in-Soil-Science/book-series/CRCADVSOILSCI.

Soil and Fertilizers

Managing the Environmental Footprint

Edited by
Rattan Lal

CRC Press
Taylor & Francis Group
Boca Raton London New York

CRC Press is an imprint of the
Taylor & Francis Group, an **informa** business

CRC Press
Taylor & Francis Group
6000 Broken Sound Parkway NW, Suite 300
Boca Raton, FL 33487-2742

First issued in paperback 2022

ISBN-13: 978-1-138-60007-2 (hbk)
ISBN-13: 978-1-03-233621-3 (pbk)
DOI: 10.1201/9780429471049

Publisher's Note

The publisher has gone to great lengths to ensure the quality of this reprint but points out that some imperfections in the original copies may be apparent.

**Visit the Taylor & Francis Web site at
http://www.taylorandfrancis.com**

**and the CRC Press Web site at
http://www.crcpress.com**

Typeset in Times
by Lumina Datamatics Limited

Contents

Preface

Global demand for three principal fertilizers (N, P_2O_5, K_2O) was about 200 million Mg per year for 2019–2020. FAO (2019) has estimated that fertilizer demand (10^6 Mg) was 184 in 2015, 186 in 2016, 191 in 2017, 194 in 2018, and 198 in 2019, and estimates it will be 202 in 2020. Of the 186 million Mg used in 2018, 57% was N, 24% P, and 19% K. Further, N and P fertilizer rates have increased by a factor of 8 and 3 between 1961 and 2013, respectively. The conversion of unreactive N_2 in the atmosphere to reactive N as ammonia for fertilizer use has increased global cereal production from 736 million Mg in 1961 to 2980 million Mg in 2017 by a factor of 4.05. In comparison, the world population increased from 3.05 billion in 1961 to 7.55 billion in 2017 by a factor of 2.48. Therefore, per capita cereal production increased from 241 kg in 1961 to 395 kg in 2017. Fertilizer use, being an integral component of the Green Revolution package (along with improved varieties, irrigation, and pesticides), has increased global food production and strengthened food security. Furthermore, judicious use of fertilizers and other inputs in the agroecosystems are also critical to advancing Sustainable Development Goal #2 (Zero Hunger) by 2030. The number of people fed because of N fertilizer is estimated at 48% of the total in 2015, or 3.54 billion people. However, the increased use of fertilizers (along with pesticides and irrigation) has adversely impacted air quality (emission of N_2O); water quality (eutrophication, algal bloom, and anoxia of coastal ecosystems); soil health and functionality (e.g., acidification, elemental cycling, and biodiversity); and overall degradation of the environment.

As Barry Commoner stated in his book *The Closing Circle*, there is no such thing as a free lunch, and everything released into nature must go somewhere. Indeed, the environmental cost (footprint) of food produced through agricultural intensification is steep and rising because of fertilizers entering into and impacting all vital components of the environment – air, water, soil, vegetation, and biodiversity. Despite its adverse impacts, the use of fertilizers is expected to increase because the increase in demand for production of victuals (i.e., food, feed, fiber, and fuel) is rising for the growing and increasingly affluent world population.

Therefore, the objective of this 14-chapter volume is to deliberate the technological options that reconcile the need for increasing the global supply of essential victuals with the urgency of restoring and sustaining the environment. The book discusses several options, including the urgency of enhancing the use efficiency of fertilizers and reducing their leakage into the environment. The use efficiency of N is hardly 30% even in the best-case scenarios. Rather than increasing the amount or rate of fertilizer use, the strategy is to double the efficiency and thus reduce the input. Using the 4-R principle (Right Rate, Right Time, Right Method, and Right Place) can enhance the efficiency and minimize its leakage into the environment.

Whereas the rate of application of fertilizer must be increased in sub-Saharan Africa and elsewhere in developing countries, it must be reduced in China and other developed countries by enhancing the use efficiency. The latter can be increased by restoring soil health (i.e., increasing soil organic carbon, concentration soil structure, plant available water capacity, and soil biodiversity) and using the strategy of integrated nutrient management. Biological N fixation, inoculation for mycorrhiza, recycling of crop and animal residues, and integration of crops with trees and livestock are among other options of minimizing the dependence on chemical fertilizers. Use efficiency of fertilizers can also be enhanced by conserving soil and water, reducing risks of erosion by water and wind, increasing plant available water capacity, and increasing the root distribution into the subsoil.

This 14-chapter book addresses these options and the appropriate policy implications. In addition to case studies, the book also contains chapters on best management practices and policies for enhancing use efficiency of fertilizers and reducing their leakage into the environment.

I thank all the authors for their outstanding contributions and for sharing their knowledge and experience with the global soil science community. Preparation of the manuscript, involving collation and synthesis of the literature and interpretation of the data from context-specific situations, is a time-consuming process that requires dedication and commitment. Thanks are also due to the editorial staff of Taylor and Francis for their timely help and prompt responses to numerous questions and queries from the editors and authors. Special thanks are due to the office staff of the Carbon Management and Sequestration Center of The Ohio State University for providing support to the flow of the manuscripts between authors and editors and making valuable contributions. In this context, special thanks and appreciation are due to Ms. Abigail Hansen, Ms. Janelle Watts, Ms. Maggie Weidner-Willis, and Ms. Gabrielle Collier, who formatted the text and prepared the final submission. While it is a major challenge to list all those who made direct and indirect contributions toward completion of this book, it is important to thank contributions of all those who supported the completion of this volume. Nonetheless, it is important to build upon the contributions of all those who study the use of fertilizer in relation to soil processes and environmental impact and share the knowledge contained in this volume with others from around the world.

Rattan Lal
The Ohio State University
Columbus, OH

Editor

Rattan Lal, PhD, is a Distinguished University Professor of Soil Science and Director of the Carbon Management and Sequestration Center, The Ohio State University, and an Adjunct Professor at the University of Iceland. His current research focus is on climate-resilient agriculture, soil carbon sequestration, sustainable intensification, enhancing use efficiency of agroecosystems, and sustainable management of soil resources of the tropics. He has received honorary degrees of Doctor of Science from Punjab Agricultural University (2001); the Norwegian University of Life Sciences, Aas (2005); Alecu Russo Balti State University, Moldova (2010); Technical University of Dresden, Germany (2015); University of Lleida, Spain (2017); Gustavus Adolphus College, Saint Peter, Minnesota (2018); and PUCV, Valparaiso, Chile (2019). He was President of the World Association of Soil and Water Conservation (1987–1990), the International Soil Tillage Research Organization (1988–1991), and the Soil Science Society of America (2005–2007), and is President of the International Union of Soil Sciences (2017–2018). He was a member of the US National Climate Assessment and Development Advisory Committee–NCADAC (2010–2013); a member of the SERDP Scientific Advisory Board of the US Department of Energy (2011–2018); Senior Science Advisor to the Global Soil Forum of the Institute for Advanced Sustainability Studies, Potsdam, Germany (2010–2015); a member of the Advisory Board of Joint Program Initiative of Agriculture, Food Security and Climate Change (FACCE-JPI) of the European Union (2013–2016); and Chair of the Advisory Board of the United Nations Institute for Integrated Management of Material Fluxes and of Resources (UNU-FLORES), Dresden, Germany (2014–2019). Professor Lal was a lead author of IPCC (1998–2000). He has mentored 115 graduate students, 54 postdoctoral researchers, and hosted 174 visiting scholars. He has authored/coauthored 950 refereed journal articles and has written 20 and edited/coedited 75 books. For six years (2014–2019), Reuters Thomson listed him among the world's most influential scientific minds and as having citations of publications among the top 1% of scientists in agricultural sciences. He is recipient of the 2018 GCHERA World Agriculture Prize, the Glinka World Soil Prize, and the 2019 Japan Prize.

Contributors

Sampson Agiyin-Birikorang
International Fertilizer Development Center
  (IFDC)
Muscle Shoals, Alabama

Akram Ahmed
ICAR Research Complex for Eastern Region
Patna, India

H. Arunakumari
Division of Resource Management
ICAR-Central Research Institute for Dryland
  Agriculture (CRIDA)
Hyderabad, India

M. Vijay Sankar Babu
Department of Soil Science
Agricultural Research Station
Anantapur, India

Mohamed Badraoui
Institut National de la Recherche Agronomique
Rabat, Morocco

Andre Bationo
International Fertilizer Development Center
  (IFDC) North & West Africa
Accra, Ghana

Bhagwati Prasad Bhatt
ICAR Research Complex for Eastern Region
Patna, India

Prem Bindraban
International Fertilizer Development Center
  (IFDC)
Muscle Shoals, Alabama

Rachid Bouabid
Ecole National d'Agriculture de Meknès
Rabat, Morocco

Michelle Anne Bunquin
International Rice Research Institute
Metro Manila, Philippines

Pauline Chivenge
International Rice Research Institute
Metro Manila, Philippines

Yash P. Dang
School of Agriculture and Food Sciences
The University of Queensland
St. Lucia, Queensland, Australia

Ekwe Dossa
International Fertilizer Development Center
  (IFDC)
Lome, Togo

Yunying Fang
NSW Department of Primary Industries
Elizabeth Macarthur Agricultural Institute
Menangle, New South Wales, Australia

V. V. Gabhane
Department of Soil Science and Agricultural
  Chemistry
Dr. Panjabrao Deshmukj Krishi Vidyapeeth
Akola, India

Y. G. M. Galal
Soil and Water Research Department
Nuclear Research Center
Egyptian Atomic Energy Authority
Cairo, Egypt

Amit Gross
Department of Environmental Hydrology and
  Microbiology
Zuckerberg Institute for Water Research
Jacob Blaustein Institutes for Desert Research
Ben Gurion University of the Negev
Beersheba, Israel

Sumanta Kundu
Division of Resource Management
ICAR-Central Research Institute for Dryland
    Agriculture (CRIDA)
Hyderabad, India

C. Subha Lakshmi
Director Cell
ICAR-National Academy of
    Agricultural Research Management
    (NAARM)
Hyderabad, India

Rattan Lal
Carbon Management & Sequestration Center
School of Environment & Natural Resources
The Ohio State University
Columbus, Ohio

Narendra Kumar Lenka
Indian Institute of Soil Science
Indian Council of Agricultural Research
Bhopal, India

Sangeeta Lenka
Indian Institute of Soil Science
Indian Council of Agricultural Research
Bhopal, India

François Lompo
International Fertilizer Development Center
    (IFDC)
Ouagadougou, Burkina Faso

Vivian Mau
Department of Environmental Hydrology and
    Microbiology
Zuckerberg Institute for Water Research
Jacob Blaustein Institutes for Desert Research
Ben Gurion University of the Negev
Beersheba, Israel

Promil Mehra
NSW Department of Primary Industries
Elizabeth Macarthur Agricultural Institute
Menangle, New South Wales, Australia

Surajit Mondal
ICAR Research Complex for Eastern
    Region
Patna, India

Sushanta Kumar Naik
ICAR Research Complex for Eastern Region
Research Centre
Ranchi, India

K. C. Nataraj
Department of Soil Science
Agricultural Research Station
Anantapur, India

Sarah R. Noack
Hart Field-Site Group
Clare, South Australia

Kathryn Page
School of Agriculture and Food Sciences
The University of Queensland
St. Lucia, Queensland, Australia

Himanshu Pathak
National Rice Research Institute
Indian Council of Agricultural Research
Cuttack, India

Karnena Koteswara Rao
ICAR Research Complex for Eastern Region
Patna, India

Amit Roy
International Fertilizer Development Center
    (IFDC)
Muscle Shoals, Alabama
and
Sasakawa Africa Association
Addis Ababa, Ethiopia

Kazuki Saito
Africa Rice Center (AfricaRice)
Abidjan, Côte d'Ivoire

Pallab K. Sarma
Department of Soil Science
All India Coordinated Research Project for
    Dryland Agriculture
Assam Agricultural University
Biswanath Chariali, India

Ayyappa Sathish
Department of Soil Science
University Agricultural Sciences
GKVK
Bengaluru, India

Kirti Saurabh
ICAR Research Complex for Eastern Region
Patna, India

Sheetal Sharma
International Rice Research Institute
New Delhi, India

Darryl D. Siemer
ISU Nuclear Engineering
Idaho Falls, Idaho

Bhupinder Pal Singh
NSW Department of Primary Industries
Elizabeth Macarthur Agricultural Institute
Menangle, New South Wales, Australia
and
University of Newcastle
Callaghan, New South Wales, Australia
and
University of New England
Armidale, New South Wales, Australia

Bijay-Singh
Department of Soil Science
Punjab Agricultural University
Ludhiana, India

Upendra Singh
International Fertilizer Development Center
   (IFDC)
Muscle Shoals, Alabama

S. M. Soliman
Soil and Water Research Department
Nuclear Research Center
Egyptian Atomic Energy Authority
Cairo, Egypt

Brahim Soudi
Institut Agronomique et Vétérinaire Hassan II
Rabat, Morocco

Ch. Srinivasarao
Director Cell
ICAR-National Academy of Agricultural
   Research Management (NAARM)
Hyderabad, India

John Wendt
East and Southern Africa Division
International Fertilizer Development Center
   (IFDC)
Muscle Shoals, Alabama

1 Effects of Fertilizers on Soil Quality and Functionality

Rattan Lal

CONTENTS

1.1 INTRODUCTION

1.1.1 GLOBAL FERTILIZER USE

Global fertilizer supply has increased drastically since 1960. The present global fertilizer supply ($N + P_2O_5 + K_2O$, 10^6 Mg) for 2015–2020 (Table 1.1) indicates an increasing trend of 1.46%/yr over the six-year period. However, the rate of increase in global fertilizer consumption was 5.5%/yr between 1960 and 1990 and 2.9%/yr between 1990 and 2020. Among all fertilizers, the annual rate of growth of N was 7.1% between 1960 and 1990, compared with 4.4% between 1990 and 2020 (Table 1.2). The annual rate of fertilizer growth between 1990 and 2020 has been especially high for Asia (i.e., China and India) but has lagged behind in sub-Saharan Africa (SSA). Thus, the agronomic yield of crops in SSA has also stagnated, and the small rate of growth in some regions is much lower than its technical potential (Lal 2017). While the growth rate and the total global consumption of fertilizers have increased, the use efficiency of fertilizer has remained low, especially that of the nitrogenous (N) fertilizer. In developing countries and also in emerging economies (i.e., India and China), the use efficiency of N fertilizers can be as low as 30%. Therefore, a large proportion of the reactive N is leaked into the environment (water and air) with dire consequences.

TABLE 1.1

Total Global Grain Production from 2008–2009 to 2018–2019

Year	World Population (Billions)	Total Grain Production (Million Tons)	Per Capita Grain Production (Kg)
2008/09	6.789	2241.6	330
2009/10	6.872	2241.5	332
2010/11	6.957	2200.4	316
2011/12	7.041	2314.4	328
2012/13	7.126	2266.2	318
2013/14	7.211	2474.7	367
2014/15	7.295	2532.0	347
2015/16	7.380	2058.0	278
2016/17	7.464	2186.0	292
2017/18	7.547	2142.0	283
2018/19	7.631	2120.0	277
2019/20	7.713		
2020/21	7.795		

Source: FAO (2017a) and UN (2019).

TABLE 1.2

World's Fertilizer Use from 1960 to 2020

Fertilizer	Fertilizer Use (10^6 Mg/yr)				Annual Growth (%)	
	1959/60	1989/90	2018	2020	1960–1990	1990–2020
Nitrogen	9.5	79.2	169.0	170.8	7.1	4.4
Phosphorus	9.7	37.5	51.2	53.1	4.5	1.1
Potash	8.1	26.9	47.2	49.5	4.0	1.2
Total	27.4	143.6	267.4	273.4	5.5	2.9

Source: Adapted and recalculated from Bumb and Baanante (1996, Columns 2, 3, and 6) and FAO (2017b, Columns 4, 5, and 6).

1.1.2 THE SOIL-WATER-AIR-QUALITY NEXUS

The low use efficiency of fertilizer has strong adverse impacts on environmental quality. There exists a strong interconnectivity between soil, water, and air (Figure 1.1). Thus, decline in the quality of one leads to decline in the quality of the other two. Soil degradation, both due to natural and anthropogenic factors, is a serious global issue. It implies a decline in quality and functionality with the attendant weakening of essential ecosystem services or even creation of some disservices, and is a global issue of the twenty-first century with severe ramifications. Already 23% of ice-free land is prone to degradation (Bai et al. 2008). Among principal types of degradation (Figure 1.2), soil physical degradation (i.e., decline in soil structure and accelerated soil erosion by water and wind) are among the ramifications of the Anthropocene (Crutzen and Steffen 2003). Soil erosion by water is causing global transport of sediments at the rate of 36 Gt per year (Walling 2008). Heavy sediment load has strong implications for water quality (e.g., nonpoint source pollution and algal bloom) and air quality (emission of greenhouse gases, especially those of CH_4 and N_2O, the particulate matter), and is associated with the increase in the frequency and intensity of dust storms caused by acceleration of the wind erosion. Soil degradation, and the attendant decline in provisioning of

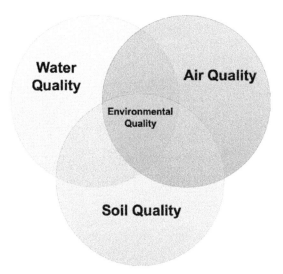

FIGURE 1.1 Constituents of environment quality: Soil, water, and air.

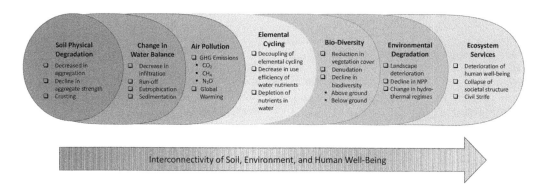

FIGURE 1.2 The cascading effects of soil degradation on the deterioration of water, air, and plants, biodiversity, environment, and human well-being.

ecosystem services and even generation of some severe disservices, also adversely impacts the use efficiency of fertilizers and uptake of nutrients and water by plant roots. Thus, reducing the risks of soil degradation and restoring degraded soils and desertified lands are high priorities. Soil degradation exacerbates contamination/eutrophication of water and pollution of air, because of the strong interconnectivity among them that leads to the cascading effect (Figures 1.1 and 1.2). Indeed, the quality of soil also determines those of water and air, and vice versa. Furthermore, quality of all three is a strong determinant of the fertilizer use efficiency, and of the use efficiency of nutrients applied and inherent in the soil.

1.1.3 OBJECTIVES AND EXPECTED OUTPUT

The objective of this book, and specifically that of this chapter, is to deliberate the interrelationship between the use of chemical fertilizers on the properties and processes of soil, and the interrelationship between soil properties and processes on the use efficiency of fertilizer in general and of essential plant nutrients in particular. The chapter and the book are based on the hypotheses that (1) fertilizer demand can be reduced and efficiency enhanced by reducing the processes of soil

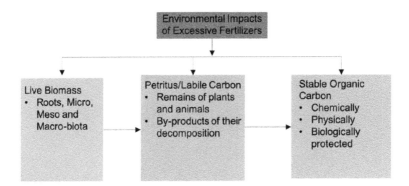

FIGURE 1.3   Constituents of soil organic carbon.

degradation (Figure 1.3); (2) an increase in soil organic carbon (SOC) concentration in strongly and severely depleted soils (SOC concentration < 1 g/kg in 0–20 cm depth) would increase fertilizer and nutrient use efficiency; (3) integrated nutrient management (INM), that is, judicious combination of organic amendments and chemical fertilizers, is the best strategy to sustain productivity and reduce the environmental footprint of agroecosystems; and (4) carbon, nitrogen, phosphorus, sulfur (CNPK) is a better recommendation for plant nutrients and soil fertility management than that of nitrogen, phosphorus, sulfur (NPK).

1.2   GLOBAL FOOD DEMAND

Global food demand is increasing because of the growing and increasingly affluent world population. Starting from the beginning of world agriculture about 10 to 20 millenia ago, the world population is now projected to reach 7.8 billion in 2020. The world population is projected to reach 8.5 billion by 2030, 9.7 billion by 2050, and 10.9 billion by 2100 (UN 2019). It is argued that between 2005 and 2050, global grain production may have to be increased by 60% and as much as doubled in some developing countries. That being the case, there is additional demand for arable land area, fertilizers, pesticides, irrigation water, and energy use (Alexandratos and Bruinsma 2012). However, rather than appropriating additional resources to be used for agricultural production, a better strategy would be to narrow the yield gap from existing lands by adopting proven technology so that resources saved (i.e., land, water, and energy) can be saved for nature (Lal 2016, 2018).

1.2.1   TRENDS IN FOOD GRAIN PRODUCTION

Global food grain production has increased drastically since 1960, and the annual rate of increase in grain production has, on average, exceeded that of the rate of population growth. Global cereal production increased by 280% between 1961 and 2014, while the world population increased by 136% (Ritchie and Roser 2020). The increase in cereal production was caused by an increase in average cereal yield of 175% between 1961 and 2014. Consequently, per capita cereal production has also increased since the 1960s. From 2008 to 2013, however, per capita grain production was almost constant at about 330 kg and then declined to ~280 kg between 2013 and 2018. The declining trend in per capita grain production since 2013, a major cause of concern, may be attributed to the drought/flood syndrome, an increase in susceptibility to degradation, a decrease in use efficiency of water and nutrients, and an increase in the incidence of plant diseases. Nonetheless, the present level of food production is adequate to meet the food demand of the projected world population of 9.7 billion by 2050 (Berners-Lee et al. 2018).

Despite the positive trends, there are two issues that require an objective and a critical evaluation. One, the increase in cereal yield is caused by an increase in the use of fertilizers and other chemicals with a strong environmental footprint. Two, the increase in intensification of agroecosystems is also associated with an increase in risks of soil degradation. Indeed, the decline in per capita food grain production since 2013 may be an indication of the widening of the gap between population and food grain production because of soil degradation. Further, the number of hungry people in the world has also increased since 2014, and has increased to 821 million by 2017 (FAO et al. 2017). In addition to malnutrition, the incidence of undernourishment (~2 billion people) is also on the rise. While the rate of fertilizer input may be increasing, especially that of N, the rate of food grain production is either stagnant or declining. Thus, an input of chemical fertilizers alone is not adequate to meet the demands of the ever-increasing world population for food and other victuals.

1.3 FERTILIZER USE AND CROP RESPONSE

1.3.1 Adverse Impacts on Soil, Water, and Air

The use of chemical fertilizers has increased since the 1960s (Table 1.2). The annual rate of application of chemical fertilizer (10^6 Mg per year) for 1959/60 and 2020, respectively, was 9.5 and 170.8 for nitrogen (N), 9.7 and 53.1 for phosphate (P_2O_5), 8.1 and 49.5 for potash (K_2O), and 27.4 and 273.4 in total ($N+P_2O_5+K_2O$). Thus, the total fertilizer input globally increased by a factor of 10 from 27.4 million Mg per year in 1959/60 to 273.4 million in 2020 (Table 1.2). Along with the input of herbicides and pesticides, the ever-increasing rate of the input of fertilizers has a strong adverse impact on the environment (Savci 2012). Some environmental impacts of excessive and indiscriminate use of fertilizers are outlined in Figure 1.4. Therefore, the strategy is to use chemical fertilizers and biocides judiciously and prudently and in a targeted manner, so that the leakage into the environment is minimal. Experiments on a temperate grasslands in the Hebei Province of China showed that application of N fertilizer increased the C:N ratio and the total amount of the above-ground biomass. However, it decreased the diversity and evenness of the plant community.

1.3.2 Soil Biodiversity and Enzymes

The long-term use of chemical fertilizer reduced the biodiversity and abundance of bacteria (Zhou et al. 2015), and the adverse effect was greater and more concentrated than that of the less-concentrated fertilizers. Continuous application of inorganic fertilizers can impact microbial community and soil health (Kumar et al. 2018). Manjunath et al. (2018) reported that inorganic control recorded the lowest values of microbial diversity indices. In the Indian mid-Himalayas, Mahanta et al. (2017) observed that industrial

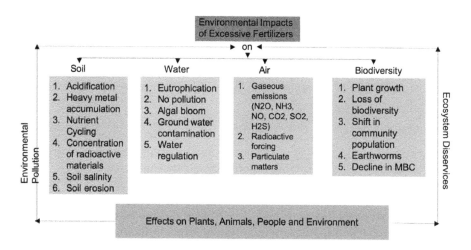

FIGURE 1.4 Environmental impacts of heavy and indiscriminate use of chemical fertilizers.

agriculture diminished soil microbial biodiversity. In a pot culture study, Hassan et al. (2013) showed that sunflowers grown in manure-amended treatments had a distinct arbuscular mycorrhizal (AM) fungal community structure with an abundance of *Rhizophagus intraradices* (B2). Beauregard et al. (2013) observed that different organic or mineral sources of P had less effect on the AM fungal communities because these are temporally stable. A study in the Mediterranean arable system indicated that the mites/collembolans ratio was higher in conventionally managed soil than in organically managed soil, probably because of rapid tillage performed in organic fields. Nelson and Spaner (2010) observed that management systems characterized by reduced tillage, diverse crop rotations and intercrops, and a low rate of input of chemical fertilizers and pesticides contained a large and diverse soil microbial community with mycorrhizal fungi. Van Eekeren et al. (2009) reported that available N in soil (NO_3, NH_4^+) strongly impacts the population of nematodes and microorganisms. A lower concentration of N in soil from inorganic fertilizers was associated with a higher root mass and higher population of herbivorous nematodes. In contrast, higher concentrations of N from organic fertilizers increased bacterial activity and bacterivorous nematodes. Further, the number of earthworm burrows was higher in treatments with organic fertilizers. Van Eekeren et al. (2009) concluded that organic fertilizers enhance the ecosystem service of water regulation more than inorganic fertilizers. Indiscriminate use of fertilizers and pesticides reduces the population of invertebrates and vertebrate groups (McLaughlin and Mineau 1995).

1.3.3 Effects on Soil Enzyme

Alternations in soil microbial community also result in a change in concentration of soil enzyme. A 29-year study in the Loess Plateau of Northwest China indicated that soil enzymes (e.g., dehydrogenase, urease, alkaline phosphatase, invertase, and glomalin) were impacted by the balanced application of fertilizer nutrients and organic manure (Hu et al. 2014). Goyal et al. (1999) observed that combined use of manure and inorganic fertilizer increased the activity of urase and phosphatase.

1.3.4 Soil Physical Properties

Long-term application of chemical fertilizers can strongly impact soil structure and its attendant physical, hydrological, and thermal properties. Based on a nine-year field experiment on a fine loamy mixed hyperthermic udic (Ustrochrept), Sarkar et al. (2003) reported that the addition of organic manures increased aggregate stability, moisture retention capacity, and infiltration rate while decreasing bulk density and improving porosity. In contrast, application of inorganic fertilizers produced 10%–17% higher grain yield of rice but not that of lentils. In Nigeria, Obi and Ebo (1995) observed that input of poultry manure increased SOC, total porosity, infiltration rate, and available water capacity.

Improvement in soil physical properties through the use of organic fertilizers is also associated with an increase in SOC concentration. In an 11-year study, Goyal et al. (1999) reported that concentration of SOC increased with the application of farmyard manure and green manure (with *Sesbania*). Further, microbial biomass carbon also increased with manuring. Palmer et al. (2017) reported that an increase in SOC concentration increased water-holding capacity.

1.3.5 Soil Organic Carbon Concentration

Concentration of SOC, a key indicator of soil quality, is strongly improved by manuring. A modeling study conducted in a 30-year experiment in Sweden showed an increase of 30% in SOC and N concentration in a manured treatment. The increase in SOC was attributed to the rate of input of biomass-C, its lignin content, and C/N ratio. There was also a positive effect of input of N on SOC. Palmer et al. (2017) observed that an increase in SOC may enhance emission of NO_2, but not necessarily increase agronomic yield if the availability of N is not limiting. Palmer and colleagues concluded that an increase in SOC by manuring may accentuate N cycling with both positive and negative effects on ecosystem services.

1.3.6 PRODUCTIVITY EFFECTS OF ORGANIC VERSUS INORGANIC FERTILIZERS

In general, organic sources result in lower agronomic productivity than inorganic fertilizers. A field experiment in Denmark from 1997 to 2008 under diverse soil and climatic conditions indicated a yield gap of 15% and 21% in systems with and without grass/clover, respectively, and the use of catch crops reduced the yield gap by 3 and 5 percentage points in the respective systems. However, the agronomic efficiency of N in manure plots was greater than that in plots receiving inorganic fertilizers.

1.4 HOLISTIC MANAGEMENT TO REDUCE THE INPUT OF CHEMICAL FERTILIZERS

Indiscriminate and excessive use of chemical fertilizers and biocides can adversely impact soil quality and the environment. Therefore, a principal strategy is to reduce the dependency of cropping systems on the import of resources while reducing the environmental footprint. Reduction in the use of inorganic fertilizers is feasible with the adoption of a holistic and systems-based approach (Ning et al. 2017). However, continuous application of manure can also lead to accumulation of Zn, Cd, and Cr in soil. With judicious management of soil and crops, however, chemical fertilizers could be completely replaced by organic amendments while enhancing SOC stocks and improving crop yield (Li et al. 2017). In the Northeast China Plain, NPK fertilizers could be at least partially replaced by manure (Li et al. 2017). Further, C-rich biofertilizers enhance soil quality by improving soil biodiversity and strengthening ecosystem resilience against stress. Liu et al. (2016) recommend retention of crop residues as an effective strategy to enhance soil biodiversity in agroecosystems and strengthen the ecological resilience of croplands. Analysis of a 50-year experiment in Switzerland showed that SOC concentration was increased by retention of crop residue mulch and manuring as compared with the use of inorganic fertilizer. Soil microbial population was also enhanced by organic amendments. Application of farmyard manure (FYM) also increased crop yield by 3.5% as compared to mineral fertilizers (Blanchet et al. 2016). A 42-year study on a flooded-rice field in India showed that grain yield was increased significantly (+74%) compared with the application of FYM + NPK (Bhattacharyya et al. 2015).

Improvements in soil properties and agronomic yield by manuring are often associated with increases in SOC concentration in the rootzone. In the Loess Plateau of China, Zhang et al. (2015) reported that with 20 years of stimulated manuring, SOC stock in the region increased by 1.15 PgC. Improvement in SOC was associated with a higher annual precipitation, which also increased agronomic yield, and supplemental irrigation. In central Italy, Migliorini et al. (2014) reported that after 16 years of comparison, comparable grain yield was observed among three agroecosystems. However, the young organic agroecosystem was the most effective in terms of the long-term increase of SOC storage (+13%), but the oldest organic agroecosystem was the most efficient in soil N storage (+9%). Migliorini and colleagues (2014) concluded that adoption of organic farming methods can enhance environmental sustainability in stockless arable systems. Walsh et al. (2012) observed that crop yields were equal or enhanced with the application of digestate generated from the anaerobic digestion of slurry. Walsh and colleagues concluded that replacing inorganic fertilizer with digestate may maintain grassland productivity but with a minimal impact on the environment. Thorup-Kristensen et al. (2012) explored some novel approaches to reduce the reliance on input of external resources. Adoption of fertility-building crops could strongly reduce the import of inputs. Efthimiadou et al. (2010) observed that combined organic and inorganic fertilization enhanced SOC concentration and improved yield of maize. In Pingliang, Gansu, China, Liu et al (2010) also observed that combined application of organic manure and inorganic fertilizers had the most beneficial effects on grain yield and soil quality. Similar observations were reported from a three-year experiment on sugarcane in India. Singh et al. (2007) observed that strategic planning involving an integrated application of manures with inorganic chemicals reduced dependence on chemical fertilizers and sustained productivity. On the basis of long-term soil fertility experiments in Gongzhuling Northeast China Plain, Li et al. (2017) concluded that NPK fertilizers could be at least partially replaced by manure to sustain high yield and enhance the SOC stock to the threshold level of 42 MgC/ha.

Therefore, it is possible to reconcile the need for advancing food and nutritional security with the absolute necessity of improving the environment by a strategic planning of combining the use of manure and organic amendments with the supplemental use of chemical fertilizers. Integrated nutrient management is the best option.

REFERENCES

Alexandratos, N., and J. Bruinsma. 2012. "World Agriculture towards 2030/2050: The 2012 Revision." ESA Working Paper 12-03, Agricultural Development Economic Division, Food and Agriculture Organization of the United Nations (FAO), Rome, Italy.

Bai, Z. G., D. L. Dent, L. Olsson, and M. E. Schaepman. 2008. "Proxy Global Assessment of Land Degradation." *Soil Use and Management* 24(3): 223–234.

Beauregard, M. S., M.-P. Gauthier, C. Hamel, T. Zhang, T. Welacky, C. S. Tan, and M. St-Arnaud. 2013. "Various Forms of Organic and Inorganic P Fertilizers Did Not Negatively Affect Soil- and Root-Inhabiting AM Fungi in a Maize–Soybean Rotation System." *Mycorrhiza* 23(2): 143–154.

Berners-Lee, M., C. Kennelly, R. Watson, and C. N. Hewitt. 2018. "Current Global Food Production Is Sufficient to Meet Human Nutritional Needs in 2050 Provided There Is Radical Societal Adaptation." *Elementa Science Anthropocene* 6: 52.

Bhattacharyya, P., A. K. Nayak, M. Shahid, et al. 2015. "Effects of 42-Year Long-Term Fertilizer Management on Soil Phosphorus Availability, Fractionation, Adsorption–Desorption Isotherm and Plant Uptake in Flooded Tropical Rice." *Crop Journal* 3(5): 387–395.

Blanchet, G., K. Gavazov, L. Bragazza, and S. Sinaj. 2016. "Responses of Soil Properties and Crop Yields to Different Inorganic and Organic Amendments in a Swiss Conventional Farming System." *Agriculture, Ecosystems & Environment* 230: 116–126.

Bumb, B. L., and C. A. Baanante 1996. "World Trends in Fertilizer Use and Projections." International Food Policy Research Institute 2020 Brief 38, Washington, DC, October 1996.

Crutzen, P. J., and W. Steffen. 2003. "How Long Have We Been in the Anthropocene Era?" *Climatic Change* 61(3): 251–257.

Efthimiadou, A., D. Bilalis, A. Karkanis, and B. Froud-Williams. 2010. "Combined Organic/Inorganic Fertilization Enhance Soil Quality and Increased Yield, Photosynthesis and Sustainability of Sweet Maize Crop." *Australian Journal of Crop Science* 4: 722–729.

FAO (Food and Agriculture Organization). 2017a. *Statistical Yearbook of the Food and Agriculture Organization*. Rome, Italy: FAO.

FAO (Food and Agriculture Organization). 2017b. *World Fertilizer Trends and Outlook to 2020*. Rome, Italy: FAO.

FAO, IFAD, UNICEF, WFP, and WHO 2017. *The State of Food Insecurity and Nutrition in the World 2017. Building Resilience for Peace and Food Security*. Rome, Italy: FAO.

Goyal, S., K. Chander, M. C. Mundra, and K. K. Kapoor. 1999. "Influence of Inorganic Fertilizers and Organic Amendments on Soil Organic Matter and Soil Microbial Properties under Tropical Conditions." *Biology and Fertility of Soils* 29(2): 196–200.

Hassan, S. E. D., A. Liu, S. Bittman, T. A. Forge, D. E. Hunt, M. Hijri, and M. St-Arnaud. 2013. "Impact of 12-Year Field Treatments with Organic and Inorganic Fertilizers on Crop Productivity and Mycorrhizal Community Structure." *Biology and Fertility of Soils* 49(8): 1109–1121.

Hu, W., Z. Jiao, F. Wu, Y. Liu, M. Dong, X. Ma, T. Fan, L. An, and H. Feng. 2014. "Long-Term Effects of Fertilizer on Soil Enzymatic Activity of Wheat Field Soil in Loess Plateau, China." *Ecotoxicology* 23(10): 2069–2080.

Kumar, U., A. Kumar Nayak, M. Shahid, et al. 2018. "Continuous Application of Inorganic and Organic Fertilizers over 47 Years in Paddy Soil Alters the Bacterial Community Structure and Its Influence on Rice Production." *Agriculture, Ecosystems & Environment* 262: 65–75.

Lal, R. 2016. "Feeding 11 Billion on 0.5 Billion Hectare of Cropland." *Food and Energy Security Journal* 5(4): 239–251.

Lal, R. 2017. "Improving Soil Health and Human Protein Nutrition by Pulses-Based Cropping Systems." *Advances in Agronomy* 145: 167–204.

Lal, R. 2018. "Promoting '4 per Thousand' and 'Adapting African Agriculture' by South-South Cooperation: Conservation Agriculture and Sustainable Intensification." *Soil & Tillage Research* 188: 27–34.

Li, H., W. Feng, X. He, P. Zhu, H. Gao, N. Sun, and M. Xu. 2017. "Chemical Fertilizers Could Be Completely Replaced by Manure to Maintain High Maize Yield and Soil Organic Carbon (SOC) When SOC Reaches a Threshold in the Northeast China Plain." *Journal of Integrative Agriculture* 16(4): 937–946.

Liu, E., C. Yan, X. Mei, W. He, S. H. Bing, L. Ding, Q. Liu, S. Liu, and T. Fan. 2010. "Long-Term Effect of Chemical Fertilizer, Straw, and Manure on Soil Chemical and Biological Properties in Northwest China." *Geoderma* 158(3–4): 173–180.

Liu, T., X. Chen, F. Hu, W. Ran, Q. Shen, H. Li, and J. K. Whalen. 2016. "Carbon-Rich Organic Fertilizers to Increase Soil Biodiversity: Evidence from a Meta-Analysis of Nematode Communities." *Agriculture, Ecosystems & Environment* 232: 199–207.

Mahanta, D., R. Bhattacharyya, P. K. Mishra, et al. 2017. "Influence of a Six-Year Organic and Inorganic Fertilization on the Diversity of the Soil Culturable Microorganisms in the Indian Mid-Himalayas." *Applied Soil Ecology* 120: 229–238.

Manjunath, M., U. Kumar, R. B. Yadava, A. B. Rai, and B. Singh. 2018. "Influence of Organic and Inorganic Sources of Nutrients on the Functional Diversity of Microbial Communities in the Vegetable Cropping System of the Indo-Gangetic Plains." *Comptes Rendus Biologies* 341(6): 349–357.

McLaughlin, A., and P. Mineau. 1995. "The Impact of Agricultural Practices on Biodiversity." *Agriculture, Ecosystems & Environment* 55(3): 201–212.

Migliorini, P., V. Moschini, F. Tittarelli, C. Ciaccia, S. Benedettelli, C. Vazzana, and S. Canali. 2014. "Agronomic Performance, Carbon Storage and Nitrogen Utilisation of Long-Term Organic and Conventional Stockless Arable Systems in Mediterranean Area." *European Journal of Agronomy* 52: 138–145.

Nelson, A. G., and D. Spaner. 2010. "Cropping Systems Management, Soil Microbial Communities, and Soil Biological Fertility." In *Genetic Engineering, Biofertilisation, Soil Quality and Organic Farming*, edited by E. Lichtfouse, 217–242. Dordrecht, the Netherlands: Springer.

Ning, C., P. Gao, B. Wang, W. Lin, N. Jiang, and K. Cai. 2017. "Impacts of Chemical Fertilizer Reduction and Organic Amendments Supplementation on Soil Nutrient, Enzyme Activity and Heavy Metal Content." *Journal of Integrative Agriculture* 16(8): 1819–1831.

Obi, M. E., and P. O. Ebo. 1995. "The Effects of Organic and Inorganic Amendments on Soil Physical Properties and Maize Production in a Severely Degraded Sandy Soil in Southern Nigeria." *Bioresource Technology* 51(2–3): 117–123.

Palmer, J., P. J. Thorburn, J. S. Biggs, et al. 2017. "Nitrogen Cycling from Increased Soil Organic Carbon Contributes Both Positively and Negatively to Ecosystem Services in Wheat Agro-Ecosystems." *Frontiers in Plant Science* 8: 731.

Ritchie, H., and M. Roser. 2020. Crop Yields. Published online at ourworldindata.org. Retrived from https://ourworldindata.org/crop-yields (online resource).

Sarkar, S., S. R. Singh, and R. P. Singh. 2003. "The Effect of Organic and Inorganic Fertilizers on Soil Physical Condition and the Productivity of a Rice–Lentil Cropping Sequence in India." *Journal of Agricultural Science* 140(4): 419–425.

Savci, S. 2012. "Investigation of Effect of Chemical Fertilizers on Environment." *APCBEE Procedia* 1: 287–292.

Singh, K. P., A. Suman, P. Singh, and M. Lal. 2007. "Yield and Soil Nutrient Balance of a Sugarcane Plant–Ratoon System with Conventional and Organic Nutrient Management in Sub-tropical India." *Nutrient Cycling in Agroecosystems* 79: 209–219.

Thorup-Kristensen, K., D. B. Dresbøll, and H. L. Kristensen. 2012. "Crop Yield, Root Growth, and Nutrient Dynamics in a Conventional and Three Organic Cropping Systems with Different Levels of External Inputs and N Re-cycling through Fertility Building Crops." *European Journal of Agronomy* 37(1): 66–82.

UN Department of Economic and Social Affairs. 2019. *World Population Prospects 2019: Highlights*. ST/ESA/SER.A/423. New York: UN.

Van Eekeren, N., H. de Boer, J. Bloem, T. Schouten, M. Rutgers, R. de Goede, and L. Brussaard. 2009. "Soil Biological Quality of Grassland Fertilized with Adjusted Cattle Manure Slurries in Comparison with Organic and Inorganic Fertilizers." *Biology and Fertility of Soils* 45(6): 595–608.

Walling, D. 2008. "Studying the Impact of Global Change on Erosion and Sediment Dynamics: Current Progress and Future Challenges." Discussion document, ISI Workshop, IRTCES, Beijing, November 6.

Walsh, J. J., D. L. Jones, G. Edwards-Jones, and A. P. Williams. 2012. "Replacing Inorganic Fertilizer with Anaerobic Digestate May Maintain Agricultural Productivity at Less Environmental Cost." *Journal of Plant Nutrition and Soil Science* 175(6): 840–845.

Zhang, F., C. Li, Z. Wang, S. Glidden, D. S. Grogan, X. Li, Y. Cheng, and S. Frolking. 2015. "Modeling Impacts of Management on Farmland Soil Carbon Dynamics along a Climate Gradient in Northwest China during 1981–2000." *Ecological Modelling* 312: 1–10.

Zhou, J., D. Guan, B. Zhou, B. Zhao, M. Ma, J. Qin, X. Jiang, et al. 2015. "Influence of 34-Years of Fertilization on Bacterial Communities in an Intensively Cultivated Black Soil in Northeast China." *Soil Biology and Biochemistry* 90: 42–51.

2 Treatment of Wet Organic Waste by Hydrothermal Carbonization

Vivian Mau and Amit Gross

CONTENTS

2.1 INTRODUCTION

Wet organic waste, including animal manures, sewage and activated sludge, and the organic portion of municipal solid waste, represents both a burden that must be dealt with as well as a material source for fertilizer and compost. Depending on how these wastes are treated, their properties and behavior when applied to soils can vary significantly. The high water-fraction present in wet organic waste is commonly a disadvantage in various treatment techniques such as composting and pyrolysis, requiring drying or the addition of dry matter. Similarly, a low C:N of some of these sources (e.g., poultry manure and domestic sludge) may be challenging for commonly used biological treatments. Hydrothermal carbonization (HTC) is a technology that, as suggested by its name, involves the presence of large water fractions; thus it is ideally suitable for the treatment of wet organic waste. This chapter will introduce the HTC process, providing an overview of the main chemical conditions and reactions involved in the conversion of wet organic waste. Then, the various applications of the generated products will be discussed, with a particular focus on the addition to soils.

2.2 HYDROTHERMAL CARBONIZATION PROCESS

Hydrothermal carbonization is a technology that can potentially treat wet organic waste such as animal manures and sewage sludge efficiently and with minimum environmental pollution. During HTC, wet organic matter is converted primarily into a carbon-concentrated solid product referred to as hydrochar, as well as aqueous and gaseous phases. The product distribution depends on process conditions and feedstock used, but in general, on a weight basis, 50% to 80% of the original feedstock is present in the hydrochar, 5% to 20% in the aqueous phase, and 2% to 5% in the gaseous phase (Libra et al. 2011). HTC takes place under a typical temperature range of 180°C–250°C and reaction times varying from minutes to several hours (Funke and Ziegler 2010). The process is conducted in a closed vessel; therefore, the internal pressure naturally increases with the increase in temperature. Anoxic or low oxygen concentrations are preferred to reduce oxidation of carbon to CO_2, or, in other words, to recover more carbon as hydrochar, thus improving efficiency. Due to the high temperature and pressure conditions, the water in the feedstock stays in the aqueous phase, avoiding the energy-intensive step of evaporation that is required in pyrolysis to generate biochar (Peterson et al. 2008). For this reason, it could be more energetically efficient to convert wet organic waste to hydrochar instead of biochar (Mau and Gross 2018). Table 2.1 provides a general comparison between hydrochar and biochar. The high HTC treatment temperatures and pressures also ensure that the process destroys pathogens and their DNAs, including antibiotic-resistant genes, as well as pharmaceutically active compounds (Libra et al. 2011; Ducey et al. 2017). Hence, the final products are considered sterile and likely prevent the spread of many of hazardous micropollutants.

The fundamental scientific knowledge base of HTC was compiled by Friederich Bergius starting in 1913 (Ramke et al. 2009). This process was found to mimic the natural process of conversion of organic matter into coal. Interest in the field was regained in the beginning of the 2000s when HTC experiments were undertaken for the production of carbon spheres from sugar and glucose (Titirici et al. 2007). Since then, HTC has been considered for the production of nanostructured materials, carbon sequestration, adsorbent material, soil amendment, energy production, nutrient recycling, and waste treatment (Libra et al. 2011). A general scheme of the HTC process and some environmental applications of the generated products are presented in Figure 2.1.

TABLE 2.1
General Comparison of Formation Process and Properties of Hydrochar and Biochar

	Hydrochar	Biochar	References
Formation process	Hydrothermal carbonization	Slow pyrolysis	Kambo and Dutta (2015)
Surrounding medium	Water	Gases	Libra et al. (2011)
Production temperature (°C)	180–260	300–650	Kambo and Dutta (2015)
Production pressure	Autogenous	Atmospheric	Libra et al. (2011)
Production time	5 min–12 h	5 min–12 h	Kambo and Dutta (2015)
Production heating rate (°C/min)	10–30	5–10	Kambo and Dutta (2015)
Char yield (%)	50–80	35	Libra et al. (2011)
pH	Acidic	Basic	Wiedner et al. (2013)
Electric conductivity	Lower	Higher	Liu et al. (2017)
Organic composition	Alkyls	Aromatics	Wiedner et al. (2013)
Ash content	Lower than feedstock	Higher than feedstock	Kambo and Dutta (2015)
H:C ratio	Higher	Lower	Kambo and Dutta (2015)
N concentration	Higher	Lower	Wiedner et al. (2013)
Thermal stability	Lower	Higher	Gascó et al. (2018)
Surface area	Lower	Higher	Kambo and Dutta (2015)
Porosity	Lower	Higher	Kambo and Dutta (2015)
Mean pore size	Higher	Lower	Liu et al. (2017)

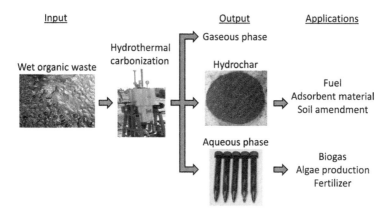

FIGURE 2.1 Scheme of hydrothermal carbonization process and environmental applications.

2.2.1 Subcritical Water

Hydrothermal carbonization is possible due to the presence of subcritical water, liquid water below the critical point of 374°C and 22 MPa (217 atm.). Subcritical water is considerably different from ambient water, facilitating many processes necessary for the transformation of organic matter.

The structure of water near the critical point is altered due to the loss of the tetrahedral coordination of water molecules present in ambient conditions (Akiya and Savage 2002). The infinite network of hydrogen bonds observed in ambient liquid water is disrupted by a decrease in hydrogen bonding between the molecules. As a result, water molecules are present in small clusters approaching gas structure, allowing more translational and rotational motions, thus increasing the self-diffusivity of water (Akiya and Savage 2002). With increasing temperature water density decreases, and consequently, the relative static dielectric constant decreases as well. These changes transform water into fairly nonpolar molecules, increasing its affinity for organic hydrocarbons (Kruse and Dahmen 2015). The dissociation constant of water increases up to 4 orders of magnitude with increasing temperature (Figure 2.2, from Peterson et al. 2008). High dissociation constant at the hydrothermal carbonization temperature range increases the ionization of water and facilitates more acid–base-catalyzed reactions (Shah 2014). As a result of these changes, water becomes a good solvent for typically nonpolar and hydrophobic hydrocarbons. Subcritical water acts as a reactant and a catalyst, with the ability to carry out condensation, cleavage, complete oxidation and decomposition of organic materials, various chemical syntheses, and to affect selective ionic chemistry, which are more compatible with organic reactions (Shah 2014).

2.2.2 Reaction Mechanisms

Organic matter is converted into hydrochar through two main pathways: (1) solid–solid conversion, in which the structural elements and morphology of the feedstock are present in the hydrochar; (2) aqueous phase degradation of the organic matter followed by polymerization and precipitation into a solid phase (Lucian et al. 2018). Figure 2.3 summarizes the chemical reaction pathways involved in the HTC process.

Hydrolysis, the cleavage of a chemical bond by the addition of a water molecule, is the initial reaction for both pathways in hydrochar formation (Kruse et al. 2013). Hydrolysis reactions during the HTC process mainly break down ether and ester bonds, resulting in a wide range of products that include saccharides of cellulose and phenolic fragments of lignin (Kruse et al. 2013). The hydrolysis temperature varies depending on the component: hemicellulose is hydrolyzed around 180°C,

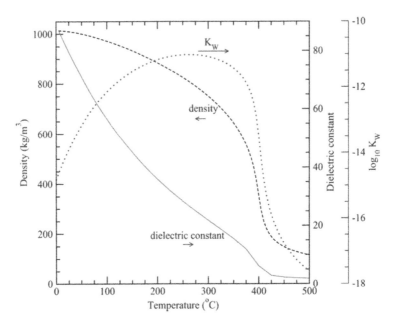

FIGURE 2.2 Changes in water properties as a function of temperature at 30 MPa. (From Peterson, A.A., et al., *Energy Environ. Sci.*, 1, 32–65, 2008.)

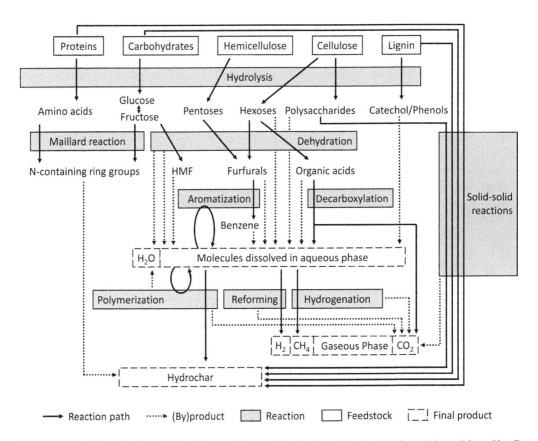

FIGURE 2.3 Principal chemical reactions pathways in hydrothermal carbonization. (Adapted from He, C. et al., *Appl. Energy*, 111, 257–266, 2013; Kruse, A., et al., *Curr. Opin. Chem. Biol.*, 17, 515–21, 2013.)

cellulose around 200°C (Funke and Ziegler 2010), and lignin around 250°C (Kim et al. 2016). The rate of biomass hydrolysis is primarily determined by diffusion and thus limited by transport phenomena within the biomass matrix (Funke and Ziegler 2010).

HTC is governed in sum by dehydration and decarboxylation reactions. Dehydration is the chemical removal of water molecules from the biomass components. It significantly carbonizes biomass by lowering the H:C and O:C ratios, through the removal of hydroxyl groups (R-OH) (Funke and Ziegler 2010). Decarboxylation is the chemical removal of carboxyl groups (R-COOH) from organic matter and release of carbon dioxide. Decarboxylation reactions are attractive in the conversion process of biomass to hydrochar because they decrease the oxygen content of the feedstock, while increasing the H:C ratio, which typically leads to higher-quality fuels (Peterson et al. 2008). Carboxyl groups rapidly degrade above 150°C (Funke and Ziegler 2010). Cellulose starts decomposing by dehydration, and decarboxylation only appears after a specific amount of water is formed (Funke and Ziegler 2010). The rate of dehydration appears to be much higher than decarboxylation during HTC (Funke and Ziegler 2010); however, some studies have shown the opposite trend (Berge et al. 2011; Mumme et al. 2011; Spitzer et al. 2018).

Other reactions present during the process are polymerization, aromatization, Maillard reaction, transformation reactions, demethylation, pyrolytic reactions, and Fischer–Tropsch-type reactions (Funke and Ziegler 2010).

2.2.3 Influence of HTC Conditions

There are several conditions that influence hydrothermal carbonization. They have been investigated at different degrees with several feedstocks. Despite the difficulty in comparing studies due to different methodologies and reporting parameters, a general understanding of the influence of process parameters is possible. The parameters investigated include temperature, reaction time, solids load, and process water characteristics.

2.2.3.1 Feedstock

The feedstock type used in HTC is an important aspect of the overall process. Several feedstock types have been investigated in recent HTC studies, such as agricultural wastes (Benavente et al. 2015; Hoekman et al. 2011; Liu et al. 2013; Reza et al. 2013), sewage sludge (Danso-Boateng et al. 2015; He et al. 2013; Zhao et al. 2014), municipal solid waste (Berge et al. 2011; Lu et al. 2012), algal residues (Broch et al. 2014; Levine et al. 2013), wet animal manures (Ekpo et al. 2016; Heilmann et al. 2014; Oliveira et al. 2013), as well as simpler feedstock that consist of single compounds such as glucose, cellulose, starch, xylose, and lignin (Falco et al. 2011; Lu and Berge 2014; Peterson et al. 2008).

Feedstock compositions and structure play a significant role during solids generation. The type and complexity of feedstock influences HTC kinetics due to the different reaction mechanisms previously discussed. The amount of cellulose, hemicellulose, lignin, sugars, and proteins present in the feedstock will highly influence the HTC process as these compounds hydrolyze at different temperatures. In general, feedstocks that are soluble in water at room temperature have an increase in solids generated with time, as the carbon moves from the liquid to solid phase, while those that are insoluble in water at room temperature have a decrease in solids (Lu and Berge 2014). Moreover, as feedstock carbon content increases, hydrochar yield (recovered solid mass of the initial feedstock mass) and carbon content increase as well (Lu and Berge 2014).

2.2.3.2 Temperature

Temperature is considered as the HTC process condition that influences most of the process as well as the composition of the final products. As discussed previously, different organic matter components hydrolyze at different temperatures; therefore, temperature is an important parameter controlling hydrothermal carbonization. Generally, an increase in temperature decreases the hydrochar yield and increases the yield of the aqueous phase and gases such as CO_2, CO, and H_2. An increase in temperature also decreases the H:C and O:C ratios in the hydrochar (Funke and Ziegler 2010), although it has a greater influence on the O:C than that on the H:C ratio (Lu et al. 2013). As temperature increases, the rate of the initial solids disappearance increases due to the intensification of the reactions involved, and also probably due to dilution of aqueous extractives present in the solids (Benavente et al. 2015). The rate and/or extent of initial feedstock solubilization is dependent on the heating rate of the reactor, and thus on the final temperature (Lu et al. 2013). In general, higher temperatures accelerate hydrothermal carbonization, but also result in higher energy consumption and higher pressure, which increases investment costs for pressure equipment (Funke and Ziegler 2010).

2.2.3.3 Reaction Time

The influence of reaction time has been investigated in various studies using different feedstocks. A longer reaction time generally increases the severity of the process and increases the hydrochar yield. This is due to ongoing polymerization of solved fragments in the aqueous phase, which finally leads to precipitation of insoluble solids (Funke and Ziegler 2010). In shorter reaction times (up to 24 h), an increase in reaction time results in a decrease of hydrochar yield and an increase in aqueous and gaseous phase yield (Benavente et al. 2015). In the initial period of reaction, there is a rapid decline in hydrochar carbon, likely due to feedstock solubilization, coupled with a simultaneous increase in aqueous and gas-phase carbon. After this initial period, there are slower and less significant changes in carbon distribution (Lu et al. 2013). Despite the trends observed, reaction time has relatively little effect on hydrochar yield and carbon content in comparison to temperature (Mumme et al. 2011).

The effect of reaction temperature and time have been combined into a single parameter called the severity factor (*SF*). The most commonly used severity factors are the ones established by Ruyter (1982) and by Overend et al. (1987). The first was developed by modeling HTC kinetics according to hydrochar oxygen loss (Ruyter 1982). Then it was calibrated with data from HTC of feedstocks ranging from cellulose to subbituminous coal treated at temperatures of 120°C–390°C and reaction times ranging from 1 min to 6 months. It has linear dependence on time and an exponential dependence on temperature following Arrhenius behavior:

$$SF = 50 \times t^{0.2} \times e^{\frac{-3500}{T}} \tag{2.1}$$

where t is the reaction time in seconds, and T is the temperature in K. The second was developed based on observations from the Kraft pulping process and steaming process used in the pulp and paper industry (Overend et al. 1987). It also has a linear dependence on time, and an exponential dependence on temperature of the first-order rate:

$$SF = \log R_o = \log\left(t \times e^{\frac{T-100}{14.75}} \right) \tag{2.2}$$

where t is the reaction time in min, and T is the temperature in °C. The fitted value of 14.75 was shown to be inversely proportional to the activation energy of hemicellulose hydrolysis (Posmanik et al. 2017). In both models the severity factor implies that the same reaction can be achieved by using different combinations of temperature and time.

2.2.3.4 Solid Load

Generally, very low concentration of biomass in water may result in very low production of hydrochar since most biomass may be dissolved in the water (Funke and Ziegler 2010). Excessive biomass load is problematic only if the water content is not sufficient to keep it submerged, which would prevent carbonization (Funke and Ziegler 2010). A high ratio of biomass enhances polymerization, and also keeps energetic losses and investment costs for pumps and heat exchangers low

(Funke and Ziegler 2010). It has been shown that hydrochar yield and carbon concentration are positively impacted by solid load (Heilmann et al. 2014). However, at high solid loads, such as solid-to-water ratios of 1:5 and 1:3, the effect on hydrochar as well as aqueous and gaseous phases composition is no longer significant (Mau et al. 2016).

2.2.3.5 Process Water

Process water includes the water content already present in the biomass as well as any other liquid component added to the feedstock prior to HTC. A neutral to weakly acidic environment appears to be necessary to achieve a simulation of natural coalification, and an acidic environment improves the overall reaction rate of HTC (Funke and Ziegler 2010). For this reason, acids are sometimes added to the process water to catalyze the reactions. Chemical properties, however, are not sufficient in explaining the influence of process water composition on carbonization. Different acids and different salt compositions have been shown to lead to different effects on process kinetics, carbon distribution, and hydrochar energy content (Lu et al. 2014).

Process water also influences HTC in a process called aqueous phase recirculation. In this process the obtained aqueous phase of the HTC treatment is reused as the water input of subsequent HTC runs. Recirculating the aqueous phase reduces fresh water use and production of HTC aqueous phase (Catalkopru et al. 2017; Kambo et al. 2018; Stemann et al. 2013; Weiner et al. 2014). Recirculation is specifically effective with relatively dry feedstocks, which requires significant water addition for HTC (Weiner et al. 2014; Catalkopru et al. 2017). Moreover, heating energy could be saved, if the process design can accommodate recirculating the aqueous phase while still hot (Stemann and Ziegler 2011). Aqueous phase recirculation significantly increases the concentration of organic C and nutrients in the aqueous phase, but does not result in major changes to hydrochar composition and calorific value (Mau et al. 2019).

2.2.4 Typical Characteristics of Generated Products

As suggested, the products' characteristics change according to substrate properties (e.g., sludge, manure, or plant biomass), but in general the direction of the process and the properties of the hydrochar as well as aqueous and gaseous phases are often similar.

Hydrochar has a skeletal structure comparable to that of natural coal, despite a higher amount of functional groups, especially oxygen-containing groups (Funke and Ziegler 2010). The largest components of hydrochar are alkyls followed by aromatics (Berge et al. 2011). The removal of hydroxyl and carbonyl groups during HTC results in a product that has a higher water repellency than the starting material (Funke and Ziegler 2010). The partition of inorganics to the hydrochar or aqueous phase depends on the nature of the element itself. In general, the concentration of inorganics in the hydrochar decreases with increasing temperature as different biomass components start to react. At higher temperatures when the hydrochar is more porous, some inorganics may be absorbed (Reza et al. 2013).

Hydrochar is often characterized in terms of proximate and ultimate analysis. The proximate analysis includes moisture, ash, volatile matter, and fixed carbon, and the ultimate analysis includes elemental composition of C, H, N, O. Nizamuddin et al. (2017) and Wang et al. (2018) recently summarized these hydrochar properties obtained using a wide range of organic waste as feedstock. The elemental composition of the hydrochar is often analyzed in terms of its H:C and O:C ratios in a Van Krevelen diagram (Figure 2.4). It enables an easy comparison between various hydrochars, as well as typical values for biomass, peat, lignite, coal, and anthracite. It also allows for a visual analysis of the extent of dehydration and decarboxylation reactions that took place during HTC. Another important property of the hydrochar is its calorific value, or heating value. Hydrochar has been considered as a potential energy source that could replace combustion of biomass and coal since interest in hydrothermal carbonization reemerged in the late 2000s (Ramke et al. 2009); thus its energy density is of high importance. A calorific value as high as 33.3 MJ/kg has been reported

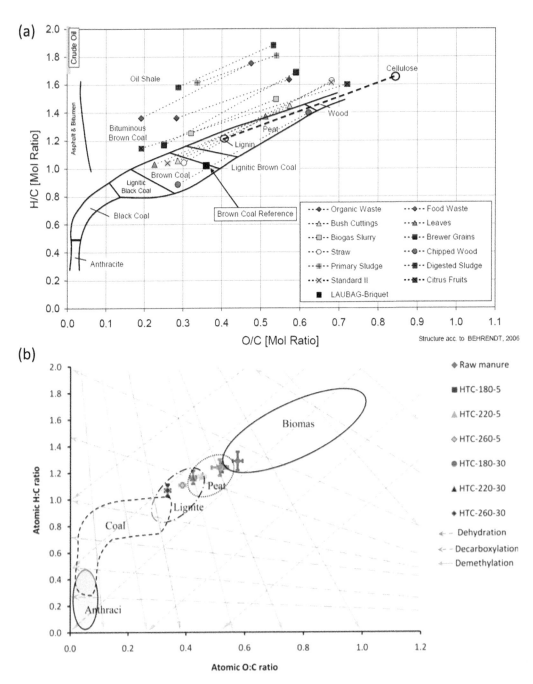

FIGURE 2.4 Van Krevelen diagram of (a) hydrochars derived from various feedstocks generated at 180°C after 12 hours. (After Ramke, H-G. et al., Hydrothermal Carbonization of Organic Waste, In *Sardinia 2009: Twelfth International Waste Management and Landfill Symposium, Sardinia, Italy, 05–09 October 2009*, edited by R. Cossu, L.F. Diaz, and R. Stegmann, CISA Publisher, Cagliari, Italy, 139–148, 2009.) (b) Cow manure hydrochar generated at temperatures of 180°C, 220°C, and 260°C after 5 and 30 min. (After Reza, M.T. et al., *Environ Prog Sustain Energy*, 35, 1002–1011, 2016.)

for hydrochar derived from maize silage at 230°C for 10 hours (Mumme et al. 2011). Typically, hydrochar has a calorific value of around 25 MJ/kg (Kambo and Dutta 2015), comparable to sub-bituminous coal, which is used for electricity production (Mau et al. 2016).

The aqueous phase has high concentrations of organics and inorganics, many of which are potentially valuable chemicals, and unless recovered become a major loss in the overall process (Funke and Ziegler 2010). The total organic carbon present in the aqueous phase is of polar and aromatic nature, and a major portion is of higher molecular weight (Stemann et al. 2013). Aldehydes, alkenes, and furanic and phenolic compounds are present in the aqueous phase as products from the decomposition of the biomass and intermediates. Acetic acid is present as a product of decomposition of hydrolysis products (Berge et al. 2011), and has been found to be the most prominent organic acid in the aqueous phase (Levine et al. 2013). Benzene, butanoic acid, pentanoic acid, propanoic acid, and Maillard products such as aldehydes, pyrroles, pyrazines, and pyridines have been identified in the aqueous phase (Danso-Boateng et al. 2015). These compounds are not detected in the untreated feedstock, confirming they arise as a result of feedstock carbonization (Danso-Boateng et al. 2015). The aqueous phase is also rich in several nutrients such as N, P, K, Na, Ca, Mg, and S, with concentrations varying depending on the original feedstock (Spitzer et al. 2018; Mau et al. 2016).

The gaseous phase formed during hydrothermal carbonization mainly consists of CO_2 with minor fractions of CO, CH_4, and H_2, and traces of CmHn (Funke and Ziegler 2010). The gas composition does not vary significantly with the carbonization of different feedstock (Berge et al. 2011). Trace gases present in the gas phase include ethylene, ethane, propene, propane, butane, and furan (Lu et al. 2013). Other gases of environmental concern that have been identified are hydrogen sulfide, probably a product of Maillard reactions; nitrogen oxides, from oxidation of nitrogenous compounds in the feedstock; nitric oxide; and ammonia (Danso-Boateng et al. 2015).

2.3 ENVIRONMENTAL USES OF HTC PRODUCTS

Hydrochar has various environmental uses, such as an energy source through combustion, an adsorbent for heavy metals and organic pollutants, and as a soil amendment (Fang et al. 2018). The uses of the aqueous phase have been investigated to a lesser extent, and the gaseous phase is considered mainly as a by-product. In this section we will present some background on environmental applications related to soils, and will focus on the use of hydrochar as a soil amendment. Hydrochar is also being investigated for other applications, such as to produce capacitors, quantum dots for medical applications, and compounds for the manufacture of chemicals in the biorefinery industry (Kambo and Dutta 2015; Fang et al. 2018). These uses are outside the scope of this chapter; accordingly, they will not be discussed.

2.3.1 AQUEOUS PHASE USES

The aqueous phase has mainly been seen as a by-product of HTC and is typically considered an environmental burden (Belete et al. 2019). Some studies have investigated its use as a valuable product, such as for enhancing biogas production in anaerobic reactors (Danso-Boateng et al. 2015; Mumme et al. 2011; Oliveira et al. 2013; Poerschmann et al. 2014) and as a nutrient source for algal biomass production (Levine et al. 2013; Du et al. 2012; Belete et al. 2019). Biogas production from the aqueous phase depends on the process parameters employed during HTC. A theoretical methane yield of up to 77.43% has been estimated following HTC of sewage sludge at 180°C for 30 minutes (Danso-Boateng et al. 2015). The production of algae from the aqueous phase enables both the generation of a product that could be used for biofuel production or as a feed, as well as a much cleaner water stream that can be more easily disposed or reused (Belete et al. 2019). Additionally, the aqueous phase could be used as a liquid fertilizer due to its high nutrient content (Vozhdayev et al. 2015). Studies have been conducted on the effect of the aqueous phase on seed germination. In general, it was found that HTC aqueous phase can inhibit seed germination (Bargmann et al. 2013; Fregolente et al. 2018; Novianti et al. 2016). However, if employed during the plant growth phase, the aqueous phase can result in plant growth similar to that achieved with fertilizers (Vozhdayev et al. 2015; Mau et al. 2019). The long-term effect on soils of using the aqueous phase as a fertilizer has not been considered.

2.3.2 HYDROCHAR AS AN ADSORBENT MATERIAL

Adsorption characteristics of hydrochar are less understood than those of biochar. Hydrochar is less aromatic, has more polar functional groups, and higher O:C ratios than biochar, resulting in better adsorption for a number of polar and nonpolar organics (Parshetti et al. 2014; Sun et al. 2011). Hydrochar has been shown to effectively remove textile dyes, bisphenol A, 17a-ethinyl estradiol, phenanthrene, lead, cadmium, uranium, and arsenic from water (Elaigwu et al. 2014; Kumar et al. 2011; Parshetti et al. 2014; Poerschmann et al. 2014; Sun et al. 2011). Hydrochar adsorption characteristics are significantly improved through treatments like alkali activation and hydrogen peroxide and acetone washing (Flora et al. 2013; Regmi et al. 2012; Xue et al. 2012). Other activation methods, such as by Fenton reaction, were also suggested and proved efficient (Belete 2019). In terms of application to soil, the success of hydrochar as an adsorbent material means it could be utilized for soil remediation. Initial experiments, however, did not obtain satisfactory results for the immobilization of zinc, copper, cadmium, and lead in contaminated soils (Wagner and Kaupenjohann 2014).

2.3.3 HYDROCHAR AS A SOIL AMENDMENT

Based on numbers of published scientific articles to date, the application of hydrochar as a soil amendment to improve soil properties and fertility has received less attention than as an energy source and as an adsorbent material. Hydrochar has been proposed as a possible way to improve soil properties due to its apparent similarities to biochar (Libra et al. 2011). However, as more research was conducted the differences between hydrochar and biochar became more noticeable and significant (Table 2.1). For example, hydrochar possesses an acidic pH and high N concentration, and is mainly composed of alkyls, whereas biochar has a basic pH, has low N concentration, contains mostly aromatics, and has high thermal stability (Wiedner et al. 2013; Gascó et al. 2018). Therefore, the behavior of hydrochar in soil cannot be directly inferred from experiments on biochar.

The application method of hydrochar to soils is considered to be similar to other amendments such as biochar, manures, and composts, involving mainly mechanical incorporation into the topsoil (Blackwell et al. 2012; Bargmann et al. 2013; Gajić and Koch 2012). Studies considered an incorporation depth of 10 to 20 cm (Bargmann et al. 2013; Gajić and Koch 2012; Melo et al. 2018). Biochar studies have investigated a wide range of application rates, from 0.5 to 135 ton/ha (Glaser et al. 2002), though most studies have focused on rates of up to 39 ton/ha (Jeffery et al. 2011). So far, studies on hydrochar have also considered a wide range of application rates, though lower rates are preferred as they are closer to the more realistic field application rates of 10 to 20 ton/ha (Bargmann et al. 2013; Gajić and Koch 2012).

The application of hydrochar to soils leads to a decrease in soil bulk density (Abel et al. 2013). The decrease in bulk density results from the lower particle density of the hydrochar compared to the soil (Abel et al. 2013; Eibisch et al. 2015). The effect on soil porosity is less clear and seems to be feedstock dependent. An increase in soil porosity was observed for maize silage hydrochar generated at 200°C applied to sand and loamy sand soils (Abel et al. 2013), but no difference was detected with the addition of woodchip hydrochar generated at 200°C and 250°C to loamy sand soil (Eibisch et al. 2015). The possible increase in soil porosity is likely due to the filling of wide soil pores by hydrochar and the small pore size of hydrochar itself (Abel et al. 2013). The role of hydrochar pore

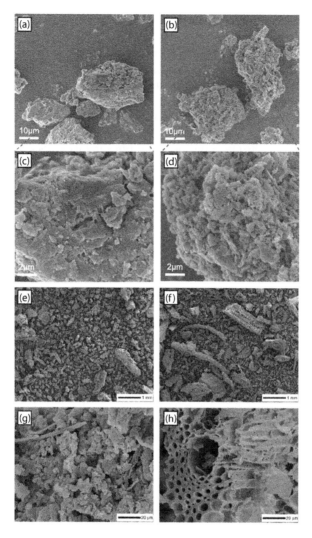

FIGURE 2.5 Scanning electron microscope images of dried sewage sludge at (a) 10 μm and (c) 2 μm; sewage sludge derived hydrochar at (b) 10 μm and (d) 2 μm; maize silage–derived hydrochar at (e) 1 mm and (g) 20 μm; maize silage–derived biochar at (f) 1 mm and (h) 20 μm. (Adapted from Gai, C., et al., *Int. J. Hydrog. Energy*, 41, 3363–3372, 2016; Abel, S., et al., *Geoderma*, 202–203, 183–191, 2013.)

size in increasing the soil's porosity became evident when application rates as high as 15% (w/w) were used (Kalderis et al. 2018). The pore structures of original feedstock, generated hydrochar, and generated biochar can be seen in Figure 2.5.

Hydrochar was shown to increase the available water capacity in sand and loamy sand soils (Abel et al. 2013). This observed increase is because the addition of hydrochar changes the pore distribution of the soil. It creates a secondary pore maximum in the range where water is held by capillary and adsorptive forces but is still available to plants (Abel et al. 2013). In other words, it increases the moisture content between field capacity and wilting point. The increase in available water capacity may also be due to increased aggregate stability (Eibisch et al. 2015). Based on these results, hydrochar was concluded to possibly be a good soil amendment, increasing plant available soil water especially for coarse-grained soils with low organic matter content (Abel et al. 2013).

Hydrochar is a highly hydrophobic material (Mau et al. 2018), and so it decreases the soil's wettability (Eibisch et al. 2015). The hydrochar's wettability was found to be more dependent on the specific surface area than the functional surface groups (Eibisch et al. 2015). Despite the significant effects on soil wettability, hydrochar addition was not found to impact soil properties (application rate of 2% w/w to loamy sand soil) such as saturation value, available water capacity, and hydraulic conductivity (Eibisch et al. 2015).

Hydrochar also affects soil microorganisms. Reports suggest both positive and negative impacts. This is not surprising as feedstock and production conditions impact the quality of the hydrochar and consequently its effect on soils. Soil type and its properties, as well as microorganisms present and their activity, also influence the hydrochar amendment effects (Reza et al. 2014). For example, hydrochar at high concentrations (20% v/v) was shown to stimulate root colonization and spore germination by arbuscular mycorrhizal symbiosis, which is known to enhance plant growth (Rillig et al. 2010). Additionally, microbial activity is stimulated by the addition of hydrochar, which has high available carbon content; however, it leads to N-immobilization, which reduces the N available to plants. It has been shown that N-immobilization depends on the hydrochar feedstock, and is diminished for hydrochar with low C/N ratio (16 versus 38) (Gajić and Koch 2012).

Significant research has been conducted on the effect of hydrochar on plants, focusing on the germination stage. These experiments demonstrated that hydrochar inhibits or delays seed germination (Busch et al. 2012, 2013; Bargmann et al. 2013, 2014). However, application rates much higher than what is practical for field application were investigated (Bargmann et al. 2013, 2014). When considering application rates suitable for agriculture, the negative effects observed in these studies will likely not be present. The inhibitory effects can be minimized by post-treatment such as composting. It was shown that hydrochar composting prior to field application removed germination inhibitory effects, and even increased plant weight at application rates as high as 30 ton/ha (Busch et al. 2013). Another proposed method to avoid inhibitory effects is to apply hydrochar several weeks prior to sowing such that microbial degradation in the soil can remove the undesirable effects (Bargmann et al. 2014).

The effect on plant growth is less clear, with both positive and negative results reported. Significant improvement in lettuce growth were obtained upon addition of poultry litter–derived hydrochar to sandy soils at 0.5% and 1% (w/w) rates (Figure 2.6). Hydrochar derived from sewage sludge applied at rates of up to 32 ton/ha did not impact plant weight at the end of the first crop harvest. At the end of the second crop harvest there was a significant increase in plant weight, indicating hydrochar acted as a slow-release fertilizer (Melo et al. 2018). Similarly, no effects on plant growth were observed when applying beet root–derived hydrochar to Albic luvisol. However, when the hydrochar application rate was increased to 10% (v/v) or higher, it resulted in decreased plant growth (Rillig et al. 2010). George et al. (2012) did not detect any significant effect on plant weight when adding hydrochar at concentrations of up to 10% (v/v); however, significant leaf tip necroses was apparent, indicating that the hydrochar did have a negative effect on the plant health.

FIGURE 2.6 Lettuce after 40 days of growth on sand mixed with poultry litter–derived hydrochar (200°C after 1 hour) at 0.5% and 1% (w/w). (Courtesy of Vivian Mau.)

2.4 CONCLUSION

In this chapter, we presented an overview of the HTC process and its main environmental applications. The intersection of HTC and soil science has still not received the adequate attention it warrants. This becomes even more apparent when considering the amount of information available on the use of biochar in soils. This is a research area that we expect will see significant growth in the future, with several research opportunities available.

REFERENCES

Abel, Stefan, Andre Peters, Steffen Trinks, Horst Schonsky, Michael Facklam, and Gerd Wessolek. 2013. "Impact of Biochar and Hydrochar Addition on Water Retention and Water Repellency of Sandy Soil." *Geoderma* 202–203: 183–191. doi:10.1016/j.geoderma.2013.03.003.

Akiya, Naoko, and Phillip E. Savage. 2002. "Roles of Water for Chemical Reactions in High-Temperature Water." *Chemical Reviews* 102(8): 2725–2750. doi:10.1021/cr000668w.

Bargmann, Inge, Matthias C. Rillig, W. Buss, Andrea Kruse, and Martin Kuecke. 2013. "Hydrochar and Biochar Effects on Germination of Spring Barley." *Journal of Agronomy and Crop Science* 199(5): 360–373. doi:10.1111/jac.12024.

Bargmann, Inge, Matthias C. Rillig, Andrea Kruse, Jörg Michael Greef, and Martin Kücke. 2014. "Initial and Subsequent Effects of Hydrochar Amendment on Germination and Nitrogen Uptake of Spring Barley." *Journal of Plant Nutrition and Soil Science* 177: 68–74. doi:10.1002/jpln.201300160.

Belete, Yonas Zeslase. 2019. "Hydrothermal Carbonization of Wet-Biomass Wastes: Utilization and Application Opportunities." MSc Thesis. Ben-Gurion University of the Negev, Midreshet Ben Gurion, Israel.

Belete, Yonas Zeslase, Stefan Leu, Sammy Boussiba, Boris Zorin, Clemens Posten, Laurenz Thomsen, Song Wang, Amit Gross, and Roy Bernstein. 2019. "Characterization and Utilization of Hydrothermal Carbonization Aqueous Phase as Nutrient Source for Microalgal Growth." *Bioresource Technology* 290: 121758. doi:10.1016/J.BIORTECH.2019.121758.

Benavente, Verónica, Emilio Calabuig, and Andres Fullana. 2015. "Upgrading of Moist Agro-Industrial Wastes by Hydrothermal Carbonization." *Journal of Analytical and Applied Pyrolysis* 113: 89–98. doi:10.1016/j.jaap.2014.11.004.

Berge, Nicole D., Kyoung S. Ro, Jingdong Mao, Joseph R. V. Flora, Mark A. Chappell, and Sunyoung Bae. 2011. "Hydrothermal Carbonization of Municipal Waste Streams." *Environmental Science and Technology* 45(13): 5696–5703. doi:10.1021/es2004528.

Blackwell, Paul, Glen Riethmuller, and Mike Collins. 2012. "Biochar Application to Soil." In *Biochar for Environmental Management: Science and Technology*, edited by Johannes Lehmann and Stephen Joseph, 207–226. London, UK: Earthscan. doi:10.4324/9781849770552.

Broch, Amber, Umakanta Jena, S. Kent Hoekman, and Joel Langford. 2014. "Analysis of Solid and Aqueous Phase Products from Hydrothermal Carbonization of Whole and Lipid-Extracted Algae." *Energies* 7(1): 62–79. doi:10.3390/en7010062.

Busch, Daniela, Claudia Kammann, Ludger Grünhage, and Christoph Müller. 2012. "Simple Biotoxicity Tests for Evaluation of Carbonaceous Soil Additives: Establishment and Reproducibility of Four Test Procedures." *Journal of Environment Quality* 41(4): 1023–1032. doi:10.2134/jeq2011.0122.

Busch, Daniela, Arne Stark, Claudia I. Kammann, and Bruno Glaser. 2013. "Genotoxic and Phytotoxic Risk Assessment of Fresh and Treated Hydrochar from Hydrothermal Carbonization Compared to Biochar from Pyrolysis." *Ecotoxicology and Environmental Safety* 97: 59–66. doi:10.1016/j.ecoenv.2013.07.003.

Catalkopru, Arzu Kabadayi, Ismail Cem Kantarli, and Jale Yanik. 2017. "Effects of Spent Liquor Recirculation in Hydrothermal Carbonization." *Bioresource Technology* 226: 89–93. doi:10.1016/J.BIORTECH.2016.12.015.

Danso-Boateng, Eric, Gilbert Shama, Andrew D. Wheatley, Simon J. Martin, and Richard G. Holdich. 2015. "Hydrothermal Carbonisation of Sewage Sludge: Effect of Process Conditions on Product Characteristics and Methane Production." *Bioresource Technology* 177: 318–327. doi:10.1016/j.biortech.2014.11.096.

Du, Zhenyi, Bing Hu, Aimin Shi, Xiaochen Ma, Yanling Cheng, Paul Chen, Yuhuan Liu, Xiangyang Lin, and Roger Ruan. 2012. "Cultivation of a Microalga Chlorella Vulgaris Using Recycled Aqueous Phase Nutrients from Hydrothermal Carbonization Process." *Bioresource Technology* 126: 354–357. doi:10.1016/J.BIORTECH.2012.09.062.

Ducey, Thomas F., Jessica C. Collins, Kyoung S. Ro, Bryan L. Woodbury, and D. Dee Griffin. 2017. "Hydrothermal Carbonization of Livestock Mortality for the Reduction of Pathogens and Microbially-Derived DNA." *Frontiers of Environmental Science and Engineering* 11(3): 1–8. doi:10.1007/s11783-017-0930-x.

Eibisch, Nina, Wolfgang Durner, Michel Bechtold, Roland Fub, Robert Mikutta, Susanne K. Woche, and Mirjam Helfrich. 2015. "Does Water Repellency of Pyrochars and Hydrochars Counter Their Positive Effects on Soil Hydraulic Properties?" *Geoderma* 245–246: 31–39. doi:10.1016/j.geoderma.2015.01.009.

Ekpo, Ugo, Andrew Ross, Miller Alonso Camargo-Valero, and P. T. Williams. 2016. "A Comparison of Product Yields and Inorganic Content in Process Streams Following Thermal Hydrolysis and Hydrothermal Processing of Microalgae, Manure and Digestate." *Bioresource Technology* 200: 951–960. doi:10.1016/j.biortech.2015.11.018.

Elaigwu, Sunday E., Vincent Rocher, Georgios Kyriakou, and Gillian M. Greenway. 2014. "Removal of Pb^{2+} and Cd^{2+} from Aqueous Solution Using Chars from Pyrolysis and Microwave-Assisted Hydrothermal Carbonization of Prosopis Africana Shell." *Journal of Industrial and Engineering Chemistry* 20(5): 3467–3473. doi:10.1016/j.jiec.2013.12.036.

Falco, Camillo, Niki Baccile, and Maria-Magdalena Titirici. 2011. "Morphological and Structural Differences between Glucose, Cellulose and Lignocellulosic Biomass Derived Hydrothermal Carbons." *Green Chemistry* 13(11): 3273–3281. doi:10.1039/c1gc15742f.

Fang, June, Lu Zhan, Yong Sik Ok, and Bin Gao. 2018. "Minireview of Potential Applications of Hydrochar Derived from Hydrothermal Carbonization of Biomass." *Journal of Industrial and Engineering Chemistry* 57: 15–21. doi:10.1016/j.jiec.2017.08.026.

Flora, Justine F. R., Xiaowei Lu, Liang Li, Joseph R. V. Flora, and Nicole D. Berge. 2013. "The Effects of Alkalinity and Acidity of Process Water and Hydrochar Washing on the Adsorption of Atrazine on Hydrothermally Produced Hydrochar." *Chemosphere* 93(9): 1989–1996. doi:10.1016/j.chemosphere.2013.07.018.

Fregolente, Laís Gomes, Thaiz Batista Azevedo Rangel Miguel, Emilio de Castro Miguel, Camila de Almeida Melo, Altair Benedito Moreira, Odair Pastor Ferreira, and Márcia Cristina Bisinoti. 2018. "Toxicity Evaluation of Process Water from Hydrothermal Carbonization of Sugarcane Industry By-Products." *Environmental Science and Pollution Research*. doi:10.1007/s11356-018-1771-2.

Funke, Axel, and Felix Ziegler. 2010. "Hydrothermal Carbonization of Biomass: A Summary and Discussion of Chemical Mechanisms for Process Engineering." *Biofuels Bioproducts and Biorefining* 4(3): 160–177. doi:10.1002/bbb.198.

Gai, Chao, Yanchuan Guo, Tingting Liu, Nana Peng, and Zhengang Liu. 2016. "Hydrogen-Rich Gas Production by Steam Gasification of Hydrochar Derived from Sewage Sludge." *International Journal of Hydrogen Energy* 41(5): 3363–3372. doi:10.1016/j.ijhydene.2015.12.188.

Gajić, Ana, and Heinz-Josef Koch. 2012. "Sugar Beet (*Beta vulgaris* L.) Growth Reduction Caused by Hydrochar Is Related to Nitrogen Supply." *Journal of Environment Quality* 41(4): 1067–1075. doi:10.2134/jeq2011.0237.

Gascó, G., Jorge Paz-Ferreiro, Maria Luisa Álvarez, Antonie Saa, and Ana Méndez. 2018. "Biochars and Hydrochars Prepared by Pyrolysis and Hydrothermal Carbonisation of Pig Manure." *Waste Management* 79: 395–403. doi:10.1016/j.wasman.2018.08.015.

George, Carmen, Marcel Wagner, Martin Kücke, and Matthias C. Rillig. 2012. "Divergent Consequences of Hydrochar in the Plant-Soil System: Arbuscular Mycorrhiza, Nodulation, Plant Growth and Soil Aggregation Effects." *Applied Soil Ecology* 59: 68–72. doi:10.1016/j.apsoil.2012.02.021.

Glaser, Bruno, Johannes Lehmann, and Wolfgang Zech. 2002. "Ameliorating Physical and Chemical Properties of Highly Weathered Soils in the Tropics with Charcoal – A Review." *Biology and Fertility of Soils* 35(4): 219–230. doi:10.1007/s00374-002-0466-4.

He, Chao, Apostolos Giannis, and Jing Yuan Wang. 2013. "Conversion of Sewage Sludge to Clean Solid Fuel Using Hydrothermal Carbonization: Hydrochar Fuel Characteristics and Combustion Behavior." *Applied Energy* 111: 257–266. doi:10.1016/j.apenergy.2013.04.084.

Heilmann, Steven M., Joseph S. Molde, Jacobe G. Timler, Brandon M. Wood, Anthony L. Mikula, Georgiy V. Vozhdayev, Edward C. Colosky, Kurt A. Spokas, and Kenneth J. Valentas. 2014. "Phosphorus Reclamation through Hydrothermal Carbonization of Animal Manures." *Environmental Science & Technology* 48: 10323–10329. doi:10.1021/es501872k.

Hoekman, S. Kent, Amber Broch, and Curtis Robbins. 2011. "Hydrothermal Carbonization of Lignocellulosic Biomass." *Bioresource Technology* 25: 1802–1810. doi:10.1016/j.biortech.2012.05.060.

Jeffery, Simon, Frank G. A. Verheijen, Marjin van der Velde, and Ana Catarina Bastos. 2011. "A Quantitative Review of the Effects of Biochar Application to Soils on Crop Productivity Using Meta-Analysis." *Agriculture, Ecosystems and Environment* 144(1): 175–187. doi:10.1016/j.agee.2011.08.015.

Kalderis, Dimitrios, George Papameletiou, and Berkant Kayan. 2018. "Assessment of Orange Peel Hydrochar as a Soil Amendment: Impact on Clay Soil Physical Properties and Potential Phytotoxicity." *Waste and Biomass Valorization*, 1–14. doi:10.1007/s12649-018-0364-0.

Kambo, Harpreet Singh, and Animesh Dutta. 2015. "A Comparative Review of Biochar and Hydrochar in Terms of Production, Physico-chemical Properties and Applications." *Renewable and Sustainable Energy Reviews* 45: 359–378. doi:10.1016/j.rser.2015.01.050.

Kambo, Harpreet Singh, Jamie Minaret, and Animesh Dutta. 2018. "Process Water from the Hydrothermal Carbonization of Biomass: A Waste or a Valuable Product?" *Waste and Biomass Valorization* 9(7): 1181–1189. doi:10.1007/s12649-017-9914-0.

Kim, Daegi, Kwanyong Lee, and Ki Young Park. 2016. "Upgrading the Characteristics of Biochar from Cellulose, Lignin, and Xylan for Solid Biofuel Production from Biomass by Hydrothermal Carbonization." *Journal of Industrial and Engineering Chemistry* 42: 95–100. doi:10.1016/J.JIEC.2016.07.037.

Kruse, Andrea, and Nicolaus Dahmen. 2015. "Water – A Magic Solvent for Biomass Conversion." *Journal of Supercritical Fluids* 96: 36–45. doi:10.1016/j.supflu.2014.09.038.

Kruse, Andrea, Axel Funke, and Maria Magdalena Titirici. 2013. "Hydrothermal Conversion of Biomass to Fuels and Energetic Materials." *Current Opinion in Chemical Biology* 17(3): 515–521. doi:10.1016/j.cbpa.2013.05.004.

Kumar, Sandeep, Vijay A. Loganathan, Ram B. Gupta, and Mark O. Barnett. 2011. "An Assessment of U(VI) Removal from Groundwater Using Biochar Produced from Hydrothermal Carbonization." *Journal of Environmental Management* 92(10): 2504–2512. doi:10.1016/j.jenvman.2011.05.013.

Levine, Robert B., Christian O. Sambolin Sierra, Ryan Hockstad, Wassim Obeid, Patrick G. Hatcher, and Phillip E. Savage. 2013. "The Use of Hydrothermal Carbonization to Recycle Nutrients in Algal Biofuel Production." *Environmental Progress & Sustainable Energy* 32(4): 962–975. doi:10.1002/ep.11812.

Libra, Judy A., Kyoung S. Ro, Claudia Kammann, Axel Funke, Nicole D. Berge, York Neubauer, Maria-Magdalena Titirici, et al. 2011. "Hydrothermal Carbonization of Biomass Residuals: A Comparative Review of the Chemistry, Processes and Applications of Wet and Dry Pyrolysis." *Biofuels* 2(1): 89–124. doi:10.4155/bfs.10.81.

Liu, Yuxue, Shuai Yao, Yuying Wang, Haohao Lu, Satinder Kaur Brar, and Shengmao Yang. 2017. "Bio- and Hydrochars from Rice Straw and Pig Manure: Inter-comparison." *Bioresource Technology* 235: 332–337. doi:10.1016/j.biortech.2017.03.103.

Liu, Zhengang, Augustine Quek, S. Kent Hoekman, and R. Balasubramanian. 2013. "Production of Solid Biochar Fuel from Waste Biomass by Hydrothermal Carbonization." *Fuel* 103: 943–949. doi:10.1016/j.fuel.2012.07.069.

Lu, Xiaowei, and Nicole D. Berge. 2014. "Influence of Feedstock Chemical Composition on Product Formation and Characteristics Derived from the Hydrothermal Carbonization of Mixed Feedstocks." *Bioresource Technology* 166: 120–131. doi:10.1016/j.biortech.2014.05.015.

Lu, Xiaowei, Joseph R. V. Flora, and Nicole D. Berge. 2014. "Influence of Process Water Quality on Hydrothermal Carbonization of Cellulose." *Bioresource Technology* 154: 229–239. doi:10.1016/j.biortech.2013.11.069.

Lu, Xiaowei, Beth Jordan, and Nicole D. Berge. 2012. "Thermal Conversion of Municipal Solid Waste via Hydrothermal Carbonization: Comparison of Carbonization Products to Products from Current Waste Management Techniques." *Waste Management* 32(7): 1353–1365. doi:10.1016/j.wasman.2012.02.012.

Lu, Xiaowei, Perry J. Pellechia, Joseph R. V. Flora, and Nicole D. Berge. 2013. "Influence of Reaction Time and Temperature on Product Formation and Characteristics Associated with the Hydrothermal Carbonization of Cellulose." *Bioresource Technology* 138: 180–190. doi:10.1016/j.biortech.2013.03.163.

Lucian, Michela, Maurizio Volpe, Lihui Gao, Giovanni Piro, Jillian L. Goldfarb, and Luca Fiori. 2018. "Impact of Hydrothermal Carbonization Conditions on the Formation of Hydrochars and Secondary Chars from the Organic Fraction of Municipal Solid Waste." *Fuel* 233: 257–268. doi:10.1016/j.fuel.2018.06.060.

Mau, Vivian, Gilboa Arye, and Amit Gross. 2018. "Wetting Properties of Poultry Litter and Derived Hydrochar." *PLoS One* 13(10): 1–15. doi:10.1371/journal.pone.0206299.

Mau, Vivian, and Amit Gross. 2018. "Energy Conversion and Gas Emissions from Production and Combustion of Poultry-Litter-Derived Hydrochar and Biochar." *Applied Energy* 213: 510–519. doi:10.1016/j.apenergy.2017.11.033.

Mau, Vivian, Juliana Neumann, Bernhard Wehrli, and Amit Gross. 2019. "Nutrient Behavior in Hydrothermal Carbonization Aqueous Phase Following Recirculation and Reuse." *Environmental Science & Technology* 53(17): 10426–10434. doi:10.1021/acs.est.9b03080.

Mau, Vivian, Julie Quance, Roy Posmanik, and Amit Gross. 2016. "Phases' Characteristics of Poultry Litter Hydrothermal Carbonization under a Range of Process Parameters." *Bioresource Technology* 219: 632–642. doi:10.1016/j.biortech.2016.08.027.

Melo, Tatiane Medeiros, Michael Bottlinger, Elke Schulz, Wilson Mozena Leandro, Adelmo Menezes de Aguiar Filho, Hailong Wang, Yong Sik Ok, and Jörg Rinklebe. 2018. "Plant and Soil Responses to Hydrothermally Converted Sewage Sludge (Sewchar)." *Chemosphere* 206: 338–348. doi:10.1016/j.chemosphere.2018.04.178.

Mumme, Jan, Lion Eckervogt, Judith Pielert, Mamadou Diakité, Fabian Rupp, and Jürgen Kern. 2011. "Hydrothermal Carbonization of Anaerobically Digested Maize Silage." *Bioresource Technology* 102(19): 9255–9260. doi:10.1016/j.biortech.2011.06.099.

Nizamuddin, Sabzoi, Humair Ahmed Baloch, Greg J. Griffin, N. Mujawar Mubarak, Abdul Waheed Bhutto, Rashid Abro, Shaukat Ali Mazari, and Brahim Si Ali. 2017. "An Overview of Effect of Process Parameters on Hydrothermal Carbonization of Biomass." *Renewable and Sustainable Energy Reviews* 73 (December 2016): 1289–1299. doi:10.1016/j.rser.2016.12.122.

Novianti, Srikandi, Anissa Nurdiawati, Ilman Nuran Zaini, Hiroaki Sumida, and Kunio Yoshikawa. 2016. "Hydrothermal Treatment of Palm Oil Empty Fruit Bunches: An Investigation of the Solid Fuel and Liquid Organic Fertilizer Applications." *Biofuels* 7(6): 627–636. doi:10.1080/17597269.2016.1174019.

Oliveira, Ivo, Dennis Blöhse, and Hans Günter Ramke. 2013. "Hydrothermal Carbonization of Agricultural Residues." *Bioresource Technology* 142: 138–146. doi:10.1016/j.biortech.2013.04.125.

Overend, Rolph P., Esteban Chornet, and J. A. Gascoigne. 1987. "Fractionation of Lignocellulosics by Steam-Aqueous Pretreatments [and Discussion]." *Philosophical Transactions of the Royal Society A: Mathematical, Physical and Engineering Sciences* 321(1561): 523–536. doi:10.1098/rsta.1987.0029.

Parshetti, Ganesh K., Shamik Chowdhury, and Rajasekhar Balasubramanian. 2014. "Hydrothermal Conversion of Urban Food Waste to Chars for Removal of Textile Dyes from Contaminated Waters." *Bioresource Technology* 161: 310–319. doi:10.1016/j.biortech.2014.03.087.

Peterson, Andrew A., Frédéric Vogel, Russell P. Lachance, Morgan Fröling, Michael J. Antal Jr., and Jefferson W. Tester. 2008. "Thermochemical Biofuel Production in Hydrothermal Media: A Review of Sub- and Supercritical Water Technologies." *Energy & Environmental Science* 1(1): 32–65. doi:10.1039/b810100k.

Poerschmann, Juergen, Barbara Weiner, Harald Wedwitschka, I. Baskyr, R. Koehler, and F. D. Kopinke. 2014. "Characterization of Biocoals and Dissolved Organic Matter Phases Obtained upon Hydrothermal Carbonization of Brewer's Spent Grain." *Bioresource Technology* 164: 162–169. doi:10.1016/j.biortech.2014.04.052.

Posmanik, Roy, Danilo A. Cantero, Arnav Sunil Malkani, Deborah L. Sills, and Jefferson W. Tester. 2017. "Biomass Conversion to Bio-Oil Using Sub-critical Water: Study of Model Compounds for Food Processing Waste." *Journal of Supercritical Fluids* 119: 26–35. doi:10.1016/j.supflu.2016.09.004.

Ramke, Hans-Günter, Dennis Blöhse, Hans-Joachim Lehmann, and Joachim Fettig. 2009. "Hydrothermal Carbonization of Organic Waste." In *Sardinia 2009: Twelfth International Waste Management and Landfill Symposium*, Sardinia, Italy, 5–9 October 2009, edited by R. Cossu, L. F. Diaz, and R. Stegmann, 139–148. Cagliari, Italy: CISA.

Regmi, Pusker, Jose Luis Garcia Moscoso, Sandeep Kumar, Xiaoyan Cao, Jingdong Mao, and Gary Schafran. 2012. "Removal of Copper and Cadmium from Aqueous Solution Using Switchgrass Biochar Produced via Hydrothermal Carbonization Process." *Journal of Environmental Management* 109: 61–69. doi:10.1016/j.jenvman.2012.04.047.

Reza, M. Toufiq, Janet Andert, Benjamin Wirth, Daniela Busch, Judith Pielert, Joan G. Lynam, and Jan Mumme. 2014. "Hydrothermal Carbonization of Biomass for Energy and Crop Production." *Applied Bioenergy* 1(1): 11–29. doi:10.2478/apbi-2014-0001.

Reza, M. Toufiq, Ally Freitas, Xiaokun Yang, Sage Hiibel, Hongfei Lin, and Charles J. Coronella. 2016. "Hydrothermal Carbonization (HTC) of Cow Manure: Carbon and Nitrogen Distributions in HTC Products." *Environmental Progress & Sustainable Energy* 35(4): 1002–1011. doi:10.1002/ep.12312.

Reza, M. Toufiq, Joan G. Lynam, M. Helal Uddin, and Charles J. Coronella. 2013. "Hydrothermal Carbonization: Fate of Inorganics." *Biomass and Bioenergy* 49: 86–94. doi:10.1016/j.biombioe.2012.12.004.

Rillig, Matthias C., Marcel Wagner, Mohamed Salem, Pedro M. Antunes, Carmen George, Hans-Günter Ramke, Maria-Magdalena Titirici, and Markus Antonietti. 2010. "Material Derived from Hydrothermal Carbonization: Effects on Plant Growth and Arbuscular Mycorrhiza." *Applied Soil Ecology* 45(3): 238–242. doi:10.1016/j.apsoil.2010.04.011.

Ruyter, Herman P. 1982. "Coalification Model." *Fuel* 61(12): 1182–1187. doi:10.1016/0016-2361(82)90017-5.

Shah, Yatish T. 2014. "Hydrothermal Processes in Subcritical Water." In *Water for Energy and Fuel Production*, 113–155. Boca Raton, FL: CRC Press. doi:10.1201/b16904.

Spitzer, Reut Yahav, Vivian Mau, and Amit Gross. 2018. "Using Hydrothermal Carbonization for Sustainable Treatment and Reuse of Human Excreta." *Journal of Cleaner Production* 205: 955–963. doi:10.1016/j. jclepro.2018.09.126.

Stemann, Jan, Anke Putschew, and Felix Ziegler. 2013. "Hydrothermal Carbonization: Process Water Characterization and Effects of Water Recirculation." *Bioresource Technology* 143: 139–146. doi:10.1016/j.biortech.2013.05.098.

Stemann, Jan, and Felix Ziegler. 2011. "Hydrothermal Carbonisation (HTC): Recycling of Process Water." In *European Biomass Conference and Exhibition Proceedings*, 1894–1899. Florence, Italy: ETA-Florence Renewable Energies. doi:10.5071/19theubce2011-oc8.2.

Sun, Ke, Kyoung Ro, Mingxin Guo, Jeff Novak, Hamid Mashayekhi, and Baoshan Xing. 2011. "Sorption of Bisphenol A, 17a-Ethinyl Estradiol and Phenanthrene on Thermally and Hydrothermally Produced Biochars." *Bioresource Technology* 102(10): 5757–5763. doi:10.1016/j.biortech.2011.03.038.

Titirici, Maria-Magdalena, Arne Thomas, and Markus Antonietti. 2007. "Back in the Black: Hydrothermal Carbonization of Plant Material as an Efficient Chemical Process to Treat the CO_2 Problem?" *New Journal of Chemistry* 31(6): 787–789. doi:10.1039/b616045j.

Vozhdayev, Georgiy V., Kurt A. Spokas, Joseph S. Molde, Steven M. Heilmann, Brandon M. Wood, and Kenneth J. Valentas. 2015. "Response of Maize Germination and Growth to Hydrothermal Carbonization Filtrate Type and Amount." *Plant and Soil* 396(1–2): 127–136. doi:10.1007/s11104-015-2577-3.

Wagner, Anne, and Martin Kaupenjohann. 2014. "Suitability of Biochars (Pyro- and Hydrochars) for Metal Immobilization on Former Sewage-Field Soils." *European Journal of Soil Science* 65: 139–148. doi:10.1111/ejss.12090.

Wang, Tengfei, Yunbo Zhai, Yun Zhu, Caiting Li, and Guangming Zeng. 2018. "A Review of the Hydrothermal Carbonization of Biomass Waste for Hydrochar Formation: Process Conditions, Fundamentals, and Physicochemical Properties." *Renewable and Sustainable Energy Reviews* 90 (December 2016): 223–247. doi:10.1016/j.rser.2018.03.071.

Weiner, Barbara, Juergen Poerschmann, Harald Wedwitschka, Robert Koehler, and Frank Dieter Kopinke. 2014. "Influence of Process Water Reuse on the Hydrothermal Carbonization of Paper." *ACS Sustainable Chemistry and Engineering* 2(9): 2165–2171. doi:10.1021/sc500348v.

Wiedner, Katja, Cornelia Rumpel, Christoph Steiner, Alessandro Pozzi, Robert Maas, and Bruno Glaser. 2013. "Chemical Evaluation of Chars Produced by Thermochemical Conversion (Gasification, Pyrolysis and Hydrothermal Carbonization) of Agro-industrial Biomass on a Commercial Scale." *Biomass and Bioenergy* 59: 264–278. doi:10.1016/j.biombioe.2013.08.026.

Xue, Yingwen, Bin Gao, Ying Yao, Mandu Inyang, Ming Zhang, Andrew R. Zimmerman, and Kyoung S. Ro. 2012. "Hydrogen Peroxide Modification Enhances the Ability of Biochar (Hydrochar) Produced from Hydrothermal Carbonization of Peanut Hull to Remove Aqueous Heavy Metals: Batch and Column Tests." *Chemical Engineering Journal*, 673–680. doi:10.1016/j.cej.2012.06.116.

Zhao, Peitao, Yafei Shen, Shifu Ge, and Kunio Yoshikawa. 2014. "Energy Recycling from Sewage Sludge by Producing Solid Biofuel with Hydrothermal Carbonization." *Energy Conversion and Management* 78: 815–821. doi:10.1016/j.enconman.2013.11.026.

3 Crop Residue Management for Improving Soil Carbon Storage, Nutrient Availability, and Fertilizer Use Efficiency

Bhupinder Pal Singh, Bijay-Singh, Sarah R. Noack, Yunying Fang, Promil Mehra, Kathryn Page, and Yash P. Dang

CONTENTS

3.1   INTRODUCTION

Soil fertility is fundamental to crop production and depends in part on soil and crop management strategies that improve soil organic carbon (SOC) and enhance the supply of nutrients to crops (Lal 1995; Kumar and Goh 1999; Chen et al. 2014a). Poor soil fertility is a major constraint to crop productivity. Hence, investment in practices that improve and conserve soil resources, such as soil organic matter (SOM), is likely to provide a large economic return to farmers with minimal impact on the environment (Lal 1995; Lal et al. 2004). The appropriate management of crop residues is an example of one practice that can help maintain soil quality and provide carbon (C) and nutrients in soils. In this chapter, crop residue is defined as any aboveground plant biomass that is generated after crop harvesting in agricultural fields.

Globally, total crop residue production (in 2013) was estimated at ~5 billion tons per year, of which 3.6 billion tons was from cereals, 0.6 billion tons from sugar crops, 0.4 billion tons from legumes, 0.3 billion tons from oil crops, and 0.1 billion tons from tubers (Cherubin et al. 2018). Almost all crop residues serve as sources of major nutrients (such as nitrogen [N], phosphorus [P], sulfur [S], and potassium [K]) for subsequent crops over the short and long term; the range of nutrient concentrations reported for a selection of major crops is provided in Table 3.1. Many of these nutrients (N, P, and S) are released in plant-available forms during biological decomposition (a process known as "mineralization"), or in the case of K, via leaching. Some soil-mineral-bound nutrients such as P (and S to a lesser extent) may also be released in plant-available forms via dissolution and desorption reactions of decomposing crop residues with soil minerals (Guppy et al. 2005; Keiluweit et al. 2015; Sarker et al. 2018b, 2019). Nevertheless, nutrient immobilization via microbial utilization or clay fixation/metal complexation may also occur simultaneously (Singh et al. 2006; Noack et al. 2014b; Sarker et al. 2018b, 2019). In addition, some studies have reported that the input of crop residues in soil can enhance the decomposition of native SOM ("positive priming") (Guenet et al. 2010; Qiu et al. 2016; Fang et al. 2018a; Lenka et al. 2019), and has the potential to enhance the quantity of plant-available nutrients from SOM reserves (Sarker et al. 2018b, 2019).

Overall, the net release of available nutrients from crop residues will be dependent on the balance between nutrient mineralization–immobilization and sorption–desorption processes in a particular soil. Consequently, it is challenging to estimate the quantity of nutrients that will become available during the growing season due to these complex biochemical processes. Moreover, there is limited information on the levels of residue- and SOM-derived C mineralization and nutrient release in plant-available forms. This requires much attention to guide predictions of soil C storage, nutrient

TABLE 3.1

Nutrient Content (g kg⁻¹) in Dry Mass of Different Crop Residues

Crop Residue	N	P	K	S	References
	span (g kg⁻¹)				
Wheat	3.0–9.6	0.3–1.8	4.0–19.2	0.3–0.9	van Duivenbooden et al. (1996), Schomberg and Steiner (1999), Mertens et al. (2002), Reddy et al. (2002), Singh and Rengel (2007), Sarker et al. (2018b)
Maize	3.7–11.0	0.3–1.5	6.8–18.8	0.40–0.90	van Duivenbooden et al. (1996), Schomberg and Steiner (1999), Sahrawat et al. (2008), Cherubin et al. (2018)
Rice	4.4–8.2	0.5–1.9	11.8–27.0	0.5–1.0	van Duivenbooden et al. (1996), Dobermann and Fairhurst (2002)
Sorghum	2.2–12.2	0.3–1.5	5.7–20.0	0.49–0.92	van Duivenbooden et al. (1996), Schomberg and Steiner (1999), Sahrawat et al. (2008)
Sugarcane	4.4–12.3	0.8–1.3	3.1–17.2	0.7–4.0	Cherubin et al. (2018)
Canola	3.4–9.1	2.1	13.8	1.4–2.0	Lupwayi et al. (2006), Singh et al. (2006), Sarker et al. (2018b)
Lupin	7.8–10.5	0.3–0.5	9.3–15.6	2.5–2.8	Schultz and French (1978), Singh and Rengel (2007)

supply, and nutrient use efficiency under different crop production systems with variations in the type of tillage, residue, and soil.

This chapter provides a summary of the pros and cons (relating to soil quality and crop yield) associated with key crop residue management strategies under different tillage systems. In addition, the importance of integrated fertilizer management, and how residue management strategies influence soil C and plant-available nutrient dynamics, is discussed. A better understanding of the components of residue management is important for improving soil fertility, meeting crop nutrient demands, enhancing agricultural sustainability, and minimizing nutrient losses (e.g., via leaching or soil erosion).

3.2 CROP RESIDUE MANAGEMENT STRATEGIES: AGRONOMIC IMPLICATIONS

Farmers employ three main strategies to manage crop residues, namely, removal, incorporation, or retention. The removal of crop residues typically occurs via burning, animal grazing, or export off site for use as animal feed, building material, or fuel. Residue incorporation involves mixing residues into the soil using tillage operations, while surface retention requires that residues are left on the soil surface without incorporation and is generally practiced in conjunction with no-till or minimal tillage, in a practice known as conservation tillage. Crop residue retention on the soil surface has been considered a key management practice to increase and/or maintain SOM levels (Singh and Rengel 2007; Mehra et al. 2018a, 2018b; Singh et al. 2018). Residue incorporation may also maintain SOM levels, particularly when practiced in conjunction with an optimal supply of nutrients (Kirkby et al. 2016; Sarker et al. 2017; Bijay-Singh 2018; Fang et al. 2018a). All residue management strategies have advantages and disadvantages in terms of soil and moisture conservation, agricultural operations, and nutrient management, as summarized in Figure 3.1 and Table 3.2 and as discussed below. Overall, the variation in residue management practices observed in the field reflects the diverse nature of cropping systems and individual farmer approaches.

3.2.1 Soil and Moisture Conservation

One of the greatest advantages with residue retention compared to incorporation and removal is the protection it provides to the soil from wind and water erosion. This occurs due to physical protection of the soil surface by residue material, increased aggregate stability attributed to

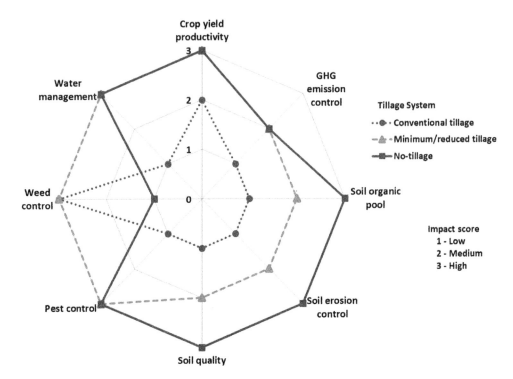

FIGURE 3.1 The radar chart illustrates the impact of different types of tillage on different soil functions and agronomic aspects. These impacts are scored as following: 1 low, 2 medium, and 3 high. Three different levels of tillage practices are conventional tillage, minimum/reduced tillage, and no-till. A conventional tillage system involves the tillage having exhaustive soil disturbance, with or without removal of crop residue. Minimum/reduced tillage is tillage with minimum disturbance and minimum to moderate removal of crop residue. Finally, no-till indicates no tillage and no removal of crop residue. (Modified from Stavi, I. et al., *Agron. Sustain. Dev.*, 36, 32, 2016.)

TABLE 3.2
Advantages and Disadvantages of Different Residue Management Strategies

Residue Management	Advantages	Disadvantages
Surface retention	↓ erosion ↑ soil moisture ↑ soil organic matter and nutrient reserves ↑ soil physical and biological quality ↓ prevalence of some weed and disease species	↓ ease of planting and crop establishment ↓ nutrient availability due to stratification and/or immobilization ↓ soil temperatures ↑ in some weed and disease species ↓ effectiveness of preemergence herbicides
Incorporation	↑ ease of seeding operations ↑ speed of nutrient cycling and crop availability ↓ nutrient stratification ↓ prevalence of some weed and disease species ↑ effectiveness of preemergence herbicides	↑ rates of organic matter decomposition ↓ soil physical and biological quality ↓ soil moisture ↑ erosion
Removal	↑ ease of seeding operations ↓ prevalence of some weed and disease species ↑ effectiveness of preemergence herbicides	↑ nutrient loss ↓ soil physical and biological quality ↓ soil moisture ↑ erosion

additional organic-C/microbiological activity, and decreased runoff intensity as a result of greater infiltration rates (Chan et al. 2003; Mandal et al. 2004; Blanco-Canqui and Lal 2009; Sahu et al. 2015). These characteristics are also commonly reported to increase stored soil moisture, which is important in dryland cropping areas where yield is limited by water availability (Chan et al. 2003; Mandal et al. 2004; Blanco-Canqui and Lal 2009; Avci 2011; Page et al. 2013b; Ranaivoson et al. 2017). The increased soil moisture and insulation of the soil surface in residue retained systems can also decrease soil temperatures. In cooler temperate climates, this decrease in soil temperature has been associated with reduced or slower germination, frost damage, and overall decrease in grain yield (Lyon et al. 2004; Bruce et al. 2005; Blanco-Canqui and Lal 2009; Ma and Rivero 2010).

3.2.2 Agricultural Operations

Retaining crop residues within the farming system has a number of implications for machinery choice (e.g., harvester, seeder) and pest management (e.g., weeds, insects, disease). Retaining residue at the surface adds many physical problems as it can interfere and cause blockages with sowing machinery, whereas incorporation and removal are much more effective at clearing fields after harvest to facilitate smooth planting and sowing operations (Lyon et al. 2004; Scott et al. 2010; Avci 2011; Sahu et al. 2015). Particularly during sowing with no-till implements, high stubble loads (unless pushed away from the seed row) can reduce seed emergence and plant establishment, and thus subsequently affect the yield (Dean and Merry 2015). Studies like Scott et al. (2010) and Carvalho et al. (2017) indicate that under a high stubble load scenario, residues may also decrease the effectiveness of preemergence herbicides, as they become bound to the residues, which may result in poor weed control.

The incidence of weeds, pests, and plant diseases can also be affected by residue management, although impacts can vary depending on the environment favored by the weed/pest/disease in question. For example, some weed species (*Chenopodium album, Digitaria sanguinalis, Portulaca oleracea* L., *Phalaris minor*, etc.) can be suppressed by residue retention due to residues physically impeding weed growth, inhibiting weed germination due to allelopathic effects, and increasing weed seed predation (Scott et al. 2010; Chauhan et al. 2012; Ranaivoson et al. 2017). However, the soil disturbance associated with residue incorporation/removal can also help to reduce other species by physically disturbing the weed location and/or burying weed seeds deeper in the soil profile, preventing further germination (Page et al. 2013b). Similarly, particular pests and diseases can either be suppressed or encouraged by residue retention, depending on their required environmental conditions.

3.2.3 Nutrient Management

The most common management practices that conserve crop residues (retention and incorporation) are generally associated with increased organic C and plant nutrients relative to systems practicing residue removal (Figure 3.1) (Mandal et al. 2004; Govaerts et al. 2007; Carvalho et al. 2017; Ranaivoson et al. 2017; Cherubin et al. 2018). Consequently, the retention of residues may lead to reduced fertilizer requirements over the long term, depending on the quality and quantity of residue (Scott et al. 2010; Page et al. 2013b; Sahu et al. 2015; Ranaivoson et al. 2017). However, it should be noted that in some instances the capacity of residues to meet the nutrient requirements of the subsequent crop may be limited – for example, some studies have observed that, when wheat residues are incorporated, this has the capacity to fulfill only 1%–6% of the N requirement (Kirkegaard et al. 2018). In addition, where residues have high C-to-nutrient ratios, nutrient immobilization can occur, and, in the short term, fertilizer will be required to meet the crop demand.

Compared to surface retention, the incorporation of residues increases the rate of decomposition, which can accelerate the release of plant-available nutrients (Bailey and Lazarovits 2003;

Yadvinder-Singh et al. 2005; Sarker et al. 2019). In some instances, while there may be productivity benefits to the current or subsequent crops, enhanced nutrient cycling may lead to losses via leaching or gaseous pathways if plant growth is not synchronized with nutrient release (Mubarak et al. 2002). The soil disturbance created by crop residue incorporation also accelerates the breakdown of native SOM (Balesdent et al. 2000; Christensen 2001), which can lead to overall declines in SOM, as well as soil physical and biological quality (Chan et al. 2003; Page et al. 2013a). However, where incorporation and soil mixing does not occur, the stratification of less mobile plant nutrients (such as P and K) can occur at the soil surface, creating problems in drier regions when plants are unable to extract nutrients from the surface soil at times of low soil moisture (Bockus and Shroyer 1998; Chan et al. 2003; Scott et al. 2010; Dang et al. 2015). The implications of crop residue and fertilizer management for soil C and nutrient cycling and fertilizer use efficiency are discussed further below.

3.3 CROP RESIDUE MANAGEMENT AND SOIL CARBON DYNAMICS

Net SOC storage in croplands is related to the balance between organic C inputs and outputs. This balance is impacted by many agricultural practices, such as the rate of crop residue return, tillage system, and fertilizer inputs. Other processes, such as microbial decomposition and erosion, can also have an impact on soil C dynamics (Van Wesemael et al. 2010; Powlson et al. 2014; Singh et al. 2018; Zhao et al. 2018). Increasing SOC storage via best management practices that translate the field-scale input of organic C resources (such as crop residues) into stable SOC pools is important for mitigating climate change (Dignac et al. 2017; Chenu et al. 2018; Singh et al. 2018). In terms of soil health and fertility, both stable and labile pools of SOM also contribute to improving soil water retention capacity (Rawls et al. 2003), soil structure (Six et al. 1999, 2000; Sarker et al. 2018a), and plant-available nutrients (Sarker et al. 2018b, 2019), with implications for improving crop yields.

3.3.1 Crop Residue Management, Tillage Systems and Soil Carbon Storage

Crop residue retention combined with conservation tillage is a key strategy that minimizes SOC loss, increases plant C input, and helps to maintain or increase SOC storage at the soil surface (Powlson et al. 2014; Zhao et al. 2018). In contrast, crop residue incorporation using conventional tillage enhances SOC loss (both residue-derived and native SOC) by disturbing soil structure and increasing organic C mineralization (Raiesi 2006; Lal et al. 2007; Sarker et al. 2018a, 2018b). From a global database of 67 long-term agricultural experiments, West and Post (2002) found higher SOC levels under no-till than conventional tillage systems, and calculated that there was an average SOC sequestration rate under no-till of 0.57 t C ha^{-1} yr^{-1} up to 30 cm depth. However, it should be noted that the results of West and Post (2002) were only derived from analysis of the top 30 cm of the profile. When the distribution of SOC is considered in the soil profile, meta-analysis and long-term studies have identified that conservation tillage mainly results in SOC gains on the soil surface, with no change or loss of SOC in deeper soil layers and vice versa for residue incorporation by tillage (Angers and Eriksen-Hamel 2008; Luo et al. 2010; Aguilera et al. 2013; Powlson et al. 2014). For example, a meta-analysis by Luo et al. (2010) compared 69 sets of paired data for no-till and conventional tillage and showed a net gain in SOC stocks in the 0–10 cm layer under no-till, relative to conventional tillage. However, a net loss of SOC under no-till was found in the 10–40 cm layer, while SOC stocks were similar between these contrasting tillage systems in deeper (40–60 cm) soil layers (Luo et al. 2010). Since the soil inversion by tillage would move surface SOC to lower depths (Olson and Al-Kaisi 2015), there may not be any overall change in SOC stocks in the entire soil profile (Powlson et al. 2014). This highlights the importance of considering the entire profile (e.g., 0–60 cm) to thoroughly assess the influence of residue management practices on SOC gains or losses (Powlson et al. 2014).

In some agricultural systems, the practice of residue retention has been threatened by difficulties in controlling pests, and the nutrient stratification that often emerges due to an absence of soil mixing (Dang et al. 2015). These concerns have prompted the use of strategic tillage (an occasional onetime tillage event in a continuous no-till managed system) to help mitigate some of the negative impacts of no-till systems. Although the adoption of strategic tillage is still in its infancy, some recent studies have indicated that tillage can be occasionally introduced for pest and weed control, without any long-term negative impacts on crop yield and soil quality (Dang et al. 2018; Mehra et al. 2018a; Conyers et al. 2019). It should be noted that as conventional tillage operations are known to enhance SOC loss, it is possible that the concept of introducing strategic tillage could lead to overall declines in SOC (Stockfisch et al. 1999). However, there is currently limited research to fully understand the effect of strategic tillage on SOC stocks in long-term no-till systems.

3.3.2 INTEGRATED CROP RESIDUE AND FERTILIZER MANAGEMENT FOR SOIL CARBON STORAGE

Fertilizer addition can increase crop yield while also providing nutrients for building SOC (Paustian et al. 1992). Recently, crop residue input in combination with an adequate supply of nutrients (termed integrated nutrient management; see Fang et al. 2018a) has been suggested as a strategy to meet crop demand, while promoting humification (stabilization) of residue-C into a stable fine fraction SOC (Kirkby et al. 2013). For example, Kirkby et al. (2014) reported an increase of net humification of wheat straw by two- to eight-fold when inorganic-N, -P, and -S were added along with wheat straw to four soils with different textures (sand, sandy loam, sandy clay loam, and clay loam). Another field-based study (~5 years) showed that when supplementary nutrients were applied together with incorporated wheat or canola residues to a sandy loam soil (10–15 cm), SOC stocks to 1.6 m depth increased by 5.5 t C ha^{-1}. Where no supplementary nutrients were added, SOC stocks decreased by 3.2 t C ha^{-1} (Kirkby et al. 2016). Similarly, using a meta-analysis based on 257 published studies, Lu et al. (2011) revealed that despite increased soil respiration, fertilizer N application caused a 3.5% increase in SOC storage in croplands, possibly linked with increased belowground biomass C inputs. It should be noted, however, that negative impacts of long-term integrated crop residue and fertilizer N inputs on SOC stocks have also been reported. For example, long-term field trials (up to 50 years) in North America have shown that crop residue and N fertilizer inputs decreased SOC stocks, likely via increasing decomposition of residue and SOM (Russell et al. 2005; Khan et al. 2007). Nevertheless, in a recent review, Bijay-Singh (2018) concluded that fertilizer N, when applied as per the need of field crops in a balanced proportion along with other nutrients (such as P and K) or organic manures, could increase SOC storage in agricultural systems.

3.3.2.1 Priming Effect on Native Soil Organic Matter under Integrated Nutrient and Residue Management

The uncertainty surrounding the impact of crop residues on SOC stocks can partly be attributed to the fact that crop residues can enhance the microbial mineralization of native SOM, an effect known as positive "priming effect" (PE) (Kuzyakov et al. 2000; Fontaine et al. 2003). A positive PE induced by crop residue input has been reported by several workers after incorporation in soil under laboratory conditions (Li et al. 2013; Chen et al. 2014b; Fang et al. 2018a). As crop residues add a large amount of labile organic C to soils, and may enhance enzyme activities and change microbial communities (Fang et al. 2018a), there can be a strong, short-term increase in the mineralization of native SOC following residue addition (Guenet et al. 2010; Kirkby et al. 2014; Liu et al. 2014). The growth of some microbial groups (such as oligotrophs) may facilitate the use of nutrients from native SOM, leading to "microbial nutrient mining" after the input of low-quality crop residues, for example, with high C-to-nutrient ratios (Fierer et al. 2007; Kaiser et al. 2014). Moreover, although exogenous nutrient supply (such as N, P, or S) may alleviate the positive PE, nutrient supply in soils amended with a sufficient amount (20 g kg^{-1} soil) of crop residues may accelerate mineralization of both residue-C and native SOM-C via increasing growth of soil microbial communities (Fang et al. 2018a). This process may contribute to a longer phase of positive PE, via supporting "microbial stoichiometry decomposition" (Fierer et al. 2007; Blagodatskaya and Kuzyakov 2008; Fontaine et al. 2011; Ramirez et al. 2012; Chen et al. 2014b).

3.3.2.2 Impact of Fertilizer Nutrient Inputs on Microbial Use Efficiency and Formation of Soil Organic Matter

There are currently two different concepts on the formation of SOM. The first of these asserts that recalcitrant fractions of plant-derived materials (such as lignin and lipids) are selectively preserved and become a substantial portion of stable SOM (Angst et al. 2017; Almeida et al. 2018). However, some decomposition studies using plant residues enriched with ^{13}C and ^{15}N observed that, at an early stage of decomposition, readily decomposable residues (such as plant leaves) contribute more to the accumulation of soil C and N than slowly decomposable residues (such as stems and roots) (Bird et al. 2008; Rubino et al. 2010). This prompted the theory that SOM formation and stabilization occurs through a microbial-driven pathway, mainly as a result of microbial products (e.g., necromass and exudates) produced during crop residue decomposition (Cotrufo et al. 2015). It is believed that these microbial products are incorporated into soil mineral fractions, or act as binding agents to form aggregates, and are thereby protected from decomposition (Bradford et al. 2013; Cotrufo et al. 2013; Cotrufo et al. 2015; Kallenbach et al. 2016). In other words, compared to relatively recalcitrant plant residues, readily decomposable residues may be more likely to contribute to stable SOC pools due to physicochemical protection (Cotrufo et al. 2015).

The degree to which products from microbial decomposition are able to contribute to SOM can depend on a range of factors. The soil microbial biomass generally has a much lower C:N:P ratio than crop residue (Xu et al. 2013; Lal 2014). As such, the nutrient demand (such as N and P) of the microbial community can control microbial growth and the use of crop residues. The large stoichiometric difference between the decomposers and the externally applied substrates (such as crop residues) drives the immobilization of nutrients (such as N and P) in the soil (Manzoni et al. 2008; Zechmeister-Boltenstern et al. 2015). Mineralization of residue-C increases as the mineral-N to residue-C ratio increases (Recous et al. 1995; Henriksen and Breland 1999), which can potentially increase the incorporation of residue-C in stable SOC pools. A wide range of C:N values for the conversion of residue-C to stable SOC have been reported, ranging from as low as 0.053 up to 0.43 (Kirkby et al. 2013).

Microbial carbon-use efficiency (CUE) can also affect the conversion of residues to SOM. Microbial CUE is normally defined as the fraction of microbial assimilation of substrates that is allocated to growth (via biosynthetic processes) over substrate C uptake (growth + total respiration) (Manzoni et al. 2012). Microbial CUE relates to the conversion of plant-produced C into microbial products that contribute to stabilized SOM and has been used as an indicator of stable SOM formation (Cotrufo et al. 2015). As integrated residue and nutrient management increases microbial biomass and CUE, this may lead to an increase in SOC storage (Bradford et al. 2013; Cotrufo et al. 2015; Fang et al. 2018b).

3.3.3 OTHER FACTORS IMPACTING SOIL CARBON DYNAMICS IN AGRICULTURAL SYSTEMS

In addition to tillage practices and integrated crop residue and fertilizer management, the formation and storage of SOC in agricultural systems can also be influenced by other factors such as residue quality, soil texture/mineralogy, and climate, as discussed below.

3.3.3.1 Crop Residue Quality

Crop residue quality can impact SOC formation and stabilization dynamics (Adair et al. 2008). High-quality residues are easily decomposable, characterized by low C:N ratios and low phenol/lignin concentrations (Castellano et al. 2015). While the influence of residue quality on SOM formation is still under debate, a review of 13 publications by Castellano et al. (2015) concluded that residue quality was not correlated with the stabilization of residue-C in SOM. This may be due to several reasons. First, during the conversion of plant residue to SOM, the residue may mainly contribute to free and occluded particulate organic matter in soil aggregates, whereas microbial residues may be dominant in the mineral-associated (stable) SOM pool over time. Second, the effect of litter quality on SOM stabilization is modulated by the extent of soil C saturation (i.e., soils have a finite capacity to store C within relatively stable pools in the mineral soil matrix) (Castellano et al. 2015). Third, there could be higher positive PE by residues with high C:N ratio (compared to low C:N), possibly induced by N limitation.

3.3.3.2 Soil Type and Climate

The interaction of crop residue–derived C with clay minerals (Fe-, Al-, Mn-oxides, phyllosilicates) and metal ions can limit SOC accessibility and enhance SOC storage (von Lützow et al. 2006; Cotrufo et al. 2015). Hence, soil type and mineralogy can play an important role in SOC stabilization (Golchin et al. 1994; Sollins et al. 1996; Torn et al. 1997; von Lützow et al. 2006). For instance, in long-term (8–10 years) field experiments, Jenkinson (1977) found that the incorporation of residue-derived C into soil increased with increasing soil clay content. Another study found a greater retention of residue-derived C in subsoil that had higher clay content, possibly from stabilization of leached soluble substrates and their microbial products (Ladd et al. 1985).

Similar to soil type, crop residue management and tillage practices and climatic conditions may also have an important impact on SOC storage in agricultural systems with a diversity of best management practices (Ogle et al. 2015; Fujisaki et al. 2018). For example, under humid and subhumid climates, no-till or reduced tillage (compared to conventional tillage) has been found to increase SOC storage (Francaviglia et al. 2017), whereas under semiarid regions, different tillage and crop rotation systems have had minimal impact on SOC storage (Fang et al. 2016; Francaviglia et al. 2017; Sarker et al. 2018a). In a meta-analysis of studies worldwide, Ogle et al. (2015) observed greater increases in SOC upon conversion from conventional tillage to no-till (estimated after a 20 year period) in tropical moist (23% increase) > tropical dry (17% increase) > temperate moist (16% increase) > temperate dry (10% increase) climates. Hence, agricultural management impacts on SOC storage and dynamics can be sensitive to climatic conditions (moist versus dry) in different agroregions (tropical versus temperate), which may be further driven by plant-derived C inputs, particularly in tropical croplands (Fujisaki et al. 2018), with a greater influence on SOC priming (Lenka et al. 2019).

3.4 CROP RESIDUE MANAGEMENT AND SOIL NITROGEN DYNAMICS

Most crop residues are a poor source of N, which is a key nutrient that limits crop production in agricultural systems. When crop residues with high C:N ratio are incorporated into soil, microbial N immobilization and a temporary decrease in plant-available N typically occurs. This initial phase of net N immobilization, which lasts for several weeks after residue incorporation, is followed by net N mineralization (Yadvinder-Singh et al. 2005; Roy et al. 2011; Chen et al. 2014a; Turmel et al. 2015). Factors such as the decomposition rate, residue quality, and environmental conditions determine the duration of net N immobilization and the net supply of N from crop residues and native SOM to the subsequent crop. The application of crop residues without supplemental fertilizer N will not generally meet crop N demand, and thus may lead to yield decline. However, the return of crop residues over the long term may lead to a buildup of readily mineralized organic soil N, and potentially a reduction in N fertilizer requirements.

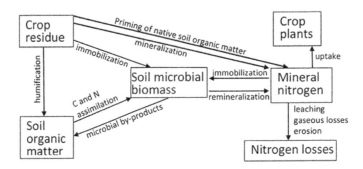

FIGURE 3.2 Schematic diagram showing different pathways of nitrogen in crop residue amended soils.

A schematic diagram showing the flows of N within a cropping system is depicted in Figure 3.2. Although it is evident that N is released from crop residues in both organic and inorganic forms, most organic N is not available to plants directly. While a small portion of crop residue N may be mineralized immediately after application, a larger portion will become immobilized in the soil microbial pool, later to be mineralized or transformed into other SOM pools as microbial by-products (Heijboer et al. 2016; Kopittke et al. 2018; Sarker et al. 2018b). This mineralized N may be taken up by crop plants, recycled in the microbial biomass, or lost from the soil-plant system via leaching, erosion, or in gaseous forms (Mary et al. 1996). A portion of the crop residue N may enter the complex SOM pools (Jansson and Persson, 1982) or organomineral fractions (Lehmann and Kleber 2015). According to Kindler et al. (2009), residue-derived N is introduced into SOM pools as microbial residues after cell death. In fact, N in microbial residual products such as empty hyphae, dead microbial cell residues and cell exudates (Müller et al. 1998; Nishio and Oka 2003; Shindo and Nishio 2005) may end up in different organic N fractions, which are further assimilated by the soil microbial biomass, resulting in additional mineralization (Myers et al. 1994). Immobilization occurs simultaneously with mineralization, and the rate at which crop residue and native SOM–derived N becomes available in the inorganic N pool (mineral N) depends on the net balance of the N mineralization–immobilization turnover (NMIT). The priming effect of crop residues on the turnover of native SOM may also influence this process (see details in Sections 3.3.2.1 and 3.4.1).

3.4.1 Nitrogen Mineralization–Immobilization Turnover in Crop Residue Amended Soils

The NMIT revolves around the growth and death of the soil biota. Mineralization is a sequence of enzymatic reactions that convert organic N into inorganic forms, the first of which is ammonium (NH_4^+). This NH_4^+ may be used by the soil microorganisms conducting mineralization for growth or excreted from the cell into the soil solution where it is free to be used by plants or other micro-organisms (Ladd and Jackson 1982). Where organic material contains insufficient N for growth, microorganisms will assimilate the required N from the soil (immobilization). The preferred inorganic N source for assimilation by most bacteria and fungi is NH_4^+, although nitrate (NO_3^-) is also used (Marzluf 1997; Myrold and Posavatz 2007). When microbes die, the immobilized N in microbial residues/products transfers to either stable SOM pools (Kopittke et al. 2018) or is converted back to inorganic N (referred to as remineralization).

The C:N ratio of residues is one of the most important factors controlling the NMIT balance, although environmental factors such as climate and soil type can also influence the rate of the transfer between the pools. In general, residues with a wide C:N ratio decompose more slowly than those with a narrow C:N ratio, and plant residues with high N content show high decomposition rates and release of mineral N (Janzen and Kucey 1988; Lynch et al. 2016). Recently, Chen et al. (2014a) attempted to define the C:N ratios at which mineralization, immobilization–mineralization, and

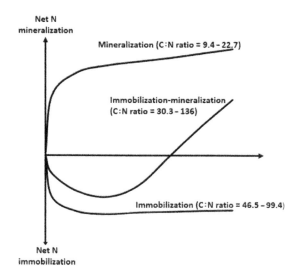

FIGURE 3.3 Schematic diagram showing dynamics of immobilization and mineralization processes as a function of time and C:N ratios of crop residues applied to the soil at zero time. (Modified from Chen, B. et al., *Agron. Sustain. Dev.*, 34, 429–442, 2014a.)

immobilization occur (Figure 3.3). Chen et al. (2014a) determined that the addition of residues with C:N ratios between 9.4 and 22.7 would lead to mineralization of N, immobilization–mineralization would occur at ratios between 30 and 136, and net immobilization would occur at ratios between 47 and 99. The dividing line between immobilization–mineralization process and immobilization is not discrete and can depend on the time period over which monitoring occurs. For example, Mohanty et al. (2010) observed only N immobilization with wheat straw having a C:N ratio of 79, but Hadas et al. (2004) reported the occurrence of an immobilization–mineralization process curve with wheat residues having a C:N ratio of 136. This occurred primarily because Hadas et al. (2004) studied NMIT for a longer period.

In annual cropping systems (based on 2 or 3 crops), the remineralization of N following immobilization may often not be observed during the growing period of the subsequent crop. Indeed, a net N immobilization phase followed by a net remineralization phase, as shown in Figure 3.3, is often observed when residues of cereal crops like rice and wheat are applied to the soil (Müller et al. 1988; Yadvinder-Singh et al. 1988; Mishra et al. 2001a, 2001b). According to Yadvinder-Singh et al. (2004), if a crop of wheat was to be planted after rice, for example, in the Indo-Gangetic plain in South Asia, rice straw decomposition over a period of 20 days or more is required before planting wheat to avoid low N levels in soil up to the crown root initiation phase. Nitrogen transformations, in general, and crop residue NMIT markedly differ in upland (aerobic) and lowland (anaerobic) soils, likely due to the difference in the activity of microorganisms under these contrasting conditions. Nitrogen in crop residues is released in larger quantities under anaerobic conditions, although the release rate may be slower (Liu et al. 1996). Thus, the long-term incorporation of rice residue under anaerobic soil conditions can increase readily mineralizable organic N that can translate into reduced fertilizer N rates for optimal rice yield (Bird et al. 2001; Eagle et al. 2003).

Nitrogen mineralization from residues may also be influenced by soil type. For example, in soils with higher clay content, N mineralization may be lowered due to the adsorption of organic N by clays (Kopittke et al. 2018). It should be noted, however, that in flooded soils, clay content may have a positive effect on N mineralization, with Becker et al. (1994) recording almost twice the mineralization of N from crop residues in a flooded clayey soil compared to a sandy one. Differences in the population and C:N ratio of microorganisms and microfauna between different soil types may also impact N mineralization (Cabrera et al. 2005). Soil acidity can also be influential, with strongly acidic conditions retarding both the decomposition of crop residues and the mineralization of residue N. For example, Fu et al. (1987) reported increased mineralization of N in crop residues when soil pH increased from 5 to 7. Thus it is important to consider differences between soil types, as they can have significant implications in terms of N cycling and thus fertilizer N requirements.

3.4.2 PLANT AVAILABILITY OF NITROGEN IN SOIL AS INFLUENCED BY CROP RESIDUE MANAGEMENT

The synchronism of the NMIT dynamics with plant N demand plays a big role in influencing N nutrition of crops in residue amended soils. Obviously, if synchronization is poor, N assimilation by the crop is adversely affected. Field crops generally remove 70% to 80% of their total N requirement from the soil during the vegetative growth stage. Thus, yields will be adversely affected if not enough N is available during this stage of crop growth. The addition of residues from rice, wheat, barley, maize, and other small grains with high C:N ratios will lead to initial N immobilization, which will adversely affect N nutrition and yield of the crop (Christensen 1986; Cassman et al. 1997). The availability of N from crop residues and native SOM to subsequent crops is also determined by the decomposition rate, residue quality, and soil management as well environmental conditions (Sarker et al. 2018b, 2018c).

A study by Singh et al. (2007) found that synchrony between N release and plant uptake was best achieved in soil receiving wheat residues (high C:N ratio) along with green manure (low C:N ratio). This is because a temporary lag in N mineralization (from green manure) brought by N immobilization (from wheat residue) provided an N conserving mechanism for the system. Similarly, Kaewpradit et al. (2009) demonstrated that mixing residues of groundnut and rice could delay N release during the prerice lag phase, leading to an improved synchrony in N demand/supply and increased growth and yield of the succeeding rice and reduced N losses from the soil–plant system. In another approach, Gentile et al. (2008) applied a combination of crop residue and fertilizer N to four differently textured soils and observed that the interactive effect of combining fertilizer with residue on mineral N changed from negative to positive with increasing residue quality. Although, in practical terms, it is hard for farmers to apply residue mixes, capitalizing on the interactions between fertilizer and organic residues can allow for the development of sustainable nutrient management practices. Thus, even if there occurs net immobilization as a part of the NMIT, the uptake of N by the crops will not necessarily be reduced if there is sufficient synchronization between the changing soil inorganic N (via a mix of residue type and fertilizer inputs) and the crop N demand.

3.4.2.1 Crop Residue Management and N Nutrition of Crops Grown under Aerobic (Upland) Soil Conditions

The way in which crop residues are managed can often have a significant influence on the N nutrition of crops grown under aerobic conditions. The length of time between residue addition and crop growth, in particular, can have a significant impact. For example, Ichir and Ismaili (2002) observed a five-month N immobilization period in the 0–15 cm soil layer when high C:N ratio wheat residues were returned to the soil, leading to reduced dry matter wheat yield and N accumulation. However, this N immobilization decreased in the order of 61.6, 46.4, and 30.0 mg kg^{-1} when residues were returned at seeding, and 15 and 30 days before seeding, respectively. Similarly, Thomsen and Christensen (1998) reported that incorporation of low-quality (high C:N ratio) crop residues reduced the yield and N uptake of the first barley crop, but not of sugar beet due to its longer growth period. To reduce immobilization of applied N, crop residues should be allowed to decompose for some time before sowing. For example, in the rice–wheat cropping system in South Asia, in situ incorporation of rice straw into the soil about three weeks before sowing of wheat is recommended (Yadvinder-Singh et al. 2004).

Many studies have also observed that residue quality can influence immobilization and the degree of synchronization between N release and crop needs, thus affecting subsequent crop yield

(Trinsoutrot et al. 2000; Kumar and Goh 2002; Gentile et al. 2009). For example, Trinsoutrot et al. (2000) observed that crop residues with low N concentrations resulted in net N immobilization (to the extent of −22 to −14 mg N g^{-1} of added C), whereas residues with high N concentrations induced very small net immobilization or mineralization (−3 to +4 mg N g^{-1} of added C), thereby offering improved synchronization with plant N needs. Similarly, Gentile et al. (2009) showed that application of Tithonia residues resulted in an early season N release of 22 kg N ha^{-1}, whereas maize residues led to an immobilization of 34 kg N ha^{-1}, due to differences in their C:N ratios. Interestingly, Hemwong et al. (2008) reported that application of high C:N ratio sugarcane straw did not significantly affect the growth of legumes due to their N$_2$ fixation capabilities.

3.4.2.2 Crop Residue Management and N Nutrition of Rice Grown under Anaerobic (Lowland) Soil Conditions

The role of crop residues in the N nutrition of lowland rice grown under anaerobic soil conditions can differ from that observed for upland crops. Rice can take up more organic N than any other crops because (1) it can take up NH$_4^+$, amino acids, or relatively large organic N molecules; (2) it can compete better than other crops with soil microorganisms for N uptake; (3) its roots secrete organic substances that support the multiplication of microbes, thereby leading to the rapid decomposition of crop residues; and (4) it has high Km (Michaelis constant), Vmax (maximum uptake velocity), and Cmin (minimum concentration of a nutrient) for N uptake (Yamagata and Ae 1996). Thus, rice is expected to respond to crop residue application better than upland crops, although as in upland soils, the application of crop residues has been shown to depress the NH$_4^+$-N concentration in soil and flood water due to N immobilization leading to lower N uptake by rice (Huang and Broadbent 1988; Nagarajah et al. 1989). In general, however, the continuous application of crop residues to lowland rice builds up SOM and ensures high N content and uptake and reduced fertilizer N requirement (Kosuge and Zulkarnani 1981; Eagle et al. 2000; Witt et al. 2000; Dobermann and Fairhurst 2002). For example, Eagle et al. (2003) applied ^{15}N labeled rice straw and observed no effects on rice grain yield, but N uptake increased when straw was incorporated for five years. It was concluded that fertilizer N rates could be reduced by at least 12 kg N ha^{-1} when straw was incorporated. Similarly, Cao et al. (2018) studied the fate of ^{15}N labeled urea in rice in the presence and absence of wheat straw (C:N ratio 144). Their results suggested that the addition of wheat straw can increase plant N uptake by contributing to N immobilization early in the season and subsequent mineralization later in the season. In a rice-rice cropping system in the Philippines, Witt et al. (2000) observed that early residue incorporation improved the congruence between soil N supply and crop demand, and grain yields of rice were 13%–20% greater with early (63 days before transplanting) compared to late (14 days before transplanting) residue incorporation. In irrigated systems it may also be possible to achieve greater release of N by using alternate wetting and drying of the soil to hasten residue decomposition (Kongchum et al. 2006).

3.4.2.3 Placement of Crop Residues in the Soil and N Availability to Crops

Crop residue placement changes the microclimate at the soil-crop residue interface, while the enhanced soil-residue contact that occurs when residues are incorporated influences several soil biological, chemical, and physical parameters that can modify residue decomposition (Fruit et al. 1999). Incorporation of crop residues is associated with bacterial-based food webs that support fast rates of litter decomposition and nutrient mineralization compared to surface-applied residues, which are decomposed via slower, fungal-based systems (Beare et al. 1996). The slower rates of decomposition in surface-applied residues can lead to lower rates of N immobilization (Coppens et al. 2007; Li et al. 2013), although the time period over which immobilization occurs can be longer. For example, Schomberg et al. (1994) reported that immobilization of N from surface-applied cereal residues continued for more than a year, while immobilization only lasted for about four months for soil incorporated residues. The location of residues can also result in differences in the distribution of N in the soil profile. Coppens et al. (2007) revealed that surface retention of residues can lead to greater mineral

N accumulation near the soil surface compared to residue incorporation, and that this may increase the risks of nitrate leaching, particularly where soil water content is greater due to residue retention.

3.4.2.4 Losses of Inorganic Nitrogen from the Soil-Plant System Amended with Crop Residues

Changes in mineral N in the soil following the application of crop residues not only affect the availability of N to crop plants, but also the losses of N from the soil-plant system. Crop residues with high C:N ratios can strongly immobilize N released during decomposition, thus reducing N losses during leaching, or as nitrous oxide emissions during nitrification/denitrification (Gentile et al. 2009; Muhammad et al. 2011; Chen et al. 2013). This immobilized N may then be conserved until a later crop growth phase (Sugihara et al. 2012). However, N released during the decomposition of low C:N ratio residues (e.g., <45) may lead to losses of N from the soil-plant system via leaching and gaseous emissions if the release of N is not synchronized with crop demand (Thomsen and Christensen 1998; Chen et al. 2013).

3.4.3 FERTILIZER NITROGEN MANAGEMENT UNDER DIFFERENT TILLAGE AND CROP RESIDUE MANAGEMENT SCENARIOS

In intensive cropping systems, fertilizer N management needs to help supplement the release of N from crop residues so that soil N supply matches the temporal demand of the crop (Cassman et al. 2002). In cereal cropping systems with returned crop straw, a substantial portion of the fertilizer N would probably be needed during early crop growth to counter N immobilization, while less is required at later stages when increased N release during residue decomposition occurs. Leaf color charts and chlorophyll meters can be used to help fine-tune the supply of fertilizer N and optimally supplement N supplied from mineralization of crop residues (Witt et al. 2005; Varinderpal-Singh et al. 2011, 2012; IRRI 2018).

3.4.3.1 Changes in Fertilizer N Requirements with Residue Management

The retention of cereal crop residues is often observed to have only a limited capacity to supply the N demands of subsequent crops due to their relatively low N contents and high C:N ratios. For example, using a litter bag decomposition technique, Yadvinder-Singh et al. (2004) studied the in situ N release dynamics during the wheat growing season under upland conditions for 7.1 t ha^{-1} rice residues containing 40 kg N ha^{-1}. In this study it was revealed that the total amount of N released during the life span of the wheat crop (~150 days) only ranged from 6 to 9 kg N ha^{-1}. Indeed, in studies where incorporated residues were added along with adequate amounts of N, P, and K fertilizers, it has been observed that yield increases are rarely observed in residue amended soils, and where they are, they are generally not related to the contribution of N contained in the residues (Bijay-Singh et al. 2008). For example, Thuy et al. (2008) observed that while the net supply of plant-available N from rice and wheat residues ranged from 3 to 14 kg N ha^{-1}, the return of crop residues had no net benefit to crop yield when fertilizer N was supplied at recommended rates. A recent study has also shown that wheat or canola crop residues can result in a positive PE on native SOM in soil aggregates, which indicates that in certain circumstances residue addition may lead to an increased release of N from SOM (Sarker et al. 2018b). However, in this study no increase in plant-available N was observed after 126 days, possibly because of the dominance of microbial N immobilization relative to mineralization induced by the nutrient-poor residues (Sarker et al. 2018b). Overall, it is clear that because the amounts of N released from the incorporated residues with low C:N ratios are small, significant savings in fertilizer N input are unlikely in these systems. It should, however, be noted that when Robertson and Thorburn (2007) studied the effect of returning sugarcane harvest residues to the soil for up to six years they found that SOC increased by 8%–15%, total soil N increased by 9%–24%, and inorganic soil N increased by 37 kg ha^{-1} year^{-1}. Long-term N balance calculations from this study suggested that fertilizer N application should not be reduced in the first six years, but small reductions may be possible in the longer term (>15 years).

While the net amount of N released from crop residues may often have a limited impact on crop N supply, residue addition is capable of affecting the agronomic efficiency of fertilizer N (AE_N). For example, both increases and decreases and no change in AE_N have been reported for rice when crop residues were incorporated rather than removed (Bijay-Singh et al. 2008). In Central Thailand, Phongpan and Mosier (2003a, 2003b, 2003c) recorded a 20% to 50% reduction in AE_N when 70 kg N ha^{-1} was applied to dry-season rice grown in a residue amended soil. Thuy et al. (2008) observed no change in AE_N in rice due to incorporation of rice and wheat residues in a three-year study at two locations in China, and in an experiment in the Philippines, AE_N was significantly increased when rice residues were incorporated 20 days before transplanting dry-season rice (Buresh et al., unpublished, as cited by Bijay-Singh et al. 2008). This increase in AE_N was observed due to the increased response of rice to fertilizer N with incorporation of residue, as the incorporation of residue without fertilizer N resulted in a significant reduction in yield due to immobilization.

Whether crop residues are incorporated into the soil or placed on the soil surface can have important implications in terms of fertilizer N management in residue amended soils. A reduction in fertilizer N losses due to crop residue mulch on the soil surface should result in enhanced recovery of applied N by the crop. For example, using a vented chamber method under field conditions, Yang et al. (2015) could measure 7.7% less losses via ammonia volatilization by following stalk mulch–based conservation tillage practices than under conventional tillage (no mulch). Gao et al. (2009) recorded higher wheat grain yields for straw mulch (4.5 t ha^{-1} straw equivalent to half the average wheat straw production in a year spread evenly between wheat rows) than in no mulch treatment when high levels of fertilizer N (120 kg N ha^{-1} or more) were applied. Similar results have been reported by Fan et al. (2005). Rahman et al. (2005) reported a significantly larger fertilizer N recovery efficiency when straw from the previous crop was allowed to remain on the soil surface as mulch rather than when it was removed. As reviewed by Bijay-Singh et al. (2008), the higher frequency of yield gains with mulching than incorporation of residue suggests that these trends were due to increased fertilizer N use efficiency when residues are placed on the soil surface, rather than incorporated into the soil.

3.4.3.2 Timing of Fertilizer N Application to Crops Grown in Residue Amended Soils

Because the incorporation of crop residues often increases N supply to the crop at later growth stages, adjustments in the timing and rate of fertilizer N can help optimize N supply to crops receiving residues. For example, Jiang et al. (1998) reported that N utilization by wheat in the presence of wheat straw (4.5 t ha^{-1}) was the highest when fertilizer N was applied in three equal split doses at sowing, tillering, and stem elongation. Xu et al. (2007) showed that fertilizer N use efficiency in rice could be markedly increased when incorporation of residues from a previous crop of wheat was combined with timing and rates of fertilizer N that better matched the needs of the rice crop. Similarly, Gupta et al. (2017) and Kirkegaard et al. (2018) recommended application of more N (5 kg N for each t ha^{-1} of cereal residue) and early in residue retained no-till systems so as to avoid impacts of N immobilization on crop yield and protein.

3.4.3.3 Fertilizer N Placement Effects in Crops Grown under Conservation Tillage Systems

In conservation tillage systems, the appropriate placement of fertilizers plays an important role in improving nutrient acquisition by the crop and yield. Recent developments in fertilizer application equipment allow fertilizers to be effectively placed below the thatch layer (i.e., between 5 and 10 cm and more than 10 cm) with minimal residue and soil disturbance, thereby reducing the potential for immobilization of N during decomposition of crop residues in the surface layers of the soil (Patel and Murthy 2017). A number of studies have reported greater yields and/or fertilizer use efficiency when using subsurface fertilizer injection in conservation tillage systems, including those producing corn (Vetsch and Randall 2000) and wheat (Kelley and Sweeney 2005, 2007; Yadvinder-Singh and Sidhu 2014). These yield differences are primarily related to the better utilization of subsurface-placed N compared to surface-broadcast N. Machines that allow one-pass seeding and fertilizing as a side band or mid-row band are preferred by farmers for managing N fertilizers and achieving high N use efficiency in no-till systems (Malhi et al. 2001). However, these cannot handle the high fertilizer N rates used in wheat in South Asia. As application of more than 130 kg diammonium phosphate ha^{-1} along with 56 kg urea ha^{-1} adversely affected germination and grain yield of wheat, new-generation planters for sowing wheat in rice residues are being perfected to allow drilling of up to 80% of the recommended fertilizer N dose (120 kg N ha^{-1}) in a separate band between two rows of wheat (Yadvinder-Singh and Sidhu 2014).

3.5 CROP RESIDUE MANAGEMENT AND SOIL PHOSPHORUS DYNAMICS

3.5.1 CROP RESIDUE PHOSPHORUS

The release of P from crop residues can be important to support the growth and yield of subsequent crops (Blair and Boland 1978; McLaughlin et al. 1988a). The most commonly used measure of P in crop residues is total P, although this can be broken down into measures of inorganic and organic components. The C:P ratio of crop residues has also been widely used to predict potential P immobilization or mineralization.

Phosphorus is known to occur in several forms in living plants and crop residues (Bieleski 1973; Mengel and Kirkby 1982). Recently, the speciation of P in fresh and mature plant materials has been conducted using ^{31}P NMR Spectroscopy (Makarov et al. 2002; Makarov et al. 2005; Noack et al. 2012; Noack et al. 2014a). These studies have generally found the main forms of P in residues are orthophosphate, phospholipids, and nucleic acids (e.g., RNA). Phytate has also been detected in chaff residues (up to 45% of total P), but minor amounts (<1%) are found in stem and leaf residues. It is expected that differences in the chemical composition of P among residues plays an important role in how P derived from residues will cycle in soils, and the processes involved. Crop residue P can be released to soil as soluble P (organic and inorganic), assimilated by microorganisms (microbial P), or it can contribute to more chemically stable P pools in soil (Figure 3.4). Crop residue speciation plays an important role in determining the partitioning of residue-derived P into these three pools. The net effect of crop residues on soil P availability will depend on the balance between these processes, although exactly how P availability will be affected is often unclear due to the complexity of P cycling in soil.

3.5.1.1 Release of Soluble Residue Phosphorus

In general, 40%–85% of the P in crop residues is soluble in water and weak acid, although this varies significantly between studies and residue types (Bromfield 1960; Birch 1961; Barr and Ulrich 1963; Floate 1970a; Martin and Cunningham 1973; White and Ayoub 1983; McLaughlin et al. 1988a; Iqbal 2009; Noack et al. 2012). However, the reported inorganic P concentrations in plant material may

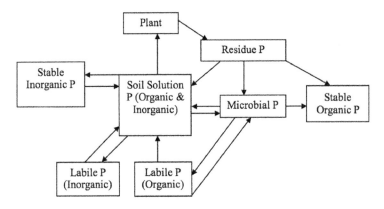

FIGURE 3.4 Role of residue P in plant-soil P cycling. (Modified from Stewart, J. W. B., and Tiessen, H., *Biogeochemistry*, 4, 41–60, 1987.)

be overestimated because some organic P may be hydrolyzed in acid extracts or through the release of enzymes in water extracts (Crowther and Westman 1954; Tadano et al. 1993; Bishop et al. 1994). Nevertheless, this suggests that the bulk of crop residue P has the potential to be brought into solution under field conditions after the first significant rainfall. When this soluble P is released into soil solution, it can be utilized by plant roots and soil microorganisms, sorbed onto soil particles, or precipitated with cations.

3.5.1.2 Microbial Processing of Crop Residues

Microbial processes in soil influence the distribution of P between various inorganic and organic P forms and consequently affect the potential availability of P for plant acquisition. For the soil microbial population to proliferate there needs to be adequate supply of C, N, and P to assimilate and drive growth. As was the case for N, net release of inorganic P to the soil (mineralization) will occur if microbial demand for P is less than the quantity of P contained in the residue, but immobilization will occur if microbial demand for P exceeds the quantity of P present.

3.5.1.2.1 Immobilization of Phosphorus in Crop Residues by the Microbial Biomass

Where P is limiting and C is nonlimiting, soil microorganisms can immobilize residue and soil P during decomposition of low-P plant materials. In a growth chamber study, McLaughlin and Alston (1986) found that P held in the microbial biomass was considerably higher in soils that had received medic residues. In a subsequent field experiment, McLaughlin et al. (1988a) found 29% of medic P was incorporated into microbial biomass 95 days after residue addition. The authors suggested that much of the P initially present in decomposing plant material was not available for plant uptake, as the microbial biomass responded quickly to the change in environmental conditions and rapidly assimilated the P released from residues. These observations are consistent with a wide body of literature (Bünemann et al. 2004; Iqbal 2009; Almeida et al. 2012; Noack et al. 2014b) that shows significant amounts of residue P will be immobilized by the microbial biomass. These studies also show that soil microorganisms are highly efficient in obtaining P to meet their own requirements and that soil microbial populations are more likely to be limited by the availability of C rather than P.

3.5.1.2.2 Mineralization of Microbial Phosphorus

The time needed to complete the process of decomposition and mineralization may range from days to years, depending on the many factors that affect microbial growth, including moisture and temperature. There has been little success in establishing a clear relationship between the turnover of microbial P and P availability to plants (Fabre et al. 1996; Kwabiah et al. 2003). Indeed, radioactive tracer studies have shown that exchange of P between microbes and the soil solution can occur without any net change in the size of the microbial P pool, demonstrating the constant flux of P through this pool (Oehl et al. 2001).

Using mature wheat, canola and pea residues incorporated into a red-brown Chromosol, Iqbal (2009) demonstrated that P present in added residues was not available to crop plants for at least six weeks after addition. Residues were added at a rate of 10 mg P kg^{-1} soil, with three crop growth periods of wheat, each lasting for 28 days. Microbial P was much higher for the residue amended soils and decreased with time as the residue was broken down. As the microbial P pool turned over, some of this P contributed to an increase in available P to plants when grown in soil 56–84 days after mature wheat residue was applied. It was suggested that the rest of the P released from the microbial pool was absorbed by the wheat, precipitated, or sorbed to soil particles.

3.5.1.2.3 Predicting Immobilization and Mineralization of P from Crop Residues

During decomposition and mineralization, different rates of microbial transformation of P are commonly observed and are assumed to reflect the substrate quality. Total P and the C:P ratio of crop residues have been widely suggested as indices of residue quality in terms of its potential to provide P to subsequent crops (Fuller et al. 1956; Enwezor 1976; Singh and Jones 1976; Kwabiah et al. 2003;

TABLE 3.3
Critical Phosphorus Concentration (%) and C:P Ratio beyond Which Immobilization of P Occurred

Authors	Residue/Plant Type	Residue % P	Residue C:P
Fuller et al. (1956)	Wheat, barley, flax straw, tomato, lettuce, alfalfa, clover, and bean tops	0.2%	200
Barrow (1960b)	Phalaris and lucerne	0.2%	55
Floate (1970b)	*Nardus* grass and *Agrostis-Festuca* grass	0.1%	
Blair and Boland (1978)	White clover		150–300
White and Ayoub (1983)	Faba bean		123–251
Kwabiah et al. (2003)	18 plant species		156–252
Iqbal (2009)	Wheat, canola, lupin, pea	0.2%	253

Iqbal 2009). Numerous experiments with crop residues have reported different P concentrations below which P immobilization occurred. Some of these are summarized in Table 3.3. It is apparent that a wide range of critical levels of total P and C:P ratios have been proposed; this is especially the case for C:P ratio, with critical values ranging from 55:1 to 500:1.

In a recent review, Damon et al. (2014) developed an empirical model for the mobilization of P from crop residues to provide a simple predictive tool. This model used a two-stage approach to C and P release, whereby an initial rapid release of soluble inorganic P was followed by a slower release phase of plant-available P from recalcitrant P pools. The model's ability to predict residue P release was largely unaffected by climatic factors but was governed by the microbial biomass (Damon et al. 2014). However, the variability in residue P cycling and crop uptake (due to factors such as residue management and soil pH) mean there are still no simple methods which can be used to advise farmers on residue P release and fertilizer recommendations.

There are also diverging views about the link between C and P cycling (McGill and Cole 1981; Smeck 1985) and predicting when P from crop residues will be released based on these measures has proven more difficult than analogous predictions of N release based on N concentrations or C:N ratio (Kirkby et al. 2011). One difference between P and N is that there is a substantial amount of inorganic P in crop residues (Birch 1961; Jones and Bromfield 1969; Martin and Cunningham 1973; McLaughlin et al. 1988a), whereas the vast majority of N in crop residues is organic. Thus, although the release of mineral N from crop residues almost exclusively involves microbial decomposition of organic forms, release of considerable amounts of mineral P can occur without microbial decomposition.

A second difference between P and N in crop residues is the diversity of organic forms present. Most organic N is present in the amide linkages of proteins (Smernik and Baldock 2005), whereas P is present in a variety of chemically diverse molecules including phospholipids, nucleic acids and inositol phosphates (Bieleski 1973; Noack et al. 2012). Reliable determination of P speciation in crop residues is needed to better predict their contribution to P cycling in soil.

3.5.1.3 Sorption of Phosphorus Released from Crop Residues in Soils

Phosphorus released from crop residues (through leaching or mineralization) into the soil solution is rapidly adsorbed onto clay surfaces (Singh and Jones 1976; White and Ayoub 1983; Friesen and Blair 1988; Bah et al. 2006). Consequently, Umrit and Friessen (1994) found that residue P has a greater effect on nutrient availability in soils with lower P-sorbing capacities. In higher P-fixing soils, soluble P released from the residue and from microbial decomposition was rapidly sorbed to soil rather than remaining available for plants. Similarly, Singh and Jones (1976) used several organic materials, incubated them with soil for 150 days and conducted P sorption and desorption measurements. From the sorption isotherms they showed that, in general, organic materials that contained >0.31% P decreased the amount of P sorbed by the soil and increased the level of P in solution. Hence, if fertilizer was applied after residue incorporation, more of the fertilizer would be available for plant uptake as the soil's P sorbing capacity has been decreased by occupation of P sorption sites by residue-derived P. Whether this effect is significant would depend on the sorption capacity of the soil and the concentration of P in the residue.

3.5.2 CROP RESIDUE MANAGEMENT AND PHOSPHORUS AVAILABILITY IN SOIL

3.5.2.1 Timing and Quantity of Phosphorus Release from Crop Residues

The most important issues for farmers when considering the effect of residue management on P nutrition are whether residue management influences the supply of P to crops, how much P residues supply, and when this P will be available to plants during the growing season. Generally, studies using residues with or without fertilizer have found that 1%–45% of the shoot P uptake comes from residues (Blair and Boland 1978; Till and Blair 1978; McLaughlin and Alston 1986; McLaughlin et al. 1988b; Nachimuthu et al. 2009; Noack et al. 2014b).

Inconsistencies in the timing and quantities of P released from crop residues in the literature reflect differences in residue type, placement, moisture supply, and rate applied. Jones and Bromfield (1969) showed that hayed-off pasture (*Phalaris tuberose* L., now known as *Phalaris aquatica*) under sterile conditions lost 80% of P via leaching (mainly as inorganic P) in the first two weeks across four different treatments (three leaching events with one to four weeks between leaching events). Little additional P was lost after the third leaching event. Under nonsterile conditions, where microbial immobilization was likely to occur, most P was retained in the residues during decomposition and leaching. This data shows that the intensity and duration of the first rainfall affects the amount of inorganic P released into the soil and the amount left in the crop residues.

Friesen and Blair (1988) found 50% of ^{32}P from labeled oat (*Avena Sativa*) residues in the inorganic P soil pool 11 days after addition. In contrast to the previous study, residues in this study were thoroughly incorporated into soil, and pots were only watered to field capacity and not leached. Both studies suggest that the majority of residue P can be released within the first week of addition. In both studies, however, the residues were ground to less than 2 mm, increasing the surface area available for P leaching or microbial attack, which would have increased the release of soluble P.

More recent studies suggest that the release of residue P can be slower than the two previous studies suggested. Noack et al. (2014b) reported that, at day 80 of a glasshouse experiment, >60% of added P from field pea residue was measured in plant, microbial, and resin P pools. Surface or incorporated residues contribute equivalent to 0.6–2.0 kg P ha^{-1} to the subsequent wheat crop at day 80. Similarly, at the end of a 52-week field study, Lupwayi et al. (2007) found that mature pea residues released 27% (0.4 kg P ha^{-1}) and mature canola residues 33% (0.8 kg P ha^{-1}) of total residue P using surface-placed litter bags. Crop residues in the field are likely to experience more frequent extremes in drying and wetting cycles, as well as variable temperatures, leading to the slower release of residue P.

3.5.2.2 Effect of Tillage and Residue Management on the Release of Phosphorus from Crop Residues and Native Soil Reserves

Many studies have demonstrated some effects of tillage and residue management on the distribution of P through the soil profile (Zibilske and Bradford 2003; Bünemann et al. 2006; Noack et al. 2014b). The clearest demonstrations of such effects arise from the comparison of soils under full tillage (full inversion tillage using a plough) versus no-till systems (direct seeding, e.g., using discs and knifepoints). Under no-till systems, crop residues are left on the soil surface and decompose more slowly because less surface area is available for microorganisms to attack, compared to residues that are fully incorporated into soil. Furthermore, decomposition may be limited by environmental conditions as the moisture and temperature in the surface soil (e.g., the top 2 mm) are often warmer and drier than for those for residues incorporated into the top 10–15 cm of soil.

The adoption of no-till systems also has implications for the transfer of residue-derived nutrients such as P into the soil compared with traditional full tillage systems (Buchanan and King 1993; Deubel et al. 2011). The minimization of soil mixing under no-till results in elevated concentrations of nutrients in the top soil layers (e.g., 0–15 cm), compared to the rest of the soil profile. This is commonly referred to as nutrient stratification and results from increased levels of organic matter and fertilizer left on the surface (Crozier et al. 1999; Saavedra et al. 2007). Mobile nutrients like N and S can move deeper into the profile; however, immobile nutrients like P tend to be concentrated in the top soil layers. This poses an issue for predicting the quantity of P available to a subsequent crop, as the rate of P cycling in these enriched surface layers will differ from that of P cycled in the remainder of the soil profile.

The mechanisms of P release may also differ between surface and incorporated residues. For example, P concentrations in surface-placed residues experience more frequent fluctuations in response to wetting and drying events compared to residues incorporated below the soil surface (Jones and Bromfield 1969; Sharpley and Smith 1989). Jones and Bromfield (1969) found that inorganic P was readily leached from crop residues when microbial activity was inhibited. Wetting events, therefore, may be the dominant mechanism contributing to loss of P from surface-placed residues, given the likely lower levels of microbial activity at the surface and the high solubility of P in crop residues. Similarly, in an incubation study using six residue types, Sharpley and Smith (1989) found greater amounts of inorganic P were leached from surface-placed residues compared with incorporated residues. However, although residue P has been shown to be more readily leached in no-till systems, the ability of a subsequent crop to access this residue-derived P has received little attention. A better understanding of stratification and leaching of P from residues under no-till management would support better fertilizer recommendations.

The cycling of SOM and nutrients from crop residues will also be affected by strategic tillage operations. Tillage practices that invert the soil are likely to address the issue of nutrient stratification, although less aggressive forms of tillage, such as chisel ploughs, have been observed to have limited ability to redistribute nutrients from the surface into the subsoil (Dang et al. 2018). Wortmann et al. (2010) found that strategic tillage had no significant effect on stratification of soil P, crop yield, soil aggregation, SOC, bulk density, or soil microbial biomass, with values measured five years post cultivation, similar to that of long-term no-till treatments. Asghar et al. (1996) conducted a onetime tillage (up to 10 cm depth) in an eight-year no-till system where nutrients (e.g., P and K) were stratified. They reported uniform distribution of the nutrients in the 0–10 cm depth after one-time tillage. Overall, the varying outcomes in these studies show the effects of strategic tillage on residue P cycling are still unknown. Further evaluation of strategic tillage practices is required to understand the possible P nutrition benefits in a cropping rotation.

A study by Sarker et al. (2018b) highlighted the impact that tillage or stubble management and soil type can have on the rates and quantities of plant-available P. Within this study, the main influence on P immobilization was soil type, with the Vertisol (smectitic-rich) binding more P compared to the Luvisols (kaolinitic-rich). In a separate incubation study using the same tillage systems and soils, the input of crop (canola or wheat) residues enhanced the release of SOM-bound nutrients. This may have occurred via positive priming, or due to the mobilization of mineral-bound nutrients in soil aggregates, with tillage intensity and soil-type interactions modulating these processes (Sarker et al. 2018b). As low molecular weight organic acids may be released during the decomposition of crop residue (Guppy et al. 2005; Kumari et al. 2008), these could also decrease the nutrient sorption capacity of clay minerals and displace adsorbed P, depending on clay content and soil mineralogy (Bolan et al. 1994; Guppy et al. 2005; Fink et al. 2016). Further, organic acids may enhance the release of mineral-bound SOM through dissolution of protective mineral phases, potentially enhancing nutrient release from SOM and stimulating microbial activity (Keiluweit et al. 2015). Clearly, there are challenges in predicting the level of P release, not only from crop residues but also from native soil reserves under various managements due to the complexity of P cycling in different soil types.

3.5.3 FERTILIZER PHOSPHORUS MANAGEMENT UNDER TILLAGE AND RESIDUE MANAGEMENT SYSTEMS

Fertilizer application method is critical for efficient fertilizer use due to the low mobility of P in soils. In Australia, regardless of tillage-residue management, most seeding systems apply the fertilizer with the seed or at depth, and application may be split banded. Where "deep placement" occurs, the exact depth of placement varies between seeders and studies. For example, deep placement is defined as 8 mm by Officer et al. (2009), 10 mm by Dunbabin et al. (2009), and 12–15 mm by Alston (1980). Liquid fertilizer applications are also becoming increasingly common, in particular on Calcareous soils to improve crop access to fertilizer P (Holloway et al. 2001).

In general, banding P fertilizer close to the seed (<40 mm) increases P uptake in the early stages of crop growth (Officer et al. 2009; McLaughlin et al. 2011; Nkebiwe et al. 2016). The main benefit of deep placement is attributed to plant roots being able to access fertilizer P at depth when the top soil is drying out. However, there are also studies that have shown no advantage from deep placement when the soil profile remains wet during the growing season (Alston 1980).

3.6 CROP RESIDUE MANAGEMENT AND SOIL SULFUR AND POTASSIUM DYNAMICS

3.6.1 SULFUR

Sulfur (S) is an essential plant nutrient that is becomingly increasingly deficient in many soils due to decreases in S deposition from industrial sources, decreased S contamination in pesticides and fertilizers, loss of SOM, and increased S removal due to agricultural intensification (Scherer 2001; Eriksen 2009; Kopittke et al. 2016). In most aerobic surface soils, >95% of S is present in organic forms, with smaller amounts present as SO_4^{2-} in the soil solution/adsorbed onto clay surfaces, and as precipitated Ca, Mg, or Na SO_4^{2-} (Scherer 2001; Eriksen 2009).

Crop residues can contain between 0.4 and 2.0 g S kg^{-1} for species such as wheat, rice, maize, canola, sugarcane, and soybean (Dobermann and Fairhurst 2002; Mertens et al. 2002; Reddy et al. 2002; Singh et al. 2006; Cherubin et al. 2018; Sarker et al. 2018b). Consequently, management strategies that retain residues are important to help maintain soil S stocks and decrease the need for S fertilizers (Sharma et al. 2016; Sarker et al. 2018b). Strategies that remove residues off site represent a complete loss of residue S from the soil system and will result in depletion over time. Practices such as burning lead to partial loss, with 50%–60% of the S contained in crop residues lost during burning operations (Dobermann and Fairhurst 2002).

Where residues are returned on site (surface retained or incorporated), most of crop residue–derived S must be mineralized to SO_4^{2-} before it is available for plant uptake (Fitzgerald 1976; Eriksen 2009). This mineralization of residue-derived S can occur via two distinct pathways – biochemical mineralization and biological mineralization, with both contributing to the breakdown of residues in soil (McGill and Cole 1981; Eriksen 2005; Eriksen 2009). Biochemical mineralization involves the release of S from sulfate esters (C-OS) by enzymes (McGill and Cole 1981; Eriksen 2009). These enzymes are usually released in soil when there is insufficient inorganic S for microbial reactions, and their production tends to be inhibited by high concentrations of SO_4^- (McGill and Cole 1981; Eriksen 2009). Bacteria and fungi are the main sources of sulfatatse enzymes in soil, although plant roots and mammalian urine may act as secondary sources (Fitzgerald 1976). Biological mineralization occurs when SO_4^- is released as a by-product following the oxidation of organic C by microbial biomass to obtain energy for growth (McGill and Cole 1981; Eriksen 2009). Mineralization of S via this pathway is primarily governed by the C:S ratio of the organic material and soil biological activity (Barrow 1960a; Wu et al. 1995; Chapman 1997). The C:S ratio at which net immobilization occurs following residue addition varies from study to study. Results of Barrow (1960a), who after testing a range of organic materials observed net S release at a C:S ratio of <200, with complete immobilization occurring at >420, are often used as a rough guide to predict when immobilization is likely to occur.

Limited work has been conducted regarding the relative importance of biochemical versus biological mineralization of S, and consequently the effect that various tillage and residue management practices have on these two processes is poorly understood. However, when total S mineralization rates are considered, it is clear that soils receiving large amounts of crop residue with low C:S ratios may experience net S immobilization to the detriment of the subsequent crop, and in these instances the addition of extra inorganic S is required to ensure adequate plant nutrition (Wu et al. 1993; Chapman 1997; Chowdhury et al. 2000). Sulfur immobilized in microbial biomass will eventually become incorporated into SOM and then into physically protected sites inside aggregates (Wu et al. 1995; Eriksen 2009). The release of S from the general SOM pool is a slow process (Sarker et al. 20108b), although this can be accelerated by management practices such as residue incorporation and tillage to breakup soil aggregates and expose protected SOM (Balesdent et al. 2000; Christensen 2001; Sarker et al. 2018b). Consequently, residue incorporation can lead to greater release of S due to acceleration in the breakdown of added residues after mixing with the soil (Singh et al. 2006), and possibly also due to accelerated release of S from the native SOM pool (Sarker et al. 2018b).

Once released from organic materials, S exists primary as SO_4^{2-} (Singh et al. 2006) and is fairly mobile in soils due to its relatively weak adsorption to soil surfaces (Curtin and Syers 1990). The mobility of SO_4^- can present problems if the release during residue breakdown does not occur at the same time as crop demand and losses via leaching occur (Konboon et al. 2000; Eriksen 2009). Management practices that increase the movement of water through the soil profile, such as no-till and irrigation, are thus also likely to affect SO_4^- losses via leaching.

3.6.2 POTASSIUM

Potassium (K) is one of the major nutrients required by plants, and where it is deficient, crop growth is generally reduced (Marschner 1995). Significant declines in soil K can occur following agricultural intensification, particularly following increases in crop production as a result of N fertilization without corresponding additions of K (Singh et al. 2002). The recycling of K in plant residues is particularly important for the maintenance of soil K reserves, as a high proportion of the K extracted by the plant is often retained in residue material (Whitbread et al. 2000; Yadvinder-Singh et al. 2005; Cherubin et al. 2018). On the other hand, management practices that remove crop residues (e.g., oaten hay) generally result in negative K balances (Daliparthy et al. 1992; Prasad et al. 1999; Whitbread et al. 2000; Bijay-Singh et al. 2002; Bijay-Singh et al. 2004; Rafique et al. 2012). It should, however, be noted that relatively little K is lost during residue burning (Prasad et al. 1999; Dobermann and Fairhurst 2002). Overall, studies generally observe that residue retention decreases requirements for fertilizer application (Bijay-Singh et al. 2008; Sui et al. 2015; Carvalho et al. 2017; Sui et al. 2017), and the amount of K that can be supplied by residues should be considered in calculating K fertilizer requirements to increase fertilizer efficiency (Sui et al. 2015; Venkatesh et al. 2017). Values for the concentration of K in crop residues have been reported to range between 3 and 17 g kg^{-1} for species such as wheat, rice, maize, and sugarcane, and will typically vary depending on plant species and crop nutrition (Schomberg and Steiner 1999; Dobermann and Fairhurst 2002; Reddy et al. 2002; Cherubin et al. 2018).

Potassium in soil may be present in the soil solution, adsorbed onto the soil surface, held in a nonexchangeable form between the platelets of shrink-swell clay minerals, and held within crystalline mineral structures (Thomas and Hipp 1968; Sardans and Peñuelas 2015). As it is not a constituent of biomolecules, only very small amounts of K are present in SOM (Sardans and Peñuelas 2015). Soil solution and surface adsorbed K is the pool immediately available for plant uptake, although K may diffuse from the interlayer of clay minerals into the exchangeable pool in response to negative concentration gradients at a rate sufficient to supply K to crops during the growing season (Thomas and Hipp 1968; Sardans and Peñuelas 2015). It may also be released more slowly from mineral structures during weathering (Thomas and Hipp 1968; Sardans and Peñuelas 2015). When concentrations of K in the soil solution are high (such as after fertilizer addition or rapid release from plant residues) uptake, or fixation, of K into the interlayer of clay minerals may also occur (Bijay-Singh et al. 2004). However, this will often later be re-released into the exchangeable fraction following uptake of K from the soil solution by the growing crop (Bijay-Singh et al. 2004).

The cycling and release of potassium from residues is not as controlled by biological processes as it is for N, P, and S. Similar to P, it can be easily washed from crop residues and its release is less dependent on microbial decomposition, and thus residue quality (Schomberg and Steiner 1999; Yadvinder-Singh et al. 2005; Lupwayi et al. 2006). A large proportion of K (60%–80%) is generally released from residues in the first few weeks following addition to the soil due to rainfall events and incorporation activities, with further release occurring as the organic material continues to break down (Rosolem et al. 2005; Bijay-Singh et al. 2008). Potassium immobilization is not generally an issue reported during residue decomposition studies (Lupwayi et al. 2006).

In lighter-textured soil with low cation exchange capacity (CEC), K is subject to leaching, and significant amounts can be lost from the soil profile via this pathway (Bijay-Singh et al. 2004; Calonego and Rosolem 2013; Ranaivoson et al. 2017). Management practices that affect soil water movement, such as no-till and irrigation, are thus likely to affect K reserves in these types of soils. However, where CEC is higher and rates of water movement lower, K is generally retained strongly in the soil profile. Indeed, in systems practicing no-till where limited mixing of the soil occurs, stratification of K in the surface of the profile can become an issue affecting crop yield (Vyn and Janovicek 2001; Yin and Vyn 2003). In these circumstances, the application of additional fertilizer in bands below the soil surface may be required to help improve K nutrition (Vyn and Janovicek 2001; Yin and Vyn 2003).

3.7 CONCLUSIONS AND FUTURE RESEARCH NEEDS

The management of crop residues on-farm is continually changing and evolving. All residue management strategies are associated with advantages and disadvantages. However, residue retention combined with conservation tillage is generally viewed as a more sustainable residue management option relative to complete residue removal or incorporation with intensive tillage. This is due to its ability to mitigate soil erosion and increase soil moisture retention, and its greater capacity to maintain soil health and nutrient reserves. There are, however, some issues associated with residue retention, particularly around nutrient immobilization and the proliferation of certain pest species. Nevertheless, research to identify ways to overcome these issues under different cropping systems is likely to lead to benefits in terms of improved crop yield and environmental sustainability.

Presently, the majority of research on crop residue management has focused on understanding the differences between conservation and conventional tillage systems and their impact on soil properties and functions. However, we are increasingly seeing the adoption of new harvesting techniques and residue management strategies that have changed the amount and size of crop residues remaining on the soil surface. For example, newer harvesting equipment like Harrington seed destructor or the "seed terminator" may lead to only minor amounts of chaff/fine material being returned to the soil, while certain harvest weed-seed control technologies "mill" the chaff/stalks into fine particles to destroy weed seeds. We currently have limited understanding of how the use of these new techniques will affect the breakdown of residue and release of major nutrients. Similarly, our understanding of new approaches to tillage management, such as strategic tillage, and the extent to which this practice will influence C and nutrient cycling from crop residues over the longer term is also in its infancy and requires further investigation. Even our understanding of more traditional management strategies, such as stubble burning, is incomplete, and greater research is required to increase our understanding of the effect of burning on nutrient content, forms, and therefore availability.

It is well established that the surface retention of residues can contribute to increasing SOC content, relative to residue incorporation. However, studies are still needed to identify the impact of crop residues on SOC stocks under different integrated residue and nutrient management systems, soil types, and environments. By studying the different labile and stable SOC fractions of not only the bulk soil, but also soil aggregates, more detailed knowledge can be obtained regarding the impact of crop residues and their management practices on SOC and nutrient dynamics and their relationships with soil microbial communities (their composition and biomass). In particular, future research is needed regarding: (1) the long-term effect of integrated residue and nutrient management on microbial C use efficiency and conversion rates of residue to stabilized SOC; and (2) the qualitative changes in SOC and nutrient availability over time under field conditions, with implications for nutrient use efficiency.

In term of N dynamics, crop residues are generally classified on the basis of C:N ratios for their involvement in different NMIT processes in the soil. But the rate of decomposition, the quality of residues in terms of N, lignin and polyphenol contents, and environmental conditions also determine the net supply of N from crop residues, and our understanding of the effect of these factors is less complete. In addition, the studies on NMIT have been carried out under laboratory conditions, and to what extent these results can be extrapolated to field conditions is not clear. Also, studies on the effect of climate on NMIT and net N mineralization from crop residues and their return practices are rare. There is an urgent need to understand the interaction between temperature and water content in incorporated and surface-applied residues to predict the effect of climate on N nutrition of crops for both upland (aerobic) and lowland (anaerobic) environments. In addition, improved understanding of the optimum timing, rate, and placement of fertilizer N to minimize N tie-up by immobilization, while optimizing N supply and fertilizer N use efficiency in crops grown in residue amended soils, is required. Recently, the role of crop residues relating to positive priming of SOM decomposition has been highlighted by several workers, but it still remains to be seen how much SOM priming can contribute to improving nutrient (N, P, S) availability in the soil.

Our understanding of many aspects of P dynamics in response to residue management is also incomplete and requires further research. Currently, the complexity of residue P cycling in soil means there are no simple methods that can be used to advise farmers on residue P release and fertilizer recommendations. Our understanding of how P cycling is affected by residue type/composition, residue placement, and environmental conditions such as soil type and soil moisture regimes (particularly under field conditions) is incomplete and requires improvement to better predict residue contributions to P cycling in soil. A better understanding of stratification and leaching of P from residues under no-till management would also support better fertilizer recommendations.

Similar to N and P, a sound understanding of the magnitude of residue-based S and K stocks is important when determining crop fertilizer requirements in order to maximize fertilizer use efficiency; as is an understanding of C:S ratios in order to prevent plant available S being immobilized in the microbial biomass. While a number of short-term studies to determine the effect of residue management on soil S and K stocks have been conducted, the effect of longer-term management is less well researched, and our understanding of the effect of different residue management strategies would be enhanced by greater work in this area. Greater work to develop strategies to best deal with nutrient stratification in no-till systems would also benefit the management of K reserves, which can become concentrated in the surface soils with higher CEC.

Overall, while many studies on nutrient dynamics in response to crop residue and tillage management have been conducted, there is currently a gap between this mechanistic research and field-based problems. There is a need for simple predictive models to help farmers and advisers determine likely rates of nutrients release from crop residues and SOM and make more informed decisions about fertilizer requirements.

REFERENCES

Adair, E. C., W .J. Parton, S. J. Del Grosso, W. L. Silver, M. E. Harmon, S. A. Hall, I. C. Burke, and S. C. Hart. 2008. "Simple Three-pool Model Accurately Describes Patterns of Long-term Litter Decomposition in Diverse Climates." *Global Change Biology* 14(11): 2636–2660.

Aguilera, E., L. Lassaletta, A. Gattinger, and B. S. Gimeno. 2013. "Managing Soil Carbon for Climate Change Mitigation and Adaptation in Mediterranean Cropping Systems: A Meta-Analysis." *Agriculture, Ecosystems and Environment* 168: 25–36.

Almeida, L. F., L. C. Hurtarte, I. F. Souza, E. M. Soares, L. Vergütz, and I. R. Silva. 2018. "Soil Organic Matter Formation as Affected by Eucalypt Litter Biochemistry – Evidence from an Incubation Study." *Geoderma* 312: 121–129.

Almeida, M., A. McNeill, C. Tang, and P. Marschner. 2012. "Changes in Soil P Pools during Legume Residue Decomposition." *Soil Biology and Biochemistry* 49: 70–77.

Alston, A. 1980. "Response of Wheat to Deep Placement of Nitrogen and Phosphorus Fertilizers on a Soil High in Phosphorus in the Surface Layer." *Australian Journal of Agricultural Research* 31(1): 13–24.

Angers, D., and N. Eriksen-Hamel. 2008. "Full-Inversion Tillage and Organic Carbon Distribution in Soil Profiles: A Meta-Analysis." *Soil Science Society of America Journal* 72(5): 1370–1374.

Angst, G., K. E. Mueller, I. Kögel-Knabner, K. H. Freeman, and C. W. Mueller. 2017. "Aggregation Controls the Stability of Lignin and Lipids in Clay-Sized Particulate and Mineral Associated Organic Matter." *Biogeochemistry* 132(3): 307–324.

Asghar, M., D. W. Lack, B. A. Cowie, and J. C. Parker. 1996. "Effects of Surface Soil Mixing after Long-Term Zero Tillage on Soil Nutrient Distribution and Wheat Production." In *Agronomy – Science with Its Sleeves Rolled Up: Proceedings of the 8th Australian Agronomy Conference.* University of Southern Queensland, Toowoomba, Queensland. pp. 88–91.

Avci, M. 2011. "Conservation Tillage in Turkish Dryland Research." *Agronomy for Sustainable Development* 31(2): 299–307.

Bah, A. R., A. Zaharah, and A. Hussin. 2006. "Phosphorus Uptake from Green Manures and Phosphate Fertilizers Applied in an Acid Tropical Soil." *Communications in Soil Science and Plant Analysis* 37(13–14): 2077–2093.

Bailey, K. L., and G. Lazarovits. 2003. "Suppressing Soil-Borne Diseases with Residue Management and Organic Amendments." *Soil and Tillage Research* 72(2): 169–180.

Balesdent, J., C. Chenu, and M. Balabane. 2000. "Relationship of Soil Organic Matter Dynamics to Physical Protection and Tillage." *Soil and Tillage Research* 53(3–4): 215–230.

Barr, C. E., and A. Ulrich. 1963. "Phosphorus Status of Crops, Phosphorus Fractions in High and Low Phosphate Plants." *Journal of Agricultural and Food Chemistry* 11(4): 313–316.

Barrow, N. J. 1960a. "A Comparison of the Mineralization of Nitrogen and of Sulphur from Decomposing Organic Materials." *Australian Journal of Agricultural Research* 11(6): 960–969.

Barrow, N. J. 1960b. "Stimulated Decomposition of Soil Organic Matter during the Decomposition of Added Organic Materials." *Australian Journal of Agricultural Research* 11(3): 331–338.

Beare, M. H., W. R. Cookson, and P. E. Wilson. 1996. "Effects of Straw Residue Management Practices on the Composition and Activity of Soil Microbial Communities and Patterns of Residue Decomposition." In *Proceedings of the ASSS and NZSSS National Soils Conference*, 11–12. Melbourne, Australia: .

Becker, M., J. Ladha, I. Simpson, and J. Ottow. 1994. "Parameters Affecting Residue Nitrogen Mineralization in Flooded Soils." *Soil Science Society of America Journal* 58(6): 1666–1671.

Bieleski, R. L. 1973. "Phosphate Pools, Phosphate Transport, and Phosphate Availability." *Annual Review of Plant Physiology* 24(1): 225–252.

Bijay-Singh, B. 2018. "Are Nitrogen Fertilizers Deleterious to Soil Health?" *Agronomy* 8(4): 48.

Bijay-Singh, Y. H. Shan, S. E. Johnson-Beebout, Yadvinder-Singh, and R. J. Buresh. 2008. "Crop Residue Management for Lowland Rice-Based Cropping Systems in Asia." *Advances in Agronomy* 98: 117–199.

Bijay-Singh, Yadvinder-Singh, P. Imas, and X. Jian-chang. 2004. "Potassium Nutrition of the Rice–Wheat Cropping System." *Advances in Agronomy* 81: 203.

Birch, H. F. 1961. "Phosphorus Transformations during Plant Decomposition." *Plant and Soil* 15(4): 347–366.

Bird, J. A., W. R. Horwath, A. J. Eagle, and C. Van Kessel. 2001. "Immobilization of Fertilizer Nitrogen in rice: Effects of Straw Management Practices." *Soil Science Society of America Journal* 65(4): 1143–1152.

Bird, J. A., M. Kleber, and M. S. Torn. 2008. "^{13}C and ^{15}N Stabilization Dynamics in Soil Organic Matter Fractions during Needle and Fine Root Decomposition." *Organic Geochemistry* 39(4): 465–477.

Bishop, M. L., A. Chang, and R. Lee. 1994. "Enzymatic Mineralization of Organic Phosphorus in a Volcanic Soil in Chile." *Soil Science* 157(4): 238–243.

Blagodatskaya, E., and Y. Kuzyakov. 2008. "Mechanisms of Real and Apparent Priming Effects and Their Dependence on Soil Microbial Biomass and Community Structure: Critical Review." *Biology and Fertility of Soils* 45(2): 115–131.

Blair, G. J., and O. W. Boland. 1978. "The Release of Phosphorus from Plant Material Added to Soil." *Soil Research* 16(1): 101–111.

Blanco-Canqui, H., and R. Lal. 2009. "Crop Residue Removal Impacts on Soil Productivity and Environmental Quality." *Critical Reviews in Plant Science* 28(3): 139–163.

Bockus, W. W., and J. P. Shroyer. 1998. "The impact of Reduced Tillage on Soilborne Plant Pathogens." *Annual Review of Phytopathology* 36(1): 485–500.

Bolan, N. S., R. Naidu, S. Mahimairaja, and S. Baskaran. 1994. "Influence of Low-Molecular-Weight Organic Acids on the Solubilization of Phosphates." *Biology and Fertility of Soils* 18(4): 311–319.

Bradford, M. A., A. D. Keiser, C. A. Davies, C. A. Mersmann, and M. S. Strickland. 2013. "Empirical Evidence That Soil Carbon Formation from Plant Inputs Is Positively Related to Microbial Growth." *Biogeochemistry* 113(1–3): 271–281.

Bromfield, S. M. 1960. "Some Factors Affecting the Solubility of Phosphates during the Microbial Decomposition of Plant Material." *Australian Journal of Agricultural Research* 11(3): 304–316.

Bruce, S. E., J. A. Kirkegaard, J. E. Pratley, and G. N. Howe. 2005. "Impacts of Retained Wheat Stubble on Canola in Southern New South Wales." *Australian Journal of Experimental Agriculture* 45(4): 421–433.

Buchanan, M., and L. D. King. 1993. "Carbon and Phosphorus Losses from Decomposing Crop Residues in No-Till and Conventional Till Agroecosystems." *Agronomy Journal* 85(3): 631–638.

Bünemann, E. K., D. P. Heenan, P. Marschner, and A. M. McNeill. 2006. "Long-Term Effects of Crop Rotation, Stubble Management and Tillage on Soil Phosphorus Dynamics." *Soil Research* 44(6): 611–618.

Bünemann, E. K., F. Steinebrunner, P. C. Smithson, E. Frossard, and A. Oberson. 2004. "Phosphorus Dynamics in a Highly Weathered Soil as Revealed by Isotopic Labeling Techniques." *Soil Science Society of America Journal* 68(5): 1645–1655.

Cabrera, M. L., D. E. Kissel, and M. F. Vigil. 2005. "Nitrogen Mineralization from Organic Residues." *Journal of Environmental Quality* 34(1): 75–79.

Calonego, J. C., and C. A. Rosolem. 2013. "Phosphorus and Potassium Balance in a Corn–Soybean Rotation under No-Till and Chiseling." *Nutrient Cycling in Agroecosystems* 96(1): 123–131.

Cao, Y., H. Sun, J. Zhang, et al. 2018. "Effects of Wheat Straw Addition on Dynamics and Fate of Nitrogen Applied to Paddy Soils." *Soil and Tillage Research* 178: 92–98.

Carvalho, J. L. N., R. C. Nogueirol, L. M. S. Menandro, et al. 2017. "Agronomic and Environmental Implications of Sugarcane Straw Removal: A Major Review." *Global Change Biology Bioenergy* 9(7): 1181–1195.

Cassman, K. G., A. Dobermann, and D. T. Walters. 2002. "Agroecosystems, Nitrogen-Use Efficiency, and Nitrogen Management." *AMBIO: A Journal of the Human Environment* 31(2): 132–140.

Cassman, K. G., S. Peng, and A. Dobermann. 1997. "Nutritional Physiology of the Rice Plants and Productivity Decline of Irrigated Rice Systems in the Tropics." *Soil Science and Plant Nutrition* 43(supl): 1101–1106.

Castellano, M. J., K. E. Mueller, D. C. Olk, J. E. Sawyer, and J. Six. 2015. "Integrating Plant Litter Quality, Soil Organic Matter Stabilization, and the Carbon Saturation Concept." *Global Change Biology* 21(9): 3200–3209.

Chan, K. Y., D. P. Heenan, and H. B. So. 2003. "Sequestration of Carbon and Changes in Soil Quality under Conservation Tillage on Light-Textured Soils in Australia: A Review." *Australian Journal of Experimental Agriculture* 43(4): 325–334.

Chapman, S. J. 1997. "Barley Straw Decomposition and S Immobilization." *Soil Biology and Biochemistry* 29(2): 109–114.

Chauhan, B. S., R. G. Singh, and G. Mahajan. 2012. "Ecology and Management of Weeds under Conservation Agriculture: A Review." *Crop Protection* 38: 57–65.

Chen, H., X. Li, F. Hu, and W. Shi. 2013. "Soil Nitrous Oxide Emissions following Crop Residue Addition: A Meta-Analysis." *Global Change Biology* 19: 2956–2964.

Chen, B., E. Liu, Q. Tian, C. Yan, and Y. Zhang. 2014a. "Soil Nitrogen Dynamics and Crop Residues: A Review." *Agronomy for Sustainable Development* 34(2): 429–442.

Chen, R., M. Senbayram, S. Blagodatsky, O. Myachina, K. Dittert, X. Lin, E. Blagodatskaya, and Y. Kuzyakov. 2014b. "Soil C and N Availability Determine the Priming Effect: Microbial N Mining and Stoichiometric Decomposition Theories." *Global Change Biology* 20(7): 2356–2367.

Chenu, C., D. A. Angers, P. Barré, D. Derrien, D. Arrouays, and J. Balesdent. 2018. "Increasing Organic Stocks in Agricultural Soils: Knowledge Gaps and Potential Innovations." *Soil and Tillage Research*. doi:10.1016/j.still.2018.04.011

Cherubin, M. R., D. M. D. S. Oliveira, B. J. Feigl, et al. 2018. "Crop Residue Harvest for Bioenergy Production and Its Implications on Soil Functioning and Plant Growth: A Review." *Scientia Agricola* 75(3): 255–272.

Chowdhury, M. A. H., K. Kouno, T. Ando, and T. Nagaoka. 2000. "Microbial Biomass, S Mineralization and S Uptake by African Millet from Soil Amended with Various Composts." *Soil Biology and Biochemistry* 32(6): 845–852.

Christensen, B. T. 1986. "Barley Straw Decomposition under Field Conditions: Effect of Placement and Initial Nitrogen Content on Weight Loss and Nitrogen Dynamics." *Soil Biology and Biochemistry* 18(5): 523–529.

Christensen, B. T. 2001. "Physical Fractionation of Soil and Structural and Functional Complexity in Organic Matter Turnover." *European Journal of Soil Science* 52(3): 345–353.

Conyers, M., V. van der Rijt, A. Oates, G. Poile, J. Kirkegaard, and C. Kirkby. 2019. "The Strategic Use of Minimum Tillage within Conservation Agriculture in Southern New South Wales, Australia." *Soil and Tillage Research* 193: 17–26.

Coppens, F., P. Garnier, A. Findeling, R. Merckx, and S. Recous. 2007. "Decomposition of Mulched versus Incorporated Crop Residues: Modelling with PASTIS Clarifies Interactions between Residue Quality and Location." *Soil Biology and Biochemistry* 39(9): 2339–2350.

Cotrufo, M. F., J. L. Soong, A. J. Horton, et al. 2015. "Formation of Soil Organic Matter via Biochemical and Physical Pathways of Litter Mass Loss." *Nature Geoscience* 8(10): ngeo2520.

Cotrufo, M. F., M. D. Wallenstein, C. M. Boot, K. Denef, and E. Paul. 2013. "The Microbial Efficiency-Matrix Stabilization (MEMS) Framework Integrates Plant Litter Decomposition with Soil Organic Matter Stabilization: Do Labile Plant Inputs Form Stable Soil Organic Matter?" *Global Change Biology* 19(4): 988–995.

Crowther, J. P., and A. Westman. 1954. "The Hydrolysis of the Condensed Phosphates: I. Sodium Pyrophosphate and Sodium Triphosphate." *Canadian Journal of Chemistry* 32(1): 42–48.

Crozier, C. R., G. C. Naderman, M. R. Tucker, and R. E. Sugg. 1999. "Nutrient and pH Stratification with Conventional and No-Till Management." *Communications in Soil Science and Plant Analysis* 30(1–2): 65–74.

Curtin, D., and J. K. Syers. 1990. "Extractability and Adsorption of Sulphate in Soils." *Journal of Soil Science* 41(2): 305–312.

Daliparthy, J., B. N. Chatterjee, and S. S. Mondal. 1992. "Changes in Soil Nitrogen, Phosphorus, and Potassium under Intensive Cropping in Sub-humid Tropics of India." *Communications in Soil Science and Plant Analysis* 23(15–16): 1871–1884.

Damon, P. M., B. Bowden, T. Rose, and Z. Rengel. 2014. "Crop Residue Contributions to Phosphorus Pools in Agricultural Soils: A Review." *Soil Biology and Biochemistry* 74: 127–137.

Dang, Y. P., A. Balzer, M. Crawford, et al. 2018. "Strategic Tillage in Conservation Agricultural Systems of North-Eastern Australia: Why, Where, When and How?" *Environmental Science and Pollution Research* 25(2): 1000–1015.

Dang, Y. P., N. P. Seymour, S. R. Walker, M. J. Bell, and D. M. Freebairn. 2015. "Strategic Tillage in No-Till Farming Systems in Australia's Northern Grains-Growing Regions: I. Drivers and Implementation." *Soil and Tillage Research* 152: 104–114.

Dean, G. J., and A. Merry. 2015. "Comparison of Stubble Management Strategies in the High Rainfall Zone." In *Proceedings of the 17th Australian Society of Agronomy Conference*, 1–4. 20–24 September 2015, Hobart, Australia.

Deubel, A., B. Hofmann, and D. Orzessek. 2011. "Long-Term Effects of Tillage on Stratification and Plant Availability of Phosphate and Potassium in a Loess Chernozem." *Soil and Tillage Research* 117: 85–92.

Dignac, M.-F., D. Derrien, P. Barré, et al. 2017. "Increasing Soil Carbon Storage: Mechanisms, Effects of Agricultural Practices and Proxies: A Review." *Agronomy for Sustainable Development* 37(2): 14.

Dobermann, A., and T. H. Fairhurst. 2002. "Rice Straw Management." *Better Crops International* 16(1): 7–11.

Dunbabin, V., R. Armstrong, S. Officer, and R. Norton. 2009. "Identifying Fertiliser Management Strategies to Maximise Nitrogen and Phosphorus Acquisition by Wheat in Two Contrasting Soils from Victoria, Australia." *Soil Research* 47(1): 74–90.

Eagle, A. J., J. A. Bird, J. E. Hill, W. R. Horwath, and C. van Kessel. 2003. "Nitrogen Dynamics and Fertilizer Use Efficiency in Rice following Straw Incorporation and Winter Flooding." In *Management of Crop Residues for Sustainable Crop Production*, 81–97. IAEA-TECDOC-1354. Vienna, Austria: International Atomic Energy Agency.

Eagle, A. J., J. A. Bird, W. R. Horwath, B. A. Linquist, S. M. Brouder, J. E. Hill, and C. van Kessel. 2000. "Rice Yield and Nitrogen Utilization Efficiency under Alternative Straw Management Practices." *Agronomy Journal* 92: 1096–1103.

Enwezor, W. O. 1976. "The Mineralization of Nitrogen and Phosphorus in Organic Materials of Varying C: N and C: P Ratios." *Plant and Soil* 44(1): 237–240.

Eriksen, J. 2005. "Gross Sulphur Mineralization–Immobilization Turnover in Soil Amended with Plant Residues." *Soil Biology and Biochemistry* 37(12): 2216–2224.

Eriksen, J. 2009. "Soil Sulfur Cycling in Temperate Agricultural Systems." *Advances in Agronomy* 102: 55–89.

Fabre, A., G. Pinay, and C. Ruffinoni. 1996. "Seasonal Changes in Inorganic and Organic Phosphorus in the Soil of a Riparian Forest." *Biogeochemistry* 35(3): 419–432.

Fan, M., R. Jiang, X. Liu, et al. 2005. "Interactions between Non-flooded Mulching Cultivation and Varying Nitrogen Inputs in Rice–Wheat Rotations." *Field Crops Research* 91(2–3): 307–318.

Fang, Y., B. P. Singh, W. Badgery, and X. He. 2016. "In Situ Assessment of New Carbon and Nitrogen Assimilation and Allocation in Contrastingly Managed Dryland Wheat Crop–Soil Systems." *Agriculture, Ecosystems and Environment* 235: 80–90.

Fang, Y., L. Nazaries, B. K. Singh, and B. P. Singh. 2018a. "Microbial Mechanisms of Carbon Priming Effects Revealed during the Interaction of Crop Residue and Nutrient Inputs in Contrasting Soils." *Global Change Biology* 24(7): 2775–2790.

Fang, Y., B. P. Singh, D. Collins, B. Li, J. Zhu, and E. Tavakkoli. 2018b. "Nutrient Supply Enhanced Wheat Residue-Carbon Mineralization, Microbial Growth, and Microbial Carbon-Use Efficiency When Residues Were Supplied at High Rate in Contrasting Soils." *Soil Biology and Biochemistry* (under review).

Fierer, N., M. A. Bradford, and R. B. Jackson. 2007. "Toward an Ecological Classification of Soil Bacteria." *Ecology* 88(6): 1354–1364.

Fink, J. R., A. V. Inda, J. Bavaresco, V. Barrón, J. Torrent, and C. Bayer. 2016. "Phosphorus Adsorption and Desorption in Undisturbed Samples from Subtropical Soils under Conventional Tillage or No-Tillage." *Journal of Plant Nutrition and Soil Science* 179(2): 198–205.

Fitzgerald, J. W. 1976. "Sulfate Ester Formation and Hydrolysis: A Potentially Important yet Often Ignored Aspect of the Sulfur Cycle of Aerobic Soils." *Bacteriological Reviews* 40(3): 698.

Floate, M. J. S. 1970a. "Decomposition of Organic Materials from Hill Soils and Pastures: III. The Effect of Temperature on the Mineralization of Carbon, Nitrogen and Phosphorus from Plant Materials and Sheep Faeces." *Soil Biology and Biochemistry* 2(3): 187–196.

Floate, M. J. S. 1970b. "Decomposition of Organic Materials from Hill Soils and Pastures: IV. The effects of Moisture Content on the Mineralization of Carbon, Nitrogen and Phosphorus from Plant Materials and Sheep Faeces." *Soil Biology and Biochemistry* 2(4): 275–283.

Fontaine, S., C. Henault, A. Aamor, et al. 2011. "Fungi Mediate Long Term Sequestration of Carbon and Nitrogen in Soil through Their Priming Effect." *Soil Biology and Biochemistry* 43(1): 86–96.

Fontaine, S., A. Mariotti, and L. Abbadie. 2003. "The Priming Effect of Organic Matter: A Question of Microbial Competition?" *Soil Biology and Biochemistry* 35(6): 837–843.

Francaviglia, R., C. Di Bene, R. Farina, and L. Salvati. 2017. "Soil Organic Carbon Sequestration and Tillage Systems in the Mediterranean Basin: A Data Mining Approach." *Nutrient Cycling in Agroecosystems* 107(1): 125–137.

Friesen, D. K., and G. J. Blair. 1988. "A Dual Radiotracer Study of Transformations of Organic, Inorganic and Plant Residue Phosphorus in Soil in the Presence and Absence of Plants." *Soil Research* 26(2): 355–366.

Fruit, L., S. Recous, and G. Richard. 1999. "Plant Residue Decomposition: Effect of Soil Porosity and Particle Size." In *Effect of Mineral-Organic-Microorganism Interaction on Soil and Freshwater Environments*, edited by J. Berthelin, P. M. Huang, J.-M. Bollag, and F. Andreux, 189–196. New York: Kluwer Academic.

Fu, M. H., X. C. Xu, and M. A. Tabatabai. 1987. "Effect of pH on Nitrogen Mineralization in Crop-Residue-Treated Soils." *Biology and Fertility of Soils* 5(2): 115–119.

Fujisaki, K., T. Chevallier, L. Chapuis-Lardy, A. Albrecht, T. Razafimbelo, D. Masse, Y. B. Ndour, and J.-L. Chotte. 2018. "Soil Carbon Stock Changes in Tropical Croplands Are Mainly Driven by Carbon Inputs: A Synthesis." *Agriculture, Ecosystems and Environment* 259: 147–158.

Fuller, W. H., D. R. Nielsen, and R. W. Miller. 1956. "Some Factors Influencing the Utilization of Phosphorus from Crop Residues." *Soil Science Society of America Journal* 20(2): 218–224.

Gao, Y., Y. Li, J. Zhang, et al. 2009. "Effects of Mulch, N Fertilizer, and Plant Density on Wheat Yield, Wheat Nitrogen Uptake, and Residual Soil Nitrate in a Dryland Area of China." *Nutrient Cycling in Agroecosystems* 85(2): 109–121.

Gentile, R., B. Vanlauwe, P. Chivenge, and J. Six. 2008. "Interactive Effects from Combining Fertilizer and Organic Residue Inputs on Nitrogen Transformations." *Soil Biology and Biochemistry* 40(9): 2375–2384.

Gentile, R., B. Vanlauwe, C. Van Kessel, and J. Six. 2009. "Managing N Availability and Losses by Combining Fertilizer-N with Different Quality Residues in Kenya." *Agriculture, Ecosystems and Environment* 131(3–4): 308–314.

Golchin, A., J. M. Oades, J. O. Skjemstad, and P. Clarke. 1994. "Study of Free and Occluded Particulate Organic Matter in Soils by Solid State ^{13}C CP/MAS NMR Spectroscopy and Scanning Electron Microscopy." *Soil Research* 32(2): 285–309.

Govaerts, B., M. Mezzalama, Y. Unno, et al. 2007. "Influence of Tillage, Residue Management, and Crop Rotation on Soil Microbial Biomass and Catabolic Diversity." *Applied Soil Ecology* 37(1): 18–30.

Guenet, B., C. Neill, G. Bardoux, and L. Abbadie. 2010. "Is There a Linear Relationship between Priming Effect Intensity and the Amount of Organic Matter Input?" *Applied Soil Ecology* 46(3): 436–442.

Guppy, C. N., N. Menzies, P. W. Moody, and F. Blamey. 2005. "Competitive Sorption Reactions between Phosphorus and Organic Matter in Soil: A Review." *Soil Research* 43(2): 189–202.

Gupta, V., J. Kirkegaard, T. McBeath, et al. 2017. "The Effect of Stubble on Nitrogen Tie-Up and Supply – The Mallee Experience 2017." Accessed 21 July 2018. http://www.msfp.org.au/wp-content/uploads/The-effect-of-stubble-on-nitrogen-tie-up-and-supply-Vadakattu-Gupta-Full.pdf.

Hadas, A., L. Kautsky, M. Goek, and E. E. Kara. 2004. "Rates of Decomposition of Plant Residues and Available Nitrogen in Soil, Related to Residue Composition through Simulation of Carbon and Nitrogen Turnover." *Soil Biology and Biochemistry* 36(2): 255–266.

Heijboer, A., H. F. ten Berge, P. C. de Ruiter, H. B. Jørgensen, G. A. Kowalchuk, and J. Bloem. 2016. "Plant Biomass, Soil Microbial Community Structure and Nitrogen Cycling under Different Organic Amendment Regimes; A ^{15}N Tracer-Based Approach." *Applied Soil Ecology* 107: 251–260.

Hemwong, S., G. Cadisch, B. Toomsan, V. Limpinuntana, P. Vityakon, and A. Patanothai. 2008. "Dynamics of Residue Decomposition and N_2 Fixation of Grain Legumes upon Sugarcane Residue Retention as an Alternative to Burning." *Soil and Tillage Research* 99(1): 84–97.

Henriksen, T. M., and T. A. Breland. 1999. "Nitrogen Availability Effects on Carbon Mineralization, Fungal and Bacterial Growth, and Enzyme Activities during Decomposition of Wheat Straw in Soil." *Soil Biology and Biochemistry* 31(8): 1121–1134.

Holloway, R., I. Bertrand, A. Frischke, D. Brace, M. J. McLaughlin, and W. Shepperd. 2001. "Improving Fertiliser Efficiency on Calcareous and Alkaline Soils with Fluid Sources of P, N and Zn." *Plant and Soil* 236(2): 209–219.

Huang, Z.-W., and F. E. Broadbent. 1988. "The Influences of Organic Residues on Utilization of Urea N by Rice." *Fertilizer Research* 18(3): 213–220.

Ichir, L. L., and M. Ismaili. 2002. "Decomposition and Nitrogen Dynamics of Wheat Residues and Impact on Wheat Growth Stages." *Comptes Rendus Biologies* 325(5): 597–604.

Iqbal, S. M. 2009. "Effect of Crop Residue Qualities on Decomposition Rates, Soil Phosphorus Dynamics and Plant Phosphorus Uptake." School of Earth and Environmental Sciences, University of Adelaide, Adelaide, Australia.

IRRI. 2018. Site-Specific Nutrient Management. http://www.knowledgebank.irri.org/ericeproduction/IV.4_SSNM.htm (Accessed 15 August 2018).

Jansson, S., and J. Persson. 1982. "Mineralization and Immobilization of Soil Nitrogen." In *Nitrogen in Agricultural Soils*, edited by F. J. Stevenson. Madison, WI: American Society of Agronomy.

Janzen, H. H., and R. M. N. Kucey. 1988. "C, N, and S Mineralization of Crop Residues as Influenced by Crop Species and Nutrient Regime." *Plant and Soil* 106(1): 35–41.

Jenkinson, D. S. 1977. "Studies on the Decomposition of Plant Material in Soil. V. The Effects of Plant Cover and Soil Type on the Loss of Carbon from ^{14}C Labelled Ryegrass Decomposing under Field Conditions." *Journal of Soil Science* 28(3): 424–434.

Jiang, Y. F., X. H. Jiang, and L. Zhou. 1998. "Techniques for Return of Straw to Fields and Combined Application of Nitrogen Fertilizer." *Jiangsu Journal of Agricultural Sciences* 4: 43–45.

Jones, O. L., and S. M. Bromfield. 1969. "Phosphorus Changes during the Leaching and Decomposition of Hayed-Off Pasture Plants." *Australian Journal of Agricultural Research* 20(4): 653–663.

Kaewpradit, W., B. Toomsan, G. Cadisch, et al. 2009. "Mixing Groundnut Residues and Rice Straw to Improve Rice Yield and N Use Efficiency." *Field Crops Research* 110(2): 130–138.

Kaiser, C., O. Franklin, U. Dieckmann, and A. Richter. 2014. "Microbial community Dynamics Alleviate Stoichiometric Constraints during Litter Decay." *Ecology Letters* 17(6): 680–690.

Kallenbach, C. M., S. D. Frey, and A. S. Grandy. 2016. "Direct Evidence for Microbial-Derived Soil Organic Matter Formation and Its Ecophysiological Controls." *Nature Communications* 7: 13630.

Keiluweit, M., J. J. Bougoure, P. S. Nico, J. Pett-Ridge, P. K. Weber, and M. Kleber. 2015. "Mineral Protection of Soil Carbon Counteracted by Root Exudates." *Nature Climate Change* 5(6): 588–595.

Kelley, K. W., and D. W. Sweeney. 2005. "Tillage and Urea Ammonium Nitrate Fertilizer Rate and Placement Affects Winter Wheat following Grain Sorghum and Soybean." *Agronomy Journal* 97(3): 690–697.

Kelley, K. W., and D. W. Sweeney. 2007. "Placement of Preplant Liquid Nitrogen and Phosphorus Fertilizer and Nitrogen Rate Affects No-Till Wheat following Different Summer Crops." *Agronomy Journal* 99(4): 1009–1017.

Khan, S. A., R. L. Mulvaney, T. R. Ellsworth, and C. W. Boast. 2007. "The Myth of Nitrogen Fertilization for Soil Carbon Sequestration." *Journal of Environmental Quality* 36(6): 1821–1832.

Kindler, R., A. Miltner, M. Thullner, H.-H. Richnow, and M. Kästner. 2009. "Fate of Bacterial Biomass Derived Fatty Acids in Soil and Their Contribution to Soil Organic Matter." *Organic Geochemistry* 40(1): 29–37.

Kirkby, C., J. Kirkegaard, A. Richardson, L. Wade, C. Blanchard, and G. Batten. 2011. "Stable Soil Organic Matter: A Comparison of C: N: P: S Ratios in Australian and Other World Soils." *Geoderma* 163(3–4): 197–208.

Kirkby, C. A., A. E. Richardson, L. J. Wade, et al. 2014. "Nutrient Availability Limits Carbon Sequestration in Arable Soils." *Soil Biology and Biochemistry* 68: 402–409.

Kirkby, C. A., A. E. Richardson, L. J. Wade, G. D. Batten, C. Blanchard, and J. A. Kirkegaard. 2013. "Carbon-Nutrient Stoichiometry to Increase Soil Carbon Sequestration." *Soil Biology and Biochemistry* 60: 77–86.

Kirkby, C. A., A. E. Richardson, L. J. Wade, M. Conyers, and J. A. Kirkegaard. 2016. "Inorganic Nutrients Increase Humification Efficiency and C-Sequestration in an Annually Cropped Soil." *PLoS One* 11(5): e0153698.

Kirkegaard, J., T. Swan, G. Vadakattu, J. Hunt, and K. Jones. 2018. "The Effects of Stubble on Nitrogen Tie-Up and Supply." GRDC Update Papers. https://grdc.com.au/resources-and-publications/grdc-update-papers/tab-content/grdc-update-papers/2018/02/the-effects-of-stubble-on-nitrogen-tie-up-and-supply (Accessed 23 July 2018).

Konboon, Y., G. Blair, R. Lefroy, and A. Whitbread. 2000. "Tracing the Nitrogen, Sulfur, and Carbon Released from Plant Residues in a Soil/Plant System." *Soil Research* 38(3): 699–710.

Kongchum, M., P. Bollich, W. Hudnall, R. DeLaune, and C. Lindau. 2006. "Decreasing Methane Emission of Rice by Better Crop Management." *Agronomy for Sustainable Development* 26(1): 45–54.

Kopittke, P. M., R. C. Dalal, and N. W. Menzies. 2016. "Sulfur Dynamics in Sub-tropical Soils of Australia as Influenced by Long-Term Cultivation." *Plant and Soil* 402(1): 211–219.

Kopittke, P. M., M. C. Hernandez-Soriano, R. C. Dalal, et al. 2018. "Nitrogen-Rich Microbial Products Provide New Organo-Mineral Associations for the Stabilization of Soil Organic Matter." *Global Change Biology* 24(4): 1762–1770.

Kosuge, N., and I. Zulkarnaini. 1981. "Effect of the Rice Straw Application to the Paddy Field in Indonesia." *Bulletin of the Hokuriku National Agricultural Experiment Station* 23: 167–186.

Kumar, K., and K. M. Goh. 1999. "Crop Residues and Management Practices: Effects on Soil Quality, Soil Nitrogen Dynamics, Crop Yield, and Nitrogen Recovery." *Advances in Agronomy* 68: 197–319.

Kumar, K., and K. M. Goh. 2002. "Management Practices of Antecedent Leguminous and Non-leguminous Crop Residues in relation to Winter Wheat Yields, Nitrogen Uptake, Soil Nitrogen Mineralization and Simple Nitrogen Balance." *European Journal of Agronomy* 16(4): 295–308.

Kumari, A., K. K. Kapoor, B. S. Kundu, and R. Kumari Mehta. 2008. "Identification of Organic Acids Produced during Rice Straw Decomposition and Their Role in Rock Phosphate Solubilization." *Plant Soil and Environment* 54(2): 72.

Kuzyakov, Y., J. K. Friedel, and K. Stahr. 2000. "Review of Mechanisms and Quantification of Priming Effects." *Soil Biology and Biochemistry* 32(11–12): 1485–1498.

Kwabiah, A. B., C. A. Palm, N. C. Stoskopf, and R. P. Voroney. 2003. "Response of Soil Microbial Biomass Dynamics to Quality of Plant Materials with Emphasis on P Availability." *Soil Biology and Biochemistry* 35(2): 207–216.

Ladd, J. N., M. Amato, and J. M. Oades. 1985. "Decomposition of Plant Material in Australian Soils. III. Residual Organic and Microbial Biomass C and N from Isotope-Labelled Legume Material and Soil Organic Matter, Decomposing under Field Conditions." *Soil Research* 23(4): 603–611.

Ladd, J. N., and R. B. Jackson. 1982. "Biochemistry of Ammonification." In *Nitrogen in Agricultural Soils*, edited by F. J. Stevenson, 173–227. Agronomy Monograph 22. Madison, WI: American Society of Agronomy.

Lal, R. 1995. "The Role of Residues Management in Sustainable Agricultural Systems." *Journal of Sustainable Agriculture* 5(4): 51–78.

Lal, R. 2014. "Societal Value of Soil Carbon." *Journal of Soil and Water Conservation* 69(6): 186a–192a.

Lal, R., J. Apt, L. Lave, and M. Morgan. 2004. "Managing Soil Carbon." *Science* 304: 393.

Lal, R., D. C. Reicosky, and J. D. Hanson. 2007. "Evolution of the Plow over 10,000 Years and the Rationale for No-Till Farming." *Soil and Tillage Research* 93(1): 1–12.

Lehmann, J., and M. Kleber. 2015. "The Contentious Nature of Soil Organic Matter." *Nature* 528(7580): 60–68.

Lenka, S., P. Trivedi, B. Singh, B. P. Singh, E. Pendall, A. Bass, and N. K. Lenka. 2019. "Effect of Crop Residue Addition on Soil Organic Carbon Priming as Influenced by Temperature and Soil Properties." *Geoderma* 347: 70–79.

Li, L.-J., X.-Z. Han, M.-Y. You, Y.-R. Yuan, X.-L. Ding, and Y.-F. Qiao. 2013. "Carbon and Nitrogen Mineralization Patterns of Two Contrasting Crop Residues in a Mollisol: Effects of Residue Type and Placement in Soils." *European Journal of Soil Biology* 54: 1–6.

Liu, C., M. Lu, J. Cui, B. Li, and C. Fang. 2014. "Effects of Straw Carbon Input on Carbon Dynamics in Agricultural Soils: A Meta-Analysis." *Global Change Biology* 20(5): 1366–1381.

Liu, Z., Z. Gao, R. Shi, Z. Dai, and K. Feng. 1996. "Effect of Carbon and C/N on the Characteristics of Soil Nitrogen Supply under Aerobic and Anaerobic Incubation." *Journal-Nanjing Agricultural University* 19: 70–74.

Lu, M., X. Zhou, Y. Luo, Y. Yang, C. Fang, J. Chen, B. Li. 2011. "Minor Stimulation of Soil Carbon Storage by Nitrogen Addition: A Meta-Analysis." *Agricultural, Ecosystems and Environment* 140: 234–244.

Luo, Z., E. Wang, and O. J. Sun. 2010. "Can No-Tillage Stimulate Carbon Sequestration in Agricultural Soils? A Meta-Analysis of Paired Experiments." *Agriculture, Ecosystems and Environment* 139(1–2): 224–231.

Lupwayi, N. Z., G. W. Clayton, J. T. O'Donovan, K. N. Harker, T. K. Turkington, and Y. K. Soon. 2006. "Potassium Release during Decomposition of Crop Residues under Conventional and Zero Tillage." *Canadian Journal of Soil Science* 86(3): 473–481.

Lupwayi, N. Z., G. W. Clayton, J. T. O'Donovan, K. N. Harker, T. K. Turkington, and Y. K. Soon. 2007. "Phosphorus Release during Decomposition of Crop Residues under Conventional and Zero Tillage." *Soil and Tillage Research* 95(1–2): 231–239.

Lynch, M., M. Mulvaney, S. Hodges, T. Thompson, and W. Thomason. 2016. "Decomposition, Nitrogen and Carbon Mineralization from Food and Cover Crop Residues in the Central Plateau of Haiti." *Springerplus* 5(1): 973.

Lyon, D., S. Bruce, T. Vyn, and G. Peterson. 2004. "Achievements and Future Challenges in Conservation Tillage." *Proceedings of the 4th International Crop Science Congress.* Queensland, Australia. www.cropscience.org.au.

Ma, B. L., and M. C. Rivero. 2010. "Fertilization and Crop Residue Management in No-Till Corn Production." In *No-Till Farming*, edited by E. T. Nardali. New York: Nova Science.

Makarov, M. I., L. Haumaier, and W. Zech. 2002. "Nature of Soil Organic Phosphorus: An Assessment of Peak Assignments in the Diester Region of ^{31}P NMR Spectra." *Soil Biology and Biochemistry* 34(10): 1467–1477.

Makarov, M., L. Haumaier, W. Zech, O. Marfenina, and L. Lysak. 2005. "Can 31P NMR Spectroscopy Be Used to Indicate the Origins of Soil Organic Phosphates?" *Soil Biology and Biochemistry* 37(1): 15–25.

Malhi, S. S., C. A. Grant, A. M. Johnston, and K. S. Gill. 2001. "Nitrogen Fertilization Management for No-Till Cereal Production in the Canadian Great Plains: A Review." *Soil and Tillage Research* 60(3–4): 101–122.

Mandal, K. G., A. K. Misra, K. M. Hati, K. K. Bandyopadhyay, P. K. Ghosh, and M. Mohanty. 2004. "Rice Residue-Management Options and Effects on Soil Properties and Crop Productivity." *Journal of Food Agriculture and Environment* 2: 224–231.

Manzoni, S., R. B. Jackson, J. A. Trofymow, and A. Porporato. 2008. "The Global Stoichiometry of Litter Nitrogen Mineralization." *Science* 321(5889): 684–686.

Manzoni, S., P. Taylor, A. Richter, A. Porporato, and G. I. Ågren. 2012. "Environmental and Stoichiometric Controls on Microbial Carbon–Use Efficiency in Soils." *New Phytologist* 196(1): 79–91.

Marschner, H. 1995. *Mineral Nutrition of Higher Plants.* 2nd ed. San Diego, CA: Academic Press.

Martin, J. K., and R. B. Cunningham. 1973. "Factors Controlling the Release of Phosphorus from Decomposing Wheat Roots." *Australian Journal of Biological Sciences* 26(4): 715–728.

Mary, B., S. Recous, D. Darwis, and D. Robin. 1996. "Interactions between Decomposition of Plant Residues and Nitrogen Cycling in Soil." *Plant and Soil* 181(1): 71–82.

Marzluf, G. A. 1997. "Genetic Regulation of Nitrogen Metabolism in the Fungi." *Microbiology and Molecular Biology Reviews* 61(1): 17–32.

McGill, W. B., and C. V. Cole. 1981. "Comparative Aspects of Cycling of Organic C, N, S and P through Soil Organic Matter." *Geoderma* 26(4): 267–286.

McLaughlin, M. J., and A. Alston. 1986. "The Relative Contribution of Plant Residues and Fertilizer to the Phosphorus Nutrition of Wheat in a Pasture Cereal System." *Soil Research* 24(4): 517–526.

McLaughlin, M. J., A. M. Alston, and J. K. Martin. 1988a. "Phosphorus Cycling in Wheat Pasture Rotations. I. The Source of Phosphorus Taken Up by Wheat." *Soil Research* 26(2): 323–331.

McLaughlin, M. J., A. M. Alston, and J. K. Martin. 1988b. "Phosphorus Cycling in Wheat-Pasture Rotations. III. Organic Phosphorus Turnover and Phosphorus Cycling." *Soil Research* 26(2): 343–353.

McLaughlin, M. J., T. M. McBeath, R. Smernik, S. P. Stacey, B. Ajiboye, and C. Guppy. 2011. "The Chemical Nature of P Accumulation in Agricultural Soils – Implications for Fertiliser Management and Design: An Australian Perspective." *Plant and Soil* 349(1–2): 69–87.

Mehra, P., J. Baker, R. E. Sojka, N. Bolan, J. Desbiolles, M. B. Kirkham, C. Ross, and R. Gupta. 2018a. "A Review of Tillage Practices and Their Potential to Impact the Soil Carbon Dynamics." *Advances in Agronomy* 150: 185–230.

Mehra, P., B. P. Singh, A. Kunhikrishnan, A. L. Cowie, and N. Bolan. 2018b. "Soil Health and Climate Change: A Critical Nexus." In *Fundamentals*, edited by D. Reicosky. Vol. 1 of *Managing Soil Health for Sustainable Agriculture*. Cambridge, UK: Burleigh Dodds Science.

Mengel, K., and E. A. Kirkby. 1982. *Principles of Plant Nutrition*. Worblaufen-Bern, Switzerland: International Potash Institute.

Mertens, J., G. Verlinden, and M. Geypens. 2002. "The Influence of Organic Material on the S Mineralization in Sandy and Loamy Soils." In *Improving Sulphur Nutrition in Plants – Molecular and Agronomic Approaches. Proceedings of Working Groups I and III*. Harpenden, Herts, UK: IARC-Rothamsted. https://www.plantsulfur.org/PreviousMeetings?action=AttachFile&do=get&target=Rothamsted-Sulfur-Workhop.pdf (Accessed 15 August 2018).

Mishra, B., P. K. Sharma, and K. F. Bronson. 2001a. "Decomposition of Rice Straw and Mineralization of Carbon, Nitrogen, Phosphorus and Potassium in Wheat Field Soil in Western Uttar Pradesh." *Journal of the Indian Society of Soil Science* 49(3): 419–424.

Mishra, B., P. K. Sharma, and K. F. Bronson. 2001b. "Kinetics of Wheat Straw Decomposition and Nitrogen Mineralization in Rice Field Soil." *Journal of the Indian Society of Soil Science* 49(2): 249–254.

Mohanty, M., M. E. Probert, K. S. Reddy, R. C. Dalal, A. S. Rao, and N. W. Menzies. 2010. "Modelling N Mineralization from High C: N Rice and Wheat Crop Residues." In *Proceedings of the 19th World Congress of Soil Science*, edited by R. Gilkes and N. Prakongkep, 28–31. Brisbane, Australia: International Union of Soil Sciences.

Mubarak, A. R., A. B. Rosenani, A. R. Anuar, and D. Siti Zauyah. 2002. "Effect of Incorporation of Crop Residues on a Maize–Groundnut Sequence in the Humid Tropics. I. Yield and Nutrient Uptake." *Journal of Plant Nutrition* 26(9): 1841–1858.

Muhammad, W., S. M. Vaughan, R. C. Dalal, and N. W. Menzies. 2011. "Crop Residues and Fertilizer Nitrogen Influence Residue Decomposition and Nitrous Oxide Emission from a Vertisol." *Biology and Fertility of Soils* 47(1): 5–23.

Müller, M. M., V. Sundman, O. Soininvaara, and A. Merilainen. 1988. "Effect of Chemical Composition on the Release of Nitrogen from Agricultural Plant Materials Decomposing in Soil under Field Conditions." *Biology and Fertility of Soils* 6: 78–83.

Müller, T., L. S. Jensen, N. Nielsen, and J. Magid. 1998. "Turnover of Carbon and Nitrogen in a Sandy Loam Soil following Incorporation of Chopped Maize Plants, Barley Straw and Blue Grass in the Field." *Soil Biology and Biochemistry* 30(5): 561–571.

Myers, R. J. K., C. A. Palm, E. Cuevas, I. U. N. Gunatilleke, and M. Brossard. 1994. "The Synchronisation of Nutrient Mineralization and Plant Nutrient Demand." In *The Biological Management of Tropical Soil Fertility*, edited by. P. L. Woomer and M. J. Swift, 81–116. Chichester, UK: John Wiley & Sons.

Myrold, D. D., and N. R. Posavatz. 2007. "Potential Importance of Bacteria and Fungi in Nitrate Assimilation in Soil." *Soil Biology and Biochemistry* 39(7): 1737–1743.

Nachimuthu, G., C. Guppy, P. Kristiansen, and P. Lockwood. 2009. "Isotopic Tracing of Phosphorus Uptake in Corn from ^{33}P Labelled Legume Residues and ^{32}P Labelled Fertilizers Applied to a Sandy Loam Soil." *Plant and Soil* 314(1–2): 303–310.

Nagarajah, S., H. U. Neue, and M. C. R. Alberto. 1989. "Effect of Sesbania, Azolla and Rice Straw Incorporation on the Kinetics of NH_4, K, Fe, Mn, Zn and P in Some Flooded Rice Soils." *Plant and Soil* 116(1): 37–48.

Nishio, T., and N. Oka. 2003. "Effect of Organic Matter Application on the Fate of ^{15}N-Labeled Ammonium Fertilizer in an Upland Soil." *Soil Science and Plant Nutrition* 49(3): 397–403.

Nkebiwe, P. M., M. Weinmann, A. Bar-Tal, and T. Müller. 2016. "Fertilizer Placement to Improve Crop Nutrient Acquisition and Yield: A Review and Meta-Analysis." *Field Crops Research* 196: 389–401.

Noack, S. R., M. J. McLaughlin, R. J. Smernik, T. M. McBeath, and R. D. Armstrong. 2012. "Crop Residue Phosphorus: Speciation and Potential Bio-availability." *Plant and Soil* 359(1–2): 375–385.

Noack, S. R., T. M. McBeath, M. J. McLaughlin, R. J. Smernik, and R. D. Armstrong. 2014a. "Management of Crop Residues Affects the Transfer of Phosphorus to Plant and Soil Pools: Results from a Dual-Labelling Experiment." *Soil Biology and Biochemistry* 71: 31–39.

Noack, S. R., M. J. McLaughlin, R. J. Smernik, T. M. McBeath, and R. D. Armstrong. 2014b. "Phosphorus Speciation in Mature Wheat and Canola Plants as Affected by Phosphorus Supply." *Plant and Soil* 378(1–2): 125–137.

Oehl, F., A. Oberson, M. Probst, A. Fliessbach, H.-R. Roth, and E. Frossard. 2001. "Kinetics of Microbial Phosphorus Uptake in Cultivated Soils." *Biology and Fertility of Soils* 34(1): 31–41.

Officer, S., R. Armstrong, and R. Norton. 2009. "Plant Availability of Phosphorus from Fluid Fertiliser Is Maintained under Soil Moisture Deficit in Non-calcareous Soils of South-Eastern Australia." *Soil Research* 47(1): 103–113.

Ogle, S. M., F. J. Breidt, and K. Paustian. 2005. "Agricultural Management Impacts on Soil Organic Carbon Storage under Moist and Dry Climatic Conditions of Temperate and Tropical Regions." *Biogeochemistry* 72(1): 87–121.

Olson, K. R., and M. M. Al-Kaisi. 2015. "The Importance of Soil Sampling Depth for Accurate Account of Soil Organic Carbon Sequestration, Storage, Retention and Loss." *Catena* 125: 33–37.

Page, K. L., R. C. Dalal, M. J. Pringle, et al. 2013a. "Organic Carbon Stocks in Cropping Soils of Queensland, Australia, as Affected by Tillage Management, Climate, and Soil Characteristics." *Soil Research* 51(8): 596–607.

Page, K., Y. Dang, and R. Dalal. 2013b. "Impacts of Conservation Tillage on Soil Quality, including Soil-Borne Crop Diseases, with a Focus on Semi-Arid Grain Cropping Systems." *Australasian Plant Pathology* 42(3): 363–377.

Patel, T., and G. R. R. Murthy. 2017. "Precision Fertilizer Applicator: Localized and Subsurface Placement." In *Engineering Practices for Agricultural Production and Water Conservation*, edited by M. R. Goyal and R. K. Sivanappan, 139–152. Waretown, NJ: Apple Academic Press.

Paustian, K., W. J. Parton, and J. Persson. 1992. "Modeling Soil Organic Matter in Organic-Amended and Nitrogen-Fertilized Long-Term Plots." *Soil Science Society of America Journal* 56(2): 476–488.

Phongpan, S., and A. R. Mosier. 2003a. "Effect of Crop Residue Management on Nitrogen Dynamics and Balance in a Lowland Rice Cropping System." *Nutrient Cycling in Agroecosystems* 66(2): 133–142.

Phongpan, S., and A. R. Mosier. 2003b. "Effect of Rice Straw Management on Nitrogen Balance and Residual Effect of Urea-N in an Annual Lowland Rice Cropping Sequence." *Biology and Fertility of Soils* 37(2): 102–107.

Phongpan, S., and A. R. Mosier. 2003c. "Impact of Organic Residue Management on Nitrogen Use Efficiency in an Annual Rice Cropping Sequence of Lowland Central Thailand." *Nutrient Cycling in Agroecosystems* 6 (3): 233–240.

Powlson, D. S., C. M. Stirling, M. Jat, et al. 2014. "Limited Potential of No-Till Agriculture for Climate Change Mitigation." *Nature Climate Change* 4(8): 678.

Prasad, R., B. Gangaiah, and K. C. Aipe. 1999. "Effect of Crop Residue Management in a Rice–Wheat Cropping System on Growth and Yield of Crops and on Soil Fertility." *Experimental Agriculture* 35(4): 427–435.

Qiu, Q., L. Wu, Z. Ouyang, et al. 2016. "Priming Effect of Maize Residue and Urea N on Soil Organic Matter Changes with Time." *Applied Soil Ecology* 100: 65–74.

Rafique, E., M. Mahmood-ul-Hassan, A. Rashid, and M. F. Chaudhary. 2012. "Nutrient Balances as Affected by Integrated Nutrient and Crop Residue Management in Cotton-Wheat System in Aridisols. III. Potassium." *Journal of Plant Nutrition* 35(4): 633–648.

Rahman, M. A., J. Chikushi, M. Saifizzaman, and J. G. Lauren. 2005. "Rice Straw Mulching and Nitrogen Response of No-Till Wheat following Rice in Bangladesh." *Field Crops Research* 91(1): 71–81.

Raiesi, F., 2006. "Carbon and N Mineralisation as Affected by Soil Cultivation and Crop Residue in a Calcareous Wetland Ecosystem in Central Iran." *Agriculture, Ecosystem and Environment* 112: 13–20.

Ramirez, K. S., J. M. Craine, and N. Fierer. 2012. "Consistent Effects of Nitrogen Amendments on Soil Microbial Communities and Processes across Biomes." *Global Change Biology* 18(6): 1918–1927.

Ranaivoson, L., K. Naudin, A. Ripoche, F. Affholder, L. Rabeharisoa, and M. Corbeels. 2017. "Agro-ecological Functions of Crop Residues under Conservation Agriculture: A Review." *Agronomy for Sustainable Development* 37(4): 26.

Rawls, W., Y. A. Pachepsky, J. Ritchie, T. Sobecki, and H. Bloodworth. 2003. "Effect of Soil Organic Carbon on Soil Water Retention." *Geoderma* 116(1–2): 61–76.

Recous, S., D. Robin, D. Darwis, and B. Mary. 1995. "Soil Inorganic N Availability: Effect on Maize Residue Decomposition." *Soil Biology and Biochemistry* 27(12): 1529–1538.

Reddy, S. K., Singh, M., Swarup, A., Rao, S. A., Singh, K. N. 2002. "Sulfur Mineralization in Two Soils Amended with Organic Manures, Crop Residues, and Green Manures." *Journal of Plant Nutrition and Soil Science* 165: 167–171.

Robertson, F. A., and P. J. Thorburn. 2007. "Management of Sugarcane Harvest Residues: Consequences for Soil Carbon and Nitrogen." *Soil Research* 45(1): 13–23.

Rosolem, C. A., J. C. Calonego, and J. S. S. Foloni. 2005. "Potassium Leaching from Millet Straw as Affected by Rainfall and Potassium Rates." *Communications in Soil Science and Plant Analysis* 36(7–8): 1063–1074.

Roy, M. D., P. Chhonkar, and A. Patra. 2011. "Mineralization of Nitrogen from [15]N Labeled Crop Residues at Varying Temperature and Clay Content." *African Journal of Agricultural Research* 6(1): 102–106.

Rubino, M., J. Dungait, R. Evershed, et al. 2010. "Carbon Input Belowground Is the Major C Flux Contributing to Leaf Litter Mass Loss: Evidences from a [13]C Labelled-Leaf Litter Experiment." *Soil Biology and Biochemistry* 42(7): 1009–1016.

Russell, A. E., D. Laird, T. B. Parkin, and A. P. Mallarino. 2005. "Impact of Nitrogen Fertilization and Cropping System on Carbon Sequestration in Midwestern Mollisols." *Soil Science Society of America Journal* 69(2): 413–422.

Saavedra, C., J. Velasco, P. Pajuelo, F. Perea, and A. Delgado. 2007. "Effects of Tillage on Phosphorus Release Potential in a Spanish Vertisol." *Soil Science Society of America Journal* 71(1): 56–63.

Sahrawat, K., T. Rego, S. Wani, and G. Pardhasaradhi. 2008. "Sulfur, Boron, and Zinc Fertilization Effects on Grain and Straw Quality of Maize and Sorghum Grown in Semi-arid Tropical Region of India." *Journal of Plant Nutrition* 31(9): 1578–1584.

Sahu, A., S. Bhattacharjya, M. Manna, and A. Patra. 2015. "Crop Residue Management: A Potential Source for Plant Nutrients." *JNKVV Research Journal* 49(3): 301–311.

Sardans, J., and J. Peñuelas. 2015. "Potassium: A Neglected Nutrient in Global Change." *Global Ecology and Biogeography* 24(3): 261–275.

Sarker, J. R., B. P. Singh, X. He, Y. Fang, G. D. Li, D. Collins, and A. L. Cowie. 2017. "Tillage and Nitrogen Fertilization Enhanced Belowground Carbon Allocation and Plant Nitrogen Uptake in a Semi-arid Canola Crop–Soil System." *Scientific Reports* 7(1): 10726.

Sarker, J. R., B. P. Singh, A. L. Cowie, Y. Fang, D. Collins, W. Badgery, and R. C. Dalal. 2018a. "Agricultural Management Practices Impacted Carbon and Nutrient Concentrations in Soil Aggregates, with Minimal Influence on Aggregate Stability and Total Carbon and Nutrient Stocks in Contrasting Soils." *Soil and Tillage Research* 178: 209–223.

Sarker, J. R., B. P. Singh, A. L. Cowie, Y. Fang, D. Collins, W. J. Dougherty, and B. K. Singh. 2018b. "Carbon and Nutrient Mineralization Dynamics in Aggregate-Size Classes from Different Tillage Systems after Input of Canola and Wheat Residues." *Soil Biology and Biochemistry* 116: 22–38.

Sarker, J. R., B. P. Singh, W. J. Dougherty, Y. Fang, W. Badgery, F. C. Hoyle, R. C. Dalal, and A. L. Cowie. 2018c. "Impact of Agricultural Management Practices on the Nutrient Supply Potential of Soil Organic Matter under Long-Term Farming Systems." *Soil and Tillage Research* 175: 71–81.

Sarker, J. R., B. P. Singh, Y. Fang, A. L. Cowie, W. J. Dougherty, D. Collins, R. C. Dalal, and B. K. Singh. 2019. "Tillage History and Crop Residue Input Enhanced Native Carbon Mineralisation and Nutrient Supply in Contrasting Soils under Long-Term Farming Systems." *Soil and Tillage Research* 193: 71–84.

Scherer, H. W. 2001. "Sulphur in Crop Production." *European Journal of Agronomy* 14(2): 81–111.

Schomberg, H. H., and J. L. Steiner. 1999. "Nutrient Dynamics of Crop Residues Decomposing on a Fallow No-Till Soil Surface." *Soil Science Society of America Journal* 63(3): 607–613.

Schomberg, H. H., J. L. Steiner, and P. W. Unger. 1994. "Decomposition and Nitrogen Dynamics of Crop Residues: Residue Quality and Water Effects." *Soil Science Society of America Journal* 58(2): 372–381.

Schultz, J., and R. French. 1978. "The Mineral Content of Cereals, Grain Legumes and Oilseed Crops in South Australia." *Australian Journal of Experimental Agriculture* 18(93): 579–585.

Scott, B. J., P. L. Eberbach, J. Evans, and L. J. Wade. 2010. *Stubble Retention in Cropping Systems in Southern Australia: Benefits and Challenges.* EH Graham Centre Monograph No. 1, edited by E. H. Clayton and H. M. Burns. Orange, New South Wales, Australia: Industry and Investment NSW. Accessed 15 August 2018. https://www.csu.edu.au/__data/assets/pdf_file/0007/922723/stubble-retention.pdf.

Sharma, K., B. Ramachandrappa, D. S. Chandrika, et al. 2016. "Effect of Organic Manure and Crop Residue Based Long-Term Nutrient Management Systems on Soil Quality Changes under Sole Finger millet (*Eleusine coracana* (L.) Gaertn.) and Groundnut (*Arachis hypogaea* L.)–Finger Millet Rotation in Rainfed Alfisol." *Communications in Soil Science and Plant Analysis* 47(7): 899–914.

Sharpley, A. N., and S. J. Smith. 1989. "Mineralization and Leaching of Phosphorus from Soil Incubated with Surface-Applied and Incorporated Crop Residue." *Journal of Environmental Quality* 18(1): 101–105.

Shindo, H., and T. Nishio. 2005. "Immobilization and Remineralization of N following Addition of Wheat Straw into Soil: Determination of Gross N Transformation Rates by ^{15}N-Ammonium Isotope Dilution Technique." *Soil Biology and Biochemistry* 37(3): 425–432.

Singh, B. B., and J. P. Jones. 1976. "Phosphorous Sorption and Desorption Characteristics of Soil as Affected by Organic Residues." *Soil Science Society of America Journal* 40(3): 389–394.

Singh, B., and Z. Rengel. 2007. "The Role of Crop Residues in Improving Soil Fertility." In *Nutrient Cycling in Terrestrial Ecosystems*, edited by P. Marschner and Z. Rengel, 183–214. Berlin, Germany: Springer.

Singh, B., Z. Rengel, and J. Bowden. 2006. "Carbon, Nitrogen and Sulphur Cycling following Incorporation of Canola Residue of Different Sizes into a Nutrient-Poor Sandy Soil." *Soil Biology and Biochemistry* 38(1): 32–42.

Singh, B. P., R. Setia, M. Wiesmeier, and A. Kunhikrishnan. 2018. "Agricultural Management Practices and Soil Organic Carbon Storage." In *Soil Carbon Storage: Modulators, Mechanisms and Modeling*, edited by B. Singh, 207–244. London, UK: Academic Press.

Singh, M., V. P. Singh, and D. Damodar Reddy. 2002. "Potassium Balance and Release Kinetics under Continuous Rice–Wheat Cropping System in Vertisol." *Field Crops Research* 77(2): 81–91.

Singh, S., N. Ghoshal, and K. Singh. 2007. "Synchronizing Nitrogen Availability through Application of Organic Inputs of Varying Resource Quality in a Tropical Dryland Agroecosystem." *Applied Soil Ecology* 36(2–3): 164–175.

Six, J., E. Elliott, and K. Paustian. 1999. "Aggregate and Soil Organic Matter Dynamics under Conventional and No-Tillage Systems." *Soil Science Society of America Journal* 63(5): 1350–1358.

Six, J., E. T. Elliott, and K. Paustian. 2000. "Soil Macroaggregate Turnover and Microaggregate Formation: A Mechanism for C Sequestration under No-Tillage Agriculture." *Soil Biology and Biochemistry* 32(14): 2099–2103.

Smeck, N. E. 1985. "Phosphorus Dynamics in Soils and Landscapes." *Geoderma* 36(3–4): 185–199.

Smernik, R. J., and J. A. Baldock. 2005. "Solid-State ^{15}N NMR Analysis of Highly ^{15}N-Enriched Plant Materials. *Plant and Soil* 275(1–2): 271–283.

Sollins, P., P. Homann, and B. A. Caldwell. 1996. "Stabilization and Destabilization of Soil Organic Matter: Mechanisms and Controls." *Geoderma* 74(1–2): 65–105.

Stavi, I., G. Bel, and E. Zaady. 2016. "Soil Functions and Ecosystem Services in Conventional, Conservation, and Integrated Agricultural Systems: A Review." *Agronomy for Sustainable Development* 36(2): 32.

Stewart, J. W. B., and H. Tiessen. 1987. "Dynamics of Soil Organic Phosphorus." *Biogeochemistry* 4(1): 41–60.

Stockfisch, N., T. Forstreuter, and W. Ehlers. 1999. "Ploughing Effects on Soil Organic Matter after Twenty Years of Conservation Tillage in Lower Saxony, Germany." *Soil and Tillage Research* 52(1–2): 91–101.

Sugihara, S., S. Funakawa, and T. Kosaki. 2012. "Effect of Land Management on Soil Microbial N Supply to Crop N Uptake in a Dry Tropical Cropland in Tanzania." *Agriculture, Ecosystems and Environment* 146(1): 209–219.

Sui, N., C. Yu, G. Song, et al. 2017. "Comparative Effects of Crop Residue Incorporation and Inorganic Potassium Fertilisation on Apparent Potassium Balance and Soil Potassium Pools under a Wheat–Cotton System." *Soil Research* 55(8): 723–734.

Sui, N., Z. Zhou, C. Yu, et al. 2015. "Yield and Potassium Use Efficiency of Cotton with Wheat Straw Incorporation and Potassium Fertilization on Soils with Various Conditions in the Wheat–Cotton Rotation System." *Field Crops Research* 172: 132–144.

Tadano, T., K. Ozawa, H. Sakai, M. Osaki, and H. Matsui. 1993. "Secretion of Acid Phosphatase by the Roots of Crop Plants under Phosphorus-Deficient Conditions and Some Properties of the Enzyme Secreted by Lupin Roots." *Plant and Soil* 155: 95–98.

Thomas, G. W., and B. W. Hipp. 1968. "Soil Factors Affecting Potassium Availability." In *The Role of Potassium in Agriculture*, edited by V. J. Kilmer, S. E. Younts, and N. C. Brady. 269–291. Madison, WI: American Society of Agronomy, Crop Science Society of America, Soil Science Society of America.

Thomsen, I. K., and B. T. Christensen. 1998. "Cropping System and Residue Management Effects on Nitrate Leaching and Crop Yields." *Agriculture, Ecosystems and Environment* 68(1–2): 73–84.

Thuy, N. H., Y. Shan, K. Wang, Z. Cai, and R. J. Buresh. 2008. "Nitrogen Supply in Rice-Based Cropping Systems as Affected by Crop Residue Management." *Soil Science Society of America Journal* 72(2): 514–523.

Till, A. R., and G. J. Blair. 1978. "The Utilization by Grass of Sulphur and Phosphorus from Clover Litter." *Australian Journal of Agricultural Research* 29(2): 235–242.

Torn, M. S., S. E. Trumbore, O. A. Chadwick, P. M. Vitousek, and D. M. Hendricks. 1997. "Mineral Control of Soil Organic Carbon Storage and Turnover." *Nature* 389(6647): 170–173.

Trinsoutrot, I., S. Recous, B. Bentz, M. Lineres, D. Cheneby, and B. Nicolardot. 2000. "Biochemical Quality of Crop Residues and Carbon and Nitrogen Mineralization Kinetics under Nonlimiting Nitrogen Conditions." *Soil Science Society of America Journal* 64: 918–926.

Turmel, M.-S., A. Speratti, F. Baudron, N. Verhulst, and B. Govaerts. 2015. "Crop Residue Management and Soil Health: A Systems Analysis." *Agricultural Systems* 134: 6–16.

Umrit, G., and D. K. Friesen. 1994. "The Effect of C: P Ratio of Plant Residues Added to Soils of Contrasting Phosphate Sorption Capacities on P Uptake by *Panicum maximum* (Jacq.)." *Plant and Soil* 158(2): 275–285.

Van Duivenbooden, N., C. De Wit, and H. Van Keulen. 1996. "Nitrogen, Phosphorus and Potassium Relations in Five Major Cereals Reviewed in Respect to Fertilizer Recommendations using Simulation Modelling." *Fertilizer Research* 44(1): 37–49.

Van Wesemael, B., K. Paustian, J. Meersmans, E. Goidts, G. Barancikova, and M. Easter. 2010. "Agricultural Management Explains Historic Changes in Regional Soil Carbon Stocks." *Proceedings of the National Academy of Sciences* 107(33): 14926–14930.

Varinderpal-Singh, Bijay-Singh, Yadvinder-Singh, H. S. Thind, A. Kumar, and M. Vashistha. 2012. "Establishment of Threshold Leaf Colour Greenness for Need-Based Fertilizer Nitrogen Management in Irrigated Wheat (*Triticum aestivum* L.) Using Leaf Colour Chart." *Field Crops Research* 130: 109–119.

Varinderpal-Singh, Yadvinder-Singh, Bijay-Singh, H. S. Thind, A. Kumar, and M. Vashistha. 2011. "Calibrating the Leaf Colour Chart for Need Based Fertilizer Nitrogen Management in Different Maize (*Zea mays* L.) Genotypes." *Field Crops Research* 120(2): 276–282.

Venkatesh, M. S., K. K. Hazra, P. K. Ghosh, et al. 2017. "Long-term Effect of Crop Rotation and Nutrient Management on Soil–Plant Nutrient Cycling and Nutrient Budgeting in Indo–Gangetic Plains of India." *Archives of Agronomy and Soil Science* 63(14): 2007–2022.

Vetsch, J. A., and G. W. Randall. 2000. "Enhancing No-Tillage Systems for Corn with Starter Fertilizers, Row Cleaners, and Nitrogen Placement Methods." *Agronomy Journal* 92(2): 309–315.

von Lützow, M., I. Kögel-Knabner, K. Ekschmitt, et al. 2006. "Stabilization of Organic Matter in Temperate Soils: Mechanisms and Their Relevance under Different Soil Conditions – A Review. *European Journal of Soil Science* 57(4): 426–445.

Vyn, T. J., and K. J. Janovicek. 2001. "Potassium Placement and Tillage System Effects on Corn Response following Long-Term No Till Research." *Agronomy Journal* 93(3): 487–495.

West, T. O., and W. M. Post. 2002. "Soil Organic Carbon Sequestration Rates by Tillage and Crop Rotation." *Soil Science Society of America Journal* 66(6): 1930–1946.

Whitbread, A. M., G. J. Blair, and R. D. B. Lefroy. 2000. "Managing Legume Leys, Residues and Fertilizers to Enhance the Sustainability of Wheat Cropping Systems in Australia: 1. The Effects on Wheat Yields and Nutrient Balances." *Soil and Tillage Research* 54(1): 63–75.

White, R. E., and A. T. Ayoub. 1983. "Decomposition of Plant Residues of Variable C/P Ratio and the Effect on Soil Phosphate Availability." *Plant and Soil* 74(2): 163–173.

Witt, C., K. Cassman, D. Olk, et al. 2000. "Crop Rotation and Residue Management Effects on Carbon Sequestration, Nitrogen Cycling and Productivity of Irrigated Rice Systems." *Plant and Soil* 225(1–2): 263–278.

Witt, C., J. M. C. A. Pasuquin, R. Mutters, and R. J. Buresh. 2005. "New Leaf Color Chart for Effective Nitrogen Management in Rice." *Better Crops* 89(1): 36–39.

Wortmann, C. S., R. A. Drijber, and T. G. Franti. 2010. "One-Time Tillage of No-Till Crop Land Five Years Post-Tillage." *Agronomy Journal* 102(4): 1302–1307.

Wu, J., A. G. O'Donnell, and J. K. Syers. 1993. "Microbial Growth and Sulphur Immobilization following the Incorporation of Plant Residues into Soil." *Soil Biology and Biochemistry* 25(11): 1567–1573.

Wu, J., A. G. O'Donnell, and J. K. Syers. 1995. "Influences of Glucose, Nitrogen and Plant Residues on the Immobilization of Sulphate-S in Soil." *Soil Biology and Biochemistry* 27(11): 1363–1370.

Xu, X., P. E. Thornton, and W. M. Post. 2013. "A Global Analysis of Soil Microbial Biomass Carbon, Nitrogen and Phosphorus in Terrestrial Ecosystems." *Global Ecology and Biogeography* 22(6): 737–749.

Xu, G., C. Wu, H. Liu, Z. Wang, M. Zhang, and J. Yang. 2007. "Effects of Wheat Residue Incorporation and Nitrogen Management Techniques on Formation of the Grain Yield of Rice." *Acta Agronomica Sinica* 33: 284–291.

Yadvinder-Singh, Bijay-Singh, J. K. Ladha, C. S. Khind, T. S. Khera, and C. Bueno. 2004. "Effects of Residue Decomposition on Productivity and Soil Fertility in Rice–Wheat Rotation." *Soil Science Society of America Journal* 68(3): 854–864.

Yadvinder-Singh, Bijay-Singh, M. S. Maskina, and O. P. Meelu. 1988. "Effect of Organic Manures, Crop Residues and Green Manure (Sesbania aculeata) on Nitrogen and Phosphorus Transformations in a Sandy Loam at Field Capacity and under Waterlogged Conditions." *Biology and Fertility of Soils* 6(2): 183–187.

Yadvinder-Singh, and H. S. Sidhu. 2014. "Management of Cereal Crop Residues for Sustainable Rice-Wheat Production System in the Indo-Gangetic Plains of India." Proceedings of the *Indian National Science Academy* 80: 95–114.

Yadvinder-Singh, Bijay-Singh, and J. Timsina. 2005. "Crop Residue Management for Nutrient Cycling and Improving Soil Productivity in Rice-Based Cropping Systems in the Tropics." *Advances in Agronomy* 85: 269–407.

Yamagata, M., and N. Ae. 1996. "Nitrogen Uptake Response of Crops to Organic Nitrogen." *Soil Science and Plant Nutrition* 42(2): 389–394.

Yang, Y., C. Zhou, N. Li, et al. 2015. "Effects of Conservation Tillage Practices on Ammonia Emissions from Loess Plateau Rain-Fed Winter Wheat Fields." *Atmospheric Environment* 104: 59–68.

Yin, X., and T. J. Vyn. 2003. "Potassium Placement Effects on Yield and Seed Composition of No-Till Soybean Seeded in Alternate Row Widths." *Agronomy Journal* 95(1): 126–132.

Zechmeister-Boltenstern, S., K. M. Keiblinger, M. Mooshammer, et al. 2015. "The Application of Ecological Stoichiometry to Plant–Microbial–Soil Organic Matter Transformations." *Ecological Monographs* 85(2): 133–155.

Zhao, Y., M. Wang, S. Hu, et al. 2018. "Economics- and Policy-Driven Organic Carbon Input Enhancement Dominates Soil Organic Carbon Accumulation in Chinese Croplands." *Proceedings of the National Academy of Sciences* 115(16): 4045–4050.

Zibilske, L. M., and J. M. Bradford. 2003. "Tillage Effects on Phosphorus Mineralization and Microbial Activity." *Soil Science* 168(10): 677–685.

4 Improving Soil Fertility through Fertilizer Management in Sub-Saharan Africa

*Andre Bationo, Upendra Singh, Ekwe Dossa,
John Wendt, Sampson Agiyin-Birikorang,
François Lompo, and Prem Bindraban*

CONTENTS

4.1 INTRODUCTION

Sub-Saharan Africa (SSA) has been identified as a future hotspot for food shortage due to low agricultural yields and high variability in yield, cultivated acreage, and total production. Several African countries are food insecure and have persistently been unable to feed their population (NEPAD 2003). Haggblade et al. (2004) note that, over the past 40 years or so, agriculture production has increased at a rate of 2.5% per year in Africa compared to 2.9% in Latin America and 3.5% in developing Asia. As a result of this situation, Africa is a net food-importing region. Food imports in Africa rose from USD 15 billion in the 1990s to about USD 40 billion in 2007 (Rakotoarisoa et al. 2011). The number of chronically undernourished people increased from 168 million in 1990–1992 to 224 million in 2016 (FAO 2017). Agricultural productivity in Africa lags all other continents. NEPAD (2014) notes that productivity per agricultural worker has improved by a factor of only 1.6 in Africa over the past 30 years, compared to 2.5 in Asia. While cereal crop yields in Asia have doubled or quadrupled since the 1960s, they have stagnated in Africa (Haggblade et al. 2004), and as populations have increased, food production per capita has been declining in Africa for the past three decades. This leaves African families with ever less opportunity to feed themselves and their children. As a result, malnutrition remains shockingly common in Africa, increasing from 160 million people in 1990 to 205 million in 2015 (FAO 2019).

SSA uses 13 kg per hectare of nitrogen, phosphorus and potassium (NPK), fertilizers on croplands in SSA, and this usage remains low compared to the world average of 100 kg NPK ha^{-1} with consumption in 2016 in Eastern, Middle, Western, and Southern Africa of 15.8, 4.3, 9.6, and 53.4 kg ha^{-1}, up from 12.3, 2.6, 3.9, and 53.7, respectively, in 2002 (FAO 2019). Despite the increase, the consumption of fertilizer per capita is stagnant – five times lower than Asia (Figure 4.1).

As a consequence of the low use of fertilizers and suboptimal soil fertility management, crop yields have stagnated in the past 50 years, causing food insecurity and encouraging encroachment upon the remnant forests to meet the food needs of an overgrowing population. Meanwhile, Africa has considerable fertilizer resources. Seventy percent of the world's phosphate rock resources (Figure 4.2) and significant deposits of nitrogen and potash resources are found in Africa (Figure 4.3).

Although significant progress has been made in research in developing principles, methodologies, and technologies for combating soil fertility depletion, soil infertility still remains the fundamental biophysical cause for the declining per capita food production in SSA over the last 3–5 decades (Vanlauwe 2004). This is evident from the huge gap between actual and potential crop yields (FAO 1995).

During the period of 30 years from 1960 to 1990, soil fertility depletion has been estimated at an average of 660 kg N ha^{-1}, 75 kg P ha^{-1}, and 450 kg K ha^{-1} from about 200 million ha of cultivated land in 37 African countries (Vanlauwe 2004). Stoorvogel et al. (1993) estimated annual net

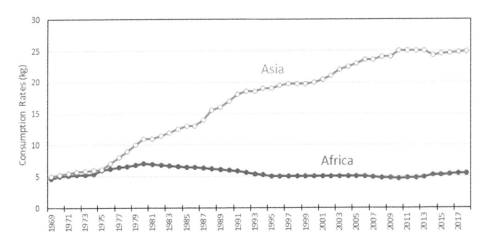

FIGURE 4.1 Evolution of fertilizer consumption per capita (kg) in Africa. Based on FAO (2019) and FAO (2016).

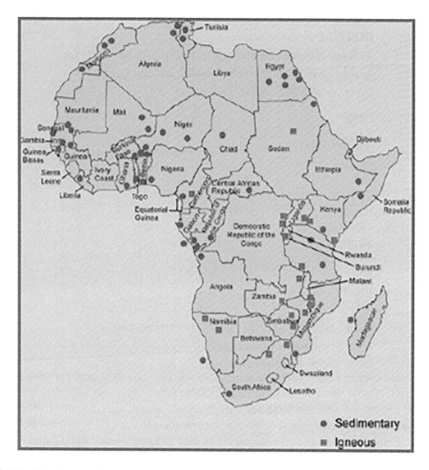

FIGURE 4.2 Phosphate rock deposits in Africa. (Source: IFDC.)

FIGURE 4.3 Nitrogen and potash deposits in Africa. (Source: IFDC.)

TABLE 4.1

Average Nutrient Balance of N, P, and K (k ha^{-1} yr^{-1}) for Arable Land for Some Sub-Saharan African Countries (Average of 1982–1984)

Country	N	P	K
Botswana	0	1	0
Mali	−8	−1	−7
Senegal	−12	−2	−10
Benin	−14	−1	−9
Cameroon	−20	−2	−12
Tanzania	−27	−4	−18
Zimbabwe	−31	−2	−22
Nigeria	−34	−4	−24
Ethiopia	−41	−6	−26
Kenya	−42	−3	−29
Rwanda	−54	−9	−47
Malawi	−68	−10	−44

Source: Stoorvogel, J.J., et al., *Fertil. Res.*, 35, 227–235, 1993.

depletion of nutrients in excess of 30 kg N and 20 kg K ha^{-1} of arable land per year in Ethiopia, Kenya, Malawi, Nigeria, Rwanda, and Zimbabwe (Table 4.1).

Given the current low levels of fertilizer use, green and animal manures are insufficient to sustain soil health and the nutrient balances remain negative for many cropping systems, indicating that farmers are mining their soils of nutrient reserves of over 50 kg ha^{-1} year^{-1} of N, P, and K combined (e.g., Lesschen et al. 2007; Cobo et al. 2010).

The development of the modern fertilizer industry had a tremendous positive impact on agricultural productivity and global food security, and it is widely acknowledged that fertilizers are responsible for at least half of the global food supply (Erisman et al. 2008). The data in Table 4.2 gives the yield of selected crops in farmer's field without the use of fertilizers as compared to the potential yields obtained on station with the use of fertilizers. It is evident from the data that for some crops it is possible to increase the yield up to five times by improving soil nutrients management.

TABLE 4.2

Yield Potential with Fertilizer Use in Africa

Crop	Actual Yields in Farms without Fertilizer (kg/ha)	Potential Yield on station with Fertilizer (kg/ha)	Increase over Control (%)
Irrigated rice West Africa	3,000	8,000	167
Upland rice West Africa	1,000	4,000	500
Cassava West Africa	8,000	47,000	487
Maize West Africa	800	6,000	650
Sorghum West Africa	600	3,000	400
Millet West Africa	300	2,000	567
Maize East Africa	1,500	8,000	433
Maize Southern Africa	1,500	8,000	433

Source: Bationo, A., unpublished data.

4.2 CHANGES IN PARADIGMS FOR SOIL FERTILITY MANAGEMENT IN SUB-SAHARAN AFRICA

Since the 1960s, the paradigms underlying soil fertility management research and development efforts have undergone substantial changes because of experiences gained with specific approaches and changes in the overall social, economic, and political environment faced by various stakeholders. During the 1960s and 1970s, an external input paradigm characterized by increased use of improved germplasm and fertilizer significantly led to a rapid increase in food production, commonly referred to as the "Green Revolution," especially in Asia and Latin America. This paradigm put little if any significance on the set of organic resources as sources of nutrients for soil health.

The impacts of the Green Revolution strategy resulted only in minor achievements in SSA. The environmental degradation resulting from massive and injudicious applications of fertilizers and pesticides observed in Asia and Latin America between the mid-1980s and early 1990s (Theng 1991) and the abolition of the fertilizer subsidies in SSA (Smaling 1993), imposed by structural adjustment programs, led to a renewed interest in organic resources in the early 1980s. The balance shifted from mineral inputs to low input sustainable agriculture (LISA) where organic resources were believed to enable sustainable agricultural production (Vanlauwe 2004). The adoption of LISA technologies, such as alley cropping or live mulch systems, was constrained by both technical (e.g., lack of sufficient organic resources) and socioeconomic factors (e.g., labor-intensive technologies) (Vanlauwe 2004). This led to the second paradigm, integrated nutrient management (INM), which emphasized the need for the judicious use of both mineral and organic inputs to sustain crop production (Vanlauwe 2004).

A further shift in paradigm in the mid-1980s to the mid-1990s advanced the combined use of organic and mineral inputs accompanied by a shift in approaches toward involvement of the various stakeholders in the research and development process, mainly driven by the "participatory" movement. One of the important lessons learned was that the farmers' decision-making process was not merely driven by soil and climate but by a whole set of factors cutting across the biophysical, socioeconomic, and political domain. The integrated natural resource management (INRM) research approach was thus formulated, aimed at developing interventions that take all the above aspects into account (Izac 2000).

Past paradigms of soil fertility management focused on fertilizer or "low-input" methods, but rarely on both, and ignored the essential scientific fact that fertilizers are most effective and efficient in the presence of soil organic matter (SOM) and well-conserved soil structure. This dichotomy is resolved by the integrated soil fertility management (ISFM) framework. The framework entails applying locally adapted soil fertility management practices to optimize the agronomic efficiency of fertilizer and organic inputs in crop production. Large-scale adoption of ISFM will promote soil fertility management practices, which include the use of mineral fertilizers, organic inputs, improved germplasm, and knowledge of their local adaptation. Such practices would maximize agronomic use efficiency of applied nutrients and improve crop productivity (Figure 4.4). Widespread adoption of ISFM is crucial in harnessing healthy soils, given that inorganic fertilizer provides most of the nutrients and organic fertilizer increases SOM status, soil structure, and buffering capacity of the soil in general. Moreover, use of both inorganic and organic fertilizers has proven to result in synergy, improving efficiency of both nutrient and water use.

In addition, the ISFM concept also takes into account other socioeconomic factors such as land tenure, input-output markets, access to credit, and institutional support, among others, in a value chain approach. ISFM, therefore, seeks to develop competitive commodity chains by strengthening the technical and managerial competencies of the various actors involved, particularly the farmers and local entrepreneurs (including inputs dealers, processors, stockists, and traders) at the grassroots level. Past paradigms had ignored such a holistic approach to agricultural development.

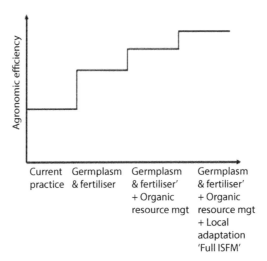

Current   Germplasm   Germplasm   Germplasm
practice   & fertiliser   & fertiliser'   & fertiliser'
                                + Organic   + Organic
                                resource mgt   resource mgt
                                                   + Local
                                                   adaptation
                                                   'Full ISFM'

FIGURE 4.4   The ISFM paradigm. (Source: Vanlauwe, B. et al., *Outlook Agr.*, 39, 17–24, 2010.)

4.3   PRODUCTIVITY OF AFRICAN SOILS

Africa covers an area of about 3.01×10^9 hectares, out of which about 230×10^6 hectares represent water bodies. Relative to soil in other parts of the world. African soils have an inherently poor fertility because they are very old and because lack of volcanic rejuvenation has caused the continent to undergo various cycles of weathering, erosion, and leaching, leaving soils poor in nutrients (Smaling 1995). Inappropriate land use, poor soil fertility management, and lack of input have led to a decline in productivity.

At least 485 million Africans are affected by land degradation, making land degradation one of the continent's urgent development issues with significant costs: Africa is burdened with a $9.3 billion annual cost of desertification. While the cumulative loss of crop productivity from worldwide land degradation between 1945 and 1990 has been estimated at 5%, as much as 6.2% of productivity has been lost in SSA. An estimated $42 billion in income and 6 million hectares of productive lands are lost every year due to land degradation and declining agricultural productivity (UNDP/GEF 2004). Globally, Africa suffered a net loss of forests exceeding 4 million hectares per year between 2000 and 2005, according to FAO (2006). This was mainly due to conversion of forest lands to agriculture. During a period of 30 years, soil fertility depletion has been estimated at an average of 660 kg N ha^{-1}, 75 kg P ha^{-1}, and 450 kg K ha^{-1} from about 200 million ha of cultivated land in 37 African countries. Africa loses $4 billion per year due to soil nutrient mining (Smaling 1993). Soil fertility depletion in smallholder farms is a fundamental biophysical root cause of the declining per capita food production; it has largely contributed to poverty and food insecurity. Over 132 million tons of N, 15 million tons of P, and 90 million tons of K have been lost from cultivated land in 37 African countries in 30 years (Smaling 1993). Nutrient loss is estimated to be 4.4 million t N, 0.5 million t P, and 3 million t K every year from cultivated land (Vanlauwe 2004). These rates are higher than Africa's annual fertilizer consumption (excluding South Africa) of 2.4 million tons N, 0.40 million tons P, and 0.5 million tons K (FAO, 2016). The loss is equivalent to 1400 kg ha^{-1} urea, 375 kg ha^{-1} Triple Super-phosphate (TSP), and 896 kg ha^{-1} KCl during the period of three decades.

N and P are the most deficient macronutrients in SSA, but the crop responses to these two nutrients is quite variable due to the high heterogeneity of the soils in their initial nutrient status, crop nutrient uptake, and response to applied fertilizer vary with soil type, as indicated in Table 4.3. Whereas P was limiting nutrient in the Nitisol, N was the most limiting nutrient in the Vertisol. These results point to the need to effectively tailor fertilizer to soil fertilizer status to ensure use efficiency of use by crops on the different soil types.

TABLE 4.3

Yields and NPK Uptake of Maize on Three Kenyan Soils as a Function of Soil Type and Fertilizer Treatment (Long Rainy Season, 1990)

Soil	Treatment	Yield (ton/ha)	Nutrient Uptake		
			N	P	K
			(kg/ha)		
Nitisol (red, clayey)	$N_0 P_0$	2.1	42	5	30
	$N_{50} P_0$	2.3	50	6	36
	$N_0 P_{22}$	4.9	79	12	58
Vertisol (black, clayey)	$N_0 P_0$	4.5	63	24	95
	$N_{50} P_0$	6.3	109	35	126
	$N_0 P_{22}$	4.7	70	23	106

Source: Smaling, E.M.A., et al., *Ambio*, 25, 492–496, 1996.
Note: N – kg/ha as CAN; P – kg/ha as TSP.

A major issue preventing effective utilization of fertilizers by crop has been the "pan-territorial/ blanket" nature of the recommendations that fail to take into account differences in farmers' resource endowment (soil type, labor capacity, climate risk, etc.). Past fertilizer recommendations have been based on single major cash crops such as maize, tea, and cotton and did not take into account complex farming systems involving crop rotations, intercropping, and conservation agriculture that are characteristic of most smallholder farming systems in Africa. Different fertilizer responses have been observed in different parts of the same field due to the existing within-farm soil fertility gradients. Research has shown that the use of high-yielding cereal varieties, along with the increasing use of fertilizers containing major nutrients (N, P, K), but without micronutrients through inorganic or organic fertilizers, dramatically increases food production under intensified systems. However, in the long run as a result of depletion of micronutrient reserves in the soil, this practice results in a number of nutrient disorders and associated nutrient imbalances. Micronutrients are required by plants in small quantities, but they limit plant growth and substantially lower yields when deficient. In SSA, only a few studies (Schutte 1954; Sillanpaa 1982; Kang and Osiname 1985) have documented the micronutrient status of soils in the region, as compared to the enormous amount of literature available on macronutrients. The study by Sillanpaa (1982) showed that copper, zinc, and molybdenum deficiencies are common in many coarse-textured acid soils of Ethiopia, Ghana, Malawi, Nigeria, Sierra Leone, Tanzania, and Zambia. Recently, Kihara et al. (2017), using a meta-analysis of published articles on crop response to secondary and micronutrients in Africa, concluded that S, Zn, Cu, and Fe induced positive and significant crop response following their addition to the major NPK fertilizers. In many SSA countries, replenishment of micronutrients through fertilizers or other amendments is still in its infancy. Additions of soil micronutrients can improve the yield response to macronutrients (N and P) on deficient soils. Nutrients such as Zn, B, S, and Mg can often be included relatively cheaply in existing fertilizer blends; when targeted to deficient soils, these nutrients can dramatically improve fertilizer use efficiency and crop yield. During the 1950s, 1960s, and 1970s, S, Mg, and (less commonly) Zn and B deficiencies were detected for maize on sandy soils in Zimbabwe (Grant 1981). Enhanced yields were obtained by including selected micronutrients in fertilizer blends (Grant 1981). Recent experience in Malawi provides a striking example of how N fertilizer efficiency for maize can be raised by providing appropriate micronutrients on a location-specific basis.

Soil organic carbon is an exhaustible natural resource capital, and, like the negative nutrient balances, its decline threatens soil productivity. The concentration of organic carbon in the top soil is reported to average 12 mg kg^{-1} for the humid zone, 7 mg kg^{-1} for the subhumid zone, 4 mg kg^{-1}

in the semiarid zone, and 2 mg kg^{-1} for the drier semiarid zone (Windmeijer and Andriesse 1993). Most African soils are inherently low in organic carbon (<20 to 30 mg kg^{-1}). This is due to the low root growth of crops and natural vegetation, but also to the rapid turnover rates of organic materials with high soil temperature, persistent bush burning, and intense faunal activity, particularly termites (Bationo et al. 2003). There is much evidence for rapid decline of soil organic C levels with continuous cultivation of crops in Africa (Bationo and Buerkert 2001). Results from long-term soil fertility trials indicate that losses of up to 0.69 tons carbon ha^{-1} yr^{-1} in the soil surface layers is common in Africa even with high levels of organic inputs (Nandwa 2003). There is much evidence for a rapid decline of Corg levels with continuous cultivation of crops (Bationo et al. 1995). For the sandy soils, average annual losses in Corg, often expressed by the k-value (calculated as the percentage of organic carbon lost per year), may be as high as 4.7%, whereas for sandy loam soils, reported losses seem much lower, with an average of 2% (Pieri 1989; Table 4). Topsoil erosion may lead to significant increases in annual Corg losses, such as from 2% to 6.3% at the Centre de Formation des Jeunes Agriculteurs (CFJA) in Burkina Faso (Table 4.4). However, such declines

TABLE 4.4
Annual Loss Rates of Soil Organic Carbon Measured at Selected Research Stations in the Sub-Saharan West Africa

Place and Source	Dominant Cultural Succession	Observations	Clay + Silt (%) (0–0.2 m)	Annual Loss Rates of Soil Organic Carbon (k)	
				Number of Years of Measurement	k (%)
Burkina Faso		With tillage			
Saria, INERA-IRAT	Sorghum monoculture	Without fertilizer	12	10	1.5
	Sorghum monoculture	Low fertilizer (lf)	12	10	1.9
	Sorghum monoculture	High fertilizer (hf)	12	10	2.6
	Sorghum monoculture	lf + crop residues	12	10	2.2
CFJA, INERA-IRCT	Cotton–cereals	Eroded watershed	19	15	6.3
Senegal		With tillage			
Bambey, ISRA-IRAT	Millet–groundnut	Without fertilizer	3	5	7.0
	Millet–groundnut	With fertilizer	3	5	4.3
	Millet–groundnut	Fertilizer + straw	3	5	6.0
Bambey, ISRA-IRAT	Millet monoculture	with PK fertilizer + tillage	4	3	4.6
Nioro-du-Rip, IRAT-ISRA	Cereal–leguminous	F0T0	11	17	3.8
	Cereal–leguminous	F0T2	11	17	5.2
	Cereal–leguminous	F2T0	11	17	3.2
	Cereal–leguminous	F2T2	11	17	3.9
	Cereal–leguminous	F1T1	11	17	4.7
Chad		With tillage, high fertility soil			
Bebedjia, IRCT-IRA	Cotton monoculture		11	20	2.8
	Cotton – cereals			20	2.4
	+ 2 years fallow			20	1.2
	+ 4 years fallow			20	0.5

Source: Pieri, C., *Fertilité des terres de savane. Bilan de trente ans de recherche et de développement agricoles au sud du Sahara.* Ministère de la Coopération, CIRAD, Paris, France, 1989.

F0 = No fertilizer, F1 = 200 kg ha^{-1} of NPK fertilizer, F2 = 400 kg ha^{-1} of NPK fertilizer + Taiba phosphate rock, T0 = manual tillage, T1 = light tillage, T2 = heavy tillage.

are site specific and heavily depend on management practices such as the choice of the cropping system, soil tillage, and the application of mineral and organic soil amendments. Data from Chad show that crop rotation and fallow management can minimize Corg losses. Thus k-values in cotton–cereal rotations were 2.4%, compared to 2.8% in a continuous cotton system. Also, four years of fallow after 16 years of cultivation led to large increases in Corg and a reduction of annual Corg losses to 0.5%.

The soil patterns in the five major agroecological zones of Africa are determined by differences in age, parent material, physiography, and present and past climatic conditions. In the humid zones, dominant soils are Ferralsols and the Acrisols. Less important in this zone are Arenosols, Nitosols, and Lixisols. The subhumid zone is characterized by the dominance of Ferralsols and Lixisols and, to a lesser extent, Acrisols, Arenosols, and Nitosols. In the semiarid zone Lixisols have the larger share followed by sandy Arenosols and Vertisols (Deckers 1993). A map showing the distribution of major soils in Africa is shown in Figure 4.5, and Table 4.5 gives the constraints of the main soils and countries covered.

Table 4.5 summarizes the extent of the main soil types, constraints, and countries covered (Bationo et al. 2007a).

Land can be classified as prime, high, medium, and low potential, and unsuitable. Figure 4.6 gives the agricultural potential of African soils. Prime land comprises those soils with deep, permeable layers and with an adequate supply of nutrients, and these soils generally do not have significant periods of moisture stress. The soils are without impermeable layers, textures are loamy to clayey with good tilth characteristics, and the land is generally level to gently undulating. They occupy about 9.6% of Africa, and they occupy significant areas in West Africa south of the Sahel, in East Africa mainly in Tanzania, and in Southern African countries of Zambia, Zimbabwe, South Africa, and Mozambique.

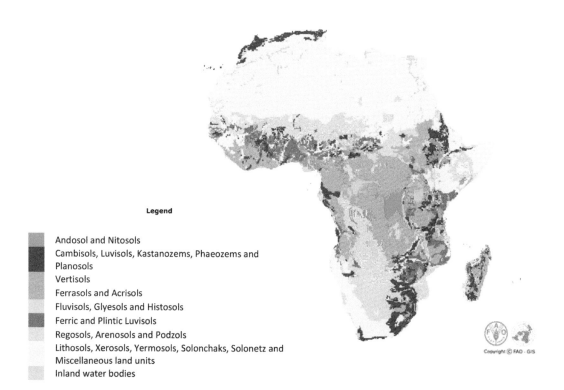

Legend

Andosol and Nitosols
Cambisols, Luvisols, Kastanozems, Phaeozems and Planosols
Vertisols
Ferrasols and Acrisols
Fluvisols, Glyesols and Histosols
Ferric and Plintic Luvisols
Regosols, Arenosols and Podzols
Lithosols, Xerosols, Yermosols, Solonchaks, Solonetz and Miscellaneous land units
Inland water bodies

Copyright © FAO - GIS

FIGURE 4.5   Major soil types in Africa.

TABLE 4.5

Main Soil Types, Extent, Constraints, and Countries Covered

Soil Type	Hectare (ha)	Percentage of Total Land (%)	Main Constraints	Main Countries Covered
Andosol and Nitosols	117,123,121	3.8	Fertile (volcanic ash), high P-fixation, Mn toxicity, medium water and nutrient retention	Rift valley (Ethiopia, Kenya, Tanzania, Zaire
Cambisols, Luvisols, Kastanozems, Phaeozems, and Planosols	211,348,146	6.8	Low to moderate nutrients content	Mediterranean countries
Vertisols	98,985,811	3.2	Heavy soils, medium mineral reserves, erodibility and flooding	Sudan, Ethiopia, South Africa, Lesotho
Ferrasols and Acrisols	500,910,947	16.2	Low nutrients content, weathered, Al and Mn toxicity, high P-fixation, low nutrient and water retention, susceptible to erosion	DRC, Angola, Zambia, Rwanda, Burundi, Uganda, Sudan, Central Afr. Rep., Cameroon, Liberia, Sierra Leone and Madagascar, subhumid zone of West Africa
Fluvisols, Glyesols and Histosols	132,037,611	4.3	Poor to moderate drainage	West, Central, and southern Africa
Ferric and Plintic Luvisols	179,786,479	5.8	Low nutrients content	Western and Southern Africa
Regosols, Arenosols and Podzols	579101963	18.7	Mainly quartz, low water and nutrient holding capacity, wind erosion, poor soils with nutrients leaching	West Africa/Sahel, Sudan, Botswana, Angola and DRC, north Africa
Lithosols, Xerosols, Yermosols, Solonchaks, Solonetz, and miscellaneous land units	1,244,513,150	40.3	Shallow soils, soils subject to drought, presence of salt	North African countries, South Africa, Namibia, Somalia, Sahel
Inland water bodies	27,230,091	0.9	Flood zones	
TOTAL	**3,091,037,319**	**100**		

High-potential soils are similar to prime soils but have some minor limitations such as extended periods of moisture stress, sandy or gravelly materials, or root-restricting layers in the soil. The high-potential lands occupy an area of about 6.7%.

Medium- and low-potential lands, which occupy 28.3% of the surface, have major constraints for low-input agriculture. These lands have a major soil constraint and one or more minor constraints that can be corrected. Constraints include adverse soil physical properties including surface soil crusting; impermeable layers; soil acidity and specifically subsoil acidity, salinity, and alkalinity; and high risks of wind and water erosion. The large contiguous areas of Central and West Africa are considered as medium potential, due to the presence of acid soils and soils that fix high amounts of P. With an inherently low soil quality, low-input agriculture can be equated to potential soil degradation. These are some of the priority areas for technical assistance and the implementation of appropriate soil management technologies.

The unsustainable class of lands are those that are considered to be fragile, easily degraded through inappropriate management, and in general are not productive or do not respond well to management. These occupy about 55% of the African continent. They are generally erodible and require high investments.

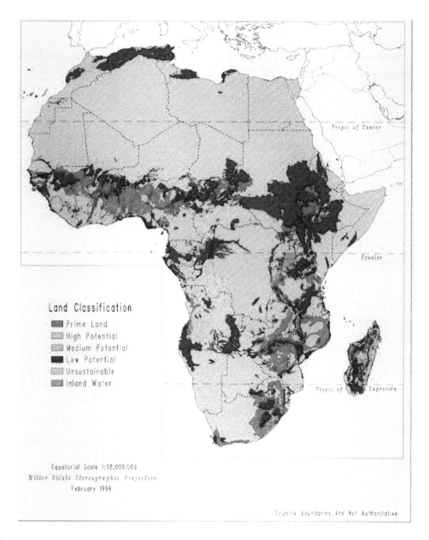

FIGURE 4.6   Land classification of African soils.

4.4   OVERVIEW OF FERTILIZER MANAGEMENT

The regional reviews by Okalebo et al. (2006), Mafongoya et al. (2006), and Schlecht et al. (2006) highlight the potential of soil fertility management technologies in Eastern, Southern, and Western Africa. These technologies include the use of mineral (soluble) fertilizers, mineral soil amendments, organic inputs, improved land use systems, and soil and water conservation.

Fertilizer was as important as seed in the Green Revolution, contributing as much as 50% of the yield growth in Asia (Hopper 1993). Several studies have found that one-third of the cereal production worldwide is due to the use of fertilizer and related factors of production (Bumb 1995, citing FAO). Van Keulen and Breman (1990) and Breman (1990) stated that the only real cure against land hunger in the West African Sahel lay in increased productivity of the arable land through the use of external inputs, mainly inorganic fertilizers. Pieri (1989), reporting on fertilizer research conducted from 1960 to 1985, confirmed that inorganic fertilizers in combination with other intensification practices had tripled cotton yields in West Africa from 310 to 970 kg ha[-1]. There are numerous cases of strong fertilizer response for maize in East and Southern Africa (Byerlee and Eicher 1997).

TABLE 4.6

Maize Response to Organic and Inorganic Fertilizer Application in Selected Sites in East, West, and Southern Africa

Site	Treatment	Yield (t/ha)	% Yield Increase
West Africa[1] (multisites 3–6 year average)	Control	1.51	–
	TSP + N + K	3.172	110
	N + K	2.319	54
	P + K	2.426	61
	P + N + K	3.765	149
	P + N + K + lime (500 kg ha^{-1} every 3 years)	3.794	151
	Crop residue (CR)	1.999	32
	Manure (10t ha^{-1} every 3 years)	2.497	65
	P + N + K + Mg + Zn	3.880	157
	P + N + K + Mg + Zn + Lime	4.006	165
	P + N + K + Mg + Zn + CR	4.083	170
	P + N + K + Mg + Zn + Manure + Lime	4.289	184
East Africa[2] (1981–1985)	Control	1.9	–
	Crop residues	2.5	32
	Manure (5 t)	3.5	84
	Fertilizer 60 kg N, 25 kg P)	4.1	116
	Manure (10 t)	4	111
	Fertilizer 120 kg N, 50 kg P)	4.6	142
	Manure 5 t + Fertilizer 60 kg N, 25 kg P)	5.2	174
Southern Africa[3]	Control	0.729	–
	N + P + K	2.194	201
	Termitaria + N + P	2.229	206
	Cattle manure + N + P	2.644	262
	Maize stover + N + P	1.575	116
	Fresh litter + N + P	2.553	250
	Crotalaria juncea + N + P	2.496	242

[1] Mokwunye et al. (1996) – Results from 6 different AEZ in Togo.
[2] Qureish (1987) – Results from Kabete, Kenya.
[3] Mtambanengwe and Mapfumo (2005) – Results from Chinyika Zimbabwe.

The data in Table 4.6 summarizes the multisite response to soil fertility improvement and clearly demonstrates the importance of fertilizers in maize yield improvement in different agro-ecological zone (AEZ) and soil types in Africa. Maize yield increase over the control due to NPK fertilizer application from six AEZ and averaged over four years was 149%, but when the soil was amended with lime and manure yield response over the control increased to 184% (Mokwunye et al. 1996). Similarly higher-yield improvements have been observed in Eastern (Qureish 1987) and Southern African countries (Mtambanengwe and Mapfumo 2005).

One lesson we learned from IFDC work in West Africa is the importance of adopting good agronomic practices before using fertilizers. In farmers' managed trials it was found that using good agronomic practices will result in higher yields in plots not using fertilizers than using fertilizers with poor agronomic practices such as planting late with poor crop density (Figure 4.7). There is a growing body

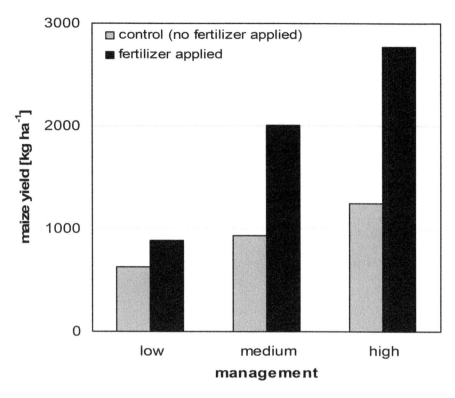

FIGURE 4.7 The effect of management intensity (planting date, crop density, and time of phosphorus application) on maize grain yield at Tinfouga, Mali. (From Bationo, A. et al., *Nutr. Cycl. Agroecosys.*, 48, 179–189, 1997).

of literature that shows fertilizers to be an essential component for sustainable yield increase in cropping systems and to be profitable in much of SSA (e.g., Droppelmann et al. 2017; Jama et al. 2017).

4.4.1 Sources and Management of Nitrogen and Phosphorus Fertilizers

The use of 15N in order to construct N balances and to determine fertilizer N uptake and losses provides an important tool for nitrogen management. IFDC undertook a systematic N balances study on nitrogen sources (urea versus calcium ammonium nitrate), time of application, and methods of application in the humid, subhumid, and semiarid zones of SSA (Mughogo et al. 1986; Bationo and Vlek 1998). One of the main findings was the high loss associated with nitrogen application in the semiarid zone (Table 4.7). The mechanism of N loss is believed to be ammonia volatilization. For all the years, losses of calcium ammonium nitrate (CAN) were less than urea because one-half of the N in CAN was in the nonvolatile nitrate form. The high losses through volatilization associated with urea N can be explained by the low cation exchange capacity of the soils and because significant rainfall does not occur shortly after N application. Although CAN has a lower N content than urea, it is attractive as an N source because of its low potential for N loss via volatilization, and its point placement will improve its special availability. At Gobery, CAN point placed outperformed urea point placed or broadcast (Figure 4.8). Nitrogen-15 data from a similar trial at Sadore illustrates the strong effect of N source and placement (Table 4.7). Plant uptake by 15N-N from point-placed CAN was almost three times that of urea applied in the same manner. A 57% reduction in fertilizer N uptake by the plant was found when the CAN was broadcast rather than point placed.

TABLE 4.7
Recovery of 15N Fertilizer by Pearl Millet Applied at Sadore, Niger, 1985

		^{15}N Recovery			
N Source	Application Method	Grain	Stover	Soil	Total
		(%)			
CAN	Point incorporated	21.3	16.8	30.0	68.1
CAN	Broadcast incorporated	10.9	10.9	42.9	64.7
Urea	Point incorporated	5.0	6.5	22.0	33.5
Urea	Broadcast incorporated	8.9	6.8	33.2	48.9
Urea	Point surface	5.3	8.6	18.0	31.9
SE		1.2	2.0	1.9	2.4

Source: Christianson, C.B., and Vlek, P.L.G., Alleviating soil fertility constraints to food production in West Africa: Efficiency of nitrogen fertilizers applied to food crops, in Mokwunye, A.U. (Ed.), *Alleviating Soil Fertility Constraints to Increased Crop Production in West Africa*, Kluwer Academic Publishers, Dordrecht, the Netherlands, 1991.

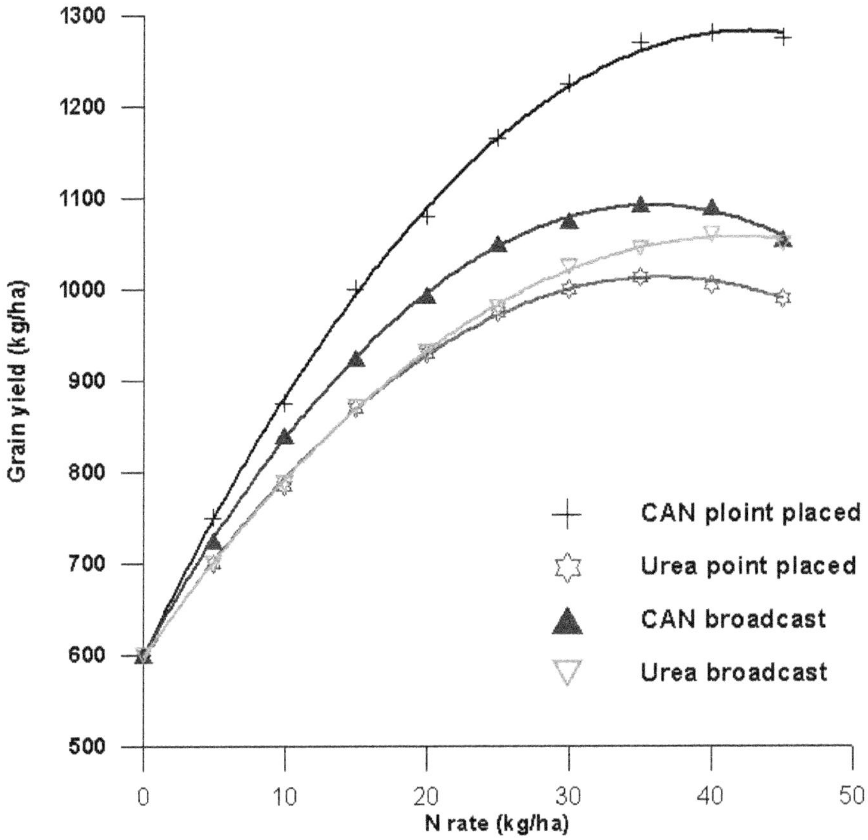

FIGURE 4.8 Effect of broadcast and point application methods for urea and CAN on grain yield of pearl millet.

4.4.2 MAINTENANCE OF SOIL ORGANIC MATTER PLAYS A KEY ROLE IN FERTILIZERS' USE EFFICIENCY AND RECOVERY

In a study in West Africa, Fofana et al. (2008) observed that grain yields across years and fertilizer treatment averaged 0.8 t ha^{-1} on outfields and 1.36 t ha^{-1} on infields (home gardens). Recovery of fertilizer N (RFN) applied varied considerably among the treatments and ranged from 17% to 23% on outfields and 34% to 37% on infields. Similarly, average recovery of fertilizer P applied (RFP) across treatments over the three-year period was 31% in the infields compared to 18% in the outfields. These results indicate higher inherent soil fertility and nutrient use efficiency in the infields (fertile fields) compared to the outfields (infertile fields) and underlines the importance of soil organic carbon and micro and secondary nutrients in improving fertilizer use efficiency. Once soils are degraded and poor in organic matter, the response to fertilizer is less and the recovery of applied fertilizers is reduced.

4.4.3 INCREASING THE LEGUME COMPONENT IN THE CROPPING SYSTEM CAN AFFECT THE EFFICIENCY OF MINERAL FERTILIZERS

Rotation of cereals and legumes can be used as means of improving soil fertility and productivity. Several researchers (Stoop and Staveren 1981; Klaij and Ntare 1995; Bagayoko et al. 1996; Bationo et al. 1998a; Bationo and Ntare 1999) have reported cereal/legume rotation effects on cereal yields A rotation of pearl millet with groundnut and cowpea resulted in significantly higher pearl millet grain yields than in the monoculture cropping of pearl millet on average over a four-year period (Figure 4.9). With no application of N fertilizer, millet grain yield after cowpea increased by 57% at Sadore, 18% at Bengou, and 87% at Tara. Even at higher levels of N, continuous cropping of millet produced the lowest yields of millet grain. One of the reasons for the improvement of the efficiency of fertilizers in the rotation system was due to the increase of soil organic carbon in this system due

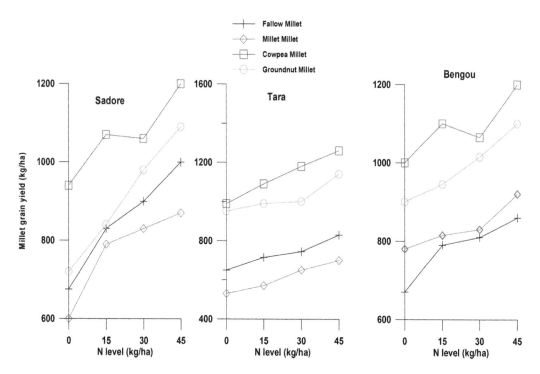

FIGURE 4.9 Effects of nitrogen and rotation on pearl millet grain yield (kg ha^{-1} average of four years) at Sadore, Tara, and Bengou.

to the dropping of legume leaves (Bationo and Ntare 1995). It has been assumed by many workers that the positive effect of rotations arises from the added N from legume in the cropping system. Some workers, however, have attributed the positive effects of rotations to the improvement of soil biological and physical properties and the solubilization of occluded P and highly insoluble calcium–bounded phosphorus by root exudates (Gardner et al. 1981; Arhara and Ohwaki 1989). Other advantages of crop rotations include soil conservation (Stoop and Staveren 1981), organic matter restoration (Spurgeon and Grissom 1965), and pest and disease control (Sunnadurai 1973).

In N15 balances trials including rotation, N use efficiency increased from 20% in continuous pearl millet cultivation to 28% when pearl millet was rotated with cowpea. Nitrogen derived from the soil is better used in rotation systems than with continuous millet (Bationo and Vlek 1998). Nitrogen derived from the soil increased from 39 kg N ha^{-1} in continuous pearl millet cultivation to 62 kg N ha^{-1} when pearl millet was rotated with groundnut.

4.4.4 UREA DEEP PLACEMENT INCREASES RICE PRODUCTIVITY AND NITROGEN USE EFFICIENCY

The main constraint to achieving high rice productivity in Africa is the chronic deficiency of major nutrients and particularly nitrogen. N is the most difficult nutrient to manage efficiently as it is very mobile in soil ecosystems and therefore is easily lost by leaching, runoff, denitrification, and volatilization. In Africa, prilled urea (PU) conventionally broadcast applied by rice farmers is inefficient in irrigated rice, largely because of serious losses (up to 60% of applied N) via ammonia volatilization and denitrification. And this calls for the development of new ergonomically efficient, economically attractive, and environmentally sound technologies aiming at improving the efficiency of input use, including improved seeds and particularly N source. Bandaogo et al. (2015) showed that through urea deep placement (UDP), the avenues for N losses in rice production system are reduced and improved N uptake by the plant is possible. In field trials conducted in Burkina Faso, the authors reported that UDP is a highly efficient soil nutrient management technology, enabling farmers to achieve higher crop yields (15%–20% higher than with broadcast application) with lower use of high-cost fertilizer (20%–30% lower than urea broadcast).

UDP is a science-based technology developed by the IFDC Research and Development Program to improve N use efficiency. It is a simple and low-cost technology, well suited to small-scale rice production. The technology is comprised of production of urea super granules (USG), compacted urea, and hand placement of USG in the puddled soil between each set of four hills of rice at a depth of 7–10 cm seven days after transplanting. It's a single basal deep placement of USG that matches more favorably the N requirement of the plant throughout the duration cycle. It significantly reduces N loss through volatilization and increases N use efficiency and paddy yield. Introduced in Asia in the 1980s, UDP's wide adoption in irrigated rice production systems in Asian countries (Bangladesh, Vietnam, Afghanistan, Cambodia, India, and Nepal) has brought about rice yield increases ranging from 23% to 70% and reduced requirement for urea fertilizer by 32% to 44%. UDP technology has proven to be highly effective in improving crop uptake of applied nitrogen fertilizers in irrigated rice and therefore merits to be experimented in similar production systems in Africa. More than 3500 demonstration plots were developed in 30 irrigation schemes located in the above-mentioned eight participating African countries. The demonstration scheme compares UDP with farmer practice (FP) – that is, the broadcast application of prilled urea.

Yield performance with UDP was higher than broadcast prilled urea (PU) (Figure 4.10). Mean yield advantage of USG over PU across all pilot countries was about 1,200 kg ha^{-1}. But it significantly varied among pilot countries, and the highest mean yield advantage was observed in Niger (1,880 kg ha^{-1}), followed by Nigeria (1,691 kg ha^{-1}) and Madagascar (1,498 kg ha^{-1}), with the lowest in Togo (390 kg ha^{-1}).

In order to study the effect of UDP on urea agronomic and economic efficiency, nitrogen agronomic efficiency (NAE) and value–cost ratio (VCR) were used. NAE was calculated as the additional grain yield (GY) produced by kg of fertilizer urea applied at constant NPK rates. It expresses

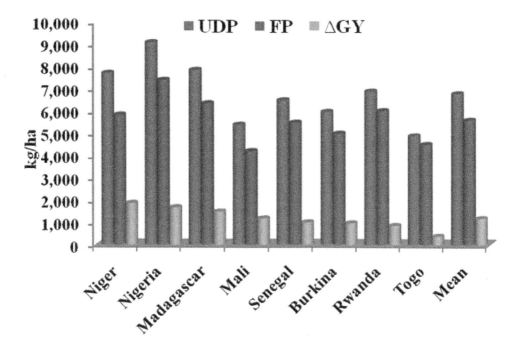

FIGURE 4.10   Efficiency of urea deep placement (UDP) and farmer practice (FP) in irrigated rice systems in the Intervention African Countries (Fofana, unpublished data).

the quantity (kg) of harvested product per kg fertilizer applied. VCR is a simple and effective indicator of nutrient effectiveness from an economic point of view. VCR was expressed as a ratio of economic yield value (with and without fertilizer application) and the total costs of fertilizer applied.

Results of NAE and VCR calculations are given in Table 4.8. The mean values of NAE and VCR varied among countries. The highest NAE (70) and VCR (8) using UDP were observed in Nigeria and Madagascar, respectively. Mean values of NAE across all target countries were 39 for PU and 56 for UDP, indicating that UDP induced an increase of NAE by 28%. Similarly, mean VCR with UDP was higher (6) than with PU (3.7), showing again an increase of 62%. These results clearly show the economic superiority of UDP over PU and suggest that under African conditions, the economics of UDP – instead of broadcast application of PU – offer an attractive N management technology that merits to be scaled up. Even under adverse conditions such as submergence-prone areas in northern Ghana, microdosing, deep placement of prilled urea, and urea briquette were more economical than farmer's practice and locally recommended practice (Table 4.9). UDP technology can be considered as a climate-smart technology because it (1) sustainably increases agricultural productivity and incomes; (2) adapts and builds resilience to climate change; and (3) reduces and/or removes greenhouse gas (GHG) emissions (Gaihre et al., 2015).

Eighty percent of the soils in Africa are P deficient, but Africa has 4,457 million tons of rock phosphate reserves representing about 75% of total world reserves of phosphate rock, and 42% of the phosphate rock used by industry to produce P fertilizers in the world is produced in six countries in Africa. SSA uses 1.6 kg P ha^{-1} of cultivated land compared to 7.9 and 14.9 kg P ha^{-1}, respectively, for Latin America and Asia.

Despite the fact that the deficiency of P is acute on the soils of Africa, local farmers use very little P fertilizers because of high costs and problems with availability. The use of locally available phosphate rock (PR) could be an alternative to imported P fertilizers. For example, Bationo et al. (1986) showed that direct application of local PR may be more economical than imported water-soluble P fertilizers. Bationo et al. (1990) showed that Tahoua PR from Niger is suitable for direct application, but Parc-W

TABLE 4.8
**Nitrogen Agronomic Efficiency and Value–Cost Ratio
for Prilled Urea and Urea Deep Placement Applied to
Irrigated Rice in 2009 in Eight African Countries**

	Technology			
	Farmer Practice – FP		Urea Deep Placement – UDP	
Country	PU – NAE	VCR	UDP – NAE	VCR
Niger	36	3.5	59	5.9
Nigeria	49	4.8	70	7.0
Madagascar	44	5.5	63	8.0
Mali	41	3.3	64	6.7
Senegal	50	3.5	69	7.3
Burkina	31	3.1	44	4.3
Rwanda	35	3.5	46	4.6
Togo	28	2.7	33	3.2
Mean	**39**	**3.7**	**56**	**6.0**

PU, Prilled Urea; NAE, Nitrogen Agronomic Efficiency; UDP, Urea Deep
Placement; VCR, Value–Cost Ratio.

TABLE 4.9
**Rice Economic Benefits under Submergence Prone Conditions in Northern Ghana with
Various Nutrient Management Technologies**

Treatment	Grain Yield (ton/ha)	Total Revenue (GH¢/ha)	Total Production Cost (GH¢/ha)	Gross Profit (GH¢/ha)	Additional Revenue for Applied N (GH¢/ha)	Additional Revenue/kg of Applied N (GH¢)
Farmer's practice – Basal NPK	1.07	1,592	1,560	32	–	–
Recommended (with urea broadcast)	2.12	3,155	1,953	1,202	1,170	7.80
Modified recommended (deep placement of prilled urea)	4.84	6,667	2,021	4,646	4,614	30.76
Microdosing	3.89	5,789	2,005	3,784	3,752	39.08
UDP (urea briquette deep placement)	5.26	7,515	2,044	5,471	5,439	48.13

from Burkina Faso has less potential for direct application. The effectiveness of local PR depends
on its chemical and mineralogical composition (Chien and Hammond 1978). Phosphate rocks can
be used as a soil amendment and the use of water-soluble P can be more profitable. Under certain
soil conditions, the direct application of PR, especially where available locally, has proved to be an
agronomically and economically sound alternative to the more expensive imported superphosphates.
Kodjari PR in Burkina Faso, Tahoua PR in Niger, and Tilemsi PR in Mali are mined and used for

soil P replenishment and crop production in the three countries. The effectiveness of PR depends on its chemical and mineral composition, soil and climatic factors, and the crops to be grown. The solubility of the three PRs in neutral Ammonium Citrate is 6.1%, 8.3%, and 10.4% for Kodjari, Tahoua, and Tilemsi, respectively. While evaluating Parc-W and Tahoua PR indigenous to Niger using field experiments, Parc-W was 48% as effective as Single Super Phosphate (SSP), whereas the effectiveness of the more reactive Tahoua rock was as high as 76% compared to that of SSP for pearl millet production (Bationo et al. 1990). In an on-station experiment in Burkina Faso, the relative effectiveness of Kodjari was found to be 68% and 39% for millet and sorghum, respectively, whereas in an on-farm experiment, the values were 54% and 50% (Lompo et al. 2018). A phosphate rock decision support system (PRDSS) has been developed by IFDC to quantify the effectiveness of various PR sources for direct application as influenced by the abovementioned factors.

The agricultural sector in Mali involves 80% of the population. However, less than 10% of the 2.7 m ha of cultivated land receives fertilizers. The 70,000 tonnes of fertilizer imported per year is applied on cash crops such as cotton, rice, and groundnuts. This amount is less than 15% of the nutrients exported by the crops. The Tilemsi reserves of PR are estimated to be 20 m tonnes, and the production in 1991 used 18,560 tonnes of PR. For many years, Tilemsi PR was evaluated in different agroecological zones in Mali. Henao and Baanante (1999a, 1999b) made a comprehensive analysis of the results, including an economic evaluation, and concluded that Tilemsi PR is practically equivalent to TSP per unit of P_2O_5.

Various ways are used to improve the agronomic effectiveness of phosphates to make the products more economically attractive. The biological means include (1) phosphocomposts, (2) inoculation of seedlings with endomyccorhizae, (3) use of phosphate-solubilizing microorganisms, and (4) use of plant genotypes. Chemical means include the partial acidulation of PR. Physical means include (1) compacted PR with water-soluble phosphate products, (2) dry mixtures of PR with water-soluble phosphate fertilizers and phosphate rock elemental sulfur assemblages. In Burkina Faso, Niger, and Mali, partial acidulation of phosphate rock (PAPR) have been tested quite extensively, but very little work has been done on other ways to improve the effectiveness of PR. PAPRs are prepared by reacting PRs, usually with H_2SO_4, in amounts less than that needed to make SSP or TSP. PARs may offer an economic means of enhancing the agronomic effectiveness of indigenous PR sources that may otherwise be unsuitable for direct application. In Burkina Faso, Lompo et al. (1994) found that the agronomic effectiveness of Kodjari PR improved from 68% to 88% for pearl millet and from 39% to 81% for sorghum when partially acidulated. In Niger, the partial acidulation of PARC-W PR increased its agronomic effectiveness from 48% to 80% for pearl millet production (Bationo et al. 1990). Recent IFDC studies have shown relatively low reactive PRs when compacted with 20% P from DAP or MAP on a wide range of soils (pH 5.0–7.9) and were 75%–100% as effective as water-soluble P.

4.4.5 Strategic Application of Fertilizer: The Microdose Technology

The strategic application of fertilizers, commonly called microdose technology, was developed for the Sahelian countries as an effective technique to increase fertilizer use efficiency and reduce investment costs for resource-poor small-scale farmers, thereby increasing crop growth and productivity (Bationo et al. 1998b; Buerkert and Hiernaux 1998). This strategic application of fertilizer is based on applying small doses of fertilizer in the hill of the target crop at planting rather than broadcasting it all over the field. The microdosing technology is affordable to the poor because of the reduced investment cost, and it gives a quick start, thus avoiding early season drought, and an earlier finish, avoiding end-of season drought while increasing crop yields (Tabo et al. 2006, 2007).

More recently, Vandamme et al. (2018) found microdosing of 3 kg P ha^{-1} in the planting hole or beneath the dry-seeded drilled rice to consistently increase early vigor and almost tripled grain yield to 3 t ha^{-1} compared to broadcast of 6 kg P ha^{-1}. A similar yield increase was obtained with microdosing of NPK compared to basal NPK application.

For many years, the microdose technology has been tested, and many countries are adopting this technology. Table 4.10 reports the most recent results of scientists from INERA achieved by

TABLE 4.10

Productivity of the Microdose Technology as Compared with the Actual Recommendation (Taonda, Unpublished Data)

Crop	Nutrient Applied with Present Recommendation (kg/ha)	Nutrient Applied with the Microdose (kg/ha)	Grain Yield in Farmers Field without Fertilizer (kg/ha)	Grain Yield with Present Recommendation (kg/ha)	Grain Yield with Microdose (kg/ha)	NUE for Microdose (kg/kg)	NUE for Present Recommendation (kg/kg)	Yield Increase with the Microdose over the Control (%)
Maize	110N, 69 P_2O_5, 42 K_2O Total = 221	41N, 29P_2O_5, 17 K_2O Total = 87	1,600	2,500	3,200	18	4	100
Sorghum	40N, 23 P_2O_5, 14 K_2O Total = 77	8.7N, 14 P_2O_5, 8.7 K_2O Total = 31	620	1,700	900	35	4	174
Millet	40N, 23 P_2O_5, 14 K_2O Total = 77	8.7N, 14 P_2O_5, 8.7 K_2O Total = 31	600	1,200	650	19	3	100
Cowpea	14N, 23 P_2O_5, 14 K_2O Total = 51	8.7N, 14 P_2O_5, 8.7 K_2O Total = 31	600	1,200	850	19	4	100
Groundnut	14N, 23 P_2O_5, 14 K_2O Total = 51	8.7N, 14 P_2O_5, 8.7 K_2O Total = 31	650	1,200	1,200	18	11	85

the microdose technology in Burkina Faso compared to the recommendation by the extension services. Whereas the mode of application of the recommended rate by the extension is broadcasting, the microdose technology consisted in applying the reduced rate of fertilizer in the hill of the crop.

The following conclusions can be made from Table 4.9:

1. The rate of the fertilizer applied with the microdose technology is reduced by more than half with the microdose technology. For example, for maize, while a total of 221 kg ha^{-1} is used as per the present rate recommended by the extension services, the microdose technology will apply only 87 kg ha^{-1} of nutrient.
2. In most of the cases, the strategic application of nutrient with the microdose resulted in higher grain yield as compared with the broadcasting of lager rates of nutrients. As an example, for pearl millet the microdose used resulted in a grain yield production of 1,700 kg ha^{-1} as compared to 900 kg ha^{-1} with the broadcasting method.
3. The yield increase over the control due to the microdose ranged from 85% to 174%. Most farmers can double their production with the adoption of the microdose technology.

4.4.6 CROP RESIDUE AND MANURE MANAGEMENT

Crop residue (CR) management can play an important role in improving crop productivity in SSA. Numerous research reports show large crop yield increases as a consequence of organic amendments in the Sahelian zone of West Africa (Abdullahi and Lombin 1978; Pieri 1986, 1989; Bationo et al. 1993, 1998a; Evéquoz et al. 1998). Bationo et al. (1995, 2007b) reported from an experiment carried out in 1985 on a sandy soil at Sadoré, Niger, that grain yield of pearl millet (*Pennisetum glaucum* L.) after a number of years had declined to only 160 kg ha^{-1} in unmulched and unfertilized control plots. However, grain yields could be increased to 770 kg ha^{-1} with a mulch application of 2 tons CR ha^{-1} and to 1,030 kg ha^{-1} with 13 kg P as SSP plus 30 kg N ha^{-1}. The combination of CR and mineral fertilizers resulted in a grain yield of 1,940 kg ha^{-1}.

In different parts of SSA, crop or organic residues applications have been shown to increase soil P availability (Kretzschmar et al. 1991), enhance PR availability (Sahrawat et al. 2001), cause better root growth (Hafner et al. 1993), improve potassium (K) nutrition (Rebafka et al. 1994), protect young seedlings against soil coverage during sand storms (Michels et al. 1995), increase water availability (Buerkert et al. 2000), and reduce soil surface resistance by 65% (Buerkert and Stern 1995) and topsoil temperature by over 4°C (Buerkert et al. 2000). These effects are especially stronger in the Sahelian zone, but weaker in other areas with lower temperatures, higher rainfall, and heavier soils (Buerkert et al. 2000). From incubation studies under controlled conditions, Kretzschmar et al. (1991) concluded that increases in P availability after crop residue (CR) application were due to a complexation of iron and aluminum by organic acids. The organic amendments have also been reported to reduce the capacity of the soil to fix P, thereby increasing P availability for uptake and hence higher P use efficiency (Buresh et al. 1997; Sahrawat et al. 2001). Availability of organic inputs in sufficient quantities and quality is one of the main challenges facing farmers and researchers today.

Manure, another farm-available soil amendment, is an important organic input in African agro-ecosystems. One of the earliest reported manure application studies in SSA was by Hartley (1937) in the Nigerian Savannah. He observed that application of 2 t ha^{-1} FYM increased seed cotton yield by 100%, equivalent to fertilizers applied at the rate of 60 kg N and 20 kg P ha^{-1}. Palm (1995) has concluded that for a modest yield of 2 t ha^{-1} of maize the application of 5 t ha^{-1} of high-quality manure can meet the N requirement, but that this cannot meet the P requirements in areas where P is deficient. Bationo and Mokwunye (1991) found no difference between applying 5 t ha^{-1} of FYM as compared to the application of 8.7 kg P ha^{-1} as Single Superphosphate, pointing to the role of manure to availability of P through complexation of iron and aluminum (Kretzschmar et al. 1991).

Other reports have shown that crop yields from the nutrient-poor West African soils can be substantially enhanced through the use of manure (McIntire et al. 1992; Sedogo 1993; Bationo and Buerkert 2001).

Several scientists have addressed the availability of manure for sustainable crop production. De Leeuw et al. (1995) reported that with the present livestock systems in West Africa the potential annual transfer of nutrients from manure is 2.5 kg N and 0.6 kg P per hectare of cropland. Although the manure rates are between 5 and 20 t ha^{-1} in most of the on-station experiments, quantities used by farmers are very low and ranged from 1,300 to 3,800 kg ha^{-1} (Williams et al. 1995). This is due to insufficient number of animals to provide the manure needed, and the problem becomes more pronounced in post-drought years (Williams et al. 1995). The amount of livestock feed and land resources available are also limited. Depending on rangeland productivity, it will require between 10 and 40 hectares of dry-season grazing land and 3–10 hectares of rangeland of wet-season grazing to maintain yields on one hectare of cropland using animal manure (Fernandez et al. 1995).

4.4.7 Managing Other Nutrients Such as Secondary Nutrients and Micronutrients

IFDC investigated whether secondary nutrients (calcium, magnesium, and sulfur) and micronutrients (zinc, boron, copper, manganese, iron, chlorine, molybdenum, nickel, and cobalt) could be limiting crop response to NPK fertilizers. Field work has been conducted in Burundi, Rwanda, Uganda, Ethiopia, Kenya, Zambia, and Mozambique to develop a "next generation" of balanced fertilizers. Over 3,000 georeferenced soil samples taken in Rwanda, Burundi, and parts of Uganda and Zambia were analyzed to map the extent of nutrient deficiencies and soil acidity constraints. These nutrient maps show large areas of S, Zn, B, and Cu deficiencies, as well as considerable regions of Ca and Mg deficiencies and soil acidity constraints. Based on this information, "best-bet" formulae were developed. Results from IFDC trials in several countries show consistently large significant yield increases on several crops due to secondary and micronutrient (SMN) additions. Yield increase over NPK fertilizers due to the addition of SMN range from 1 to 2 t ha^{-1} for cereal crops, 0.5 to 1.0 t ha^{-1} for legumes, and 5 to 10 t ha^{-1} for roots and tubers. Sometimes, yield increases due to SMNs equal yield increases due to NPK fertilizers alone (Figure 4.11). As SMNs cost less than NPKs, this represents an increased return on fertilizer investment.

With large increases in crop productivity due to the addition of SMNs, the use of fertilizers is more profitable and therefore more attractive to farmers. Fertilizer manufacturers and blenders are also keen to provide farmers with more profitable products. The data in Figure 4.4 clearly indicate that the addition of secondary and micronutrients will result in higher productivity.

In addition to increasing yield, the nutritional quality of food items is likely to simultaneously increase, with the aim to contribute to human nutrition and fight hidden hunger. The potential of "agronomic fortification," that is, the application of mineral SMN-containing NPK fertilizers to soils and/or plants' leaves, is to increase the nutrient content of edible plant parts.

Supplementation by S, Zn, B, and K have frequently been reported to increase cereal yields by 40% or more over the standard N-P recommendation alone (Wendt et al. 1994; Wendt and Rijpma 1997; Kihara and Njoroge 2013; Kihara et al. 2017).

4.4.8 Water and Nutrient Use Efficiency

Variability of rainfall is a critical factor in efficiency of fertilizers and in determining risk-aversion strategies of farmers in Africa. The variability in cultivated acreage and yield for various staple cereals is significantly related to rainfall variability.[1] Acreage and yield differences between years can reach up to 50%, and total cereal production may vary over 25% and has not declined during

---

[1] Based on unpublished data from the FAO, obtained by the author in 2019.

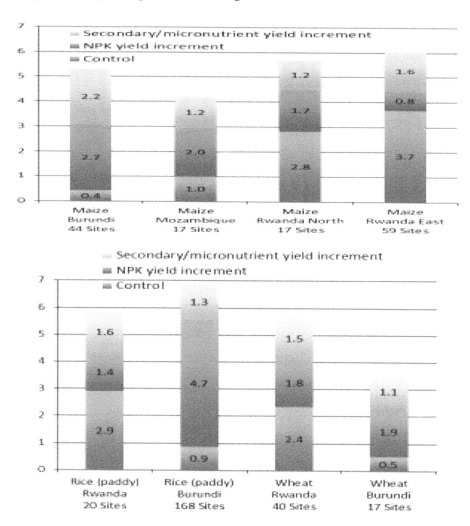

FIGURE 4.11   Impacts of secondary and micronutrients relative to NPK for maize rice and wheat in selected SSA countries (Wendt, unpublished data).

the past decades, jeopardizing food security. The tendency of African rainfall to be both spatially and temporally concentrated has important implications for fertilizer use. A survey of available data found Africa levels of available water from rainfall at 12.7 cm year^{-1} compared to North America at 25.8, South America at 64.8, and the world average at 24.9 (Brady 1990). Fertilizer is commonly thought to increase risk in dryland farming, but in some situations it may be risk neutral or even risk reducing. Phosphorus and shorter-duration millet varieties in Niger, for example, cause crops to grow hardier and mature earlier, reducing damage from and exposure to drought (ICRISAT 1985–88; Shapiro and Sanders 1998). A key constraint, though, is the availability of fertilizer and the incentive for adopting fertility-enhancing crop rotations in these zones (Thomas et al. 2004).

In the dry land of the Sahel, several scientists have reported that where the rainfall is more than 300 mm, the most limiting factor to crop production is nutrient and not water. Quantification of the attainable yields under rainfed conditions with and without fertilization using crop growth simulation models (Bindraban et al. 2000) with actual yield levels reveal great potential for productivity increase (Conijn et al. 2011) through fertilization, presuming effective water harvesting (Figure 4.12).

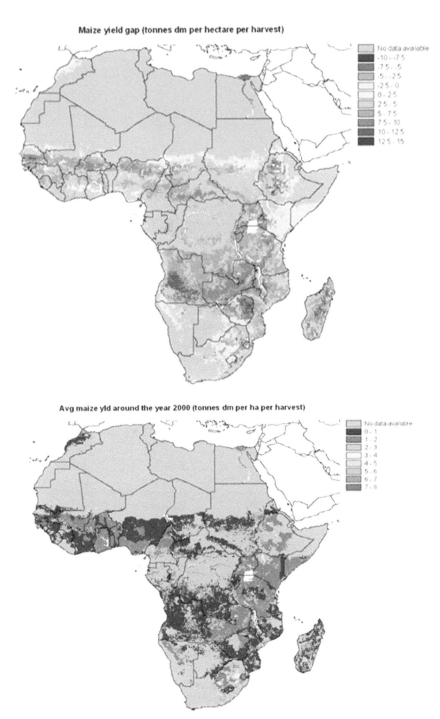

FIGURE 4.12 Average maize yield (left) around the year 2000 in tonnes dry matter per hectare harvest (Monfreda et al. 2008) and calculated yield gap with maize yield potentials under rainfed and fertilized conditions (right). (From Conijn, J.G., et al., *Agricultural Resource Scarcity and Distribution: A Case Study of Crop Production in Africa*, Plant Research International, Wageningen, the Netherlands, 2011.)

TABLE 4.11

Water Use (WU), Grain Yield (Y) and Water Use Efficiency (WUE) for Millet at Sadore and Dasso (Niger)

Treatment	Sadore			Dosso		
	WU (mm)	Y (kg ha^{-1})	WUE	WU (mm)	Y (kg ha^{-1})	WUE
Fertilizer	382	1570	4.14	400	1700	4.25
Without fertilizer	373	460	1.24	381	780	2.04

These principles have been substantiated by numerous fields studies. At Sadore, where the annual average rainfall is 560 mm, the nonuse of fertilizers resulted in a harvest of 1.24 kg of pearl millet grain per mm of water, but the use of fertilizers resulted in the harvest of 4.14 kg of millet grain per mm of water (Table 4.11).

Many development projects have invested billions of dollars in soil and water conservation. These projects mostly did not include soil fertility improvement, and the water harvested in this manner did not fulfill its full potential for productivity improvement. It is well known that fertilizers are a key to improved water use efficiency, as water harvesting can also improve fertilizer use efficiency. The Zai system is widely used in West Africa for water harvesting and soil conservation. The data in Table 4.6 indicates that the use of Zai alone will not significantly improve the productivity (only 200 kg ha^{-1} of sorghum grain), but when the Zai is associated with manure and fertilizer, large crop yield increases can be obtained (1,700 kg ha^{-1} of sorghum grain) (Table 4.12).

The mineral fertilizer microdosing (MD) technique was disseminated in the North Sudanian zone of Burkina Faso for three years using various extension tools. This study aims to analyze the economic efficiency as well as farmers' perception of the use of MD technique. Quantitative and qualitative data were collected from 60 demonstration plots conducted by innovative farmers and from 300 households using an interview guide during the focus groups. The results of the demonstration trials show that this innovation significantly increases sorghum productivity compared to farmer's practice. It even triples sorghum yields when combined with soil and water conservation (SWC) techniques and the use of improved seed varieties. The efficiency of mineral fertilizers by microdose in association with SWC techniques was assessed through the evaluation of yields and incomes from the different treatments on the demonstration plots.

Microdose fertilization resulted in significant increases in sorghum grain yields of 100% and 186% with the local seed variety and the improved seed variety, respectively, compared to the

TABLE 4.12

Effect of "Zai" on Sorghum Yields

Technology	Sorghum Yield (kg ha^{-1})	Yield Increase (%)
Only planting pits (Zai)	200	–
Zai + Cattle manure	700	250
Zai + Mineral fertilizers	1400	600
Zai + Cattle manure and fertilizers	1700	750

Source: Reij, C., et al., *Sustaining the Soil: Indigenous Soil and Water Conservation in Africa,* Earthscan, London, 1996.

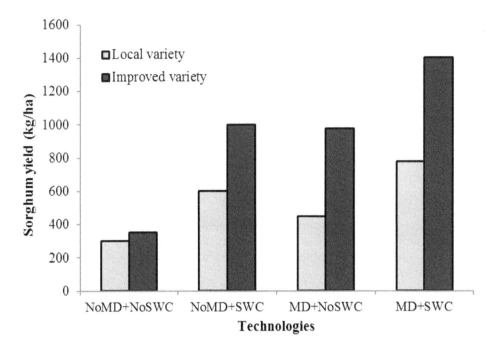

FIGURE 4.13 Response of sorghum to MD combined with SWC techniques. NoMD + NoSWC = without Microdose and without SWC techniques; NoMD+SWC = without Microdose with SWC techniques; MD+NoSWC = Microdose without SWC techniques; MD+SWC = Microdose with SWC techniques. (Source: Ouattara, B. et al., Improving agronomic efficiency of mineral fertilizers through microdose on sorghum in the sub-arid zone of Burkina Faso, in Bationo, A., Ngaradoum, D., Youl, S., Lompo, F., and Fening, J.O. (Eds.), Volume 1, *Improving the Profitability, Sustainability and Efficiency of Nutrients through Site Specific Fertilizer Recommendations in West Africa Agro-Ecosystems*, Springer, the Netherlands, 2018.)

control (Figure 4.13). Its effects are even greater when combined with SWC techniques. This combined use of techniques on the local and improved varieties yielded increases of 158% and 300%, respectively, compared to the absolute control. The yields obtained with the SWC techniques alone were higher than those of the control, but lower compared to those with the microdose alone.

4.4.9 Lessons Learned from Long-Term Experiments

Long-term experiments (LTEs) offer the best practical means of studying the effects of management or global change on soil fertility, sustainability of yield, or wider environmental issues. LTEs are used for biophysical aspects of sustainability, the impact of agriculture on environments, the impact of environment on agriculture, and the development and testing of crop models.

Most trials were designed to determine the effects of inorganic fertilizers and organic inputs on crop yields and soil properties, including other parameters such as rotations of cereal with legumes.

LTEs serve as living laboratories, providing opportunities for experimentation in which the effects of manipulation may be separated from other variables. This is clearly essential in understanding processes of soil fertility change.

The data from Figure 4.14 of the LTEs established since 1960 in Burkina Faso clearly show that the application of mineral fertilizers alone will result in crop yield decline, but when mineral fertilizers are combined with manure, sustainable higher yields can be obtained.

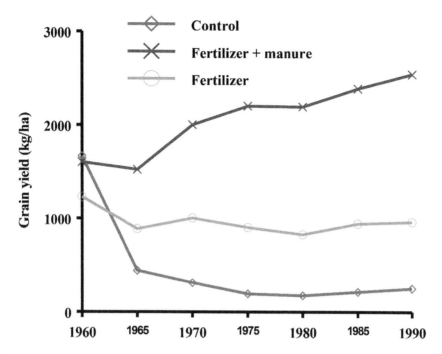

FIGURE 4.14   Sorghum grain yield as affected by mineral and organic fertilizers over time. (Source: Bationo, A. et al., Integrated nutrient management – Concepts and experience from sub-Saharan Africa, in Aulakh, M.S., and Grant, C.A. (Eds.), *Integrated Nutrient Management for Sustainable Crop Production*, CRC Press, New York, 2007a.)

4.5   THE WAY FORWARD

Improving efficiency of fertilizers will remain the center of future research, with the focus on all the different steps, as shown in Figure 4.4. Tailoring fertilizer use for mitigation and adaptation to climate change, the use of crop models, the new SMART approach develop by IFDC, and the use of balanced fertilizers to improve both food and nutritional security are proposed here as the new areas for future research focus in SSA.

4.5.1   TAILORING FERTILIZER USE FOR MITIGATION AND ADAPTATION TO CLIMATE CHANGE

There is a need for more research on the role of fertilizers on climate-smart agriculture. Tilling soils or cutting forests is well known to result in lower SOM levels and thus increase the release of CO_2 into the atmosphere. Crop intensification with the use of fertilizer will allow farmers to use less land for food production and therefore prevent carbon dioxide emission in the atmosphere.

Numerous studies suggest that managing SOM can have a profound impact on the amount of N and C released into the atmosphere, and therefore climate change. Paustian et al. (1998) estimated that atmospheric CO_2 can be reduced by 20 to 30 Pg C over the next 50 to 100 years by better SOM management. Fertilizers maintain yield and increase biomass that can be incorporated into the soil. Incorporated rates of 1–6 tons of biomass per hectare can gradually increase SOM, sequestering 200–1,000 kilograms of carbon per hectare per year.

4.5.2   THE ROLE OF CROP MODELS

Soils and climate are highly variable in Africa, calling for use crop models. Crop models offer the benefit of increasing our understanding of crop responses to management in different soil and climatic conditions. Such responses are often of a complex and nonlinear nature, given the innumerable interactions among weather, soil, crop, and management factors throughout the growing season. Crop models can also provide insights as to what might happen to productivity under various climate change scenarios, a domain beyond the reach of field experimentation. The outputs can inform key decision makers at local, national, and regional levels in order to put the appropriate measures in place. Although major advances in modeling have been made in the US, Europe, and Asia, SSA lags behind due to the limited number of soil scientists and agronomists with the skills to set up and run crop model simulations. Having a well-trained cadre of African modelers would greatly facilitate the design of best crop management and adaptation measures in the varied environments and boost agricultural productivity in the region. Based on validation trials and simulation experiment outputs, Dzotsi and colleagues (2003) developed leaflets to help farmers in southern Togo choose between varieties as a function of preferred time of sowing and expected yield range. The use of models in decision support is important as field experiments provide empirical data on responses to only a small number of possible combinations of climate, soil, and management situations. Also, existing management systems from other regions, new crops and varieties, and other technologies being developed by scientists may provide useful adaptation options. However, it is impossible to conduct experiments that cover the full range of possible management options and climate conditions to determine production systems that are more resilient to climate variability, potential changes in climate, and farmers' goals. Nix (1984) criticized the predominance of a "trial and error" approach in agricultural research for evaluating management practices. He emphasized the need for a systems approach in which: (1) experiments are conducted over a range of environments, (2) a minimum set of data is collected in each experiment, (3) cropping system models are developed and evaluated, and (4) models are used to simulate production technologies under different weather and soil conditions so as to provide a broad range of potential solutions for farmers. Nix (1984) referred to the high cost of field experiments in addition to their limited extrapolation domain because results are site specific. Yet it should be realized that models are a simplification of the real conditions, unable to simulate a large number of relevant agrotechnical interventions for the African continent. The temporal dimension of crop sequences or interactions in intercropping on water, nutrients, and weed and disease infestation are poorly understood and not or only poorly captured in models. Models accounting for nutrients are limited to N and P, with the exception of K for limited crops (Singh et al. 2007). Descriptive approach models such as QUEFTS (Janssen et al. 1990) and optimizing fertilizer recommendations in Africa (OFRA) (Wortmann and Sones 2017) capture N, P, and K responses and interactions. OFRA fertilizer optimization tools develop and fine-tune fertilizer recommendations within an integrated soil fertility management framework. While SOM dynamics is simulated by DSSAT (Hoogenboom et al. 2017) and APSIM (Keating et al. 2003) models, a better understanding of organic matter dynamics under tropical conditions with very low organic matter content is needed to improve the capability of existing simulation models. Likewise, current models do not simulate micronutrient response. Despite these limitations, crop model use improvements should be encouraged in SSA to help reinforce the existing yield gaps as well as highlight knowledge gaps. Ongoing projects such as AgMIP[2] are including researchers and policymakers from SSA and other developing regions to promote the use of crop models and model improvements relevant to the developing world.

4.5.3 The Need to Upscale IFDC's New Approach to Develop, Manufacture, and Deliver Balanced Fertilizers to Farmers: The SMaRT Approach

SMaRT is an acronym for Soil analysis, Mapping, Recommendations development, and Technology Transfer (Wendt and Muthubia, 2017). The concept behind SMaRT is to get better fertilizers to farmers for a given crop and region that substantially and sustainably outperform fertilizers currently

---

[2] http://www.agmip.org/.

used by farmers. Sustainability is addressed by using "balanced" fertilizers, which have a balance of macro, secondary, and micronutrients that address predominant nutrient deficiencies. Lime Recommendations may be part of a SMaRT recommendation when lime is required.

This approach involves:

1. **Soil sampling**. Assess the extent of nutrient and soil acidity constraints through large-scale country-wide georeferenced soil sampling, followed by chemical soil analysis by certified laboratories.
2. **Soil mapping** of nutrient deficiencies at country level. Maps of all soil nutrients create national recognition of the extent of nutrient deficiencies and soil acidity constraints. Nutrient maps generate the momentum to move away from traditional NPK fertilizers. Soil nutrient maps will guide the research community as to which constraints are most likely, and where they exist. This is the first step toward creating and validating balanced fertilizer formulations. Soil maps are a public good that guides fertilizer policy and assists the fertilizer industry to meet anticipated demand.
3. **Recommendations**. Based on soil maps, "best-bet" crop and soil-specific fertilizers are developed and tested. Best-bet formulations form the basis for evaluating yield gains and economic returns versus current formulations available to farmers, a prerequisite to investing in improved formulations. Further trials ("nutrient omission trials") are conducted to evaluate the yield and economic contribution of each nutrient in the fertilizer formulation, allowing for fine-tuning of formulations. Farmers' awareness is created through large numbers (thousands) of farmer-led demonstrations.
4. **Technology Transfer**. Recommendations alone will not deliver technology to farmers, because many private-sector operatives are in the fertilizer value chain and are collectively seen as the "go-to people" for all aspects of fertilizer knowledge, from marketing to technical support.

4.5.4 MULTIPLE FUNCTIONS FROM INNOVATIVE FERTILIZER PRODUCTS AND SECTOR TRANSFORMATION

In addition to increasing yield and resource efficiency, balanced mineral fertilizers can contribute to increasing the nutritional value of crop produce to improve human nutrition and health. They can also improve plant health to reduce biocide use, stimulate plant robustness to enhance tolerance to abiotic stresses, and increase metabolite production to improve taste and shelf life. We refrain from further elaborating these latter benefits that contribute to enhancing the resilience of production systems as they have been described by Bindraban et al. (2018).

The potential contribution of micronutrient-containing fertilizers has yet to be exploited, but initial evidence for some elements, including Zn and Se, are very promising. Micronutrient deficiencies, or "hidden hunger," affect the lives of approximately 2 billion people around the world. In developing countries, more than 10 million children under the age of five die each year; 60% of these deaths are related to malnutrition. About 1.6 billion people are anemic due to iron deficiency, vitamin A deficiency results in the death of around 1 million children each year, iodine deficiency during pregnancy contributes to the mental impairment of nearly 20 million babies annually, and deficiency in zinc is responsible for about 800,000 deaths annually from diarrhea, pneumonia, and malaria in children under five. Agriculture can play an important role in combating hidden hunger and its consequences.

Traditionally, fertilizers have been used to maintain or restore soil fertility, increase crop yield, and, to a lesser extent, improve crop quality. Their management has been progressively improved to optimize their economic return, while minimizing negative impacts on the environment. More recently, there has been increasing attention to another dimension: managing fertilizers such that they also contribute to healthy and productive lives for all. While fertilizer has been highly influential in increasing the quantity of food produced, it also holds enormous potential for improving human welfare by improving the quality of food. To eradicate the multifaceted aspects of human suffering, the future food (chain) challenge will be to continue increasing available food quantities at affordable prices, while simultaneously enhancing food quality to address hidden hunger as well.

IFDC proposes to exploit the potentials of "agronomic fortification," that is, the application of mineral SMN-containing NPK fertilizers to soils and/or plant leaves to increase nutrient contents of edible plant parts. Apart from the nutrient composition, the mode and timing of application also determines in which organ nutrients will accumulate. Based on organ-specific biological needs, foliar application could be suggested for Fe, Cu, Mn, and Zn; root application for Mo and Ni; while B, Cl, Cu, and Zn could be provided through both foliar and soil (Dimkpa and Bindraban 2016).

Protein deficiency in foodstuffs result from a lack of N, and S needed for essential amino acids like methionine and cysteine. Protein content can be increased by postanthesis nitrogen application (Worland et al. 2017). The expression of the genetic potential of biofortified crops could be maximized with adequate availability of micronutrients in the soil or from SMN fertilization (Cakmak 2008). Mandated blending of Se in fertilizers by the Finnish government increased Se levels of 125 indigenous food items including wheat, meat, and dairy products, raising the human intake of Se to sufficient intake levels (Eurola et al. 1991).

Awareness among policymakers, NGOs, and the general public about the multiple goals that fertilizers can deliver could catalyze a process of transformation of the fertilizer sector through public–private initiatives in order to unlock these potentials (Bindraban et al. 2018). More effective and efficient fertilizers that increase the economic volume in the fertilizer value chain may support the last mile delivery of smart balanced fertilizers. Appropriate site-specific fertilizers with a limited spatial reach and consequently small volumes only, and containing multiple macro- and micronutrients, call for adoption of novel business models by the fertilizer industry. The adoption of fertilizers by farmers, even when fully meeting site-specific and farmers' demand, will have to be accompanied by well-orchestrated actions by fertilizer value chain actors that provide affordable inputs and credits and reliable markets to farmers. Government incentives are imperative to support the creations of these enabling conditions, as are the efforts of knowledgeable development organizations, such as the IFDC, for training and educating chain actors in good agricultural practices and agribusiness.

REFERENCES

Abdullahi, A., and G. Lombin. 1978. *Long-Term Fertility Studies at Samaru-Nigeria: Comparative Effectiveness of Separate and Combined Applications of Mineralizers and Farmyard Manure in Maintaining Soil Productivity under Continuous Cultivation in the Savanna, Samaru.* Samaru Miscellaneous Publication No. 75. Zaria, Nigeria: Ahmadu Bello University.

Arhara, J., and Y. Ohwaki. 1989. "Estimation of Available Phosphorus in Vertisol and Alfisol in View of Root Effects on Rhizosphere Soil." XI Colloquium, Wageningen, Holland.

Bagayoko, M. S. C., M. S. Traore, and K. M. Eskridge. 1996. "Pearl Millet and Cowpea Cropping System. Yield and Soil Nutrient Levels." *African Crop Science Journal* 4: 453–462.

Bandaogo, A., F. Bidjokazo, S. Youl, E. Safo, R. Abaidoo, and A. Opoku. 2015. "Effect of Fertilizer Deep Placement with Urea Supergranule on Nitrogen Use Efficiency of Irrigated Rice in Sourou Valley (Burkina Faso). *Nutrient Cycling in Agroecosystems* 102: 79–89.

Bationo, A., E. Ayuk, D. Ballo, and M. Kone. 1997. "Agronomic and Economic Evaluation of Tilemsi Phosphate Rock in Different Agroecological Zones of Mali." *Nutrient Cycling in Agroecosystems* 48: 179–189.

Bationo, A., and A. Buerkert. 2001. "Soil Organic Carbon Management for Sustainable Land Use in Sudano-Sahelian West Africa." *Nutrient Cycling in Agroecosystems* 61: 131–142.

Bationo, A., A. Buerkert, M. P. Sedogo, B. C. Christianson, and A. U. Mokwunye. 1995. "A Critical Review of Crop Residue Use as Soil Amendment in the West African Semi-arid Tropics." In *Technical Papers: Proceedings of an International Conference, 22–26 November 1993*, edited by J. M. Powell, S.

Fernandez-Rivera, T. O. Williams, and C. Renard, 305–322. Vol. 2 of *Livestock and Sustainable Nutrient Cycling in Mixed Farming Systems of Sub-Saharan Africa*. Addis Ababa, Ethiopia: International Livestock Centre for Africa (ILCA).

Bationo, A., S. H. Chien, J. Henao, C. B. Christianson, and A. U. Mokwunye. 1990. "Agronomic Evaluation of Two Unacidulated and Partially Acidulated Phosphate Rocks Indigenous to Niger." *Soil Science Society of America Journal* 54: 1772–1777.

Bationo, A., S. H. Chien, and A. U. Mokwunye. 1986. "Chemical Characteristics and Agronomic Value of Some Phosphorus Rocks in West Africa". *Paper presented at International Symposium on Drought in sub-Saharan Africa, Nairobi*, Kenya, 19–23 May, 1986.

Bationo, A., B. C. Christianson, and M. C. Klaij. 1993. "The Effect of Crop Residue and Fertilizer Use on Pearl Millet Yields in Niger." *Fertilizer Research* 34: 251–258.

Bationo, A., J. Kihara, B. Vanlauwe, J. Kimetu, B. S. Waswa, and K. L. Sahrawat. 2007a. "Integrated Nutrient Management – Concepts and Experience from Sub-Saharan Africa." In *Integrated Nutrient Management for Sustainable Crop Production*, edited by M. S. Aulakh and C. A. Grant. New York: CRS Press.

Bationo, A., J. Kihara, B. Vanlauwe, B. S. Waswa, and J. Kimetu. 2007b. "Soil Organic Carbon Dynamics, Functions and Management in West African Agro-ecosystems." *Agricultural Systems Journal* 94: 13–25.

Bationo, A., S. Koala, and E. Ayuk. 1998a. "Fertilité des sols pour la production céréalière en zone sahélo–soudanienne et valorisation des phosphates naturels." *Cahiers Agricultures* 7: 365–371.

Bationo, A., and A. U. Mokwunye. 1991. "Role of Manures and Crop Residues in Alleviating Soil Fertility Constraints to Crop Production: With Special Reference to the Sahelian and Sudanian Zones of West Africa." *Fertilizer Research* 29: 117–125.

Bationo, A., U. Mokwunye, P. L. G. Vlek, S. Koala, and B. I. Shapiro. 2003. "Soil Fertility Management for Sustainable Land Use in the West African Sudano-Sahelian Zone. In *Soil Fertility Management in Africa: A Regional Perspective*, edited by Gichuru et al. Nairobi, Kenya: Academy Science Publishers/ TSBF-CIAT.

Bationo, A., J. Ndjeunga, C. Bielders, V. R. Prabhakar, A. Buerkert, and S. Koala. 1998b. "Soil Fertility Restoration Options to Enhance Pearl Millet Productivity on Sandy Sahelan Soils in South-West Niger." In *Proceedings of an International Workshop on the Evaluation of Technical and Institutional Options for Small Farmers in West Africa, University of Hohenheim, Stuttgart, Germany, April 21–22, 1998*, edited by P. Lawrence, G. Renard, and M. von Oppen, 93–104. Weikersheim, Germany: Margraf Verlag.

Bationo, A., and B. R. Ntare. 1999. "Rotation and Nitrogen Fertilizer Effects on Pearl Millet, Cowpea and Groundnut Yield and Soil Chemical Properties in a Sandy Soil in the Semi-arid Tropics, West Africa." *Journal of Agricultural Science* 134(3): 277–284.

Bationo, A., and P. L. G. Vlek. 1998. "The Role of Nitrogen Fertilizers Applied to Food Crops in the Sudano-Sahelian Zone of West Africa." In *Soil Fertility Management in West African Land Use Systems*, edited by G. Renard, A. Neef, K. Becker, and M. von Oppen, 41–51. Weikersheim, Germany: Margraf Verlag.

Bindraban, P. S., C. Dimkpa, S. Angle, and R. Rabbinge. 2018. "Unlocking the Multiple Public Good Services from Balanced Fertilizers." *Food Security* 10: 273–285.

Bindraban, P. S., J. J. Stoorvogel, D. M. Jansen, J. Vlaming, and J. J. R. Groot. 2000. "Land Quality Indicators for Sustainable Land Management: Proposed Method for Yield Gap and Soil Nutrient Balance." *Agriculture, Ecosystems and the Environment* 81: 103–112.

Brady, N. 1990. *The Nature and Properties of Soils*. 10th ed. New York: Macmillan. Brief No. 1. IFPRI, Washington, DC.

Breman, H. 1990. "No Sustainability without External Inputs. Sub-Saharan Africa; Beyond Adjustment." *Africa Seminar*, 124–134. The Hague, the Netherlands: Ministry of Foreign Affairs, DGIS.

Buerkert, A., and P. Hiernaux. 1998. "Nutrients in the West African Sudano-Sahelan Zone: Losses, Transfers and Role of External Inputs." *Journal of Plant Nutrition and Soil Science* 161: 365–383.

Buerkert, A., A. Bationo, and K. Dossa. 2000. "Mechanisms of Residue Mulch-Induced Cereal Growth Increases in West Africa." *Soil Science Society of America Journal* 64: 346–358.

Buerkert, A. K., and R. D. Stern. 1995. "Crop Residue and P Application Affect the Spatial Variability of Non-destructively Measured Millet Growth in the Sahel." *Experimental Agriculture* 31: 429–449.

Bumb, B. 1995. *Global Fertilizer Perspective, 1980–2000: The Challenges in Structural Transformations*. Technical Bulletin T-42. Muscle Shoals, Alabama: International Fertilizer Development Center.

Buresh, R. J., P. C. Smithson, and D. T. Hellums. 1997. "Building Soil Phosphorus Capital in Africa." In *Replenishing Soil Fertility in Africa*, edited by R. J. Buresh, et al., 111–149. SSSA Spec. Publ. 51. Madison, WI: SSSA.

Byerlee, D., and C. Eicher, eds. 1997. *Africa's Emerging Maize Revolution*. Boulder, CO: Lynn Rienner.

Cakmak, I. 2008. "Enrichment of Cereal Grains with Zinc: Agronomic or Genetic Biofortification?" *Plant Soil* 302: 1–17.

Chien, S. H., and L. L. Hammond. 1978. "A Comparison of Various Laboratory Methods for Predicting the Agronomic Potential of Phosphate Rocks for Direct Application." *Soil Science Society of America Journal* 42: 935–939.

Christianson, C. B., and P. L. G. Vlek. 1991. "Alleviating Soil Fertility Constraints to Food Production in West Africa: Efficiency of Nitrogen Fertilizers Applied to Food Crops." In *Alleviating Soil Fertility Constraints to Increased Crop Production in West Africa*, edited by A. U. Mokwunye. Dordrecht, the Netherlands: Kluwer Academic.

Cobo, J. G., G. Dercon, and G. Cadisch. 2010. "Nutrient Balances in African Land Use Systems across Different Spatial Scales: A Review of Approaches, Challenges and Progress." *Agriculture, Ecosystems and Environment* 136(1–2): 1–15.

Conijn, J. G., E. Querner, M. L. Rau, H. Hengsdijk, T. Kuhlman, G. Meijerink, B. Rutgers, and P. S. Bindraban. 2011. *Agricultural Resource Scarcity and Distribution. A Case Study of Crop Production in Africa*. Report 380. Wageningen, the Netherlands: Plant Research International.

De Leeuw, P. N., L. Reynolds, and B. Rey. 1995. "Nutrient Transfers from Livestock in West African Agricultural Systems." In *Technical Papers: Proceedings of an International Conference Held in Addis-Ababa, 22–26 November 1993*, edited by M. Powell, S. Fernandez-Rivera, T. O. Williams, and C. Renard. Vol. 2 of *Livestock and Sustainable Nutrient Cycling in Mixed Farming Systems of Sub-Saharan Africa*. Addis Ababa, Ethiopia: ILCA (International Livestock Center for Africa).

Deckers, J. 1993. "Soil Fertility and Environmental Problems in Different Ecological Zones of the Developing Countries in Sub Saharan Africa." In *The Role of Plant Nutrients and Sustainable Food Production in Sub-Saharan Africa*, edited by H. Van Reuler and W. H. Prins. Laidschendam, the Netherlands: Vereniging van Kunstmest Producenten.

Dimkpa, C., and P. S. Bindraban. 2016. "Micronutrients Fortification for Efficient Agronomic Production." *Agronomy for Sustainable Development* 36: 1–26.

Droppelmann, K. J., S. S. Snapp, and S. R. Waddington. 2017. "Sustainable Intensification Options for Smallholder Maize-Based Farming Systems in Sub-Saharan Africa." *Food Security* 9(1), 133–150.

Dzotsi, K., A. Agboh-Noaméshie, T. E. Struif Bontkes, U. Singh, and P. Dejean. 2003. "Using DSSAT to Derive Optimum Combinations of Cultivar and Sowing Date for Maize in Southern Togo." In *Decision Support Tools for Smallholder Agriculture in Sub-Saharan Africa – A Practical Guide*, edited by T. E. Struif Bontkes and M. C. S. Wopereis, 100–113. Muscle Shoals, AL: IFDC and CTA.

Erisman, J. W., M. A. Sutton, J. Galloway, Z. Klimont, and W. Winiwarter. 2008. "How a Century of Ammonia Synthesis Has Changed the World." *Nature Geoscience* 1: 636–639.

Eurola, M., P. Ekholm, M. Ylinen, P. Koivistoinen, and P. Varo. 1991. "Selenium in Finish Foods after Beginning the Use of Selenate Supplemented Fertilisers." *Journal of the Science of Food and Agriculture* 56: 57–70.

Evéquoz, M., D. K. Soumana, and G. Yadji. 1998. "Minéralisation du fumier, nutrition, croissance et rendement du mil planté dans des ouvrages anti-érosifs." In *Soil Fertility Management in West African Land Use Systems*, Niamey, Niger, 4–8 March 1997, edited by G. Renard, A. Neef, K. Becker, and M. von Oppen, 203–208. Weikersheim, Germany: Margraf Verlag.

FAO (Food and Agriculture Organization of the United Nations). 1995. *The State of Food and Agriculture*. Rome, Italy: FAO.

FAO (Food and Agriculture Organization of the United Nations). 2019. *The State of Food and Agriculture*. Rome, Italy: FAO.

FAO (Food and Agriculture Organization of the United Nations). 2006. *Statistical Yearbook 2005–2006*. Rome, Italy: FAO.

FAO (Food and Agriculture Organization of the United Nations). 2016. *World Fertilizer Trends and Outlook to 2019: Summary Report*. Rome, Italy: FAO.

FAO (Food and Agriculture Organization of the United Nations). 2017. *Regional Overview of Food Security and Nutrition in Africa 2017: The Food Security and Nutrition–Conflict Nexus: Building Resilience for Food Security, Nutrition and Peace*. Accra, Ghana: FAO.

Fernandez-Rivera, S., T. O. Williams, P. Hiernaux, and J. M. Powell. 1995. "Livestock, Feed and Manure Availability for Crop Production in Semi-arid West Africa." In *Livestock and Sustainable Nutrient Cycling in Mixed Farming Systems of Sub-Saharan Africa*, edited by J. M. Powell, S. Fernandez Rivera, T. O. Williams, and C. Renard, 149–170. Addis Ababa, Ethiopia: International Livestock Center for Africa (ILCA).

Fofana, B., M. C. S. Wopereis, A. Bationo, H. Breman, and A. Mando. 2008. "Millet Nutrient Use Efficiency as Affected by Inherent Soil Fertility in the West African Sahel." *Nutrient Cycling in Agroecosystems* 81(1): 25–36.

Gaihre, Y. K., U. Singh, S. M. M. Islam, A. Huda, M. R. Islam, M. A. Satter, J. Sanabria, Md. R. Islam, A. L. Shah. 2015. "Impacts of Urea Deep Placement on Nitrous Oxide and Nitric Oxide Emissions from Rice Fields in Bangladesh. *Geoderma* 259–260: 370–379.

Gardner, M. K., D. G. Parbery, and D. A. Barker. 1981. "Proteoid Root Morphology and Function in Legumes Alnus." *Plant Soil* 60: 143–147.

Grant, P. M. 1981. "The Fertilization of Sandy Soils in Peasant Agriculture." *Zimbabwe Agricultural Journal* 78: 169–175.

Hafner, H., J. Bley, A. Bationo, P. Martin, and H. Marschner. 1993. "Long-Term Nitrogen Balance of Pearl Millet (*Pennisetum glaucum* L.) in an Acid Sandy Soil of Niger." *Journal of Plant Nutrition and Soil Science* 156: 169–176.

Haggblade, S., P. Hazell, I. Kirsten, and R. Mkandawire. 2004. African Agriculture: Past Performance, Future Imperatives." In *Building on Successes in African Agriculture*, edited by S. Haggblade. Focus 12, Brief No. 1. 2020 Vision. Washington, DC: IFPRI.

Hartley, K. T. 1937. "An Explanation of the Effect of Farmyard Manure." *Empire Journal of experimental Agriculture* 19: 244–263.

Henao, J., and C. Baanante. 1999a. *Estimating Rates of Nutrient Depletion in Soils of Agricultural Lands in Africa*. Muscle Shoals, AL: International Fertilizer Development Centre.

Henao, J., and C. Baanante. 1999b. "Nutrient Depletion in the Agricultural Soils of Africa." 2020 Vision. Brief No. 62. Washington, DC: IFPRI.

Hoogenboom, G., C. H. Porter, V. Shelia, K. J. Boote, U. Singh, J. W. White, L. A. Hunt, et al. 2017. Decision Support System for Agrotechnology Transfer (DSSAT), Version 4.7. Gainesville, FL: DSSAT Foundation. https://DSSAT.net.

Hopper, W. D. 1993. "Indian Agriculture and Fertilizer: An Outsider's Observations." Keynote address to the FAI Seminar on Emerging Scenario in Fertilizer and Agriculture: Global Dimensions. FAI, New Delhi.

ICRISAT (International Crop Research Institute for the Semi-arid Tropics). 1985–88. *ICRISAT Sahelian Center Annual Report, 1984*. Patancheru, India: ICRISAT.

Izac, A-M. N. 2000. "What Paradigm for Linking Poverty Alleviation to Natural Resources Management?" In *Proceedings of an International Workshop on Integrated Natural Resource Management in the CGIAR: Approaches and Lessons, 21–25 August 2000*, Penang, Malaysia.

Jama, B., D. Kimani, R. Harawa, A. K. Mavuthu, and G. W. Sileshi. 2017. "Maize Yield Response, Nitrogen Use Efficiency and Financial Returns to Fertilizer on Smallholder Farms in Southern Africa." *Food Security* 9(3): 577–593.

Janssen, B. H., F. C. T. Guiking, D. van der Eijk, E. M. A. Smaling, J. Wolf, H. van Reuler. 1990. "A System for Quantitative Evaluation of the Fertility of Tropical Soils (QUEFTS)." *Geoderma* 46: 299–318.

Kang, B. T., and O.A. Osiname. 1985. "Micronutrient Problems in Tropical Africa." *Nutrient Cycling in Agroecosystems* 7(1–3): 131–150.

Keating, B. A., P. S. Carberry, G. L. Hammer, M. E. Probert, M. J. Robertson, D. Holzworth, N. I. Huth, et al. 2003. "An Overview of APSIM, a Model Designed for Farming System Simulation." *European Journal of Agronomy* 18: 267–288.

Kihara, J., and Njoroge, S. 2013. "Phosphorus Agronomic Efficiency in Maize-Based Cropping Systems: A Focus on Western Kenya." *Field Crops Research* 150: 1–8.

Kihara, J., G. W. Sileshi, G. Nziguheba, M. Kinyua, S. Zingore, and R. Sommer. 2017. "Application of Secondary Nutrients and Micronutrients Increases Crop Yields in Sub-Saharan Africa." *Agronomy for Sustainable Development* 37: 25.

Klaij, M. C., and B. R. Ntare. 1995. "Rotation and Tillage Effects on Yield of Pearl Millet (Pennisetum Glaucum) and Cowpea (Vigna Unguculata), and Aspects of Crop Water Balance and Soil Fertility in Semi-arid Tropical Environment." *Journal of Agriculture Science* 124: 39–44.

Kretzschmar, R. M., H. Hafner, A. Bationo, and H. Marschner. 1991. "Long and Short-Term of Crop Residues on Aluminium Toxicity, Phosphorus Availability and Growth of Pearl Millet in an Acid Sandy Soil." *Plant and Soil* 136: 215–223.

Lesschen, J. P., J. J. Stoorvogel, E. M. A. Smaling, G. B. M. Heuvelink, and A. Veldkamp. 2007. "A Spatially Explicit Methodology to Quantify Soil Nutrient Balances and Their Uncertainties at the National Level." *Nutrient Cycling in Agroecosystems* 78: 111–131.

Lompo, F., A. Bationo, M. P. Sedogo, V. B. Bado, V. Hien, and B. Ouattara. 2018. "Role of Local Agro-minerals in Mineral Fertilizer Recommendations for Crops: Examples of Some West Africa Phosphate Rocks." In *Improving the Profitability, Sustainability and Efficiency of Nutrients through Site Specific Fertilizer Recommendations in West Africa Agro-Ecosystems*, edited by A. Bationo, D. Ngaradoum, S. Youl, F. Lompo, and J. O. Fening. Cham, Switzerland: Springer International Publishing.

Lompo, F., M. P. Sedogo, and A. Assa. 1994. "Effets a long terme des phosphates naturels de Kodjari (Burkina Faso) sur la production du sorgho et les bilans minéraux." *Revue de Recherche en amelioration de agriculture en milieu aride* 6: 163–178.

Mafongoya, P. L., A. Bationo, J. Kihara, and B. S. Waswa. 2006. "Appropriate Technologies to Replenish Soil Fertility in Southern Africa." *Nutrient Cycling in Agroecosystems* 76: 137–151.

McIntire, J. D., Bourzat, and P. Pingali. 1992. *Crop-Livstock Interaction in Sub-Saharan Africa*. Washington DC: The World Bank.

Michels, K., M. V. K. Sivakumar, and B. E. Allison. 1995. "Wind Erosion Control Using Crop Residue>II. Effects on Millet Establishment and Yield." *Field Crops Research* 40: 111–118.

Mokwunye, A. U., A. de Jager, and E. M. A. Smaling. 1996. *Restoring and Maintaining the Productivity of West African Soils: Key to Sustainable Development*. IFDC-Africa, LEI-DLO, and SC-DLO.

Mtambanengwe, F., and P. Mapfumo. 2005. "Effects of Organic Resource Quality on Soil Profile N Dynamics and Maize Yields on Sandy Soils in Zimbabwe." *Plant and Soil* 28(1–2): 173–190.

Mughogho, S. K., A. Bationo, B. Christianson, and P. L. Vlek. 1986. "Management of Nitrogen Fertilizers for Tropical African Soils. In *Management of nitrogen and phosphorus fertilizers in sub-Saharan Africa*, edited by A. U. Mokwunye and P. L. G. Vlek, Vol. 24. Dordrecht, the Netherlands: Martinus Nijhoff. pp. 117–172.

Nandwa, S. M. 2003. "Perspectives on Soil Fertility in Africa." In *Soil Fertility Management in Africa: A Regional Perspective*, edited by Gichuru M. P., et al. Nairobi, Kenya: Academy of Science Publishers (ASP) and Tropical Soil Biology and Fertility of CIAT.

NEPAD (New Partnership for African Development). 2003. *Comprehensive Africa Agricultural Development Programme*. Pretoria: NEPAD.

NEPAD (New Partnership for African Development). 2014. *Comprehensive Africa Agricultural Development Programme*. Pretoria: NEPAD.

Nix, H. A. 1984. "Minimum Data Sets for Agro Technology Transfer. In *Proceedings of the International Symposium on Minimum Data Sets for Agro Technology Transfer*, 181–188. Patancheru, Andhra Pradesh, India: ICRISAT.

Okalebo, J. R., C. O. Othieno, P. L. Woomer, N. K. Karanja, J. R. M. Semoka, M. A. Bekunda, D. N. Mugendi, R. M. Muasya, A. Bationo, and E. J. Mukhwana. 2006. "Available Technologies to Replenish Soil Fertility in East Africa." *Nutrient Cycling in Agroecosystems* 76(2–3): 153–170.

Ouattara, B., B. Somda, I. Sermé, A. Traoré, D. Peak, F. Lompo, S. J. B. Taonda, M. P. Sedogo, and A. Bationo. 2018. "Improving Agronomic Efficiency of Mineral Fertilizers through Microdose on Sorghum in the Sub-arid Zone of Burkina Faso." In *Improving the Profitability, Sustainability and Efficiency of Nutrients through Site Specific Fertilizer Recommendations in West Africa Agro-Ecosystems*, edited by A. Bationo, D. Ngaradoum, S. Youl, F. Lompo, and J. O. Fening. Vol. 1. The Netherlands: Springer.

Palm, C. A. 1995. "Contribution of Agroforestry Trees to Nutrient Requirements of Intercropped Plants." *Agroforestry Systems* 30: 105–124.

Paustian, K., C. V. Cole, D. Sauerbeck, and N. Sampson. 1998. "CO_2 Mitigation by Agriculture: An Overview." *Climate Change* 40: 135–162.

Pieri, C. 1986. "Fertilisation des cultures vivrières et fertilité des sols en agriculture paysanne subsaharienne." *Agronomie Tropicale* 41: 1–20.

Pieri, C. 1989. *Fertilité des terres de savane. Bilan de trente ans de recherche et de développement agricoles au sud du Sahara*. Paris, France: Ministère de la Coopération, CIRAD.

Qureish, J. N. 1987. "The Cumulative Effects of N-P Fertilizers, Manure and Crop Residues on Maize Grain Yields, Leaf Nutrient Contents and Some Soil Chemical Properties at Kabete." Paper presented at the National Maize Agronomy Workshop, CIMMYT, Nairobi, Kenya, 17–19 February, 1987.

Rakotoarisoa, M. A., M. Iafrate, and M. Paschali. 2011. *Why Has Africa Become a Net Food Importer?* Rome, Italy: Trade and Markets Division, FAO.

Rebafka, F.-P., A. Hebel, A. Bationo, K. Stahr and H. Marschner. 1994. "Short- and Long-Term Effects of Crop Residues and of Phosphorus Fertilization on Pearl Millet Yield on an Acid Sandy Soil in Niger, West Africa. *Field Crops Research* 36: 113–124.

Reij, C., I. Scoones, and C. Toulmin. 1996. *Sustaining the Soil: Indigenous Soil and Water Conservation in Africa.* London, UK: Earthscan.

Sahrawat, K. L., M. K. Abekoe, and S. Diatta. 2001. "Application of Inorganic Phosphorus Fertilizer. In *Sustaining Soil Fertility in West Africa*, edited by G. Tian, F. Ishida, and D. Keatinge, 225–246. Soil Science Society of America Special Publication No. 58. Madison, WI: Soil Science Society of America and American Society of Agronomy.

Schlecht, E., A. Buerkert, E. Tielkes, and A. Bationo. 2006. "A Critical Analysis of Challenges and Opportunities for Soil Fertility Restoration in Sudano-Sahelian West Africa." *Nutrient Cycling in Agroecosystems* 76(2–3): 109–136.

Schutte, K. 1954. "A Survey of Plant Minor Element Deficiencies in Africa." *Afric Soils* 3: 285–292.

Sedogo, M. P. 1993. "Evolution des sols ferrugineux lessivés sous culture: Influences des modes de gestion sur la fertilité." Thèse de Doctorat Es-Sciences, Abidjan, Université Nationale de Côte d'Ivoire.

Shapiro, B. I., and J. H. Sanders. 1998. "Fertilizers Use in Semiarid West Africa. Profitability and Supporting Policy." *Agricultural Systems* 56: 467–482.

Sillanpaa, M. 1982. "Micronutrients and Nutrient Status of Soils: A Global Study." *FAO Soils Bulletin* 48: 343–409.

Singh, U., P. Wilkens, S. V. S. Mahla, R. Gopi and C. H. Porter. 2007. "Development and Application of a Potash Decision Support System." Conference abstract, ASA-CSSA-SSSA International Annual Meetings, New Orleans, LA, 4–8 November 2007.

Smaling, E. M. A. 1993. *An Agro-ecological Framework of Integrated Nutrient Management with Special Reference to Kenya.* Wageningen, the Netherlands: Wageningen Agricultural University.

Smaling, E. M. A. 1995. "The Balance May Look Fine When There Is Nothing You Can Mine: Nutrient Stocks and Flows in West African Soils. In *Use of Phosphate Rock for Sustainable Agriculture in West Africa. Proceedings of a Seminar on the Use of Local Mineral Resources for Sustainable Agriculture in West Africa, Held at IFDC-Africa from November 21 to 23, 1994*, edited by H. Gerner and A. U. Mokwunye.

Smaling, E. M. A., L. O. Fresco, and A. de Jager. 1996. "Classifying, Monitoring and Improving Soil Nutrient Stocks and Flows in African Agriculture." *Ambio* 25: 492–496.

Spurgeon, W. I., and P. H. Grissom. 1965. *Influence of Cropping Systems on Soil Properties and Crop Production.* Mississippi Agriculture Experiment Station Bulletin No. 710.

Stoop, W. A., and J. P. V. Staveren. 1981. "Effects of Cowpea in Cereal Rotations on Subsequent Crop Yields under Semi-arid Conditions in Upper-Volta." In *Biological Nitrogen Fixation Technology for Tropical Agriculture*, edited by P. C. Graham and S. C. Harris. Cali, Colombia: Centro International de Agricultura Tropical.

Stoorvogel, J. J., E. M. A. Smaling, and B. H. Janssen. 1993. "Calculating Soil Nutrient Balances in Africa at Different Scales. I. Supra-national Scale." *Fertilizer Research* 35: 227–235.

Sunnadurai, S. 1973. "Crop Rotation to Control Nematodes in Tomatoes." *Ghana Journal Agricultural Science* 6: 137–139.

Tabo, R., A. Bationo, G. Bruno, J. Ndjeunga, D. Marcha, B. Amadou, M. G. Annou, et al. 2007. "Improving the Productivity of Sorghum and Millet and Farmers' Income Using a Strategic Application of Fertilizers in West Africa." In *Advances in Integrated Soil Fertility Management in Sub-Saharan Africa: Challenges and Opportunities*, edited by A. Bationo, B. S. Waswa, J. Kihara, and J. Kimetu. Dordrecht, the Netherlands: Springer NL.

Tabo, R., A. Bationo, M. K. Diallo, O. Hassane, and S. Koala. 2006. *Fertilizer Micro-Dosing for the Prosperity of Small-Scale Farmers in the Sahel: Final Report.* Global Theme on Agroecosystems Report No. 23. ICRISAT Sahelian Center, Niamey, Niger.

Theng, B. K. G. 1991. "Soil Science in the Tropics – The Next 75 years." *Soil Science* 151: 76–90.

Thomas, R. J., H. El-Dessougi, and A. Tubeiley. 2004. Soil Fertility and Management under Arid and Semiarid Conditions. In *Biological approaches for sustainable soil system*, edited by N. Huphoff. New York: Marcel Decker.

UNDP/GEF. 2004. *Reclaiming the Land: Sustaining Livelihoods: Lessons for the Future.* New York: United Nations Development Fund/Global Environmental Facility.

Vandamme, E., A. Kokou, M. Leah, S. Mujuni, G. Mujawamariya, J. Kamanda, K. Senthilkumar, and K. Saito. 2018. "Phosphorus Micro-Dosing as an Entry Point to Sustainable Intensification of Rice Systems in Sub-Saharan Africa." *Field Crops Research* 222: 39–49.

Van Keulen, H., and H. Breman. 1990. "Agricultural Development in the West African Sahelian Region: A Cure against Land Hunger?" *Agriculture, Ecosystems and Environment* 32: 177–197.

Vanlauwe, B. 2004. "Integrated Soil Fertility Management Research at TSBF: The Framework, the Principles, and Their Application." In *Managing Nutrient Cycles to Sustain Soil Fertility in Sub-Saharan Africa*, edited by A. Bationo, 25–42. Nairobi, Kenya: Academy Science Publishers, A Division of African Academy of Sciences.

Vanlauwe, B., A. Bationo, J. Chianu, K. E. Giller, R. Merckx, A. U. Mokwunye, O. Ohiokpehai, et al. 2010. "Integrated Soil Fertility Management: Operational Definition and Consequences for Implementation and Dissemination." *Outlook on Agriculture* 39(1): 17–24.

Wendt, J., and L. W. Mbuthia. 2017. "A Conceptual Framework for Delivering Improved Fertilizers to Smallholder Farmers in Africa." In *Myanmar Soil Fertility and Fertilizer Management Conference Proceedings*, pp. 169–175, IFDC and DAR.

Wendt, J. W., R. B. Jones., and O. A. Itimu. 1994. "An Integrated Approach to Soil Fertility Improvement in Malawi, including Agroforestry. In *Soil Fertility and Climatic Constraints in Dry land Agriculture*, edited by E. T. Crawell and J. Simpson, 74–79. ACIAR Proceedings No. 54. Canberra, Australia: Australian Council for International Agricultural Research (ACIAR).

Wendt, J. W., and J. Rijpma. 1997. "Sulphur, Zinc, and Boron Deficiencies in the Dedza Hills and Thiwi-Lifidzi Regions in Malawi. *Tropical Agriculture* 74(2): 81–89.

Williams, T. O., J. M. Powell, and S. Fernandez-Rivera. 1995. "Manure Utilization, Drought Cycles and Herd Dynamics in the Sahel: Implications for Crop Productivity." In *Technical Papers: Proceedings of an International Conference, 22–26 November 1993*, edited by J. M. Powell, S. Fernandez-Rivera, T. O. Williams, and C. Renard, 393–409. Vol. 2 of *Livestock and Sustainable Nutrient Cycling in Mixed Farming Systems of Sub-Saharan Africa*. Addis Ababa, Ethiopia: International Livestock Centre for Africa (ILCA).

Windmeijer, P. N., and W. Andriesse. 1993. *Inland Valleys in West Africa: An Agroecological Characterization of Rice-Growing Environments*. Publication 52. Wageningen, the Netherlands: International Institute for Land Reclamation and Improvement (ILRI).

Worland, B., N. Robinson, D. Jordan, S. Schmidt, and I. Godwin. 2017. "Post-Anthesis Nitrate Uptake Is Critical to Yield and Grain Protein Content in Sorghum Bicolor." *Journal of Plant Physiology* 216: 118–124.

Wortmann, C. S., and K. Sones. 2017. *Fertilizer Use Optimization in Sub-Saharan Africa*. Nairobi, Kenya: CAB International.

5 The Storage of Organic Carbon in Dryland Soils of Africa
Constraints and Opportunities

Brahim Soudi, Rachid Bouabid, and Mohamed Badraoui

CONTENTS

5.1 INTRODUCTION

By definition, drylands include arid, semi-arid, and dry sub-humid zones. They represent about 43% of the total surface area of the African continent and cover more than 70% of agricultural land. About 50% of the population of Africa lives in these areas and is significantly vulnerable in terms of food insecurity. Climate change, which is expected to increase the frequency and severity of extreme weather events, will exacerbate the vulnerability of these lands if effective adaptation and resilience actions are not undertaken. The livelihoods of most dryland populations depend on natural resource–based activities, such as agriculture and livestock. Forced to meet urgent short-term needs, households resort to unsustainable practices, resulting in strong pressure on the natural resources, loss of biodiversity, and severe soil degradation. One of the crucial factors considered as a cause and a consequence of agroecosystem vulnerability in dry areas is soil degradation (Biancalani et al. 2015). Various interacting processes of such degradation include water and wind erosion, salinization, and loss of organic matter (OM). These processes lead to a decrease of the soil health, productivity, as well as its capacity for reducing carbon (C) emissions into the atmosphere.

Studies on the understanding of the dynamics of soil organic matter (SOM) in different soil and climate contexts of the world have been extensively addressed, and their scientific bases remain current (Hénin and Depuis 1945; Hénin et al. 1959; Jenny 1941; Laudelout 1993; Bremer et al. 1995; Janzen et al. 1997). Hénin and Dupuis (1945) developed for the first time balance equations for SOM decomposition, Hénin and Monnier (1959) addressed the physical and biochemical determinants of the dynamics of SOM, and Laudelout et al. (1960) established quantitative relations between the content of SOM and climate. These and many other research studies were motivated by the need to understand soil genesis, to improve soil properties and productivity, and to elucidate the interrelationship between plant nutrition and C and nitrogen (N) biogeochemical cycles. Studies included laboratory and field experiments involving long-term trials and modeling approaches. More recently, Campbell and Paustian (2015) conducted a fairly comprehensive literature review on developments in SOM modeling. The current concerns related to climate change drive the interest in mitigating greenhouse gases (GHG) and adaptation to climate change in soils, particularly through SOC sequestration (Lal 2004b).

This chapter discusses the status of soil organic carbon (SOC) in the agricultural soils of Africa with an attempt to debate some of the underpinning issues that remain relevant despite the abundant literature on this topic.

5.2 SOIL ORGANIC CARBON STATUS

OM plays various important roles in soil properties and behaviors. It is a major soil fertility and quality indicator and has long been closely related to the capacity of soils to sustain crop growth (FAO 2005). OM is the main source of energy for sustaining living organisms in soil and is an important source of nutrients for crops. It improves the chemical, physical, and biological properties of soil and acts as a protective factor against sealing and erosion. OM is an important soil moisture and temperature conditioner, which makes it a valuable soil component under dryland conditions (Plaza et al. 2018). In the context of global climate changes, SOM represents a major reservoir of C and is advocated as an important sink for atmospheric C sequestration (Lal 2004).

The increase and maintenance of and "adequate" amount of SOM, that is, SOC, is of major concern for farmers and scientists, especially in the context of the recent factual climatic changes. In arid and semi-arid conditions, low SOM contents are related to low biomass productivity and low returns as a result of limiting climate conditions, low water and nutrient use and efficiencies, and nonappropriate farming practices (Lal 2004; Jarecki and Lal 2003). It is estimated that SOC content in drylands is lower than 1% and in many areas does not exceed 0.5% (Lal 2002, 2004). The stock of SOC pools is mainly present in the top layer and declines rapidly with depth. Such conditions unavoidably contribute to soil fertility decline and alter the various soil functions. The level of SOC in African soils in dry areas is in general low to very low (Figure 5.1a) and exhibits a close relationship with the aridity of the climate (Figure 5.1b).

Enhancing OM levels in agricultural lands is in general a major issue, and is even more serious in dryland soils. OM decomposition depends on two natural processes, humification and mineralization. The importance of each of these processes depends greatly on climate and soil conditions, the nature of OM, and agricultural practices (Duchaufour 1976; Zech et al. 1997; Jedidi 1998; Guggenberger 2005). Humification leads to the formation of stable humus, while mineralization leads to the rapid decomposition of the labile fraction and part of the stable humus, with the release of various mineral constituents. The processes of humification and mineralization coexist, and the importance of one over the other depends on several factors, mainly soil moisture, temperature conditions, and C/N ratio.

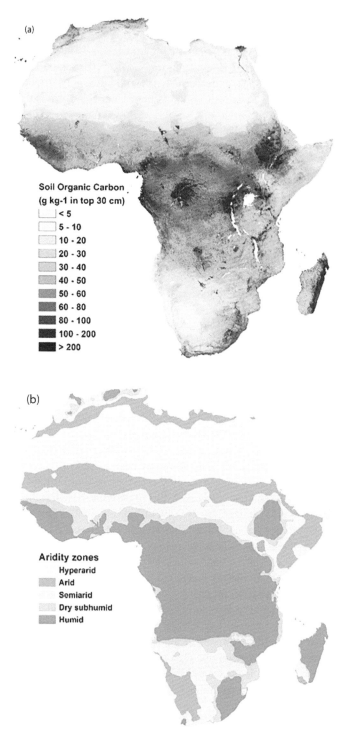

FIGURE 5.1   (a) Soil organic carbon in topsoil in Africa (drawn using data from Soil Grid/ISRIC, 2018; www.soilgrids.org); (b) aridity zones in Africa (Adapted using data from Millennium Ecosystem Assessment, 2018; www.millenniumassessment.org.).

The crucial issue for drylands lies in the predominance of mineralization over humification. The question is, How can the trend of loss of SOM occurring at variable rates and magnitudes be reversed? Scientifically, how can these antagonistic mineralization–humification processes be acted on in order to favor C sequestration, while taking into account the temperature rise (favorable to mineralization) due to climate change, extreme events (mainly droughts) affecting biomass production, low nutrient input (N and P), low residues return, etc.? Undoubtedly, it is theoretically accepted and scientifically proven that the storage of C in the soil contributes to reducing GHG emissions (Lal 2004). Although global carbon mass balances are well established, in order to move from theory to practice, and to make this concept a reality, it would be necessary to go through a drastic change in terms of farming systems as well as in terms of farmer attitudes and practices. In parallel, a series of questions yet under discussion at international level needs to be objectively addressed: (1) What is the maximum level of SOC storage that is biochemically feasible in dry agroecological zones of Africa? (2) What is the time required for C sequestration, taking into consideration the amount of fresh OM incorporated in the soil and the simultaneous halftimes of humification and mineralization processes? (3) What are the most adapted agroecological innovations in favor of C sequestration in the conditions of African agriculture? and (4) What are the perception and the degree of acceptance of small farmers with respect to these innovations?

Humification versus mineralization under dryland farming is a dilemma that needs particular attention. When the process of humification is dominant, SOM decomposition results in appreciable amounts of the stable humus that plays a major role in the definition of the soil fertility; humus buildup will contribute to C accumulation and therefore to C sequestration, depending on the rate of return to an equilibrium state that depends, in turn, on various local climatic and biophysical factors. When the process of mineralization is dominant, SOM decomposition results in high amounts of mineral constituents that can contribute to nutrient bioavailability for crops, but on the other hand, a continuous depletion of OM will take place. This is particularly true when the organic residues returned to the soil are rich in N (i.e., low C/N). The same process applies for organic amendments, such as N-rich farm manures. In drylands, where crop seasons are characterized by high temperatures and appreciable precipitations, annual SOM depletion is very important (FAO 2004).

The issue of humification versus mineralization has been raised in reaction to the 4-per-mille (4pm) initiative for soil C sequestration proposed by France in support of the Conference of Parties (COP) 21 (Paris, France, 2016) and later discussed during the COP 22 (Marrakech, Morocco, 2017). The 4PM initiative advocates for an increase of 0.4% of SOM annually. This rate of increase would undoubtedly result in significant annual accumulation of SOM. However, the pertinence and feasibility of such a rate in drylands, particularly those in Africa, is subject to debate. If the objective of a 0.4% increase is possible under humid and temperate regions, it is very difficult to achieve in the dryland agricultural soils of Africa, mainly due to the low recycling of OM in most areas, as well as to the dominance of the mineralization process.

Therefore, SOM management as a key component of soil fertility and as a potential sink of C sequestration faces major challenges in the dryland agriculture of Africa from the various dimensions of climate and pedoclimate conditions, cropping systems, and farmer practices. The interactions among these dimensions vary from one region to another and create a multitude of situations that need to be considered at the local scale. Managing SOM in arid subsistence smallholders of Africa would be completely different from that of the arid large farms of North America.

5.3 KINETICS OF SOIL ORGANIC CARBON EVOLUTION

The evolution of SOC operates according to first-order kinetics. The basic equation established in early studies (Hénin and Depuis 1945; Hénin et al. 1959; Jenny 1941) and still in use in recent models is expressed as follows:

$$dC/dt = -k \cdot C, \qquad (5.1)$$

TABLE 5.1

SOM Decomposition Rates (k) and Corresponding Half-Lives for Various Cropping Systems

k/yr⁻¹	Half-Life (years)	Cropping System/Region	Source
0.33	2.1	Maize–Rice/Zaire	Laudelout and Meyer (1954)
0.0608	11	Continuous Maize/Ohio, US	Jenny (1941)
0.052	13	Quinquennial rotation/Ohio, US	Salter and Green (1933)
0.07	10	Continuous and alternate cereals with fallow/Kansas, US	Myers et al. (1943)
0.22	3.1	Fallow/Puerto Rico	Smith et al. (1951)

Source: Laudelout, H., Bilan de la matière organique du sol: le modèle de Hénin (1945), In *Mélanges offerts à Stéphane Hénin: Sol – agronomie – environnement*. ORSTOM, Paris, pp. 117–123, 1993.

where C is carbon content, t is time, and k is the evolution rate (accumulation or depletion). The integration of this equation from t_0 to t, respectively corresponding to C_0 and C, gives the following equation:

$$A = C_0 \cdot e^{(-kt)} \tag{5.2}$$

The half-life $t_{1/2}$ corresponds to the time required for the accumulation or the depletion of 50% of the initial C_0. The log transformation of equation (5.2) allows calculating $t_{1/2}$. For a situation of depletion, at $t_{1/2}$ C = $C_0/2$, therefore:

$$\ln(C/C_0) = -k \cdot t_{1/2} \tag{5.3}$$

and

$$t_{1/2} = \ln(2)/k = 0.693/k.$$

Two important deductions can be drawn from this result: (1) the evolution of SOC, and therefore of its storage or depletion, is not a linear process but follows a first-order kinetic; (2) the theoretical humification or mineralization half-life of the initial C content (C_0) occurs in terms of years, depending on soil and climate conditions. Table 5.1 shows that values of $t_{1/2}$ vary from about 2 years for a k rate of 0.33 year⁻¹ in the conditions of Zaire for a rice (*Oryza* spp.)–maize (*Zea mays*) rotation (Laudelout and Meyer 1951), to about 13 years for a k rate of 0.0608 yr⁻¹ in the conditions of Ohio for a continuous maize (Jenny 1941). Estimated values of half-lives from other studies of steady state SOC decomposition are in the range of 5 to 20 years (Haas and Evans 1957; Mann 1986; Davidson and Ackerman 1993; Bremer et al. 1995). In drylands, the SOC evolution half-life $t_{1/2}$ would vary from 5 to 10 years depending on climatic and biophysical conditions.

5.4 SCENARIOS OF SOIL ORGANIC CARBON EVOLUTION

Three scenarios are possible for the evolution of SOC, as illustrated by Figure 5.2.

Scenario 1: This scenario corresponds to the situation where an exponential decrease in SOC takes place as a result of factors causing low return and intense mineralization. This is the situation that applies most for drylands that are characterized by high temperatures. The depletion of SOC starts when negative disturbance factors take place and tends toward a minimum steady state level ($C_{(b)}$) if practices are kept "business as usual." This evolution follows an exponential trend, and the rate of decrease following disturbance depends on the type and intensity of the depleting factors. In this case, mineralization is the dominant process, especially with favorable soil moisture and temperature conditions allowing high microbial activity. Such a trend is described and corroborated by several studies of SOM mineralization and its seasonal

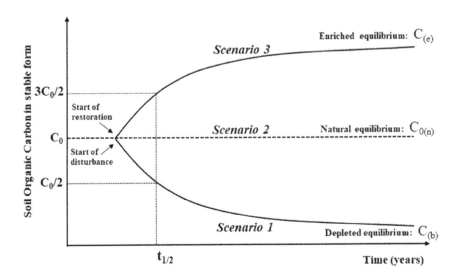

FIGURE 5.2   Scenarios of soil organic carbon evolution in soils.
$C_{(b)}$ = f(climate, soil properties, management practices business as usual)
$C_{0(n)}$ = f(climate, soil properties, ideally all management practices adopted)
$C_{(e)}$ = f(climate, soil properties, feasible management practices adopted)

variations in arid and irrigated rainfed areas (Soudi et al. 1990; Berdai et al. 2002). This scenario is the most likely to occur in dryland soils of Africa and is expected to intensify even more with global warming if appropriate measures are not undertaken to reverse it.

Scenario 2: This scenario corresponds to a situation of a "natural" equilibrium where SOC balance remains relatively stable and C stock stays in the range of the initial C_0 before disturbance. This equilibrium is driven by climate conditions, soil properties, and basic agricultural practices that are not inducive of any significant degradation. This situation corresponds to a relatively stable soil quality and to a zero growth of SOC. The amplitudes of change with time are small and the average balance $C_{0(n)}$ stays, over time, around the initial C_0. This scenario can be encountered in agroecosystems that have not been subject to much pressure or in cases where anthropic activities are conducted in harmony with available resources.

Scenario 3: In the contexts that favor the formation of stable humus, related to climate conditions, innovative and adapted agricultural practices (residues input and management, rotations, etc.), and/or to soil properties (high-charge clays and oxyhydroxides, calcium carbonates, etc.) (Lal 2016), an increase of SOC is expected. This scenario corresponds also to an exponential increase of stable SOC as a result of the dominance of humification, and a new enriched steady state (equilibrium) $C_{(e)}$ level is reached depending on the factors involved. The $C_{(e)}$ would depend also on the climatic conditions, but much more on the quantities and the nature of biomass or crop residues returned to the soil and the activities that favor their accumulation (higher humification). This situation is more likely to be found in humid temperate regions with high SOM return, but is unlikely to occur in dryland cultivated soils (such as those of Africa), where the restitution of OM is generally very low (few tons per hectare), crop residue management is often inappropriate or absent, and the rates of mineralization are high.

5.5   WHAT IS THE POSSIBLE MARGIN FOR CARBON STORAGE IN SOILS OF DRYLANDS?

The majority of dryland soils in Africa are low in SOC and are even experiencing an increasing loss of OM (Lal 2004; FAO 2004, 2005). This situation correspond to "trend a" (Figure 5.3), where the residual SOC is often stabilizing at a critical minimum ($C_{(b)}$). Assuming that a recovery of SOC is

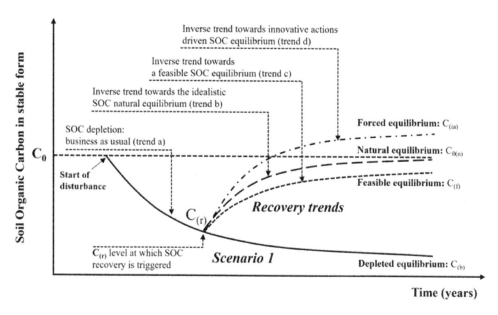

FIGURE 5.3 Idealistic and feasible options for soil organic carbon recovery and storage under dryland conditions.

$C_{(r)}$ = carbon level from which recovery is triggered

$C_{0(n)}$ = f(climate, soil properties, ideally all management practices adopted)

$C_{(b)}$ = f(climate, soil properties, management practices business as usual)

$C_{(f)}$ = f(climate, soil properties, feasible management practices adopted)

$C_{(ia)}$ = f(climate, soil properties, innovative management actions and practices adopted)

triggered from a given point on the curve of depletion, the increase of SOC (rate and maximum soil carbon) would depend on the various factors discussed earlier. This recovery is a major challenge that would require sustained efforts in terms of overcoming factors of disturbance and adopting appropriate management practices, as well as operating significant changes on farmers' attitudes and behaviors. The *recovery trend* would follow a first-order kinetics and would require several years. The reconstitution half-life of the lost C would be in general slow, from 5 to 10 years depending on the local context.

Ultimately, if all conditions are favorable, the recovery of SOC would rise back toward the "idealistic" natural initial storage capacity $C_{(n)}$ (trend b). However, this trend seems challenging, knowing that in most cases, it would be difficult to guarantee all the necessary conditions for a positive and increasing annual balance of SOC. For instance, a soil that has undergone significant degradation by water erosion as a result of land cover degradation and subsequent organic matter loss would be unlikely to restore to its initial situation despite rigorous efforts.

The alternative trend for SOC sequestration when adopting a set of "feasible" agricultural practices that favor SOC storage would be a new state of equilibrium $C_{0(f)}$ that depends on the local context (trend c), and that would be less than $C_{0(n)}$. Therefore, it is emphasized here that reversing the trends of C storage in dryland soils would depend greatly on the local conditions (climate, soil properties, and management practices). Farmers can move forward to only a set of feasible management practices according to their local context (farm size, cropping systems, level of input, productivity, income, advisory, government support, degree of adoption of innovations, etc.). The SOC would be eventually enhanced from the "business as usual" level ($C_{(b)}$) to the "maximum feasible level" ($C_{(f)}$), but would not reach the idealistic maximum level ($C_{0(n)}$).

Under special conditions relying on forced action measures, it is possible to increase SOC to a level ($C_{(ia)}$) beyond the idealistic level. This situation is optimistically achievable, but would require the introduction of innovative management practices and assumes no limiting factors to their implementation (adaptation, sufficient long-term funding, acceptance by farmers, capacity building, advisory, policy, etc.).

5.6 FACTORS DETERMINING CARBON STORAGE CAPACITY IN DRYLAND SOILS

5.6.1 QUANTITY OF CROP RESIDUES

Crop residues are the main source of SOC and improve the chemical, physical, and biological properties of the soil. According to Power and Legg (1978), degraded soils of North Africa are a good example of the effect of off-field residues exports on SOM. Ayanlaja and Sanwo (1991) showed that SOC content is proportional to the amount of residues recycled. Soudi et al. (2000) reported that the amount of residues returned to the soil in arid zones of Morocco are low (Figure 5.4), because off-field exports are used for livestock feed during the dry season. This is a typical situation in smallholders' mixed cropping in Africa (FAO 2005). Naman and Soudi (1999) and Traoré et al. (2007) underlined that improving management of crop residues is the best way to regulate the SOC levels.

5.6.2 NATURE OF CROP RESIDUES

According to Mustin (1987), the resistance to biodegradation of SOM is correlated to the biochemical nature of plant cell tissues. Similar findings were reported by studies carried out in irrigated drylands in Morocco (Naman et al. 2015; Naman et al. 2018). The biochemical fractionation of the OM returned to the soils from seven crop residues revealed that their tissues contain more cellulose, hemicellulose, and soluble fractions than lignin (Figure 5.5). Wheat residues are characterized by the highest concentration of cellulose (40.6%) compared to the other residues. Sorghum (*Sorghum*) and maize (*Zea mays*) contained moderate amounts of lignin. The hemicellulose concentration varies between 10.0% for tomato residues and 23.5% for wheat residues. The highest lignin concentration was found in sorghum residues (9.6%). In terms of soluble fractions, tomato residues had the highest fraction (65.3%), and those of wheat had the lowest fraction (29.8%). Crop residues with

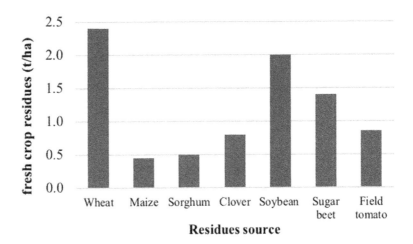

FIGURE 5.4 Quantities of residues returned by different crops in arid irrigated areas in Morocco. (From Soudi, B. et al., Problématique de gestion de la matière organique des sols: Cas des périmètres irrigués du Tadla et des Doukkala, in *Séminaire "Intensification agricole et qualité des sols et des eaux," Rabat, 2–3 Novembre 2000*, B. Soudi (ed.), pp. 25–30, 2000.)

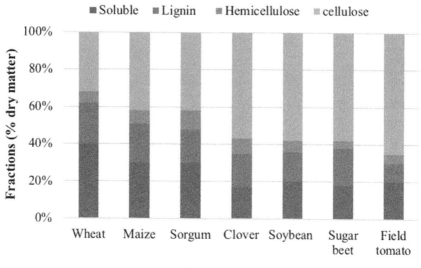

FIGURE 5.5 Biochemical fractions of crop residues in irrigated dryland soils in Morocco. (Adapted from Soudi, B. et al., Problématique de gestion de la matière organique des sols: Cas des périmètres irrigués du Tadla et des Doukkala. In *Séminaire "Intensification agricole et qualité des sols et des eaux," Rabat, 2–3 Novembre 2000*, B. Soudi (ed.), pp. 25–30, 2000.)

high concentrations of fiber generate more stable C in the soil. Lignin is degraded almost entirely to phenolic compounds with polymerization properties leading to stable humic substances (Datta et al. 2017). Organic residues decompose differently depending on their biochemical composition (Guggenberger 2005; Bouajila et al. 2014). Residues rich in lignin are difficult to decompose due to the recalcitrance of these macromolecules. Cellulose, the main constituent of the cell wall structure, is easily biodegradable, but becomes less biodegradable if coated with hemicellulose, which has a higher degree of polymerization. When hemicellulose becomes encrusted in lignin, it also becomes protected from rapid decomposition (Guggenberger 2005).

5.6.3   C/N RATIO: A STILL-PRACTICAL INDICATOR FOR ORGANIC MATTER DECOMPOSITION

The C/N ratio is an indicator of the humification potential of returned OM (plant residues, manure, etc.) to the soil. It is unanimously accepted that the higher the C/N ratio of organic residues incorporated into the soil, the slower is the decomposition and the more stable the humus produced (Waksman 1924; Jensen 1929; Allison 1955; Fog 1988). The slow decomposition is essentially attributed to the highly polymerized humus macromolecules that are difficult to degrade (Swift et al. 1979; Monties 1980).

To better understand the processes of humification and mineralization of the SOM and the consequent supply of mineral N potentially absorbed by the crops, it is essential to consider the interactions between the C and N cycles. In fact, a high C/N ratio favors the production of stable C in humus, whereas a low C/N ratio favors the mineralization of OM and the production of CO_2, mineral-N, and other mineral constituents. The breakpoint value that drives the dominance of one process over the other is related to the assimilation needs of the microorganisms present in the soil (to satisfy the C/N ratio of their tissues), and varies with their population type (Janssen 1996). C/N breakpoints varying from 20 to 25 were reported by Tate (1995) and Bengtston et al. (2003). Vigil and Kissel (1991) reported a higher value of 40. Chen et al. (2014) ranked data for C/N in relation to mineralization–immobilization for residues from various crops that showed that mineralization is dominant for C/N values less than 22.7, that concurrent mineralization–immobilization occurs for C/N values exceeding 30, while immobilization becomes dominant for high C/N values (>78).

5.6.4 MINERALIZATION VERSUS HUMIFICATION IN DRYLANDS

In drylands of Africa, the low OM maintenance in the soils, as a result of low restitution of crop residues, is often aggravated by rapid and intense mineralization because of the high temperatures favoring microbial activity. The OM mineralization rate is amplified in irrigated soils in semiarid and arid areas because of the combined effect of temperature and moisture content. Estimated rates of OM mineralization under Mediterranean conditions in Morocco range from 1.9% to 3.3% yr^{-1} (Soudi et al. 2000).

Carbon fluxes resulting from the decomposition of fresh OM added to the soil follow concurrently the process of humification with a rate of k_h and that of mineralization with a rate of k_m. The later involves primary mineralization (k_{m1}) and secondary mineralization (k_{m2}) (Figure 5.6).

In drylands, although secondary mineralization is slower compared to primary mineralization given the molecular complexity of the humic compounds, both processes greatly exceed, in terms of intensity, the humification rate. Furthermore, in these areas, the rate of mineralization is amplified when the soil moisture content is greater than 50 of the soil water-holding capacity, particularly during the warm rainy season or in the case of irrigation (Flowers and Chalagha; Soudi et al. 1990; Li 1990).

Applying equation (4) of Hénin and Dupuis (1945) to the degradation of OM:

$$dC/dt = \left(k_h{}_*fOC\right) - k_{m2}{}_*Cs, \tag{5.4}$$

where C is organic carbon, k_h is the humification rate, fOC is the fresh organic carbon in the crop residues, k_{m2} is the secondary mineralization, and Cs is the stable SOC; and using the example of data reported by Soudi (2000) and Naman et al. (2015) for Mediterranean conditions, mainly (1) an average quantity of residues returned of 1.5 Mg ha^{-1}, (2) an average humification rate (k_h) of about 15% for most crop residues, (3) an average annual OM mineralization rate (k_{m2}) of about 2.5%, and (4) an average stable SOC content of about 0.7%, the apparent annual deficit of SOC is estimated to be about 0.3 Mg ha^{-1}.

This clearly shows that current farming systems in drylands are part of a trend of SOC losses that are not expected to be counterbalanced by stable C production. A study by Naman et al. (2015) showed that the rate of compensation of C depleted by mineralization in the case of arid irrigated

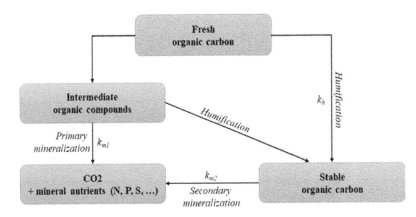

FIGURE 5.6 Soil organic carbon evolution in the soil (k_h: rate of humification; k_{m1}: rate of primary mineralization; k_{m2}: rate of secondary mineralization).

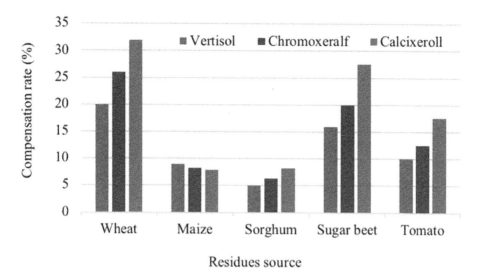

FIGURE 5.7 Compensation rate of the mineralized organic carbon by humification for selected crop residues in three different soil types. (Adapted from Naman, F. et al., *J. Mater. Environ. Sci.*, 6, 3574–3581, 2015.)

areas in Morocco varies from 4% to 32% depending on soil type and the nature of residue returned (Figure 5.7). Wheat (with the highest C/N ratio) showed the highest compensation rates among the five crops, and the calcium-rich Mollisol showed the highest values among the three soil types.

5.6.5 Effect of Soil Texture

As noted above, after going through a partial or total humification process, SOM develops close bonds with clays. Ladd et al. (1996) and Hassink et al. (1994) showed that the decomposition of fresh crop residues and the mineralization of native SOM are very fast in sandy soils compared to clay soils. This is partly attributed to a strong physical or physiochemical protection of SOM compounds by adsorption on the clay minerals surface or by their encapsulation in the very small pore of aggregates that are inaccessible to microorganisms (Elliott and Coleman 1988). In addition, the protecting effect depends on the kind of clay minerals. High-charge 2:1 clays have more bounding forces than 1:1 clays. Martin and Haider (1986) reported that smectites are very effective protectors of organic compounds, whereas kaolinites are rather weak. This protective effect is generally related to the importance of the charge, the location of the charge, and the swelling-shrinking properties. Theng et al. (1986) stated that SOM structures can be incorporated in the interlayers of clay minerals. Nguyen (1982) showed that extracellular enzymes are adsorbed by the clays and therefore are not able to reach their C substrates in the micropores. A characterization study of C contents in different particle size fractions conducted in irrigated semiarid soils in Morocco (Naman et al. 2002) showed that the three particle-size fractions, sands, silts, and clays, comprised 15% to 37%, 19% to 40%, and 24% to 66% of the total carbon, respectively.

Zech et al. (1997) studied correlations among the C stock and various soil parameters in different regions of Africa. They reported that in semiarid soils of Senegal, soil C stock was significantly correlated to the N reserves (r = 0.93***), P reserves (r = 0.63***), fine earth fraction (r = 0.51*), potential cation exchange capacity (CEC) (r = 0.8***), and clay content (r = 0.79***). Eliminating variables that were autocorrelated, the authors found that the regression can be simplified to contain only clay content as follows:

$$C \text{ stock} = 15.9 + 0.017_* Clay; (R = 0.62^{***}),$$

with C and clay contents in 10^6 g·ha^{-1} in 1 m soil depth.

5.6.6 Carbon Status and Intensive Cultivation in Dry Irrigated Areas

The history of the farming system and the degree of intensive cultivation have marked effects on the dynamics of SOC. Unsound agricultural intensification can cause negative effects on the environment. However, if intensification is accompanied by sustainable management practices (sustainable intensification), positive effects can be obtained, among which is carbon sequestration. Lal (2002) reported that in China, agricultural intensification and the adoption of a set of recommended management practices on cropland, forest land, and grazing land have a potential to sequester 59–106 Tg C yr^{-1}, with 25–37 Tg C yr^{-1} in the croplands.

Comparing two vertic calcixerolls under arid Mediterranean conditions in Morocco that have undergone similar pedogenesis, Soudi et al. (2000) observed that the soil under irrigated intensive agriculture (Tadla) is poorer in organic C, labile organic N forms, and clay-fixed (nonexchangeable) ammonium than that under rainfed agriculture (Chaouia) (Table 5.2). These differences were attributed mainly to the intensive cultivation practices that are not accompanied by adequate management of crop residues. Indeed, in most irrigated arid zones, temperature and irrigation ensure optimum thermal and moisture conditions for the mineralization process. This phenomenon is amplified by frequent tillage that increases the accessibility of OM to biodegradation. The low levels of chemically hydrolyzable-N and amino acids-N in the irrigated soil showed a tendency to deplete readily biodegradable organic-N forms. In fact, management failures of crop residues do not allow replenishment of these pools of SOM. The low nonexchangeable ammonium content in the irrigated soil, compared to that in rainfed soil, is also a worthy indicator of the effect of intensive cultivation. In fact, the intense nitrification process under irrigation and the high crop mobilization of mineral N shift the equilibrium toward the release of the clay fixed-ammonium. These results confirm that soil type alone cannot explain the trends in SOM evolution and that the degree of agricultural intensification and soil use have a significant impact.

Other studies have reported opposite trends with agricultural intensification when residues are appropriately managed. Liao et al. (2015, 2016) reported that intensification in northern China engendered a significant increase of SOC. Their results indicate that from 1982 to 2011, SOC content and C stock in the surface (0–20 cm) layer of the cropland increased from 7.8 to 11.0 g kg^{-1} (41%) and from 21 to 33.0 Mg ha^{-1} (54%), respectively. The estimated SOC stock (0–20 cm) of the farmland for the entire county increased from 0.75 to 1.2 Tg (59%). Correlation analysis revealed that under intensification conditions SOC was increased significantly with crop residues, while it was decreased with increased mean annual temperature. This shows that unless proper management practices are adopted, intensification will cause a decline of SOC.

TABLE 5.2
Comparison of Some Dynamic Parameters of Organic Matter in Topsoil Layer (0–10 cm) in Two Contrasting Arid Zones of Morocco: Chaouia (Rainfed) and Tadla (Intensive Irrigated)

Parameters	Chaouia (Rainfed)	Tadla (Intensive Irrigated)
Organic C (g/kg)	13.0	23.3
Organic N (g/kg)	1.4	2.2
Hydrolyzable-N (mg/kg)	915.6	1192
Amino-acid-N (mg/kg)	428.8	603.4
Clay fixed-ammonium (mg/kg)	71.3	120.8

Source: Soudi, B. et al., Problématique de gestion de la matière organique des sols: Cas des périmètres irrigués du Tadla et des Doukkala. In *Séminaire "Intensification agricole et qualité des sols et des eaux," Rabat, 2–3 Novembre 2000*, B. Soudi (ed.), pp. 25–30, 2000.

5.7 TREES OF CONSTRAINTS AND SOLUTION FOR CARBON SEQUESTRATION IN DRY AREAS

5.7.1 CONSTRAINTS HIERARCHY

Considering the aspects discussed above, it can be deduced that the storage of C in dryland soils in general, and in Africa in particular, is difficult and challenging, given the farming systems characterized by low input, low production, and low return of crop residues. Intense mineralization attributed to temperature and soil moisture factors during the rainy season, or to irrigation, is a driving force for organic C decline. This is aggravated when no or low exogenous organic amendments (manure, compost, etc.) are used. In addition, and as shown in Figure 5.8, the cropping systems do not always include crops generating residues with a high C/N ratio and rich in carbon compounds with high potential of carbon stabilization in humic substances.

It is important to note that while C storage is possible, it can be increased only up to a level where SOC equilibrium is established with regard to climate and appropriate farming practices. In the case of drylands in general, it would be too ambitious to strive for an increase of SOC beyond the level that is considered a feasible balance (equilibrium: C_0) as discussed earlier. The more realistic option for C storage in these

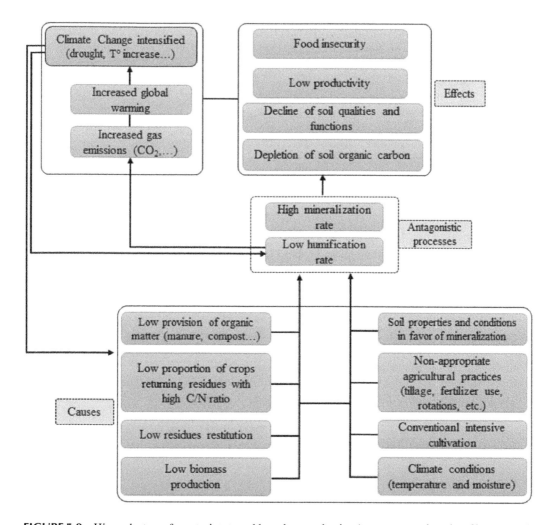

FIGURE 5.8 Hierarchy tree of constraints to stable carbon production (causes are not in order of importance).

areas would be to close the gap in the current (often deficient) level and the C_0 level (Figure 5.3) which is somehow equivalent to the maximum C storage capacity in its stable form in a given soil.

On the other hand, and as mentioned above with regard to the influence of soil type, the physical and physicochemical links between OM and soil mineral particles mean that soils do not have the same ability to store stable SOC. For instance, it is well established that the OM content of sandy soils is always lower than that of clay soils. Calcium-saturated soils (calcareous soils) have often more stable SOC than acid soils. Moreover, it is prudent to avoid comparing, in absolute terms, the values of the C or OM contents between such contrasting soils. In fact, a sandy soil with an OM content of 1% could be considered richer than a clay soil with a content of 2%. This argues further for (1) the effect of fine clay particles on the level of protection of C against mineralization, and (2) the difficulty of trying to exceed the maximum C storage capacity of the soil (i.e., natural equilibrium $C_{0(n)}$). It is somewhat comforting that these observations are based on "stable carbon" rather than "total C including that of fresh OM."

5.7.2 SOLUTIONS HIERARCHY

It is crucial to underline that even if it is difficult to close the gap toward higher levels of SOC in African soils, their potential for C storage is quantitatively important. Africa counts a vast area of arable (205 millions ha) and potentially arable (870 millions ha) land. The amount of C sequestration potential in African drylands estimated by current soil carbon stocks for continental Africa have been calculated as 80.1 Gt C for the 0–30 cm layers and 74.5 Gt C for the depth of 30–100 cm (a total of 154.6 Gt C for 0–100 cm depth) (Jones et al. 2013). By adopting feasible and viable agricultural practices as succinctly illustrated in Figure 5.8, a solutions tree is deduced by reversing the constraints in Figure 5.9.

Implementing these solutions in the real world, through agroecological innovations that promote carbon storage, requires drastic changes in African farming systems, particularly in rainfed arid and semiarid areas. Solutions need to target also changes of attitudes and behaviors, not only at the farmer level, but also at all levels of decision-making (policy, research-development, advisory, etc.). The main agricultural practices likely to maximize the socioeconomic benefits for farmers and at the same time promote the C stock in soils are briefly described below (not in order of importance).

5.7.2.1 Conservation Agriculture

Tillage is the practice that has the most influence on the mineralization of SOM. When it is too intensive, it speeds up the process. Soil tillage promotes soil aeration and increases the activity of microorganisms involved in the mineralization of SOM. It can also make the soil more susceptible to losses of OM through increased erosion (both wind and water). Therefore, the adoption of reduced tillage practices or the shift toward conservation tillage or direct seeding, with the maintenance of significant surface residues, is a measure that is advocated to contribute significantly to reducing the rapid decomposition of SOM while offering other favorable conditions for crop growth such as improved soil moisture conditions and soil fertility status.

Mulching plays an important role in preserving a permanent soil cover that contributes to soil temperature and moisture conditioning. Mulching with residues of high C/N ratio is likely to contribute to the accumulation of stable humus. Even if the amounts are low, their influence under dry conditions have been shown to be significant. The favorable conditions created with mulching have also indirect positive effects on soil moisture status, nutrient cycling, and crop growth.

5.7.2.2 Fallow

Fallow plays multiple roles in the sequestration of soil C. When the soil is left uncultivated, a protective spontaneous vegetation cover takes place. This is particularly beneficial in dry areas where the soil is exposed to erosion. Soil moisture is preserved, especially in the deep layers,

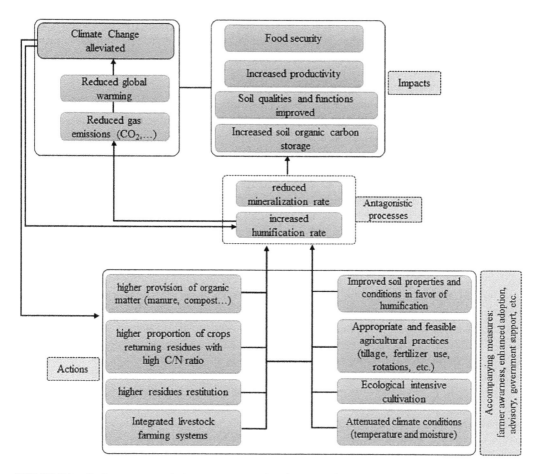

FIGURE 5.9 Carbon storage solutions tree: converting the constraints (Figure 5.8) into solutions.

and improves biomass production. The vegetation cover represents a sink for photosynthetically fixed C that becomes available for potential return and sequestration in the soil after plant senescence. In Nigeria, a study showed that deforestation caused a decline in soil carbon from 25 to 13.5 Mg ha^{-1} in seven years, but after 12–13 years of bush fallow, the soil restored its total carbon content.

5.7.2.3 Crop Rotations

The importance of crop rotation in agricultural systems is well established, and this practice has become an integral part of many soil conservation technologies. The adoption of rotations has many benefits such as improving soil structure, combating weed buildup, improving soil fertility, and diversifying crops and thus crop residue types. In many dryland areas of Africa, crop rotations are used as alternatives to fallow. The use of different crops with deep rooting facilitates the distribution of OM through the soil profile and allows C storage in deeper horizons, where it is less subject to degradation.

Crop rotations, with nitrogen-fixing (N-F) legume crops are especially used to improve soil N status, especially in low-N-fertilizer farming systems. The main concern with rotations involving N-F is their important nitrogen contribution, which can lower the overall C/N ratio of the residues and consequently favors OM mineralization at the expense of humification. Although legume crops are of paramount importance in many African countries, continuous rotation with such crops may not be in favor of storing SOC. Therefore, it is important to consider introducing other rotations with crops that leave residues of high C/N ratios that can be practiced in an alternating manner with those involving legumes.

5.7.2.4 Agroforestry and Tree Cropping

Agroforestry is a practice of introducing trees and shrubs into production and land management systems. In many areas of Africa, agroforestry systems are known to store larger amounts of C than tree-free farming systems (Somarriba et al. 2013; Unruch et al. 1993). The establishment of trees on agricultural land is relatively efficient and cost effective compared to other mitigation strategies, and it provides a range of important ancillary benefits for improving the livelihoods of small farms and for contributing to the adaptation to climate change (FAO 2010).

While it is widely accepted that agroforestry has great potential for C sequestration and for establishing buffers between primary forests and areas under anthropic pressure, the concern for many agroforestry ecosystems in Africa lies in the competition between extraction of wood for household needs and carbon sequestration (Wise and Cacho 2005). Driving the population to minimize the removal of wood to meet their needs while satisfying the goals of C sequestration projects based on agroforestry parks may result in adverse effects of more cutting of primary forests (Chomitz 1999). Therefore, in order to implement a sustainable agroforestry project in socioeconomically vulnerable areas, it is necessary to consider the factors of adoption and success (Mercer 2004). The main challenge of any project of this type is to find solutions to financial and socioeconomic constraints and to understand the importance of the cultural context (Nair 1993). The socioeconomic context is a critical determinant of C sequestration that needs to be taken into account beyond the scientific evidence of the benefits of agroforestry in carbon sequestration.

According to FAO (2003), tree crops such as cocoa (*Theobroma cacao*) plantations in Ghana, coffee (*Coffea* L) plantations in Burkina Faso, native fruit trees in South Africa, and oil palm (*Elaeis guineensis*) have increased considerably; they went from 7 million hectare (Mha) in 1965 to 43 Mha in 1990 and to 187 Mha in 2000. These crops have a positive effect on C sequestration. Olive orchards in Mediterranean North Africa, an ancient millennium practice, are a good example of soil C sequestration, especially due to their longevity. They are often associated with other crops (intercropping), which increases their benefits for high C/N residue returns as well as for additional economic value. Oladele and Braimoh (2013) estimated that C sequestration from various African cases of tree crop farming and associated alley farming vary from 1359 and 1458 kg C ha^{-1} yr^{-1}, respectively, and proposed that these practices be scaled up to more farmers due to their multiple positive effects on the environment. Ogunkunle and Eghaghara (1992) assessed the SOM contents in the surface layer (0–15 cm) of soils under different land use types, some of which were mixed cropping. The soil under the original secondary forest had the highest SOM content (61 g kg^{-1}), and the cultivated soil under the four-year-old cassava (*Manihot esculenta*) had the lowest content, 36 g kg^{-1}. However, the soil under the 10-year-old Cacao (*Theobroma cacao*) plantation mixed with yam (*Dioscorea rotunda, bulbifera*), plantain (*Musa paradisiaca*), and cocoyam (*Colocasia esculenta*) presented intermediate SOM contents of 43 g kg^{-1}.

5.7.2.5 Mixed and Intercropping

Mixed or intercropping with various field or tree crops have been reported to promote C sequestration in the soil compared to sole cropping. A well-managed mixed or intercropping may result in greater surface and belowground organic residues than sole crops and sequester more soil carbon over time due to greater input of root litter (Cong et al. 2015b). Dyer et al. (2012) found that maize–wheat intercropping enhanced soil C sequestration and reduced C emission significantly compared to the corresponding monocultures. Beedy et al. (2010) indicated that the use of intercropping had proportionally positive effects on soil fertility reconstruction as well as on C sequestration. Hu et al. (2015) found that intercropping of wheat–maize produced better yields compared to sole crops and generated less C emissions, especially with reduced tillage practice. Cong et al. (2015b) demonstrated a divergence in soil organic carbon (C) and nitrogen (N) content over a seven-year field experiment in which they compared rotational strip intercrop systems and ordinary crop rotations. Soil organic C content in the top 20 cm was 4% greater in intercrops than in sole crops, corresponding to an average difference in C sequestration rate among these systems of 184 kg C ha^{-1} yr^{-1}. Soil organic N content in the top 20 cm was 11% greater in intercrops than in sole crops, indicating also an average difference in N sequestration rate of 45 kg N ha^{-1} yr^{-1}.

Other studies reported that the mix of residues from intercropping can accelerate or inhibit decomposition depending on the nature of the resulting residue mixture, their biochemical composition, and their C/N ratio. Cong et al. (2015a) reported different decomposition rate trends depending on the crops involved in the intercropping. SOM in strip intercrop plots decomposed faster than in monocrop plots with a lower soil C/N ratio of the mixed residues. Root litter mixtures of maize (*Zea mays*) and wheat (*Triticum aestivum*) decomposed as expected from single litters, but a litter mixture of maize and faba bean (*Vicia faba*) decomposed faster than expected.

5.7.2.6 Organic Amendments and Composting

In low-input agricultural systems, the return of organic materials, such as manure, to the soil represents an important source of nutrients and contributes to enhancing SOM contents and improving soil physical, chemical, and biological properties. Vågen et al. (2005) reported that the use of manure in East Sudanian savanna in combination with crop residues results in an increase of SOC of about 0.55 Mg C ha^{-1} yr^{-1} (Table 5.3).

Composting is an important organic practice and an excellent means for adaptation to climate change. Composting involves a good mix of OM of various sources (manure, crop residues, and other organic waste) so that it can be broken down into a compost that can be used as a soil amendment. Composting can have a rapid effect on SOC increase (Calderon et al. 2017; Demelash et al. 2014). Composting has the advantage of ensuring the removal of phytopathogenic agents. The composition of composts vary widely depending on the source of organic materials involved, as their humic compounds and their organic C and N content affect their decomposition in the soil.

Most farming systems in Africa are mixed crops and livestock and often generate substantial manure and other organic wastes. Despite the low production of such outputs, their composting with residues from other sources can provide a valuable product for improving soil fertility and enhancing soil C storage. According to Lal (1999), soil amendment with compost would sequester from 0.1 to 0.3 Mg ha^{-1} yr^{-1} in dryland soils.

5.7.2.7 Fertilizer Use and Soil Nutrient Status

Fertilizer use in Africa is among the lowest worldwide, which contributes to nutrient mining (Roy and Nabhan 1999; Henao and Baanante 2006; Bouabid et al. 2018). Crop fertilization is a key to enhancing crop yields. There is overwhelming evidence for the need for increased nutrient inputs in the region, and for the necessity of fertilizer use (mainly N) if crop yields are to be enhanced (Henao and Baanante 2006; Sanchez et al. 2007). Any increase in crop production means an increase in biomass, which represents a potential residue source for C sequestration. Therefore, adequate fertilizer use combined with soil water management (in fields that are either rainfed or under irrigation) is an effective way for increasing biomass production and SOC stock. A recent review by Oladele and Braimoh (2014) showed that the average impact of fertilizer application on the carbon stock in Africa was estimated to be about 626 kg C ha^{-1} yr^{-1}.

Han et al. (2016) conducted a global meta-analysis of SOC changes under different fertilizer managements, including unbalanced application of chemical fertilizers (UCF), balanced application of chemical fertilizers (CF), chemical fertilizers with straw application (CFS), and chemical fertilizers with manure application (CFM). Their results show that topsoil organic carbon (C) increased by about 0.9, 1.7, 2.0, and 3.5 g kg^{-1} under UCF, CF, CFS and CFM, respectively, corresponding to

TABLE 5.3

Changes in SOC for Different Land Use Conversions in Sub-Saharan Africa, Based on Selected Long-Term Studies

Region	Change in Land Use From	To	Mean	Min.	Max.	SOC Gains[a]	N
			____ Mg C ha⁻¹ yr⁻¹ ____				
HS	Natural forest	Cultivated	−0.90	−1.00	−0.80		2
		Fallow	−0.57	−1.14	0.00		2
	Cultivated	Fallow	1.06	0.23	2.77	100.0	9
		Afforestation	0.12	−0.29	0.56	71.4	21
WSS	Cultivated	Cultivated CR	0.19	—	—		1
		Cultivated F	0.05	—	—		1
		Cultivated NT	0.33	−1.00	1.30	66.7	6
		Fallow	1.37	0.10	5.30	100.0	10
		Afforestation	0.07	−0.98	0.57	50.0	9
	Cultivated M	Fallow	−0.14	—	—		1
	Fallow	Cultivated	−0.11	−0.18	−0.06		2
	Savanna	Cultivated	0.05	0.00	0.15	33.3	5
		Cultivated C	−0.12	−0.15	−0.09		3
		Cultivated CR	−0.06	−0.13	0.15	20.0	2
		Cultivated F	0.09	0.00	0.20	66.7	5
		Cultivated M	0.04	−0.52	0.94	33.3	3
ESS	Savanna	Cultivated	−2.77	−5.30	−0.77		3
	Woodland	Cultivated	0.36	—	—		4
		Cultivated M	0.55	—	—		1
SA	Savanna	Cultivated	−0.82	−1.26	−0.40		4
		Fallow	0.00	—	—		1
		Pasture	0.05	−0.31	0.40	50.0	2
	Pasture	Afforestation	−0.16	−0.19	−0.13		2
	Fallow	Cultivated	−0.75	—	—		1

Source: Vågen, T.G. et al., *Land Degrad. Dev.*, 16, 53–71, 2005.

Studies sources: Agbenin and Goladi (1997); Aweto (1981); Bationo et al. (2000); Dominy and Haynes (2002); Drechsel et al. (1991); Feller et al. (1981); Glaser et al. (2001); Hartemink (1970, 1995); Impala (2001); Juo et al. (1995); Lal (2000); Manlay (2000); Materechera and Mkhabela (2001); Morris and Gray (1984); Onim et al. (1990); Pieri (1989); Solomon et al. (2000, 2002); Trouve et al. (1994).

Note: HS = Humid and Subhumid; WSS = West Sudanian Savanna; ESS = East Sudanian Savanna; SA = Southern Africa; Cultivated M = w/manure; Cultivated C = w/cover crops; Cultivated CR = w/crop residues; Cultivated F = NPK fertilizers only; Cultivated NT = no-till; N = number of observations.

[a] Percent of observations with net gain (>0).

relative SOC increases of 10%, 15.4%, 19.5%, and 36.2%, respectively. The C sequestration durations were estimated to be 28–73 years under CFS and 26–117 years under CFM, but with high variability across climatic regions. At least 2.0 Mg ha⁻¹ yr⁻¹ C input is needed to maintain the SOC in about 85% of the cases examined. They highlighted a great C sequestration potential of applying CF, and observed that adopting CFS and CFM is highly important for either improving or maintaining SOC stocks across all agroecosystems. Liu and Zhou (2017) reported that SOC built up quickly in manure and manure plus N-P fertilizer treatments compared to sole use of N-P fertilizers in newly built terraced lands in semiarid highlands of northern China. The rate of increase of SOC decreased with time and tended to stabilize at about 3.4 g kg^{-1}y^{-1} after a cumulative input of manure equivalent to 14 Mg C ha^{-1} over a six-year period.

5.7.2.8   Afforestation and Rangeland Rehabilitation

Deforestation and itinerant cultivation cause major damage to soils in Africa. Loss of SOC from forested land can occur rapidly after deforestation. Vågen et al. (2005) compiled changes in SOC for different land use conversions in different ecosystems of sub-Saharan Africa (Table 5.3). The results show that the shifts toward cultivation, especially from forest and savanna, dramatically affects the SOC stock and can attain −0.9 Mg C ha^{-1} yr^{-1}.

Restoring appreciable levels of SOC after afforestation may take years, especially in drylands. When consequent degradation is still at a reversible stage, afforestation associated with conservation techniques can help restore the degraded land and bring back SOC to a satisfactory level. Nosetto et al. (2006) reported that after 15 years of afforestation with pine (*Pinus ponderosa*) trees on degraded drylands in Argentina, more than 50% C was added to the initial ecosystem C pool, with annual sequestration rate ranging from 0.5 to 3.3 Mg C ha^{-1} year^{-1}. The C gains in afforested stands were higher above than below ground (150% vs. 32%).

Liniger et al. (2011) reported that afforestation has high potential for C sequestration and is comparable to the use of conservation agriculture. Silvopasture systems with 50 trees per hectare can store 110 to 147 tons of CO$_2$eq per hectare. Lal (2004) underlined that afforestation, through the establishment of various types of tree plantations, has great potential for C sequestration in the tropics. The SOC accumulation rate under 18-year plantation of acacia in northern Senegal was about 0.03% yr^{-1} under the tree canopy and 0.02% yr^{-1} in open ground, corresponding to SOC sequestration rates of 420 and 280 kg C ha^{-1} yr^{-1}. Afforestation actions, mainly those involving participatory approaches, have indirect effects of providing alternative income and substitutes for wood fuel (Lepetu et al. 2015). Garrity et al. (2010) indicated that introducing the "ever green approach" of afforestation (integration of particular wood tree species into the annual food crop systems) in Africa can contribute significantly to the carbon sequestration potential of agroforestry systems, enhances the quality of degraded soils, and improves the livelihood of the populations. It is estimated that "evergreen agriculture" systems can accumulate C both above and belowground in the range of 2–4 Mg C ha^{-1} yr^{-1}.

Given the vast areas of deforested land suitable for afforestation in Africa, the potential of SOC from such practice is huge, assuming necessary investment funds are deployed (Garrity et al. 2010).

Rangelands in Africa, mainly in the savannas, represent vast ecosystems with extensive biodiversity and a high potential for carbon cycling and storage. The effect of species diversity on soil C and N stocks in these natural grasslands has been attributed to positive interactions among plant communities. However, grazing and cultivation are two key factors that can disturb the functioning and services of these ecosystems and affect their potential for carbon sequestration. Conant and Paustian (2002) reported that the total grassland in Africa amounts to about 838.2 × 10^6 ha, of which about 87.7 × 10^6 (10.4%) have been subject to overgrazing. They estimate a total potential carbon sequestration of 16.7 Tt C yr^{-1} for the African continent through cessation of overgrazing and rehabilitation, of which 16.7 Tt C yr^{-1} lies in the moderate to highly degraded lands. These figures represent about 37% of global rangeland areas.

Lipper et al. (2010) recognize two main reasons for looking into the potential of sequestering carbon in West African rangelands: (1) the degradation and depletion of carbon stocks in these systems has resulted in declining rangeland and agricultural productivity, in turn reducing the livelihood of the local population and increasing their impoverishment (Batjes 2004; Tieszen et al. 2004); and (2) increasing carbon stocks in the system can be not only a way of improving the ecological health and productivity of the livestock systems, but also a significant and low-cost way of mitigating climate change (Woomer et al. 2004). The per-hectare amounts in the rangelands of western Africa are low, but aggregate potential is high. Avoiding degradation and rehabilitating slightly degraded lands are the least costly and can generate significant reductions in carbon storage and emission. Bazin (2010) considers that the rangelands of the African savannas have great potential for SOC if proper animal load and rehabilitation measures are taken.

5.7.3 COMPARATIVE BENEFITS OF SUSTAINABLE LAND MANAGEMENT TECHNOLOGIES

Figures regarding carbon sequestration from different management practices vary largely across agroecosystems. It is very difficult to make direct comparisons of their specific benefits as they are affected by many physical, environmental, and socioeconomic factors. The change for a given practice in various areas can vary considerably.

A comprehensive assessment by World Bank (2012) on carbon sequestration by agricultural soils in Africa, Asia, and Latin America compared the climate benefits of sustainable land management technologies as measured by the net rate of carbon sequestration, adjusted for emissions associated with a series of land management technologies – a measurement referred to as the "abatement rate" (AR). The emissions associated with the technologies are classified as land emissions and process emissions. Land emissions are the difference between emissions of nitrous oxides and methane by conventional and improved practices, while process emissions refer to those arising from fuel and energy use. The abatement rate is expressed in tons of carbon dioxide equivalent t CO_2e ha^{-1} yr^{-1}.

In the case of Africa (Figure 5.10) AR estimates vary from 0.29 t CO_2e ha^{-1} yr^{-1} for chemical fertilizer use to 10.3 t CO_2e ha^{-1} yr^{-1} for the use of biochar. Figure 5.10 also shows that management techniques such as improved fallow, alley farming, afforestation, and tree crop farming have high AR rates (around 7.5 t CO_2e ha^{-1} yr^{-1}), compared to mulching, rotations, no tillage, and use of manure (1.3 to 1.8 t CO_2e ha^{-1} yr^{-1}). Intercropping and mixed cropping involving trees have moderate effects on AR (4 to 5.3 t CO_2e ha^{-1} yr^{-1}).

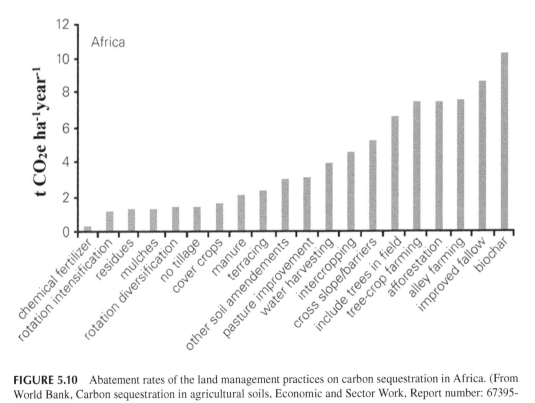

FIGURE 5.10 Abatement rates of the land management practices on carbon sequestration in Africa. (From World Bank, Carbon sequestration in agricultural soils, Economic and Sector Work, Report number: 67395-GLB, 2012.)

5.8 CONCLUSIONS

Storing C in soils or enriching it with OM is not only an option for mitigating or offsetting GHG emissions, but also for improving soil quality (water retention, aggregate stability and structure, biological activity, chelating capacity, attenuation of the alkalinity and sodicity, etc.) and in turn soil productivity. All aspects converge toward the ultimate goal of contributing to food security in the short and long terms. However, the current situation in African drylands is unanimously characterized by a low SOC content and a trend toward its decline if no measures are taken. This C decline is of particular concern in the arid and semiarid areas that are more prone to a decrease in productivity. Reversing this situation is synonymous with "triggering major changes in the farming systems" with sustained underlying supporting measures.

It is incontestable that the agricultural soils of Africa are both "victims" of climate change, but also potential "savers" in the fight against global warming, given their potential for colossal C storage at the continent scale. It is obvious that climate change, especially through a rise in temperature, can drastically modify the behavior of the soil, particularly by increasing the rate of mineralization of SOM. In addition, soil microbial dynamics can be disturbed, as extreme droughts inhibit the biological activity for C humification, while soil engorgement following heavy rain events may favor the emission of N_2O by denitrification and the loss of nitrate-nitrogen needed by crops. The alternating wetting and drying phenomena that occur with climatic disturbance can also modify the geochemical processes in the soil and lead to disturbance of its vital functions. Arid conditions can lead to salinity development and therefore negatively affect crop production and SOC accumulation. Other indirect effects of climate change on soils are reduced biomass production and therefore reduced vegetation cover and OM, both of which play an important role in protection against soil erosion.

The storage of C in the soil, as a form of C sequestration, is an irrefutable solution to compensate for GHG emissions. Theoretical mass balances are established; however, putting this concept to practice, especially in the context of African drylands, would require drastic modifications as well as introducing adapted innovations to the farming systems that take into consideration the multiple climatic, physical, and socioeconomic situations that can be encountered in this vast continent.

Considering African "soil potential," two initiatives are worth reflecting on regarding SOC sequestration and agricultural adaptation in the context of climate change: (1) The "4PM" initiative proposed by France at the COP21, which focuses on increasing soil carbon content annually by 0.4%. This is a very optimistic initiative that needs special conditions for African soils and their stakeholders. (2) The "3A initiative" (adaptation of African agriculture) proposed by Morocco at the COP22, which considers agricultural adaptation and food security as a prerequisite and an indirect way to contributing to SOC sequestration. Synergies can be established between these two initiatives, and eventually other initiatives, if the specific contexts for the various agroecological zones of Africa are taken into consideration.

Improving SOC sequestration in African dryland soils remains a challenge, as it requires answers to a series of questions, some of which were previously addressed by FAO (2012), but deserve readdressing based on the aspects and issues discussed in this chapter:

- What are the conditions that can guarantee the sustainability of C storage when agricultural practices are changed, considering the antagonistic effects of the processes of humification and mineralization?
- What is the maximum and feasible level of SOC sequestration, taking into account the natural balance of C in the conditions of dryland soils of Africa?
- How to act, in an effective way, on humification in African drylands in the present warming trends (favorable to mineralization), the extreme phenomena (droughts, floods, etc.), that limit biomass production and consequently C sequestration?

- Which cropping systems to advocate, and what is their socioeconomic feasibility?
- Which monitoring system to adopt and at what scale?
- What is the perception of small African farmers (stakeholders) with respect to proposed agroecological innovations?
- What emissions associated with increased fertilizer use, irrigation, or livestock that need to be integrated into carbon sequestration estimates?
- What is the added value for farmers of applying new farming systems in favor of carbon sequestration, and what support measures are required (access to inputs, technical advisory, access to market, etc.)?

LIST OF ABBREVIATIONS

C	carbon
CEC	Cation Exchange Capacity
FAO	Food and Agriculture Organization
GHG	greenhouse gazes
g	gram
kg	kilogram
Pg	pictogram
ha	hectare
N	nitrogen
N-F	nitrogen fixation
OM	organic matter
P	phosphorus
SOM	soil organic matter
r	coefficient of correlation

REFERENCES

Agbenin, J. O., and J. T. Goladi. 1997. Carbon, nitrogen and phosphorus dynamics under continuous cultivation as influenced by farmyard manure and inorganic fertilizers in the savanna region of northern Nigeria. *Agriculture Ecosystems and Environment* 63: 17–24.

Allison, F. E. 1955. Does nitrogen applied to crop residues produce more humus? *Soil Science Society of America, Proceedings* 19: 210–211.

Aweto, A. O. 1981. Secondary succession and soil fertility restoration in southwestern Nigeria. II. Soil fertility restoration. *Journal of Ecology* 69: 609–614.

Ayanlaja, S. A., and J. O. Sanwo. 1991. Management of soil organic matter in the farming systems of the low land humid tropics of West Africa: A review. *Soil Technology* 4(3): 265–279.

Bationo, A., S. P. Wani, C. L. Bielders, P. L. G. Vlek, and A. U. Mokwunye. 2000. Crop residue and fertiliser management to improve soil organic carbon content, soil quality and productivity in the desert margins of West Africa. In *Global Climate Change and Tropical Ecosystems*, R. Lal, J. M. Kimble, B. A. Stewart (eds.). CRC Press, Boca Raton, FL, pp. 117–145.

Batjes, N. H. 2004. Estimation of soil carbon gains upon improved management within croplands and grasslands of Africa. *Environment Development and Sustainability* 6: 133–143.

Bazin, F. 2010. Contribution de l'agriculture africaine au changement climatique et potentiel d'atténuation. *Grain de sel*, Dossier No. 49, January–March 2010.

Beedy, T. L., S. S. Snapp, F. K. Akinnifesi, and G. W. Sileshi. 2010. Impact of *Gliricidia sepium* intercropping on soil organic matter fractions in a maize-based cropping system. *Agriculture, Ecosystems & Environment* 138(3–4): 139–146.

Bengtston, G., P. Bengtson, and K. F. Mansson. 2003. Gross nitrogen mineralization-, immobilization-, and nitrification-rates as a function of soil C:N ratio and microbial activity. *Soil Biology and Biochemistry* 35(1): 143–154.

Berdai, H., N. Aghzar, F. Z. Cherkaoui, and B. Soudi. 2002. Azote minéral résiduel et son évolution pendant l'été en fonction du précédent cultural en climat méditerranéen. *Étude et Gestion des Sols (EGS)* 9: 7–23.

Biancalani, R., M. Petriet, and S. Bunning. 2015. Land use, land degradation, and sustainable and management in the drylands of sub-Saharan Africa. Unpublished document non. FAO, Rome.

Bouabid R., B. Soudi, and M. Badraoui. 2020. Nitrogen dynamics and management in rainfed drylands: Issues and challenges. In *Soils and Fertilizers: Advances in Soil Sciences*, Lal, R. (ed.). Present volume. CRC Press, Boca Raton, FL.

Bouajila, K., F. Ben Jeddi, H. Taamallah, N. Jedidi, and M. Sanaa. 2014. Effets de la composition chimique et biochimique des résidus de cultures sur leur décomposition dans un sol Limono-Argileux du semi aride [Chemical and biochemical composition's effect of crop residues on their decomposition in semi-arid limono-clay soil]. *Journal of Materials and Environmental Science* 5: 159–166.

Bremer, E., Ellert, B. H., and Janzen, H. H. 1995. Total and light-fraction carbon dynamics during four decades after cropping changes. Soil Science Society of America Journal 59(5): 1398.

Calderón, F. J., M. F. Vigil, and J. Benjamin. 2017. Compost input effects on dryland wheat and forage yields and soil quality. *Pedosphere* 28(3): 451–462.

Campbell, C. A., Y. W. Jame, and G. E. Winkleman. 1984. Mineralization rate constants and their use for estimating nitrogen mineralization in some Canadian prairie soils. *Canadian Journal of Soil Science* 64: 333–343.

Campbell, E. E., and K. Paustian. 2015. Current developments in soil organic matter modeling and the expansion of model applications: A review. *Environmental Research Letters* 10(12): 123004.

Chen, B., E. Liu, Q. Tian, C. Yan, and Y. Zhang. 2014. Soil nitrogen dynamics and crop residues: A review. *Agronomy for Sustainable Development* 34(2): 429–442.

Chomitz, K. M. 1999. Evaluating carbon offsets from forestry and energy projects: How do they compare? World Bank Policy Research Working Paper No. 2357, Washington, DC: The World Bank.

Conant, R. T., and K. Paustian. 2002. Potential soil carbon sequestration in overgrazed grassland ecosystems. *Global Biogeochemical Cycles* 16(4): 90–91.

Cong, W. F., E. Hoffland, L. Li, B. H. Janssen, and W. van der Werf. 2015a. Intercropping affects the rate of decomposition of soil organic matter and root litter. *Plant Soil* 391: 399–411.

Cong, W. F., E. Hoffland, L. Li, J. Six, J. H. Sun, X. G. Bao, F. S. Zhang, and W. Van der Werf. 2015b. Intercropping enhances soil carbon and nitrogen. *Global Change Biology* 21(4): 1715–1726.

Datta, R., Kelkar, A., Baraniya, D., Molaei, A., Moulick, A., Meena, R., and Formanek, P. 2017. Enzymatic degradation of lignin in soil: A review. *Sustainability* 9(7): 1163(1–18).

Davidson, E. A., and Ackerman, I. L. 1993. Changes in soil carbon inventories following cultivation of previously untilled soils. Biogeochemistry 20: 161–193.

Demelash, N., Bayu, W., Tesfaye, S., Ziadat, F., and Sommer, R. 2014. Current and residual effects of compost and inorganic fertilizer on wheat and soil chemical properties. Nutrient Cycling in Agroecosystems 100(3): 357–367.

Dominy, C. S., and R. J. Haynes. 2002. Influence of agricultural land management on organic matter content, microbial activity and aggregate stability in the profiles of two Oxisols. *Biology and Fertility of Soils* 36: 298–305.

Drechsel, P., B. Glaser, and W. Zech. 1991. Effect of four multipurpose tree species on soil amelioration during tree fallow in central Togo. *Agroforestry Systems* 16: 193–202.

Duchaufour, P. 1976. Dynamics of organic matter in soils of temperate regions: Its action on pedogenesis. *Geoderma* 15(1): 31–40.

Dyer, L., Oelbermann, M., and Echarte, L. 2012. Soil carbon dioxide and nitrous oxide emissions during the growing season from temperate maize-soybean intercrops. *Journal of Plant Nutrition and Soil Science* 175: 394–400.

Elliott, E. T., and D. C. Coleman. 1988. Let the soil work for us. In *Ecological Implications of Contemporary Agriculture: Proceedings of the 4th European Ecology Symposium 7–12 September 1986, Wageningen*, H. Eijsakkers and A. Qispel (eds.), Ecological Bulletins No. 39, pp. 23–32.

FAO (Food and Agriculture Organization). 2002. *Rapport sur les ressources en sols du monde – la séquestration du carbone dans le sol pour une meilleure gestion des terres: Options de gestion du sol pour la séquestration du carbone.* FAO, Rome, Italy.

FAO (Food and Agriculture Organization). 2003. *State of the World Forests.* FAO, Rome, Italy.

FAO (Food and Agriculture Organization). 2004. Carbon Sequestration in Dryland Soils. World Soils Resources Reports No. 102. FAO. 2004. Rome, Italy. 129 pp.

FAO (Food and Agriculture Organization). 2005. *The Importance of Soil Organic Matter: Key to Drought-Resistant Soil and Sustained Food Production*, A. Bot and J. Benites (eds.). FAO Soils Bulletin 80. FAO, Rome, Italy.

FAO (Food and Agriculture Organization). 2010. *Pour une agriculture intelligente face au climat: Politiques, pratiques et financements en matière de sécurité alimentaire, d'atténuation et d'adaptation.* FAO, Rome, Italy.

FAO (Food and Agriculture Organization). 2012. *Identifying Opportunities for Climate-Smart Agriculture Investments in Africa*. FAO, Rome, Italy.

Feller, C., F. Ganry, and M. Cheval. 1981. Décomposition et humification des reésidus végétaux dans un agrosystème tropical. I. Influence d'une fertilisation azotée (urée) et d'un amendement organique (compost) sur la reépartition du carbone et de l'azote dans différents compartiments d'un sol sableux. *Agronomie Tropicale* 36(1): 9–25.

Flower, T. N. and J. R. Q. Challagha. 1983. Nitrification in soils incubated with pig slurry or ammonium sulphate. *Soil Biology and Biochemistry* 15(3): 337–342.

Fog, K. 1988. The effect of added nitrogen on the rate of decomposition of organic matter. *Biological Reviews* 63(3): 433–462.

Garrity, D. P., F. K. Akinnifesi, O. C. Ajayi, S. G. Weldesemayat, J. G. Mowo, A. Kalinganire, M. Larwanou and J. Bayala. 2010. Evergreen agriculture: A robust approach to sustainable food security in Africa. *Food Security* 2: 197.

Glaser, B., J. Lehmann, M. Führböter, D. Solomon, and W. Zech. 2001. Carbon and nitrogen mineralization in cultivated and natural savanna soils of Northern Tanzania. *Biology and Fertility of Soils* 33: 301–309.

Guggenberger, G. 2005. Humification and mineralization in soil. In *Microorganisms in Soils: Roles in Genesis and Functions*, F. Buscot and A. Varma (eds.), Springer-Verlag, Berlin Heidelberg, Germany. *Soil Biology*, 3: 85–106.

Haas, H. J., and C. E. Evans. 1957. *Nitrogen and Carbon Changes in Great Plains Soils as Influenced by Cropping and Soil Treatments*. Technical Bulletin No. 1164. United States Department of Agriculture, Washington, DC.

Han, P., Zhang, W., Wang, G., Sun, W., and Huang, Y. 2016. Changes in soil organic carbon in croplands subjected to fertilizer management: A global meta-analysis. *Scientific Reports* 6(1).

Hartemink, A. E. 1995. Soil fertility decline under sisal cultivation in Tanzania. Technical Paper No. 28. ISRIC, Wageningen, the Netherlands.

Hartemink, A. E. 1997. Input and output of major nutrients under monocropping sisal in Tanzania. *Land Degradation & Development* 8(4): 305–310.

Hassink, J. 1994. Active organic matter fractions and microbial biomass as predictors of N mineralization. *European Journal of Agronomy* 3(4): 257–265.

Henao, J., and C. Baanante. 2006. *Nutrient Mining in Africa: Implications for Resource Conservation and Policy Development*. Summary report. March 2006. International Fertilizer Development Center (IFDC), Alabama, USA.

Hénin, S., and M. Dupuis. 1945. Bilan de la matière organique des sols. *Annales Agronomiques* 1: 17–29.

Hénin, S., G. Monnier, and L. Turc. 1959. Un aspect de la dynamique des matières organiques du sol. C. *Journal of the Academy of Marketing Science* 248: 138–141.

Hu, F., Q. Chai, A. Yu, W. Yin, H. Cui, and Y. Gan. 2015. Less carbon emissions of wheat – Maize intercropping under reduced tillage in arid areas. *Agronomy for Sustainable Development* 35(2): 701–711.

IMPALA. 2001. *First Annual Report of IMPALA Project, Covering the Period October 2000–December 2001*. IMPALA, Nairobi, Kenya.

Janssen, B. H. 1996. Nitrogen mineralization in relation to C:N ratio and decomposability of organic materials. *Plant Soil* 181: 47–56.

Janzen, H. H., C. A. Campbell, B. H. Ellert, and E. Bremer. 1997. Soil organic matter dynamics and their relationship to soil quality. In *Soil Quality for Crop Production and Ecosystem Health*, E. G. Gregorich and M. R. Carter (eds.), Vol. 25 of *Developments in Soil Science*. Elsevier, Amsterdam, the Netherlands, pp. 277–291.

Jarecki, M. M., and R. Lal. 2003. Crop management for soil carbon sequestration. *Critical Reviews in Plant Sciences* 22(5): 471–502.

Jedidi, N. 1998. Minéralisation et humification des amendements organiques dans un sol limono-argileux Tunisien. Thèse Ph.D., Univesité Gent, Belgique, 180pp.

Jenny, H. 1941. *Factors of Soil Formation*. McGraw-Hill, New York.

Jensen, H. L. 1929. On the influence of the carbon: Nitrogen ratios of organic material on the mineralization of nitrogen. *Journal of Agricultural Science* 19: 71–82.

Jones, A., Breuning-Madsen, H., Brossard, M., Dampha, A., Deckers, J., Dewitte, et al. (eds.) 2013. *Soil Atlas of Africa*. European Commission, Publications Office of the European Union, Luxembourg.

Juo, A. S. R., K. Franzluebbers, A. Dabiri, and B. Ikhile. 1995. Changes in soil properties during long-term fallow and continuous cultivation after forest clearing in Nigeria. *Agriculture, Ecosystems and Environment* 56: 9–18.

Ladd, J. N., M. Van Gestel, L. Jocteur Monrozier, and M. Amato. 1996. Distribution of organic 14C and 15N in particle-size fractions of soils incubated with 14C, 15N-labelled glucose/NH4, and legume and wheat straw residues. *Soil Biology and Biochemistry* 28(7): 893–905.

Lal, R. 1999. Global carbon pools and fluxes and the impact of agricultural intensification and judicious land use. In *Prevention of Land Degradation, Enhancement of Carbon Sequestration and Conservation of Biodiversity through Land Use Change and Sustainable Land Management with a Focus on Latin America and the Caribbean*. World Soil Resources Report 86. FAO, Rome, pp. 45–52.

Lal, R. 2000. Restorative effects of *Mucuna utilis* on soil organic C pool of a severely degraded Alfisol in western Nigeria. In *Global Climate Change and Tropical Ecosystems*, R. Lal, J. M. Kimble, and B. A. Stewart BA. CRC Press, Boca Raton, FL, pp. 147–165.

Lal, R. 2002. Carbon sequestration in dryland ecosystems of West Asia and North Africa. *Land Degradation and Development* 13(1): 45–59.

Lal, R. 2004a. Soil carbon sequestration for sustaining agricultural production and improving the environment with particular reference to Brazil. *Journal of Sustainable Agriculture* 26: 23–42.

Lal, R. 2004b. Soil carbon sequestration impacts on global climate change and food security. *Science* 304: 1623–1627.

Lal, R. 2016. Soil health and carbon management. *Food and Energy Security* 5(4): 212–222.

Laudelout, H. 1993. Bilan de la matière organique du sol: le modèle de Hénin (1945). In *Mélanges offerts à Stéphane Hénin: Sol – agronomie – environnement*. ORSTOM, Paris, pp. 117–123.

Laudelout, H., and J. Meyer. 1951. Temperature characteristics of the microflora of Central African soils. *Nature* 168: 791–792.

Laudelout, H., and J. Meyer. 1954. Les cycles d'éléments minéraux et de matière organique en forêt équatoriale congolaise. In *Actes et Comptes rendus, Ve Congrès International de Science du Sol*. 16–21 Août 1954, Léopoldville (Kinshasa) Zair, Volume II, pp. 267–272. International Society of Soil Science. Bruxelles, Belgique.

Laudelout, H., J. Meyer, and A. Peeters. 1960. Les relations quantitatives entre la teneur en matière organique du sol et le climat. *Agricultura* 8: 103–140.

Lepetu, J., I. Nyoka, and O. I. Oladele. 2015. Farmers' planting and management of indigenous and exotic trees in Botswana: Implications for climate change mitigation. *Environmental Economics* 6(3): 20–30.

Li, L. M. 1990. Nitrification. In *Nitrogen in Soils of China*, Z. L. Zhou and Q. X. Wen (eds.). Jiangsu Science and Technology Publishing House, Nanjing, China, pp. 94–122.

Liao, Y., Wu, W. L., Meng, F.Q., Smith, P., and Lal, R. 2015. Increase in soil organic carbon by agricultural intensification in northern China. *Biogeosciences* 12(5): 1403–1413.

Liao, Y., Wu, W. L., Meng, F. Q., and Li, H. 2016. Impact of agricultural intensification on soil organic carbon: A study using DNDC in Huantai County, Shandong Province, China. *Journal of Integrative Agriculture* 15(6): 1364–1375.

Liniger, H. P., R. Mekdaschi Studer, C. Hauert, and M. Gurtner. 2011. *Sustainable Land Management in Practice – Guidelines and Best Practices for Sub-Saharan Africa*. TerrAfrica, World Overview of Conservation Approaches and Technologies (WOCAT) and Food and Agriculture Organization of the United Nations (FAO).

Lipper, L., Dutilly-Diane, C., and McCarthy, N. 2010. Supplying carbon sequestration from West African rangelands: Opportunities and barriers. Rangeland Ecology & Management 63(1): 155–166.

Liu, C. A., and L. M. Zhou. 2017. Soil organic carbon sequestration and fertility response to newly-built terraces with organic manure and mineral fertilizer in a semi-arid environment. *Soil and Tillage Research* 172: 39–47.

Mann, L. K. 1986. Changes in soil carbon storage after cultivation. *Soil Science* 142: 279–288.

Manlay, R. J. 2000. Dynamique de la matière organique à l'échelle d'un terroir agro-pastoral de savane ouest-africaine (Sud-Sénégal). PhD thesis, Ecole Nationale du Génie Rural, des Eaux et Forêts Centre de Montpellier (ENGREF).

Martin, J. P., and Haider, K. 1971. Microbial activity in relation to soil humus formation. *Soil Science* 111: 54–63.

Materechera, S. A., and T. S. Mkhabela. 2001. Influence of land-use on properties of a ferralitic soil under low external input farming in southeastern Swaziland. *Soil and Tillage Research* 62: 15–25.

Mercer, D. E. 2004. Adoption of agroforestry innovations in the tropics: A review. *Agroforestry Systems* 61–62(1): 311–328.

Monties, B. 1980. *Les polymères végétaux. Polymères pariétaux et alimentaires non azotés*. Gauthier-Villars, Paris.

Morris, A. R., and D. C. Gray. 1984. A comparison of soil nutrient levels under grassland and two rotations of *Pinus patula* in the Usutu Forest, Swaziland. In *Proceedings of the IUFRO Symposium on Site and Productivity of Fast Growing Plantations*, A. P. G. Schonau, C. J. Schutz (eds.). South African Forest Research Institute, Pretoria, South Africa, pp. 881–892.

Mustin, M. 1987. *Le compost: Gestion de la matière organique*, François Dubux (ed.), ch. 12, pp. 743–765.

Myers, H. E., A. L. Hallsted, J. B. Kuska, and H. J. Haas. 1943. Nitrogen and carbon changes in soils under low rainfall as influenced by cropping systems and soil treatment. *Kansas Agricultural Experiment Station* 56: 1–52.

Nair, P. K. R. 1993. *An Introduction to Agroforestry*. Kluwer Academic, Dordrecht, the Netherlands.

Naman, F., B. Soudi, and C. N. Chiang 2002. Impact de l'intensification agricole sur le statut de la matière organique des sols en zones irriguées semi-arides au Maroc. *Etude et Gestion des Sols* 8(4): 269–277.

Naman, F., B. Soudi, C. El Adlouni, and C. N. Chiang. 2015. Humic balance of soils under intensive farming: The case of soils irrigated perimeter of Doukkala in Morocco. *Journal of Materials and Environmental Science* 6: 3574–3581.

Naman, F., B. Soudi, C. N. Chiang, and C. E. L. Adlouni. 2018. Evolution of carbon and nitrogen biomass of vertisol and fersiallitic soil after previous cultivation of wheat and sugar beet in the irrigated perimeter of the Doukkala in Morocco. *Journal of Materials and Environmental Science* 9(5): 1544–1550.

Nguyen, K. 1982. Mise en évidence, par étude microstructurale et microanalyse X, de l'effet stabilisateur des argiles gonflantes au cours des processus de biodégradation. *Pédologie* 32: 175–192.

Nosetto, M. D., Jobbágy, E. G., and Paruelo, J. M. 2006. Carbon sequestration in semi-arid rangelands: Comparison of Pinus ponderosa plantations and grazing exclusion in NW Patagonia. *Journal of Arid Environments* 67(1): 142–156.

Ogunkunle, A. O., and Eghaghara, O. O. 1992. Influence of land use on soil properties in a forest region of Southern Nigeria. *Soil Use and Management* 8(3): 121–125.

Oladele, O. I., and A. K. Braimoh. 2013. Climate change mitigation potential of tree crop and alley farming practices in Africa. *Asia Life Sciences* 9(20): 185–119.

Oladele, O. I., and A. K. Braimoh. 2014. Potential of agricultural land management activities for increased soil carbon sequestration in Africa – A review. *Applied Ecology and Environmental Research* 12(3): 741–751.

Onim, J. F. M., M. Mathuva, K. Otieno, and H. A. Fitzhugh. 1990. Soil fertility changes and response of maize and beans to green manures of Leucaena, Sesbania and pigeonpea. *Agroforestry Systems* 12: 197–215.

Pieri, C. 1989. *Fertilité des terres de savane. Bilan de trente ans de recherche et de développement agricoles au sud du Sahara.* Ministeere de la Coopeeration, CIRAD, Paris.

Plaza-Bonilla, D., Arrúe, J. L., Cantero-Martínez, C., Fanlo, R., Iglesias, A., and Álvaro-Fuentes, J. 2015. Carbon management in dryland agricultural systems. A review. *Agronomy for Sustainable Development* 35(4): 1319–1334.

Power, J. F., and J. O. Legg. 1978. *Crop Residue Management Systems*. In W. R. Oshwald (ed). ASA Special Publication, American Society of Agronomy, Madison, WI, pp. 85–100.

Roy, R. N., and H. Nabhan. 1999. Soil and nutrient management in sub-Saharan Africa in support of the soil fertility initiative. *Proceedings of the Expert Consultation Lusaka, Zambia 6–9 December 1999*. FAO, Rome, Italy.

Salter, R. M. and T. C. Green. 1933. Factors affecting the accumulation and loss of nitrogen and organic carbon in cropped soils. *American Society of Agronomy* 25: 622–630.

Sanchez, P., C. Palm, J. Sachs, J. Denning, G. Flor, R. Harawa, et al. 2007. The African millennium villages. *Proceedings of the National Academy of Sciences of the United States of America* 104(43): 16775–16780.

Smith, R. M., G. Samuel, and C. F. Cernuda. 1951. Organic matter and nitrogen build-ups in some Puerto Rican soil profiles. *Soil Science* 72(6): 409–427.

Solomon, D., Fritzsche, F., Lehmann, J., Tekalign, M., and Zech, W. 2002. Soil organic matter dynamics in the sub-humid agro-ecosystems of the Ethiopian Highlands: Evidence from natural 13C abundance and particle size fractionation. *Soil Science Society of America Journal* 66: 969–978.

Solomon, D., J. Lehman, and W. Zech. 2000. Land-use effects on soil organic matter properties of chromic luvisols in semi-arid northern Tanzania: Carbon, nitrogen, lignin and carbohydrates. *Agriculture, Ecosystems and Environment* 78: 203–213.

Somarriba, E., R. Cerda, L. Orozco, M. Cifuentes, H. Dávila, T. Espin, et al. 2013. Carbon stocks and cocoa yields in agroforestry systems of Central America. *Agriculture, Ecosystems and Environment* 173: 46–57.

Soudi, B., C. N. Chiang, and M. Zraouli. 1990. Seasonal variation of mineral nitrogen and the combined effect of soil temperature and moisture content on mineralization. *Actes de l'Institut Agronomique et Vétérinaire Hassan II* 10(1): 29–38, ref. 16.

Soudi, B., F. Naman and C. N. Chiang. 2000. Problématique de gestion de la matière organique des sols: Cas des périmètres irrigués du Tadla et des Doukkala. In *Actes du Séminaire "Intensification agricole et qualité des sols et des eaux"*. B. Soudi, C. Chiang, M. Badraoui, M. Agbani, and J.M. Marcoen (eds.), Rabat 2–3 November 2000. Actes Editions, Rabat, Morocco, pp. 25–30.

Swift, M. J., O. W. Heal, and M. J. Anderson. 1979. *Decomposition in Terrestrial Ecosystems*. Blackwell Scientific Publications, Oxford, UK.

Tate, R. L. 1995. *Soil Microbiology*. New York, John Wiley.

Theng, B. K. G., Churchman, G. J., and Newman, R. H. 1986. The occurrence of interlayer clay-organic complexes in two New Zealand soils. *Soil Science* 142: 262–266.

Tieszen, L. L., Tappan, G. G., and Touré, A. 2004. Sequestration of carbon in soil organic matter in Senegal: An overview. *Journal of Arid Environments* 59: 409–425.

Traoré, O., N. A. Somé, K. Traoréa, and K. Somda K. 2007. Effect of land use change on some important soil properties in cotton-based farming system in Burkina Faso. *International Journal of Biological and Chemical Sciences* 1(1): 7–14.

Trouve, C., A. Mariotti, D. Schwartz, and D. Guillet. 1994. Soil organic carbon dynamics under Eucalyptus and Pinus planted on savannahs in the Congo. *Soil Biology and Biochemistry* 26: 287–295.

Unruch, J. D., R.A Houghton, and P. A. Lefevre. 1993. Carbon storage in agroforestry: An estimate for sub-Saharan Africa. *Climate Research* 3: 39–52.

Vågen, T. G., R. Lal, and B. R. Singh. 2005. Soil carbon sequestration in sub-Saharan Africa: A review. *Land Degradation and Development* 16(1): 53–71.

Vigil, M. F., and D. E. Kissel. 1991. Equations for estimating the amount of nitrogen mineralized from crop residues. *Soil Science Society of America* 5: 757–761.

Waksman, S. A. 1924. Influence of microorganisms upon the carbon-nitrogen ratio in soil. *Journal of Agricultural Science* 14: 555–562.

Wise, R., and O. Cacho. 2005. A bioeconomic analysis of carbon sequestration in farm forestry: A simulation study of *Gliricidia sepium*. *Agroforestry Systems* 64: 237–250.

Woomer, P. L., Touré, A., and Sall, M. 2004. Carbon stocks in Senegal's Sahel transition zone. *Journal of Arid Environment* 59: 499–510.

World Bank. 2012. *Carbon Sequestration in Agricultural Soils. Economic and Sector Work*. Report 67395-GLB.

Zech, W., Senesi, N., Guggenberger, G., Kaiser, K., Lehmann, J., Miano, T. M., and Schroth, G. 1997. Factors controlling humification and mineralization of soil organic matter in the tropics. *Geoderma* 79(1–4): 117–161.

6 Manures versus Fertilizers in Rainfed Dryland Production Systems of India

Ch. Srinivasarao, Sumanta Kundu, C. Subha Lakshmi, M. Vijay Sankar Babu, V. V. Gabhane, Pallab K. Sarma, Ayyappa Sathish, K. C. Nataraj, and H. Arunakumari

CONTENTS

6.1 INTRODUCTION

India is ranked number one in the world as a rainfed agroecosystem in extent of area and value of produce. The rainfed climate is largely semiarid and dry subhumid with a short wet season followed by a long dry season. In dryland areas in general, the rainfall is low and highly variable with uneven distribution, which results in uncertain crop yields. The late onset of monsoon season in drylands delays timely sowing of crops, which ultimately leads to a reduction in crop yield. In the event of an early cessation of rains, crops experience drought during flowering and maturity stages, which results in substantial yield loss. In drylands, variations in temperature accelerate crop development, leading to forced maturity during periods of moisture stress and drought. Dryland areas are highly prone to soil degradation, especially due to soil erosion. The small size of land holdings (less than 2 ha), which are usually fragmented and scattered, frequent crop failure, poor socioeconomic condition, lack of infrastructure to boost production, and reduced access to market facilities are some of the problems associated with drylands. The rainfed agroecosystem is prone to several issues in regard to edaphic, climatic, and social factors. Frequent droughts occurring due to high variability in the quantum and distribution of rainfall, poor soil health, low and imbalanced fertilizer use, poor mechanization, low risk-bearing capacity, low credit availability, and infrastructure constraints have contributed to stagnation in productivity levels (Srinivasarao et al. 2015). Consequently, farmers are diverting from agriculture and tending to migrate to cities in search of an alternative source of livelihood. The productivity and overall net returns of rainfed crops therefore need to be boosted in order to retain the farming community in the field of agriculture to meet the growing demands of the ever-increasing population. The primary factors influencing the productivity and sustainability of these agroecosystems based on rainfall are soil health and moisture stress. The total land area coverage in India is 329 M ha, of which only 143 M ha is arable land, in which 57% of the net sown area is under rainfed agriculture system, contributing about 44% of food grains produced in the country and feeding about 40% of nation's population. Even after realization of the full irrigation potential of the country, 50% of the net sown area will continue to be under a rainfed agroecosystem (CRIDA-ORP 1997). The productivity of rainfed crops has persisted at a low level with few exemptions. Under rainfed conditions, several constraints ranging from soil related to marketing contribute to poor yields. These constraints lead to reduced biomass production, less root biomass, and low or no recycling of residues associated with low levels of soil organic carbon (SOC) concentration in soils. High variability and uncertainty of rains make agriculture in rainfed areas reliant on the amount and distribution of downpour. Moisture stress is a major challenge to rainfed agriculture, which at critical growth stages leads to low yields. Degraded soils with a high risk of accelerated erosion resulting in loss of fertile surface soil and SOC are major concerns. Depending upon soil type, vegetation, and slope gradient, the magnitude of soil loss ranges from 5 to 150 Mg ha^{-1} yr^{-1}. There have been several estimates of area of degraded/wastelands by various organizations, which have been synchronized by adopting spatial data integration using a geographic information system (GIS) for a diverse environment (Maji 2007). The total degraded area is estimated at 120.72 M ha, of which 104.19 M ha (86.3%) is arable land and 16.53 M ha (13.7%) is open forest land. Of the total degraded land area, 73.27 M ha (60.7%) is caused by water erosion, 12.40 M ha (10.3%) by wind erosion, 5.44 M ha (4.5%) by salinization, and 5.09 M ha (4.2%) by acidification of soil. Some areas are affected by more than one degradation process (Maji 2007). In comparison to irrigated crops, usage of production inputs like good-quality seeds, fertilizers, supplemental irrigation, herbicides, and pesticides are low in rainfed crops, leading to reduced yields. Fertilizer consumption pattern by major size groups of unirrigated crops against irrigated crops in India and fertilizer consumption in predominant rainfed crops in comparison with irrigated crops are shown in Figures 6.1 and 6.2, respectively. Though demonstrated amply that soils in rainfed regions are deficient in multinutrients, balanced use of these inputs in rainfed crops is seldom done. There are wide disparities of inputs between irrigated and rainfed regions. Severe incidence of diseases and insect pests contribute to low production in rainfed crops in general, and pulses in particular. Enhanced crop productivity and water and nutrient use efficiency can be accomplished with application of nutrients in synergy with profile moisture storage. Improved production technologies and varieties have been developed for rainfed regions. Nevertheless, slow adoption of rainfed management practices and the

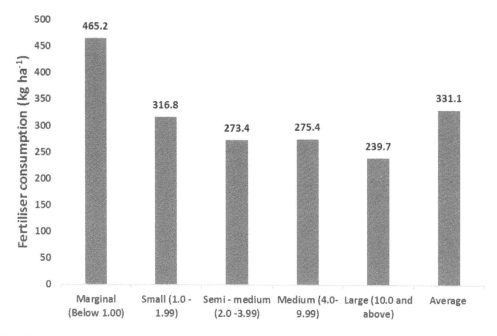

FIGURE 6.1   Fertilizer consumption pattern by major size groups in unirrigated crops against irrigated crops in India. (From Fertilizer Statistics 2017–2018, The Fertilizer Association of India, New Delhi, 63rd edition.)

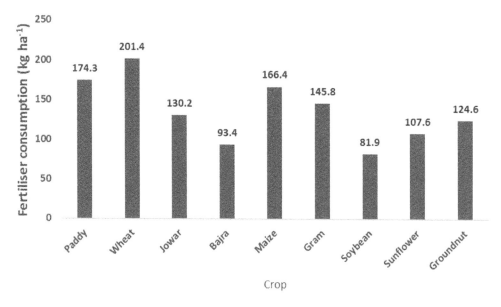

FIGURE 6.2   Fertilizer consumption in predominant rainfed crops in comparison with irrigated crops. (From Fertilizer Statistics 2017–2018, The Fertilizer Association of India, New Delhi, 63rd edition.)

dearth of focused extension programs specially aimed at rainfed regions are among the key constraints. Though coarse grains like finger millet, sorghum, pearl millet, etc. are highly nutritive and important elements of diet, lack of appropriate policy support contributes to low monetary returns, and thus farmers cultivating these crops do not emphasize application of fertilizers and input management. Soil health deterioration is another major concern affecting agricultural productivity in drylands (Srinivasarao et al. 2011b). The major causes of soil health deterioration/ degradation include nutrient mining through intensive farming, imbalanced fertilizer use, deforestation, monocropping, poor manuring, etc. Water or wind erosion also contributes largely to soil degradation. Soil degradation results in declined yields and cropping intensity. The severity of degradation caused by different factors, overall damage of a few major crops as influenced by soil degradation, and the expected land degradation scenario through 2020 is presented in Tables 6.1 and 6.2 and Figure 6.3, respectively.

TABLE 6.1
Severity of Degradation

	Severity of Degradation				
Degradation	Slight (5–10 Mg)	Moderate (11–20 Mg)	High (21–40 Mg)	Severe (>40 Mg)	Total Area (m ha)
Water erosion	5.0	24.3	107.2	12.7	148.9
Wind erosion	0.0	0.0	10.8	2.7	13.5
Loss of nutrients	0.0	0.0	3.7	0.0	3.7
Salinization	2.8	2.0	5.3	0.0	10.1
Water logging	6.4	5.2	0.0	0.0	11.6
Total	**14.2**	**31.5**	**127.0**	**15.1**	**187.7**

Source: Singh, G. et al., *J. Soil Water Conserv.*, 47, 97–99, 1990.

TABLE 6.2
Overall Damage of Some Major Crops as Influenced by Soil Degradation

Crop	Loss (Percentage)	Losses (in Billion Rs)
Sugarcane	12.1	11.7
Wheat	10.8	26.4
Cotton	10.2	2.2
Paddy	9.7	29.2
Barley	9.1	0.4
Rapeseed	8.3	3.9
Groundnut	8.1	5.8
Maize	8.0	2.4
Total	100	88.7

Source: Brandon, C., and Hommann, K., The cost of inaction: Valuing the economy wide cost of environmental degradation in India, Mimeo, Asia Environment Division, World Bank, Washington, DC, 1995.

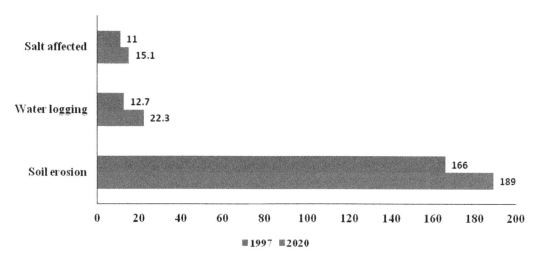

FIGURE 6.3   Expected land degradation (m ha) scenario in India through 2020.

Strategies to be adopted to improve soil health, productivity. and sustainability include fertilizer fortification, precision agriculture, nanotechnology, integrated nutrient management (INM), conservation agriculture, bioremediation or phytoremediation of contaminated soils, improving soil quality through addition of organic sources, site-specific nutrient management (SSNM), erosion control through agronomic and mechanical measures, etc. Decline in soil health coupled with inadequate and imbalanced nutrient use and limited use of organic manures leads to multinutrient deficiency in rainfed regions. Most of the soils are deficient in N and P. Because in the rainfed regions most of the soils are not only deficient in N and P but also K, Mg, S besides Zn and B which is a major constraint of higher crop yield and quality (Srinivasarao et al. 2009, 2010, 2011a). SOC is an important component for the functioning of agroecosystems, and its presence is central to the concept of sustainable maintenance of soil health (Thangavel et al. 2019). Temperature has a great influence on organic carbon depletion from soil. The majority of tropical Indian soils belong to an arid and semiarid climate, are characteristically low in SOC stock, and rarely exhibit organic carbon levels exceeding 0.6% (Virmani et al. 1982). Soil quality strongly relates to food security. Hence, enriching SOC stock and curbing depletion are vital to stabilizing agronomic productivity (Wright and Hons 2005). In this chapter, we emphasize the role of organic manure application in yield sustainability, improving nutrient use efficiency (NUE), maintaining SOC and its subsequent benefits in improving soil health, and comparative studies on the impact of application of chemical fertilizers and organic manures and their conjunctive impact on yield, NUE, and soil properties.

6.2   SCENARIOS, STATES, CROPS, FERTILIZER AND MANURE USE, AND SOIL TYPES IN RAINFED DRYLANDS

Globally, India receives the highest rainfall on a per-unit area basis due to the short duration of the rainy season. Out of 128 districts in India identified as dryland farming regions, 91 districts are spread throughout the states of Tamil Nadu, Chhattisgarh, Madhya Pradesh, and Uttar Pradesh. The remaining districts belong to the Saurashtra region of Gujarat and the rain shadow region of the Western Ghats and Central Rajasthan.

India has around 80 M ha of rainfed area, which constitutes nearly 57% of the total cultivable land. Cultivation becomes relatively challenging as it chiefly relies upon the frequency and intensity of rainfall. Crop production, therefore, in such regions is referred to as rainfed farming as there is no facility to give any irrigation, and even protective or lifesaving irrigation is not possible.

Predominantly grown dry-farming crops include millets such as finger millet, sorghum, and pearl millet; pulse crops like pigeon pea and lentil; and oilseeds like groundnut and safflower. Dryland agriculture contributes about 80% of maize and sorghum, 90% of pearl millet, nearly 95% of pulses, and 75% of oilseeds. About 70% of cotton is produced in dryland areas as well. These regions also contribute considerably to rice and wheat production. However, 66% and 33% of rice and wheat, respectively, are still rainfed. The traditional and alternate efficient crops grown in dryland regions of India are presented in Table 6.3.

Soil degradation with loss of top fertile soil with erratic and high-intensity rainfall is a common feature of dryland agroecosystems. This leads to emerging multinutrient deficiencies and crop production often affected by improper plant nutrition. In addition, an increasing number of years with deficient rainfall, mid-season droughts, and a lack of resources contribute to farming being a risky venture. These are the some of the reasons why farming in rainfed drylands is not profitable unless efficient natural resource management strategies are implemented taking into account specific location and socioeconomic considerations. In India, red soils (Alfisols), black soils (Vertisols), and arid soils (Aridisols), mostly in southern and western regions, are common soil types covering dry climates. Alfisols exist to the extent of 30%, Vertisols and Vertic subgroups up to 35%, and Entisols-Aridisols of 10% in overall rainfed dryland ecology in India (Srinivasarao et al. 2015). It was recorded that about 73 M ha land is subject to erosion, and lead soil loss up to 10 Mg ha^{-1} yr^{-1}.

Though fertilizer consumption increased during the past 70 years in India, there are wide differences in its usage depending on rainfed versus irrigated, soil type, and ecosystem characteristics. Consumption of plant nutrients in major states (kg ha^{-1}) is presented in Figure 6.4. Traditionally, rainfed dryland regions are mostly dependant on organic manures. However, in the recent past fertilizer consumption has increased significantly even in drylands, particularly in commercial crops. In general, soils in India are deficient in nitrogen, which is often associated with low SOC content (Srinivasarao et al. 2006, 2009), and low to high in available P and K, with a variable degree of deficiency in secondary and micronutrients. The fertility status of N, P, and K and micronutrient deficiencies in some states of India are presented in Tables 6.4 and 6.5, respectively. Rainfed dryland systems also have more or less similar nutrient deficiencies, but the extent of deficiency in various nutrients in soils is much in high-intensive and nutrient-exhaustive crop cultivation (Srinivasarao et al. 2014b). Fertilizer consumption during 1950–2016 in India is presented in Figure 6.5.

TABLE 6.3
Traditional and Alternate Efficient Crops in Different Dryland Regions of India

S. No.	Traditional Crop	Alternate Crop	Region
1	Rice	Ragi, groundnut, black gram	Uplands of Bihar plateau and Orissa
2	Wheat	Chickpea, safflower	Malwa plateau
3	Wheat	Mustard, Taramira	Sierozems of northwest India
4	Cotton, Wheat	Safflower	Deccan rabi season
5	Maize	Sorghum	Southeast Rajasthan
6	Maize	Soybean	North Madhya Pradesh
7	Soybean	Chickpea	Eastern Uttar Pradesh

Source: Yadav, K., Recent Advances in Dryland Agriculture, GBPUAT, Pantnagar, 2009.

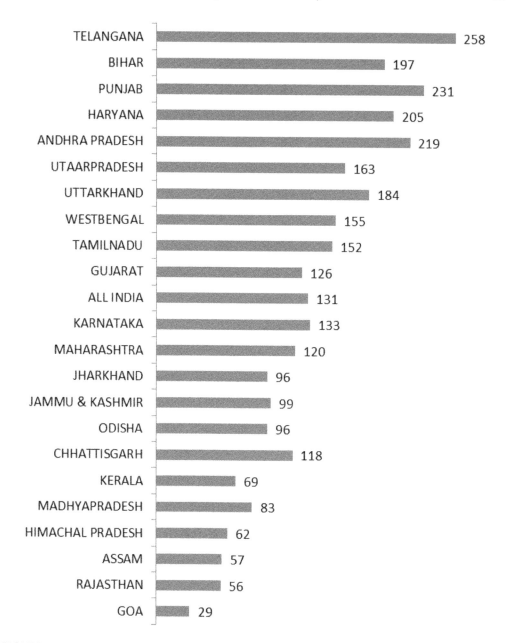

FIGURE 6.4 Consumption of plant nutrients in major states (kg ha^{-1}). (From Fertilizer Statistics 2017–2018, The Fertilizer Association of India, New Delhi, 63rd edition.)

The response of crops in rainfed drylands is not restricted to only N, P, and K. Sulfur is given through ammonium sulfate (23.7%), pyrites (53.5%), etc. In comparison to legumes, the S requirement is high in cereals. The deficiency symptoms of S are more pronounced in oilseeds and pulses due to higher removal of the nutrients. The essential micronutrient elements for crops comprise zinc (Zn), iron (Fe), manganese (Mn), copper (Cu), boron (B), and molybdenum (Mo). Further, mineral nutrients such as cobalt (Co), chlorine (Cl), nickel (Ni), and silicon (Si) are regarded as beneficial to some plants, and their universal essentiality has not been fully established. Dryland crops also show positive response to the application of micronutrients, namely, Zn, Fe, and Mn.

TABLE 6.4
Fertility Status of N, P, and K in Some States of India

State	No. of Samples Analyzed	Nitrogen			Phosphorus			Potassium		
		L	M	H	L	M	H	L	M	H
Andhra Pradesh	312,521	62	21	17	57	29	14	9	30	61
Karnataka	317,213	29	37	34	31	48	21	7	32	61
Madhya Pradesh	138,553	40	41	19	39	38	23	10	32	58
Maharashtra	93,142	67	26	7	86	12	2	8	18	74
Orissa	251,196	60	23	17	59	28	13	33	41	26
Tamil Nadu	491,657	75	16	9	24	41	35	12	36	52
Uttar Pradesh	807,424	80	15	5	71	26	3	12	55	33
India	3,650,004	63	26	11	42	38	20	13	37	50

Source: Motsara, M.R., *Fert. News*, 47, 15–21, 2002.

TABLE 6.5
Micronutrient Deficiencies in Some States of India

State	No. of Samples (Range of Samples Analyzed for Different States)	Percentage Deficiency of Available Micronutrients			
		Zn	Fe	Cu	Mn
Gujarat	29,532	24	8	4	4
UP	24,425–25,122	45	6	1	3
Karnataka	24,411–25,542	78	39	5	19
Tamil Nadu	19,559–20,580	53	15	3	8
Bihar	17,802–19,078	54	6	3	2
MP	11,204–12,000	63	3	1	3
AP	5,219–6,563	51	2	1	2

Source: Takkar, P. N., *J. Indian Soc. Soil Sci.*, 44: 563–581, 1996.

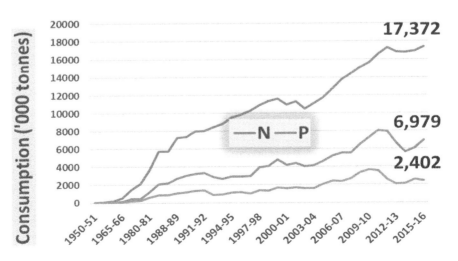

FIGURE 6.5 Fertilizer consumption during 1950–2016 in India.

Zinc is supplemented through $ZnSO_4 \cdot 2H_2O$ (35% Zn), while the chief source is $FeSO_4$ and $MnSO_4$ for Fe and Mn, respectively. Organic manures supply macro and micronutrients and at the same time improve the physical, chemical, and biological properties of soil. Farmyard manure (FYM), compost, and oil cake from castor, groundnut, mustard, mahua, safflower, etc. are used very commonly in rainfed drylands for enhancing soil fertility and accomplishing high yield. In comparison to chemical fertilizers, organic manures are bulky and slow releasing in nature. Organic manures add very few nutrients, which is not sufficient for high-responsive crops/hybrids. In that case, to derive the maximum benefits from organic sources, conjunctive use with chemical fertilizer has been emphasized and advocated.

6.3 IMPACT OF THE ADDITION OF MANURE AND FERTILIZERS ON SOIL HEALTH AND CROP PRODUCTIVITY

6.3.1 ALFISOLS

Alfisols represent one of the most prominent soil orders for food and fiber production. They are formed in semiarid to humid areas and have a clay-enriched subsoil with relatively high innate fertility, occupying around one-tenth of the earth's ice-free land surface. They are dominant in many areas such as southern and unglaciated western Europe, the Ohio River basin in the United States, the Baltic region and central European Russia, the drier parts of peninsular India, Sudan in Africa, and several parts of South America. They are generally used both in agriculture and forestry, as fertility retention of these soils is easy compared to other humid-climate soils, though those in Africa and Australia are still very deficient in nitrogen and available phosphorus. The distinctive morphology of Alfisol is characterized by prominent horizons of eluviation and illuviation. An extensive range of cereal crops (rice, maize, sorghum, and millet), root and tuber crops (yam, cassava, cocoyam, and sweet potato), and grain legumes (soybean, cowpeas, groundnuts, pigeon pea, and chickpea) are supported by Alfisols and associated soils. These soils have lower acidity and are less leached than Ultisols and Oxisols, but they exhibit high base saturation and their fertility is low to moderate. Variations in texture, depth, color, and clay mineralogy result from relief and drainage variations. The physical properties of Alfisols at three All India Coordinated Research Project for Dryland Agriculture (AICRPDA) centers is presented in Table 6.6.

TABLE 6.6
Physical Properties of Alfisols at Three AICRPDA Centers

Characters	Ananthapur 0–15	Ananthapur 15–30	Bangalore 0–15	Bangalore 15–30	Hyderabad 0–15	Hyderabad 15–30
Gravel (%)	23.6	60.0	0.00	0.40	22.9	8.8
Sand (%)	63.7	24.0	73.50	41.10	83.4	74.4
Silt (%)	3.2	2.6	6.20	8.00	5.0	4.0
Clay (%)	9.5	13.4	21.24	50.80	11.6	21.6
Bulk density (g cm^{-3})	NA	NA	1.64	1.42	1.43	1.63
Moisture at 0.033 MPa (1/3 bar)	NA	NA	14.73	18.03	12.31	17.42
Moisture at 1.5 MPa (15 bar)	NA	NA	7.57	13.09	5.84	7.49

Source: Vijayalakshmi, K., Soil Management for increasing productivity in semi-arid red soils: Physical aspects. Alfisols in the Semi-Arid Tropics, In: *Proceedings of the Consultants' Workshop on the State of the Art and Management Alternatives for Optimizing the Productivity of SAT Alfisols and Related Soils*, 1–3 December 1983, ICRISAT Center, India, 1987.

Note: NA = Not available.

Alfisols occur on gently sloping to undulating surfaces and are well drained. These soils have weak granular or subangular blocky surface A horizons, subangular blocky to prismatic B horizons, and subangular blocky C horizons. The A horizons vary from loamy sand to gravelly clay and clay, but B horizons are usually enriched with clay to give clayey or fine loamy soils. The loamy sand texture of the topsoil and predominance of kaolinite among the clay minerals make these soils structurally inert (Charreau 1977). Alfisols on peninsular gneiss are formed on very old landscapes and have 10%–15% clay content in the Ap horizon, closely followed by a well-developed argillic (Bt) horizon with a clay content of >30%. These soils are developed on the old rock system of the earth and signify ancient soils. The deposition of clay in the subsurface is caused by illuviation of clay particles. This process is active during humid past (which is being very damp and stuffy). Red soils have distinct horizons of clay enrichment that are easily visible in the field with a significant amount of amorphous material found in the profiles. The pH of Alfisols is between 5.0 and 6.0, while the base saturation must be greater than 35%, and its smectite clays (usually 2:1) have a higher cation exchange capacity (CEC) (>16 cmol kg^{-1}). Their exchange complex is dominated by a mixture of kaolinite and illite. The organic matter content ranges from very low to low. The chemical characteristics of these soils appear to reflect primarily the effect of soil-forming materials. Rainfall and drainage are the other aspects that impact these characteristics (Shankaranarayana and Sarma 1982).

In Alfisols, soil management practices should rather comprise an optimum dose of mineral fertilizer in conjunction with organic sources or lime to sustain better SOC contents and soil physical characteristics, resulting in enhanced yields. However, soils in monsoonal tropical regions tend to acidify when cultivated heavily, particularly when nitrogenous fertilizers are used. Intensive cropping with conventional tillage causes a deprivation of SOC, with a subsequent deterioration of soil physical properties. Trials conducted reveal that the decline in SOC could be prevented by the addition of organic sources and better nutrient management practices. Organic matter addition, either through crop residues or manures, augments SOC content, which ultimately impacts soil physical properties and processes like aggregation, water retention capacity (Rose 1991), hydraulic conductivity (Anderson et al. 1990; Franzluebbers 2002), bulk density (BD), and soil resistance to wind and water erosion (Zebarth et al. 1999; Celik et al. 2004). Lime application would also aid in achieving substantial improvement in crop yields. But the impact of liming on the physical properties of soil is contradictory both in temperate and tropical regions. Liming may accelerate dispersion of clay and decrease aggregate stability and infiltration rate (Tama and El-Swaify 1978; Roth and Pavan 1991). Contrarily, others have observed that application of lime can enhance aggregate stability and infiltration (Castro and Logan 1991), improve water-holding capacity (Kohn 1975), and reduce surface cracking (Hoyt 1981). To evaluate the effect of various levels of nitrogen and FYM on yield and nutrient uptake by turmeric (*Curcuma longa* L.) and residual accrual of different N forms, significant improvement in rhizome, straw yield, P and K uptake was realized up to 90 kg ha^{-1}. Nevertheless, N uptake was significant up to 150 kg N ha^{-1}, whereas the maximum NUE (34.3%) was obtained by application of 30 kg N ha^{-1}, which progressively decreased with increasing levels of nitrogen. The available N (alkaline KMnO$_4$-oxidizable N) status in postharvest soil increased successively with increasing N levels. The combined application of N with farmyard manure proved superior to inorganic N by registering significantly higher available N (Majumdar et al. 2002). Low levels of soil fertility limit groundnut cultivation in Alfisols of drylands. The dose combination of 300 kg ha^{-1} SSP – 36 + 150 kg ha^{-1} urea + cow manure 10 Mg ha^{-1} aided in achieving maximum productivity in groundnut (Suryono et al. 2015). Evaluation of the long-term effects of fertilizer, manure, and lime application on physical properties and organic carbon content under soybean–wheat rotation revealed balanced fertilizer application, in conjunction with manure or lime, improved soil water retention, soil aggregation, microporosity, and available water holding capacity and lowered BD of the soil in 0–30 cm depth in comparison to the control (Hati et al. 2008). The microbial biomass carbon and dehydrogenase activity in soil at harvest of crops was significantly increased due to the addition of fly ash and FYM in a rice–wheat cropping system in Alfisol and Vertisol (Ramteke et al. 2017). Similarly, Lal et al. (1996) reported that application of fly ash in the presence of FYM to an acid soil (Alfisols) led to increased microbial population and urease and cellulase activities than alone. Higher mean pod yield and SOC in rainfed groundnut was obtained with 50% RDF + 4 Mg ha^{-1} FYM (Srinivasarao et al. 2012). In an 18-year long-term trial conducted on red sandy loam soil in rainfed groundnut, the application of chemical fertilizers in conjunction with groundnut shells altered the sustainability yield index (SYI) to 0.320 as compared to 0.285 and 0.306 by the sole application of FYM and inorganic fertilizers, respectively (Balaguravaiah et al. 2005). The impact of fertilizers and amendments on organic matter dynamics in an acid Alfisol evaluated for 42 years indicated that continuous application of inorganic fertilizers either alone or coupled with FYM or lime considerably influenced water-soluble organic carbon (WS-OC), water-soluble carbohydrate (WS-CHO), soil microbial biomass carbon, nitrogen, phosphorus, sulfur, fulvic acid (FA), and humic acid (HA). Uninterrupted cropping avoiding fertilizers led to exhaustion in the order of 17%, 21%, 24%, 23%, 22%, 26%, 12%, and 18% in WS-OC, WS-CHO, microbial biomass carbon, nitrogen, phosphorus, sulfur, HA, and FA, respectively. Different percentages of soil organic matter (SOM) were positively and significantly correlated with straw/stover and grain yield of maize and wheat crops (Meena and Sharma 2016). A trial conducted for two successive *kharif* (rainy) seasons to calculate the rhizospheric microflora, green gram yield, and nutrient availability as influenced by organic sources, phosphorus levels, and phosphate solubilizers in Alfisols indicated that addition of organic sources considerably improved the rhizospheric microbial population (fungi, actinomycetes, and bacteria) available soil N P K production of green gram. Among organic sources, the addition of farmyard manure at 10 Mg ha^{-1} was superior over dalweed application at 15 Mg ha^{-1}. Soil inoculation with P solubilizers resulted in a substantial increase in the production of green gram. Among P solubilizers, *Pseudomonas* was found superior to Bacillus with regard to enhancing green gram yield, soil available P, and bacterial population (Chesti and Tahir 2012). Sorghum and green gram yield with conjunctive use of organic residues and fertilizer nutrients in Alfisols, as well as many soil characteristics such as available N, P, organic carbon, dehydrogenase activity, and DTPA extractable micronutrients, were positively influenced by the conjunctive use of inorganic and farm-based organic resources (Sharma et al. 2004).

6.3.2 Vertisols

Vertisols occur extensively in warm temperate zones, subtropics, and tropics, which are a group of heavy-textured soils referred to as black cotton soils, black earths, dark clays, regurs, and grumusols in other classification systems (Dudal 1965). Vertisols are predominantly found in India, Australia, Sudan, Chad, and Ethiopia; these five countries comprise above 80% of the total area of 250 M ha of Vertisols. They are regarded as significant soil orders in semiarid drylands due to their high water-holding capacity, and are suited to dryland crop production in semiarid environments with heavy and uncertain rainfall. Even though Vertisols cover only a small region of the world's land surface, and only a subdominant part of any geographical zone, they are a soil order of significance in semiarid dryland agriculture, as they are among the most productive soils in this environment. Black soils cover 257 million ha between 45" N to 45" S latitudes on a global area basis (Dudal 1965). The total area of Vertisols in the world is around 180 M ha (Swindale 1981). Vertisols and related soils in India alone occupy 73 M ha, out of which 38% (28 M ha) are Vertisols, 37% Vertic-Inceptisols, and 21% Entisols (Murthy 1981). In India, Vertisols extend over an area of 26 M ha, chiefly in the states of Madhya Pradesh (10.7 M ha), Maharashtra (5.6 M ha), Karnataka (2.8 M ha), Andhra Pradesh (2.2 M ha), and Gujarat (1.8 M ha) (Bhattacharyya et al. 2013), distributed across varying rainfall regimes ranging from 590 mm in Rajkot to 980 mm in Rewa. The predominant factor governing productivity of Vertisols in semiarid climates is their high water-retention capacity in areas of variable and uneven rainfall, sometimes heavy and often too low. The soil capacity to stock sufficient quantity of water to carry crops through periods of drought assumes utmost prominence. The major crops grown in these areas are soybean, ground-nut, maize, cotton, and pigeon pea during the rainy season and chickpea, safflower, sunflower, and sorghum during the post-rainy season.

In drier seasons, Vertisols exhibit deep cracks and crevices due to the high expansive nature of clay (montmorillonite). Alternate swelling and shrinking results in self-mulching. Vertisols have an extremely deep A horizon but no B horizon as the soil material constantly mixes itself. Generally, the texture of the surface soil is lighter and the clay content rises with increasing depth toward the subsoil. The clay content of Vertisols remains uniformly high (>35%) throughout the profile to a depth of at least 50 cm or more. In some Vertisols where the top soil is perhaps eroded, the clay content might be 40%, resulting in loam or a silty loam texture in some Vertisols in West Africa, and probably even a sandy texture in the subsurface. They are commonly referred to as deep black soils.

The high moisture-holding capacity of Vertisols is due to their clay content, which commonly lies between 40% and 60%, but it may be as high as 80% and have a profile depth of more than 90 cm (Dudal 1965; De Vos and Virgo 1969). In Vertisols throughout the profile, the clay content does not show much variation, increasing marginally with depth. They are usually calcareous, montmorillonitic, isohyperthermic soils. Because of the montmorillonitic nature of the clay minerals, Vertisols experience substantial shrinkage on drying and swelling. On drying they become hard, and sticky when wet. They have a high CEC (45–60 meq 100 g^{-1}), and an alkaline reaction with pH varying between 7.5 to 8.6, and more than 80% base saturation. The exchange complex is dominated by calcium, but on salinization these soils display distinctive characteristics of sodic-ity even when the sodium percentage is much less than 15 in the exchange complex. Due to their swelling and shrinking nature, with alterations in soil moisture content the bulk density of Vertisols differs greatly. Moisture content is the most important factor which determines the bulk density of soil. Dry soils have higher bulk density compared to wet soils (swollen stage) which may vary from 1 to 2 Mg m^{-3} due to variations in soil moisture content. $CaCO_3$ presence and high contents of bases, particularly magnesium and calcium in the profile, add to high soil pH which aids in gaseous loss of NH_3 when urea or ammonium fertilizers are surface applied (Terman 1979; Sahrawat 1980). Most of the Vertisols are calcareous, and the distribution of $CaCO_3$ may be either uniform all through the profile or may increase in the lower horizons. The content of $CaCO_3$ in Indian Vertisols may vary from nil to 10% or greater in the profile. Gypsum has been found to occur in the subsurface of the Vertisol profiles in relatively arid areas. Most of the black cotton soils of India rarely have organic matter exceeding 1.0%. Organic matter has been found to be more or less uniformly distributed in the first meter of the profile in some Indian Vertisols. Vertisols are bestowed with higher CEC owing to the dominance of 2:1-type clay minerals in the 2 μm particle size fraction, although this depends upon the actual content of clay. Calcium is the most dominant cation, accounting for 52% to 85% of the total exchange complex. Sodium is usually less than 20% of total CEC, and magnesium generally varies from 10% to 30%.

Intensive cropping of Vertisols necessitates cautious management of moisture regime and soil temperature along with INM (Srinivasarao et al. 2012). Reduced agronomic yields, low or no crop residue retention, and extended fallow periods involving uncontrolled grazing decrease the SOC concentration and stock (Srinivasarao et al. 2011a). Several researchers have reported yield augmentation to the tune of 60% by the use of organic amendments in comparison to nonamended control, whereas conjunctive use of organic amendments and N fertilizers can aid in achieving a yield increase of 114% (Chivenge et al. 2011). It is also noted that utilization of better-quality organic amendments can bring pronounced long-term residual effects. In another study, carried out by Hati et al. (2007) to evaluate the long-term effect of manure and fertilizer addition in a soybean–wheat–maize (fodder) crop rotation on SOC status and physical properties of a Vertisol in subtropical, subhumid India, when compared to the initial level, SOC increased by 22.5% and 56.3% in the treatment of 100% NPK and 100% NPK + FYM. The water retention, SOC, available water capacity, and aggregation of the soil showed substantial improvement, while the BD was decreased considerably with the application of 100% NPK + FYM. However, the use of imbalanced (100% N) and a suboptimal rate of inorganic fertilizer (50% NPK) showed no significant effect on the physical properties as compared to the unfertilized control. The study revealed that balanced application of fertilizer with organic manure could sequester SOC in the top soil layer, which sustains optimum crop productivity and improves the soil physical environment under an intensive cropping system. In a soybean–wheat (*Triticum aestivum*) sequence in Vertisols in central India, the annual rate of SOC enrichments ranged from 0.09 to 0.74 Mg C ha^{-1} for 0 to 15 cm depth and 0.05 to 0.15 Mg C ha^{-1} for 15 to 30 cm depth of soil (Kundu et al. 2001). The organic carbon content of Vertisol increased significantly (0.56% to 0.87%) with the application of manure and fertilizers in comparison to the control (0.45%) (Malewar and Hasnabade 1995).

The combined application of the recommended dose of NPK (100:50:40 for sorghum and 120:60:60 to wheat) with 10 Mg FYM ha^{-1} enhanced the available NPK and organic carbon content of soil (Sonune et al. 2003). The available NPK exhibited a positive correlation with crop yield and there was a positive balance of N and P. Thus, the balanced use of NPK fertilizers in conjunction with FYM is essential for augmenting the fertility of Vertisol in a cereal-cereal cropping sequence. The effects of long-term cropping, fertilization, and manuring, and their integration to assess the microbial community count in soil samples from five trials at Anantapur, Solapur, Bellary, Bangalore, and Coimbatore centers of different rainfed production systems revealed that the community counts are higher in treatments with conjunction of organic and inorganic fertilizers when compared to a control. Vertisols exhibited higher organic carbon levels than Alfisols. In general fungal population is higher in acid soils but continuous application of chemical fertilizers also leads to the development of fungal growth in arable soils, whereas with combined use of organic and chemical fertilizers in field results in maximum bacterial count. At most of the locations, SOC and microbial biomass carbon (MBC) exhibited significant positive correlation with microbial populations, suggesting that even under arid and semiarid tropical conditions, nutrient application through an integrated approach could improve SOC and increase microbial count (Vineela et al. 2008). Studies on the effect of organic and inorganic nutrients on yield and nutrient balance in different cropping systems indicated that the highest grain yield was obtain by application of recommended dose of fertilizer (RDF), followed by *Gliricidia* @ 1.5 Mg ha^{-1} + 25% RDF and were comparable with each other (Shirale and Khating 2009). The effect of continuous application of chemical fertilizers and manures on yield sustainability, key indicators of soil quality, soil properties, and yield sustainability under a sorghum (*Sorghum vulgare*)–wheat (*Triticum aestivum*) cropping sequence indicated that substantial improvement in the soil-quality index was observed under integrated nutrient management involving usage of 100% NPK + FYM (2.45), followed by sole application of FYM (2.16) and 150% RDF (2.15), followed by 50% RDF (1.45), while the lowest was in the control (1.14), signifying considerable enhancement in soil quality due to INM. Among different treatments, 100% RDF + FYM not only registered the highest soil-quality index but was found to be the most favorable approach from the view point of yield sustainability and maintaining higher average yields under a sorghum–wheat cropping sequence (Katkar et al. 2012). In Vertisols under semiarid conditions of Vidarbha, the integrated application of 50% RDF + FYM @ 5 Mg ha^{-1} augmented soil quality and sustained rainfed cotton productivity (Gabhane et al. 2014). In semiarid Vertisols, 25 kg N (FYM) + 25 kg N (urea) + 25 kg P ha^{-1} treatment was superior for achieving optimum SYI, NUE, RWUE, and benefit-cost (B:C) ratio, and gross and net monetary returns, beside soil fertility buildup (Maruthi Sankar et al. 2014). Studies on integrated nutrient management on soil properties under a cotton–chickpea cropping sequence in Vertisols in the Deccan plateau of India indicated that the application of manures alone or in combination with chemical fertilizers enhances the soil physical, chemical, and biological properties at the end of two crop cycles (Gudadhe et al. 2015). The conjunctive application of 100% NP + 10 kg K (inorganic) +20 kg K ha^{-1} through Gliricidia green leaf manuring at 30 days after sowing (DAS) resulted in enhanced soil fertility and yield of cotton grown in Vertisols under rainfed conditions (Naik et al. 2018). A long-term field experiment on nutrient management for a sorghum–wheat sequence on Typic Haplustert over 10 consecutive seasons (*kharif* and *rabi*) from 1988 to 1997 showed an increase in yield of sorghum and wheat and improved soil fertility with application of 100% NPK in combination with 10 Mg FYM ha^{-1} (Ravankar et al. 1999). In a rice–wheat cropping system, the highest grain yield of rice was obtained when sunhemp grown as green manure crop was incorporated along with 120 kg N ha^{-1}, followed by dhaincha and green gram. In the case of wheat, incorporated dhaincha gave the highest grain yield, followed by sunhemp and green gram. The same trend with respect to N< P< K and S uptake was observed in both crops (Chand et al. 2015). Under dryland conditions, FYM 10 Mg ha^{-1} + 100% NPK application was found to sustain finger millet crop productivity and improved soil quality. The application of 10 Mg ha^{-1} FYM + 100% NPK in rotation compared with monocropping registered higher values of SYI and SQI. Hence, the addition of FYM 10 Mg ha^{-1} + 100% NPK with crop rotation is a viable option for sustaining soil health maintenance and yield of an important staple food crop. In areas where FYM availability is a major concern due to a decline in the livestock population, compost, crop residues, green manures, and municipal biosolids may be utilized as substitutes for sustainable crop production (Sathish et al. 2016).

The importance of SOC in regulating crop yields has been well established (Lal 2004). Maintenance of SOC stock through the application of FYM and chemical fertilizers is essential to the sustainability of rainfed production systems, which need high external inputs. Any improvement in SOC enhances the available water capacity of the soil profile (Du et al. 2009). The buildup of SOC can augment the productivity of crops by improving the size of the mineralizable N and P pool; increasing soil physical properties with lower BD, improved pore connectivity, improved aggregate stability and aeration; and through amending constraints including low CEC and inappropriate soil pH. The long-term changes in SOC due to the addition of organic sources showed that on average, the organic sources increased SOC by 49% compared to an unfertilized control or 29% compared to a fertilized control (Chen et al. 2018). The application of compost significantly improves SOC levels (Eldridge et al. 2018). Crop rotation with legumes, rotational application of manures and fertilizers not only improve the yields but also enhance soil quality parameters. Crop rotations included were the soybean–wheat, rice–wheat–jute, and sorghum–wheat systems at Ranchi (Typic Haplustalf), Barrackpore (Typic Eutrochrept), and Akola (Typic Haplustert), respectively. SOC concentration in the un fertilized plot (control) reduced by 24.5%, 41.5% and 15.5% over initial values in Ranchi, Barrackpore and Akola respectively, whereas fertilized with NPK and NPK + FYM either sustained or enhanced SOC concentration over initial SOC concentration. Active fractions of soil organic carbon, namely, soil MBC, alkaline phosphatase, and dehydrogenase, increased considerably with the application of NPK and NPK + FYM. The SOC content significantly correlated with the sustainable yield index and active fractions of SOC, which encouraged better sustainable productivity (Manna et al. 2005).

The impact of conjunctive use of chemical fertilizers and the organic manures, namely, green manure, FYM, and wheat straw, in the long-term productivity of an irrigated sorghum–wheat system in permanent plots under the AICRP on cropping systems at Rahuri, Akola, and Parbhani in Maharashtra revealed the available soil S, Fe, and Mn augmented, while available Zn and Cu remained unaltered. However, the integrated nutrient supply also improved the fertility of the soil (Hegde 1996).

The effect of long-term fertilization on soil biological health in the sorghum–wheat sequence on swell-shrink soils of Central India demonstrated that combined usage of fertilizers along with wheat straw and *Leucaena* loppings showed considerable improvement in soil biological properties over the sole usage of fertilizers. The biological properties significantly declined where no organics were used continuously for 27 years (Mali et al. 2015). The highest nitrogen (3.08%) and phosphorus (0.26%) was found in *Gliricidia*, followed by sunhemp (2.95% N and 0.25% P). The potassium was found high in eupatorium leaves (2.02%), followed by *Gliricidia* (1.95%) and neem leaves (1.93%). Compared to green leaf manure crops, the NPK percentage was found at lower levels in FYM (1.00% N, 0.20% P, and 0.52% K). This study has clearly elucidated that incorporation of green leaf manures, particularly leguminous green manuring crops like *Gliricidia*, sunhemp, and dhaincha, has an added advantage for enhancing the soil nutrient status (Srinivasarao et al. 2011a; Sreeramulu and Shankar 2016).

6.3.3   Aridisols and Entisols

The arid region in India is spread over 38.7 M ha area, out of which 31.7 M ha lies in hot regions and the remaining 7 M ha lies in cold regions. The hot arid region occupies the major part of northwestern India (28.7 M ha), and the remaining 3.13 M ha is in southern India. The northwestern arid region lies between 22°30' and 32°05'N latitude and 68°05' to 75°45' E, comprising the western part of Rajasthan, northwestern Gujarat, and southwestern parts of Punjab and Haryana. In southern India it occupies parts of Andhra Pradesh, Karnataka, and Maharashtra. About 62% of the area of the arid region falls in western Rajasthan, followed by 20% in Gujarat and 7% in Punjab and Haryana. Karnataka, Maharashtra, and Andhra Pradesh together constitute around 11% area of the arid region. These soils also occur in regions where a very dry or cold climate restricts soil profile development. Entisols are the second-most-abundant soil order (after Inceptisols), covering nearly 16% of the global ice-free land area. Soils under Entisols are usually found in those parts of arid regions where high aeolian or fluvial activities are witnessed. Entisols have no diagnostic horizons, and are essentially unaltered from their parent material, which can be unconsolidated sediment or rock. This is a very diverse group of soils with one thing in common: little profile (horizon) development. Entisols are commonly found at the site of recently deposited materials (e.g., alluvium), or in parent materials resilient to weathering (e.g., sand).

The productivity potential of Entisols varies widely, from very productive alluvial soils found on floodplains to low fertility/productivity soils in sandy areas or in steep slopes. Low moisture-holding capacity coupled with low and high variable rainfall are soil-related constraints, leading to severe moisture stress of varying degrees at one or another crop growth stage, leading to reduced productivity. Low fertility status of soils and widespread multinutrient deficiencies, secondary salinization, and wind erosion are a few fertility-related constraints on realizing the productivity potential of these soils.

The potential yields recorded on Aridisols are low in comparison to that obtained on Inceptisols. They are dry soils with $CaCO_3$ (lime) accumulations, common in desert regions, light in color, and low in organic matter content. Salt and lime accumulations are common in the subsurface horizons. Some Aridisols have an argillic (clay accumulation) B horizon, likely formed during a period with a wetter climate. Deficiency of water is the main distinctive of Aridisols. Soil pH ranges from 8 to 9. A zone of accumulation of lime or lime concretions at a depth of 60 to 120 cm and presence of alkaline earth carbonate are common features of these soils. Characteristically, these soils are very low in organic matter/humus, and most of the nutrient reserve is present in unweathered mineral forms. Nutrient adsorption and retention by these soils are very low, which can be attributed to low clay and silt. Soils are usually alkaline in nature and high in soluble salts and calcium content. Reduced vegetation cover, high temperature, and coarse texture contribute to low organic matter in arid zone soils. Organic carbon content in soils below 300 mm rainfall zone varies between 0.05% and 0.2%, 0.2% and 0.3%, and 0.3% and 0.4% in coarse-textured, medium-textured, and fine-textured soils, respectively. Phosphorus is present in soils as organic and inorganic forms, but the organic form constitutes only around 10%–20% of the total phosphorus. Total phosphorus content in soils ranges between 300 and 1500 μg g^{-1} (Choudhari et al. 1979), and about 80% of the inorganic P remain bound with Ca. Al-P was higher, generally higher than Fe-P. Total and inorganic phosphorus were irregularly distributed in the soil profile of arid soils, but organic phosphorus decreased with depth (Talati et al. 1975). About 15%–20% (97–110 kg ha^{-1}) of the total P is present in organic form as lecithin, phospholipids, phytin, and other unidentified compounds (Tarafdar et al. 1989). The available phosphorus content differs widely in different soils and is around 2.4% to 3.9% of total P, and the mean content in different soil series is less than 10 μg g^{-1} (Mathur et al. 2006). Even though soils are often medium to low in available P, the response of P fertilization in arid soils is usually noticed only during periods of good rainfall (Aggarwal and Venkateswarlu 1989). An extensive survey through arid zones indicated that the P-solubilizing population of microorganisms is comparatively low and differed from region to region (Venkateswarlu et al. 1984).

The arid soils are well provided with available potassium (70–890 kg ha^{-1}). The total K content in arid region soils ranged between 980 and 1890 mg 100 g^{-1} with an average value of 1489 mg 100 g^{-1} soil. The major proportion of total potassium in arid soil is present as mineral form followed by interlayer, nonexchangeable, and water-soluble form. These forms, beside being related to each other, are correlated with sand and silt fraction and K resistance to depletion (Choudhari and Pareek 1976). The K fixation capacity was associated with the clay content, K-saturation, and weathered K-bearing minerals (Mathur et al. 1981; Dutta and Joshi 1993). Calcium is an essential element present in varying amounts from 0.1% to 5.2% in the arid soil. However, in lower layers, most soils have a higher amount varying from 15% to 20% (as CaO). Exchangeable calcium, which is an indicator of availability, varied between 1.5 to 20 meq 100 g^{-1} soil. Fine-textured Aridisols of Pipar and Pali contain 15 to 20 meq 100 g^{-1}, whereas dune and coarse-textured Aridisols contain between 1.5 to 4.0 and 2.2 to 4.0 meq 100 g^{-1} soil. Magnesium is present in different forms in arid region soil, with mineral forms as the dominant fraction (32% to 87% of total Mg) followed by dilute acid soluble, exchangeable, organic bound, and water-soluble fractions. The total sulfur content in arid regions ranges between 280 and 500 µg g^{-1}, of which 50% to 80% remains in inorganic forms. This is in contrast to most other regions, where organic-bound sulfur predominates. In surface soils, sulfate sulfur dominates in surface soils, while nonsulfate sulfur dominates the lower horizon of arid region soils. Calcium, magnesium, and sulfur contents in most arid zone soils are generally adequate for plant growth (Kalyansundaram et al. 1993).

The practice of efficient utilization of organic and inorganic sources of nutrients together in suitable proportion not only lowers inorganic fertilizers' requirement, but also aids in improving physical conditions and enhances water-holding capacity and its availability in the soil. Besides this, soil biological properties and fertilizer use efficiency improve considerably (Sharma et al. 2008, 2009). The results of several studies have indicated that the conjunctive use of fertilizer and FYM assumes importance in rainfed areas for maintaining moderate to high yields and accomplishing higher nutrient use efficiencies. The addition of FYM, composted organic wastes, and other organics enhances the stability of yield in rainfed areas (Aggarwal and Venkateswarlu 1989; Venkateswarlu and Hegde 1992; Singh et al. 2000). Under arid conditions of Jodhpur, continuous application of sheep manure in general gave considerably higher yields than sole application of urea (Singh et al. 1981). The substitution of 50% of fertilizer requirement by FYM resulted in yield levels almost comparable to those attained with 100% fertilization (Rao and Singh 1993). The application of FYM enhances the utilization efficiency of fertilizer N; however, soil fertility improvement after FYM application is a very slow process. In rainfed regions, cereal stover is often fed to livestock, and manure is applied in field. This way of residue recycling is more advantageous for crops than direct application in field (Aggarwal et al. 1996).

The most promising route to enhancing inorganic fertilizer use efficiency in cropping systems is by applying small amounts of high-quality organic matter (having narrow C/N ratio and a low lignin percent) to soils (Ladd and Amato 1985; Snapp and Silim 2002). It offers readily available N, energy (carbon), and nutrients to the soil ecosystem, increases soil microbial activity and nutrient cycling, improves structure, and reduces losses of nutrients from denitrification and leaching (De-Ruiter et al. 1998). Soil microbes not only supply nutrients directly, but also enhance the synchrony of plant-nutrient demand with soil supply by lowering large pools of free nutrients and consequent nutrient losses from the system. Hence, microbes sustain a buffered, actively cycled nutrient supply. Crop residue incorporation and natural vegetation in soil enhance microbial activity during decomposition. Also, adhesive action of decomposed products improves hydraulic conductivity, soil aggregation, and moisture retention (Venkateswarlu 1984, 1987; Gupta 1980; Gupta and Gupta 1986). Leaving the crop residues in soil generally has a positive effect on grain yield. Crop residues are also an efficient source of nutrients that are like other organics namely, cattle manure and compost. However, there was no considerable difference in the yield of succeeding crops of pearl millet after the addition of crop residues with a wide C:N ratio, whereas it was significant after residue incorporation with a narrow C:N ratio. Significant response to the application of 80 kg N ha^{-1} occurred only in years of good rainfall. A comparison of different N fertilizers showed that maximum yields were registered with ammonium sulfate. But urea has become the major source of N in arid regions due to the high cost of ammonium sulfate even though N use efficiency by pearl millet is very low in the case of the application of urea (Aggarwal et al. 1996). The frequent occurrence of drought in arid regions leads to crop failure. Therefore the application of N fertilizers is regarded as a risky input by farmers, and hence the need for its better management. Split application of N fertilizers is hence regarded as a risk-avoiding strategy that aids in realizing better yields, as compared to their onetime application, increased utilization efficiency, and lower risk. Carbon sequestration in soil is strongly affected by root production (Matamala et al. 2003). Trees in arid zones put forth a large volume of below-ground biomass in the form of roots. The tree roots play a vital role in the addition of organic matter to the soil. Charcoal or black carbon is produced by controlled burning of biomass that can contribute to carbon sequestration. Apart from locking carbon in the soil for centuries, activated charcoal can attribute positively to the CEC and physical, chemical, and biological properties of soil (Goldberg 1985).

6.3.4 INCEPTISOLS

Inceptisols are spread over an area of nearly 925 million hectares throughout the world, and are the second-largest grouping on the FAO–UNESCO soil map of the World (FAO 1998). Contrary to many other orders, Inceptisols include soils from a varied range of environments from the Arctic to the Tropics. They lack developed features but occur in association with soils of nearly every other order. Depending on the dominant processes that occur in their particular landscape and geographic area, these soils may develop through numerous ways (Foss et al. 1983). Inceptisols (Cambisols) can develop on old or young geological material. They have profile features more weakly expressed than those of many other soils and retain a close resemblance to their parent material. The central concept of Inceptisols is that of soils developed in cool to very warm humid, and subhumid regions and that have a cambic horizon and an ochric epipedon. These are relatively young soils having features more weakly expressed than mature soils and retain close resemblance to the parent material. The Inceptisols include soils from ustic and udic regions that have altered B horizons resulting from some chemical weathering processes (Soil Survey Staff 1975). They may have any moisture regime except aridic and any temperature regime. Moisture storage capacity (mm) of these soils are 90–100/m in loamy sand, 110–140/m in sandy loam and 140–180/m in sandy loam. Out of three 110–140/m and 140–180/m depth have same soil texture. These soils have severe physical constraints due to the presence of large amounts of shrinking and swelling type of clays in the finer fractions. Inceptisols in Northwest India have 10YR hue except for horizon in the imperfectly drained soils of the alluvial terrace, which has 2.5YR hue, and value and chroma vary from 3 to 6 and 2 to 6, respectively. The texture of Inceptisols varies from sandy loam to clay loam in the middle hills (Sharma et al. 1997). These textural variations could be attributed to the in situ weathering under different rainfall conditions (Verma et al. 1987) and vegetation cover (Walia and Chamuah 1992). The Inceptisols have been reported to have fine soil texture compared with coarse-textured Podzolic soils formed in temperate regions because of slate parent material (Clayden et al. 1990). These soils are very fine in texture with clay content more than 30% (Sharma et al. 1997). These soils have strong angular blocky structure with pressure faces and slickensides. Due the presence of vertical cracks up to a depth of 30 cm and the ustic moisture regime, these soils are classified as Vertic Ustochrepts. The Inceptisols with placic horizon which is sufficient evidence to consider the soils as inceptisols and observed that the soils have properties related with Podzols and recommend strong accumulation and removal of soluble salts, Mn, Fe, and other materials. Soils have a dark organic and mineral matter (O/A) on the surface, a grey albic (E) horizon, a dark reddish-brown placic (Bs) horizon, and a brownish-yellow B horizon (Wu and Chen 2005). Inceptisols of the Arunachal hills are deep to very deep except those in high hills, which are relatively shallow. Total porosity of fine-textured Inceptisols range from 33% to 47.3% as compared to 30.5% to 41.3% in coarse-textured soils, which can be attributed to a higher number of finer particles per unit volume of soil in fine-textured soil resulting in higher surface area and hence total porosity (Walia and Chamuah 1996). The fine-textured Inceptisols are more porous than coarse-textured soils. In general, the total porosity decreased down the profile. The chemical and mineralogical changes may consist of hydrolysis, solution, reduction, segregation, or loss of free iron oxides. The fine-textured Inceptisols have not been shown to be chemically constrained for crop productivity (Sharma et al. 1997). Soil factors like pH, calcium carbonate, and organic matter and particle size fractions had a strong impact on the distribution of total and available micronutrients. The soil pH is slightly acidic, ranging from 6.0 to 6.8, and increase with depth, indicating base cations leaching or perhaps an organic matter effect. The pH of subsurface horizon of Inceptisols varies from neutral (pH 7.1) to strongly alkaline (pH 9.7), and surface horizon is almost neutral (pH 7.3) to alkaline (pH 8.6). The Inceptisols are acidic in nature with pH ranging from 4.0 to 5.3 (Sharma et al. 1997). The acidic reaction is due to high aluminum ions (Walia and Chamuah 1996). The values of electrical conductivity (EC) were low in all the profiles owing to good drainage. The total nitrogen, organic carbon, and C/N values ranged generally from moderate to high. The electrical conductivity indicates that these soils are nonsaline (Sharma et al. 1997). The organic carbon content is generally medium in surface and low in subsurface horizons in the soils of both ustic and udic moisture regimes. In these soils calcium carbonate content varied from 0 to 126 g kg^{-1}. The irregular distribution of $CaCO_3$ with depth may be due to the alluvial nature of these soils.

The CEC of these soils was low and varied from 5 to 10 c mol (p^+) kg^{-1} soil. The CEC was significantly correlated with clay content. The higher CEC values in the B horizons correlated with the clay content of the profiles. Calcium plus magnesium were dominant bases followed by sodium and potassium. The low values of CEC in proportion to their organic matter and/or clay content may be due to the dominance of illitic and kaolinitic clays (Prasad and Ram 1985; Singh and Datta 1983). Base saturation values of the profiles varied from 92.7% to 98.5%. The base saturation was low due to the predominance of aluminum and hydrogen ions on exchange complexes.

The grain yield and uptake of N, P, and K by rice and wheat were high with the application of farmyard manure, which in turn was due to increased total soil porosity, hydraulic conductivity, and mean weight diameter of the soils (Rasool et al. 2007). The positive impact of integrating inorganic fertilizers with FYM and bio inoculants of dehydrogenase, phosphatase enzyme activity, and soil microbial population showed the positive correlation with nutrient uptake and grain yield (Parewa et al. 2014). The mineralization and subsequent release of N from sewage sludge meet the needs of crop demand which further leads to maximization of NUE by wheat. Sewage sludge was collected from the Okhla sewage sludge treatment plant in Delhi, and characterized and incubated in an Inceptisol along with fertilizer N to monitor the release pattern of N. A pot culture experiment was also conducted to evaluate the efficiency of sludge for synchronization of N supply with the demand of wheat crop. An incubation experiment revealed that the sludge-amended soil released significantly higher amounts of NH_4^+-N as well as NO_3^--N than that of the control. Release of NH_4^+-N from the sludge-treated soil was slow during the first 15 days of incubation, then it increased and reached its maxima at around 45 days, after which it decreased sharply. Release of NO_3^--N was slow during the first 15 days of incubation, followed by a high release rate up to 45 days and a slow release rate beyond 45 days. A pot culture experiment revealed that maximum yield, N content, and uptake by wheat was obtained when the soil was treated with sludge along with fertilizer N @ 100 mg kg^{-1} soil, but those were statistically on par with the results obtained using sludge along with fertilizer N at 50 mg kg^{-1} soil. The same treatment, that is, sludge along with fertilizer N at 50 mg kg^{-1} soil, also showed the highest N recovery (50%). This treatment also showed significantly higher yield and N uptake over the sole fertilizer treatment, that is, N at 100 mg kg^{-1} soil along with the recommended dose of P and K. Hence, with sludge amendment, the fertilizer N requirement could be curtailed by 50% and might be implicative for better NUE (Biswas et al. 2017).

Inceptisols have a high level of aluminum content, and amendments are required to ensure a satisfactory crop stand. The practice of liming aids in neutralizing $Al^{3}+$, but on application to the topsoil its action is limited to the surface layers, and soil incorporation of lime is recommended at times. However, tillage may alter the physical properties of soil negatively. To avoid these negative effects of tillage, gypsum can be utilized as a substitute to boost $Ca^{2}+$ levels and decrease Al saturation in deeper layers. The effect of integrated nutrient management on soil properties and productivity of toria revealed that the application of 75% RDF + 5 Mg ha^{-1} vermicompost recorded significantly higher yield, benefit cost ratio, and rain water use efficiency (RWUE), along with a significant improvement in soil physical and chemical properties that remained on par with 75% RDF vermicompost at 5 Mg ha^{-1} in respect of toria productivity and soil properties (Hazarika et al. 2016).

6.4 ORGANIC FARMING AND INTEGRATED NUTRIENT MANAGEMENT

6.4.1 Organic Farming

Organic farming can be referred to as a practice that comprises growing and fostering crops involving exclusion of synthetic substances and encompassing the utilization of biological materials to sustain soil fertility and ecological balance, thus lowering wastage and pollution. It depends on ecologically stable agricultural principles like green manure, organic waste, biological pest control, mineral, rock additives, and crop rotation. The addition of organic sources contributes to improvement in the physicochemical properties of soil and regulates the release of nutrients for plant growth. Organic manures often offer an opportunity to minimize the usage of chemical fertilizers (Gopinath et al. 2013). They aid in supplying essential nutrients to the first crop and also leave a substantial residual effect on the succeeding crops in the system that often lasts for several seasons (Hegde 1998). Growing health consciousness among consumers paved the way for high demand of organically grown products that fetch lucrative prices. The potential of organic farming is suggested by the fact that the farm sector has ample organic resources like crop residue, forest litter, aquatic weeds, urban and rural solid wastes, agroindustrial waste and bioproducts, etc. (Bhattacharyya and Chakraborty 2005). Converting crop residues such as cotton stalks, pigeon-pea stalks, castor stalks. and weed residue into biochar on a large scale and its field application would enhance soil fertility and also serve as a good source of nutrient supplementation to crops (Indoria et al. 2018). Predominant organic manures and related resources used in rainfed dry land agriculture systems in India are presented in Figure 6.6.

6.4.1.1 Critical Role of Organic Manures in Rainfed Drylands

Manures are plant and animal wastes that are used as sources of plant nutrients. These manures, on mineralization, process the release of nutrients required for crop production, besides improving SOC content. Organic manures also contribute to the biological health of the soil and an overall

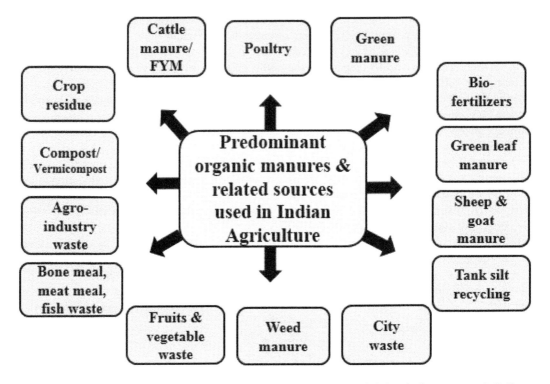

FIGURE 6.6 Predominant organic manures and related sources used in rainfed agriculture systems in India.

healthy soil system. Another essential aspect of continuous usage of organic manures in dryland farming is to retain and conserve the soil-available moisture, particularly at critical plant-growth stages. As drought, especially interdrought frequency, is increasing, retaining soil moisture and in particular protecting it from evaporation under high temperatures, the addition of organic manure has become more significant for sustainable agriculture systems in tropic regions. The critical role of adding organic manure along with fertilizers is presented in Figure 6.7, indicating that integrated use of fertilizers and manures sustained long-term finger millet production in semiarid Alfisols of southern India compared to chemical fertilizers alone (Srinivasarao et al. 2014a, 2015). Different forms of organic manures available for field application in dryland regions of India are cattle manure, compost, crop residue, vermicompost, green and green-leaf manure, sheep and goat manure, poultry manure, agroindustry wastes, bone meal, and city waste and related materials (Indoria et al. 2018).

6.4.2 INTEGRATED NUTRIENT MANAGEMENT (INM)

Reliance on organic sources alone will not aid in achieving remarkable improvement in yield, which can be attributed to their low nutrient status, while dependence on only synthetic fertilizers will result in soil health deterioration and pollution perils. Nonrenewable energy consumption by inorganic fertilizers and escalating prices also raise the utmost concern. In modern agriculture,

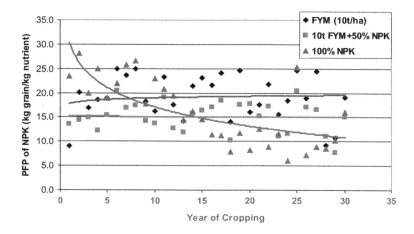

FIGURE 6.7   Critical role of soil organic matter. Partial factor productivity of NPK in finger millet in 30-year long-term experiment at Bangalore under rainfed conditions (1978–2007). (From Srinivasarao, Ch. et al., Potential and challenges of rainfed farming in India, In: Sparks, D.L. (Ed.), *Adv. Agron.*, 113–181, 2015.)

plant nutrient supply solely through chemical fertilizers or organic manures, crop residues, or biofertilizers cannot serve the entire nutrient need of crops, hence the calls for the adoption of a management strategy that combines chemical fertilizers and organic sources that is practicable, socially acceptable, ecologically sound, and economically viable. INM refers to the maintenance of soil fertility and plant nutrient supply at an ideal level for sustaining desired productivity and enhancing nutrient use efficiency utilizing all possible sources of organic, inorganic, and biological components in an integrated manner that can be regarded as a feasible option to sustain soil health and enhance productivity, eventually aiding farmers in reaping high profits (Srinivasarao et al. 2016a, 2016b, 2018).

Seed inoculation with *Azospirillum* and *Azotobacter* increased the grain yield of sorghum by 8.73% and 11.91%, respectively, compared with no inoculation. Yield attributes and grain yield were increased with increasing levels of nitrogen up to 90 kg ha^{-1}, where grain yield of 1514 kg ha^{-1} was achieved. The interaction effect between inoculation and N levels was found significant with respect to panicle length, grains/panicle, test weight, and grain yield of sorghum (Barik and Nag 2001). The effect of organics and inorganics on variation in soil water content at important crop growth periods and their effect on grain yield of sorghum under rainfed condition revealed that the combined application of FYM, vermicompost, and *Azospirillum* recorded, significantly, the highest test weight (37.16 g), grains per ear head (1642), and grain yield (3477 kg ha^{-1}) when compared to their individual application (Kalibhavi et al. 2002). Significantly higher seed cotton yield was observed with application of 50% N through *Gliricidia* + 50% N through inorganics + biofertilizers + 100% P + 25 kg K ha^{-1}, which was comparable with the application of 100% NP + biofertilizers + 25 kg K ha^{-1} (Khambalkar et al. 2017). Dryland farming zones and their treatment details (inorganic, organic, and INM), mean yield, sustainability yield index, SOC, and nitrogen use efficiency of various production systems in different soils of rainfed areas in India are presented in Tables 6.7 and 6.8, respectively.

TABLE 6.7

Dryland Farming Zones and Their Regional Characteristics, and Treatment Details of Various Production Systems in Different Soils of Rainfed Areas in India

S. No.		Location	State	Latitude and Longitude	Climate	Rainfall (mm)	Year of Initiation (Duration)	Production System	Treatment Details
1.	Alfisols	Anantapur	Andhra Pradesh	14°42′ N, 77°40′ E, 350 m	Arid	566	1985–2004 (20 years)	Groundnut	T_1 = Control (no fertilizer), T_2 = 100% recommended dose of fertilizer (RDF) (20:40:40 N, P_2O_5, K_2O), T_3 = 50% RDF + 4 Mg groundnut shells (GNS) ha^{-1}, T_4 = 50% RDF + 4 Mg FYM ha^{-1}, T_5 = 100% organic (5 Mg FYM ha^{-1})
		Bangalore	Karnataka	12°46′ N, 77°11′ E, 810 m	Semiarid	768	(1978–2004) (27 years)	Finger millet	T_1 = Control, T_2 = 10 Mg FYM ha^{-1}, T_3 = 10 Mg FYM ha^{-1} + 50% NPK, T_4 = 10 Mg FYM ha^{-1} + 100% NPK, T_5 = Recommended NPK (50:50:25 kg NPK ha^{-1} – finger millet)
		Bangalore	Karnataka	12°46′ N, 77°11′ E, 810 m	Semiarid	768	(1991–2004) (13 years)	(Finger millet– Groundnut)	T_1 = Control, T_2 = 10 Mg FYM ha^{-1}, T_3 = 10 Mg FYM ha^{-1} + 50% NPK, T_4 = 10 Mg FYM ha^{-1} + 100% NPK, T_5 = Recommended NPK (25:50:25 kg NPK ha^{-1} – groundnut; 50:50:25 kg NPK ha^{-1} – finger millet)
2.	Vertisols	Indore	Madhya Pradesh	22°71′ N, 75°85′ E	Semiarid	958	1992–2007 (15 years)	Soybean– safflower sequence	T_1 = Control, T_2 = 20 kg N+ 13 kg P, T_3 = 30 kg N+ 20 kg, T_4 = 40 kg N+ 26 kg, T_5 = 60 kg N+ 35 kg P, T_6 = 6 Mg FYM ha^{-1} + $N_{20}P_{13}$, T_7 = 5 Mg soybean residue ha^{-1} + $N_{20}P_{13}$, T_8 = 6 Mg FYM ha^{-1}, T_9 = 5 Mg soybean residues ha^{-1}

(Continued)

TABLE 6.7 (Continued)
Dryland Farming Zones and Their Regional Characteristics, and Treatment Details of Various Production Systems in Different Soils of Rainfed Areas in India

S. No.	Location	State	Latitude and Longitude	Climate	Rainfall (mm)	Year of Initiation (Duration)	Production System	Treatment Details
	Solapur	Maharashtra	17°65′ N, 75°90′ E	Semiarid	732	1985–2006 (22 years)	Rabi sorghum	T_1 = Control, T_2 = 25 kg N ha^{-1} (Urea), T_3 = 50 kg N ha^{-1} (Urea), T_4 = 25 kg N ha^{-1} through sorghum residue (CR), T_5 = 25 kg N ha^{-1} through FYM, T_6 = 25 kg N ha^{-1} (CR) + 25 kg N ha^{-1} (Urea), T_7 = 25 kg N ha^{-1} (FYM) + 25 kg N ha^{-1} (Urea), T_8 = 25 kg N ha^{-1} (CR) + 25 kg N ha^{-1} (*Leucaena* clippings), T_9 = 25 kg N ha^{-1} (*Leucaena*), T_{10} = 25 kg N ha^{-1} (*Leucaena*) + 25 kg N ha^{-1} (Urea)
3.	Inceptisols Varanasi	Uttar Pradesh	25°11′ N, 82°51′ E, 480 m	Semiarid and Subhumid	1080	1986–2007 (21 years)	Rice–lentil	T_1 = Control, T_2 = 100% RDF (inorganic), T_3 = 50% RDF (inorganic), T_4 = 100% organic (FYM), T_5 = 50% organic (FYM), T_6 = 50% RDF + 50% (foliar), T_7 = 50% organic (FYM) + 50% RDF, T_8 = Farmers' practice
4.	Aridisols SK Nagar	Gujarat	24°30′ N, 72°13′ E, 152.5 m	Semiarid/ Arid	670	1988–2006 (18 years)	Pearl millet– clusterbean– castor rotation (once in 3 years)	T_1 = Control, T_2 = 100% recommended dose of N through mineral fertilizer (RDNF), T_3 = 50% RDNF, T_4 = 50% recommended N (FYM), T_5 = 50% recommended N (fertilizer) + 50% recommended N (FYM), T_6 = Farmers' method (5 Mg of FYM ha^{-1} once in 3 years)

TABLE 6.8
Dryland Farming and Treatment Details ([Inorganic, Organic, and INM], Mean Yield, Sustainability Yield Index, Soil Organic Carbon and Nitrogen Use Efficiency) of Various Production Systems in Different Soils of Rainfed Areas in India

S. No	Soil Type (Broad Soil Group)	Location/State	Production System	Year of Initiation (Duration)	Treatmental Details	Inorganic	Mean Yield (Mg ha⁻¹)	SYI	SOC	NUE	INM	Mean Yield (Mg ha⁻¹)	SYI	SOC	NUE	Organic	Mean Yield (Mg ha⁻¹)	SYI	SOC	NUE
1.	Alfisols	Anantapur, Andhra Pradesh	Groundnut	1985–2004 (20 years)	T_1 = Control (no fertilizer), T_2 = 100% recommended dose of fertilizer (RDF) (20:40:40 N, P_2O_5, K_2O), T_3 = 50% RDF + 4 Mg groundnut shells (GNS) ha⁻¹, T_4 = 50% RDF + 4 Mg FYM ha⁻¹, T_5 = 100% organic (5 Mg FYM ha⁻¹)	T_2 = 100% recommended dose of fertilizer (RDF) (20:40:40 N, P_2O_5, K_2O)	0.98	0.32	3.3	10.0	T_3 = 50% RDF + 4 Mg groundnut shells (GNS) ha⁻¹	1.02	0.48	4.6	4.1	T_5 = 100% organic (5 Mg FYM ha⁻¹)	0.92	0.38	3.3	5.6
		Bangalore, Karnataka	Finger millet	(1978–2004) (27 years)	T_1 = Control, T_2 = 10 Mg FYM ha⁻¹, T_3 = 10 Mg FYM ha⁻¹ + 50% NPK, T_4 = 10 Mg FYM ha⁻¹ + 100% NPK, T_5 = Recommended NPK (50:50:25 kg NPK ha⁻¹ – finger millet)	T_5 = Rec. NPK (50:50:25 kg NPK ha⁻¹)	2.16	0.36	4.9	26.4	T_4 = 10 Mg FYM ha⁻¹ + 100% NPK	3.28	0.59	6.4	24.4	T_2 = 10 Mg FYM ha⁻¹	2.48	0.58	4.2	32.8

(Continued)

TABLE 6.8 (Continued)

Dryland Farming and Treatment Details ([Inorganic, Organic, and INM], Mean Yield, Sustainability Yield Index, Soil Organic Carbon and Nitrogen Use Efficiency) of Various Production Systems in Different Soils of Rainfed Areas in India

S. No	Soil Type (Broad Soil Group)	Location/State	Production System	Year of Initiation (Duration)	Treatmental Details	Inorganic	Mean Yield (Mg ha⁻¹)	SYI	SOC	NUE	INM	Mean Yield (Mg ha⁻¹)	SYI	SOC	NUE	Organic	Mean Yield (Mg ha⁻¹)	SYI	SOC	NUE
		Bangalore, Karnataka	(Finger millet Ground-nut	(1991–2004) (13 years)	T_1 = Control, T_2 = 10 Mg FYM ha⁻¹, T_3 = 10 Mg FYM ha⁻¹ + 50% NPK, T_4 = 10 Mg FYM ha⁻¹ + 100% NPK, T_5 = Recommended NPK (25:50:25 kg NPK ha⁻¹ – groundnut; 50:50:25 kg NPK ha⁻¹ – finger millet)	T_5 = Rec. NPK 50:50:25 kg NPK ha⁻¹ – finger millet; (25:50:25 kg NPK ha⁻¹ – ground nut)	2.58 for finger millet 0.72 for groundnut	0.62 for finger millet 0.16 for groundnut	4.6	35.2 for finger millet 12.8 for gnut	T_4 = 10 Mg FYM ha⁻¹ + 100% NPK	3.96 for finger millet 1.34 for ground-nut	0.76 for finger millet 0.21 for groundnut	5.7	31.4 for finger millet 12.5 for ground nut	T_2 = 10 Mg FYM ha⁻¹	3.25 for finger millet 1.10 for ground nut	0.71 for finger millet 0.23 for ground nut	5.2	48.6 for finger millet 14.0 for gnut
2.	Vertisols	Indore, Madhya Pradesh	Soybean–safflower sequence	1992–2007 (15 years)	T_1 = Control, T_2 = 20 kg N+ 13 kg P; T_3 = 30 kg N+ 20 kg. T_4 = 40 kg N+ 26 kg, T_5 = 60 kg N+ 35 kg P, T_6 = 6 Mg FYM ha⁻¹+ $N_{20}P_{13}$, T_7 = 5 Mg soybean residue ha⁻¹+ $N_{20}P_{13}$, T_8 = 6 Mg FYM ha⁻¹, T_5 = 5 Mg soybean residues ha⁻¹	T_5 = 60 kg N+ 35 kg P	1.99 for soybean 1.21 for safflower	0.48 for soybean 0.31 for safflower	4.6	15.8 for soybean 9.7 for safflower	T_6 = 6 Mg FYM ha⁻¹ $N_{20}P_{13}$	2.10 – soybean, 1.49 safflower	0.48 – soybean, 0.45 safflower	5.4	21.2 for soy bean 17.2 for safflower	T_8 = 6 Mg FYM ha⁻¹	1.86 for soybean, 1.22 for safflower	0.42 for soy-bean, 0.38 for safflower	5.1	27.3 for soy-bean 19.7 for safflower

(Continued)

TABLE 6.8 (Continued)

Dryland Farming and Treatment Details ([Inorganic, Organic, and INM], Mean Yield, Sustainability Yield Index, Soil Organic Carbon and Nitrogen Use Efficiency) of Various Production Systems in Different Soils of Rainfed Areas in India

S. No	Soil Type (Broad Soil Group) Location/State	Production System	Year of Initiation (Duration)	Treatmental Details	Inorganic	Mean Yield (Mg ha⁻¹)	SYI	SOC	NUE	INM	Mean Yield (Mg ha⁻¹)	SYI	SOC	NUE	Organic	Mean Yield (Mg ha⁻¹)	SYI	SOC	NUE
	Solapur. Maharashtra	Rabi sorghum	1985–2006 (22 years)	T_1 = Control, T_2 = 25 kg N ha⁻¹ (Urea), T_3 = 50 kg N ha⁻¹ (Urea), T_4 = 25 kg N ha⁻¹ through sorghum residue (CR), T_5 = 25 kg N ha⁻¹ through FYM, T_6 = 25 kg N ha⁻¹ (CR)+25 kg N ha⁻¹ (Urea), T_7 = 25 kg N ha⁻¹ (FYM)+25 kg N ha⁻¹ (Urea), T_8 = 25 kg N ha⁻¹ (CR)+25 kg N ha⁻¹ (Leucaena clippings), T_9 = 25 kg N ha⁻¹ (Leucaena), T_{10} = 25 kg N ha⁻¹ (Leucaena) +25 kg N ha⁻¹ (Urea)	T_3 = 50 kg N ha⁻¹ (Urea)	1.04	0.41	4.5	8.6	T_{10} = 25 kg N (Leucaena) +25 kg N ha⁻¹ (Urea)	1.19	0.44	5.6	11.6	T_{10} = 25 kg N ha⁻¹ (CR) +25 kg N ha⁻¹ (Leucaena)	0.85	0.48	4.4	4.8

(Continued)

TABLE 6.8 (Continued)

Dryland Farming and Treatment Details ([Inorganic, Organic, and INM], Mean Yield, Sustainability Yield Index, Soil Organic Carbon and Nitrogen Use Efficiency) of Various Production Systems in Different Soils of Rainfed Areas in India

S. No	Soil Type (Broad Soil Group)	Location/State	Production System	Year of Initiation (Duration)	Treatmental Details	Inorganic Mean Yield (Mg ha⁻¹)	SYI	SOC	NUE	INM Mean Yield (Mg ha⁻¹)	SYI	SOC	NUE	Organic Mean Yield (Mg ha⁻¹)	SYI	SOC	NUE
3.	Inceptisols	Varanasi, Uttar Pradesh	Rice–lentil	1986–2007 (21 years)	T$_1$ = Control, T$_2$ = 100% RDF (inorganic), T$_3$ = 50% RDF (inorganic), T$_4$ = 100% organic (FYM), T$_5$ = 50% organic (FYM), T$_6$ = 50% RDF + 50% (foliar), T$_7$ = 50% organic (FYM)+ 50%RDF, T$_8$ = Farmers' practice	1.85-rice, 0.77-lentil	0.25-rice, 0.24- lentil	1.7	12.8 for rice	1.95 for rice, 1.04-lentil	0.29-rice, 0.30-lentil	2.0	14.5 for rice	T$_4$ = 100% organic (FYM) 1.75- rice, 0.82-lentil	0.26-rice, 0.26- lentil	2.1	11.2 for rice

(Continued)

TABLE 6.8 (Continued)
Dryland Farming and Treatment Details ([Inorganic, Organic, and INM], Mean Yield, Sustainability Yield Index, Soil Organic Carbon and Nitrogen Use Efficiency) of Various Production Systems in Different Soils of Rainfed Areas in India

S. No	Soil Type (Broad Soil Group)	Location/State	Production System	Year of Initiation (Duration)	Treatmental Details	Inorganic	Mean Yield (Mg ha⁻¹)	SYI	SOC	NUE	INM	Mean Yield (Mg ha⁻¹)	SYI	SOC	NUE	Organic	Mean Yield (Mg ha⁻¹)	SYI	SOC	NUE
4.	Aridisols	SK Nagar, Gujarat	Pearl millet–cluster bean–castor rotation (once in 3 years)	1988–2006 (18 years)	T_1 = Control, T_2 = 100% recommended dose of N through mineral fertilizer (RDNF), T_3 = 50% RDNF; T_4 = 50% recommended N (FYM), T_5 = 50% recommended N (fertilizer) + 50% recommended N (FYM), T_6 = Farmers' method (5 Mg of FYM ha⁻¹ once in 3 years)	T_2 = 100% recommended dose of N through mineral fertilizer (RDNF)	0.78-pmillet, 0.45-cbean, 0.80-castor	0.24-pmillet,0.46-cbean, 0.40-castor	1.5	14.4-pmillet, 6.5-cbean, 6.0-castor	T_5 = 50% recommended N (fertilizer) + 50% recommended N (FYM)	0.81-pmillet, 0.58-cbean, 0.83-castor	0.30-pmillet, 0.69-cbean, 0.46-castor	2.2	4.8- pmillet, 13.0-cbean, 6.5- castor	T_4 = 50% recommended N (FYM)	0.55-pmillet, 0.48-cbean, 0.63-castor	0.25-pmillet, 0.46-cbean, 0.39-castor	1.9	1.5-pmillet, 8.0-cbean, 3.2-castor

6.4.3 Conservation Agriculture (CA) Systems

Conservation agriculture (CA) practices help to reduce soil losses besides soil nutrient and organic carbon losses. It also significantly contributes to drought adaptation and soil fertility enhancement (Kundu et al. 2013). Typical CA, including soil cover with crop residues, minimum tillage, and rotation with legumes, was practically not possible due to deficient rainfall or drought situations in dryland regions of India due to severe moisture stress. Harvesting rainwater, either in situ or in farm ponds, is critical for sustainable rainfed agriculture. For this soil profile, saturation first with rain in rain-dependent crops/zones with land treatments is essential. Therefore, we have proposed four principles of CA for rainfed dryland systems in India (Srinivasarao et al. 2015).

6.4.4 Role of Manure/Fertilizer Management in Climate-Resilient Villages in Dryland Ecosystems

Dryland ecosystems are the most vulnerable to climate change, and impacts and often agriculture and livelihoods are negatively affected in India and the world. Vulnerability indicators (sensitivity, exposure, and adaptation) of dryland ecosystems are very low, and a strong need was felt and implemented to build climate-resilient villages across India to reduce overall vulnerability (Srinivasarao et al. 2016b). Several location-specific technologies were developed in different rainfed dryland ecosystems of India (Srinivasarao et al. 2014b), and these technologies were implemented in climate-resilient villages in India. These model villages are being replicated in India under larger programs with good policy support. The critical role of manures and improved nutrient use efficient technologies were amply demonstrated along with provision of soil health cards in dry regions (Srinivasarao et al. 2018) and contributed to carbon sink capacity of soils and overall reduction of greenhouse gas (GHG) emissions (Srinivasarao et al. 2017a).

6.4.5 Agriculture Contingency Plans

Several technologies related to manure and fertilizer management practices that provide drought tolerance were included in district agriculture contingency plans developed for 623 districts of India. Cover cropping, mulch with manuring, green and green leaf manuring, and crop residue management technologies included in these plans not only contributed to drought adaptation but also soil fertility/health improvement (Srinivasarao et al. 2017b). However, these plans have to be implemented at the village/farm level in a much more vigorous way to harness the envisaged benefits to farming in drought-affected regions of India and elsewhere.

6.5 NATIONAL PROGRAMS ON NUTRIENT USE AND POLICY

6.5.1 Soil Health Cards

In India, the Soil Health Card (SHC) was launched by the government of India in February 2015. A SHC display the soil health indicators which is associated with clear images of soil colour which further evaluate how well soil functions since soil testing in lab often cannot be directly measured. Soil samples are analyzed in an authenticated soil laboratory and determine crops suited to that particular soil based on the availability of nutrient content in the soil as well as nutrient requirement of the crop. Based on the soil test results, the SHC will give information to farmers on the nutrient status of the soil and give essential procedures that have to be practiced in order to improve the soil fertility. In a cycle of three years, the soil will be analyzed and SHCs will be issued to all the farmers in the country so that soil properties and nutrient deficiencies are identified in order to improve the defects of the soil and indicate the status of soil health. SHC-based-nutrient application contributed to balanced nutrient management systems in dryland ecosystems of Andhra Pradesh, India, and improved crop productivity of rainfed crops (Srinivasarao et al. 2011b).

6.5.2 NATIONAL PROJECT ON ORGANIC FARMING

A central sector scheme, the National Project on Organic Farming (NPOF), was introduced in 2004, with the idea of promoting organic farming practices to lessen the huge demand for chemical fertilizers, to warrant efficient utilization of farm resources, and to meet the demands of both the domestic and the international organic market. The major objective of the NPOF is to promote organic farming. This project also enhanced the usage of four biofertilizers, namely. Azotobacter, Azospirillum, PSB, and Rhizobium, and two organic fertilizers: city waste compost and vermicompost. Production and promotion of a large number of organic inputs/products are in progress with an aim to sustain soil health. The farm waste that is crop residue waste is converted into biochar, which otherwise is burned in the fields. At the farmer level, crop residues are converted into biochar utilizing a low-cost biochar-making kiln, and its successive application enhances the percentage of organic matter and carbon sequestration process in the soil, which ultimately improves the soil physical, chemical, and biological health.

Organic farming is being encouraged through various programs and schemes, namely, the Mission for Integrated Development of Horticulture (MIDH), the National Programme on Organic Production (NPOP), Rashtriya Krishi Vikas Yojana (RKVY), National Mission on Oilseeds & Oil Palm (NMOOP), and Agricultural & Processed Food Products Export Development Authority (APEDA). Paramparagat Krishi Vikas Yojna (PKVY) was launched as a component of Soil Health Management (SHM) by the government of India in 2015. A cluster approach is being followed to implement this scheme, with every cluster consisting of 50 farmers, each having one acre for conversion to organic farming. The target was to have 10,000 such clusters in the country over a three-year period.

6.5.3 INTEGRATED FARMING SYSTEM

The integrated farming system (IFS) approach can be defined as a judicious mix of two or more components using fundamental principles of minimum competition and maximum complementarity with progressive agronomic management tools targeting sustainable and environment-friendly enhancement of farm income, family nutrition, and ecosystem services (Singh and Ravisankar 2015). The current farming situation in India demands an integrated effort to address the emerging problems and issues. The IFS approach is regarded as the most powerful tool for augmenting farming systems' profitability. Income achieved from cropping alone is not sufficient to satisfy the needs of farmers. Hence, the income of farmers be enhanced and supplemented by the adoption of effectual secondary and tertiary enterprises like horticulture (vegetables, fruits, and flowers and medicinal and aromatic plants), animal husbandry, mushroom cultivation, apiaries and fisheries, etc.

Farmers in rainfed regions practice crop–livestock mixed-farming systems, which provide stability during drought years, lessen their risk, and aid in coping with weather abnormalities. However, these traditional systems are less profitable and cannot warrant immediate livelihood security. The drop in the size of landholdings size, low investments, eroded and degraded soils with multiple nutrient deficiencies, and weather anomalies negatively affect the productivity and sustainability of farming. The farming systems approach is regarded as imperative and pertinent, particularly for marginal and small farmers, as site-specific IFS will be more adaptive and resilient to aberrations in climate. It also has the potential to overcome the diverse problems of farmers, including declining resource use efficiency, resource degradation, farm productivity, and profitability.

6.5.4 NEEM COATED UREA SCHEME

This scheme was introduced with an aim of enhancing soil health and lowering pest and disease attacks, thereby minimizing usage of plant protection chemicals, increasing the overall crop yield, and reducing the diversion of urea toward nonagricultural purposes. The soils in India are faced with an emerging crisis of multinutrient deficiencies, and the government has taken several initiatives for the reduction of organic carbon depletion and deterioration of soil health. Programs like site-specific nutrient management (SSNM) aid in the efficient use of nutrients and the application of nutrients in a balanced manner and ultimately lead to the betterment of soil health.

6.5.5 FERTILIZER POLICY

In order to accomplish enhancement in agricultural production, fertilizer plays an inevitable role. Ensuring adequate availability of fertilizers at affordable rates for farmers across the country is of utmost concern. Hence, fertilizer has been declared as an essential commodity and the Fertilizer Control Order (FCO) was passed in 1985 to regulate sale, price, and quality of fertilizers. The FCO provides for compulsory registration of fertilizer manufacturers, importers, and dealers; specification of all fertilizers sold in the country; regulation of the manufacture of fertilizer mixtures; packing and marking on fertilizer bags; the appointment of enforcement agencies; quality-control laboratory setup; and prohibition of the manufacture, import, and sale of nonstandard, adulterated, or spurious fertilizers. The escalating cost of chemical fertilizers restricts farmers from its usage, which makes clear the need for the implementation of policies and schemes that will aid in the easy procurement and usage of fertilizers. Various policies have been implemented from time to time. Different policies introduced during different periods with their objectives are presented in Table 6.9.

TABLE 6.9
Fertilizer Policies Introduced During Different Periods in India

S. No.	Policy	Period	Objective
1.	New Pricing Scheme (NPS) – I	1 April 2003 to 31 March 2004	50% and 75% of urea was to be introduced under the terms of the Essential Commodities Act of 1955.
2.	New Pricing Scheme (NPS) – II	1 April 2004 to 31 September 2006	Urea distribution to be completely decontrolled on the basis of Phase I evaluation and Ministry of Agriculture (MoA) recommendations.
3.	New Pricing Scheme (NPS) – III	1 October 2006 to 1 April 2014	Increasing efficiency of urea and its distribution in the country. Promote the usage of natural gas, which is an efficient feedstock for production of urea.
4.	Modified New Pricing Scheme (NPS) – III	2 April 2014 to 31 May 2015	Compensation of fixed cost and variable cost, namely, cost of bag, water charges, and electricity.
5.	New Urea Policy	1 June 2015 to 31 March 2019	Maximizing indigenous urea production, promoting energy efficiency in urea production, and rationalizing subsidy burden on the government.
6.	New Investment Policy	2 January 2012	Enable fresh investments, make India self-sufficient, and lower import dependency in urea sector.
7.	Nutrient Based Subsidy (NBS) Policy	February 2010	Implementing subsidy on P and K fertilizers.

6.6 CONCLUSIONS AND WAY FORWARD

Rainfed dryland agroecosystems have huge untapped potential that can contribute to a major portion of the global food basket if available technological interventions are implemented intensively in more pragmatic way. This requires a strong dynamic interrelationship between the technological needs of a particular rainfed dryland ecosystem and the implementation process. One of the important interventions urgently needed is soil health restoration, which is the basis of overall food production systems' sustainability. Declining soil fertility and the emergence of multinutrient deficiencies in Indian drylands are serious productivity constraints in targeting higher crop productivity. Over the last five decades of crop intensification in rainfed regions, the amount of fertilizer application has increased tremendously, and the corresponding partial-factor productivity of fertilizer inputs has declined gradually. During this period, the dependency of plant nutrition systems moved from traditional organic manure application to the use of chemical fertilizers alone. This led to negative ecological and environmental consequences such as higher input costs and loss of nutrients or higher GHG emissions. On the other hand, available organic resources are not used in agriculture in the form of mulch cum manuring or composting and so on, and the same crop residue is burned in the field itself. Several technologies have evolved in the judicious use of inorganic chemical fertilizers with locally available organic manures; however, these technology implementations in Indian dryland ecosystems require an acceptance/faith from farmers side towards the adaptation of this technology. Consistent application of organic amendments and residue recycling is imperative to sustain productivity and boost the benefits of fertilizer addition. The massive quantities of organic sources are available in the country that can be made available by the action of soil microbes and recycling of farm organic materials results in supply of nutrients to the soil as well as crop. As sole application of organic amendments cannot address the whole requirement of crop nutrition, an integrated nutrient management approach that involves utilization of inorganic fertilizers along with organic sources is the solution.

A comprehensive concerted effort in the utilization of locally available components of INM comprising rational and appropriate use of fertilizers and organics will go a long way in providing a sustainable crop nutrition management. Moreover, a number of nutrient elements in soil, namely, Mg, S, Zn, and B, are found deficient in sole chemical fertilizer treated fields compared to INM. In addition, fertilizer management should be based on cropping systems rather than sole crops for higher nutrient use efficiency and economics. Hence, the integrated plant nutrient system has emerged as a necessity for the sustainability of agriculture in India.

Evaluation of INM technologies (with secondary/micronutrients) ought to be made only after a thorough inventory of the available resources in a region, including the components of production, namely, tillage practices, water management, moisture conservation practices, managing crops with location-specific technology, biotic and abiotic stresses, and the cropping/farming system. Developing awareness among farmers through extension activities about deteriorating soil health, unsustainable production, and environmental pollution due to reduced or no use of organics in overall soil health management in rainfed dryland ecosystems, coupled with conservation agriculture practices, replication of climate-resilient villages, and intensive implementation of district contingency plans in the country, will result in developing resilient dryland production systems. At the same time, policy implementation in support of sustainable soil health management and incentive support will be crucial for successful village-level adoption of these technologies.

REFERENCES

Aggarwal, R. K., and J. Venkateswarlu. 1989. Long-term effect of manures and fertilizers on important cropping systems of arid region. *Fertilizer News* 34: 67–70.

Aggarwal, R. K., Praveen-Kumar, and J. F. Power. 1996. Use of crop residue and manure to conserve water and enhance nutrient availability and pearl millet yield in an arid tropical region. *Soil and Tillage Research* 41: 43–51.

Anderson, S. H., C. J. Gantzer, and J. R. Brown. 1990. Soil physical properties after 100 years of continuous cultivation. *Journal of Soil and Water Conservation* 45: 117–121.

Balaguravaiah, D., G. Adinarayana, S. Prathap, et al. 2005. Influence of long-term use of inorganic and organic manures on soil fertility and sustainable productivity of rainfed groundnut in Alfisols. *Journal of the Indian Society of Soil Science* 53(4): 608–611.

Barik, A. K., and P. Nag. 2001. Effect of bio fertilizers with nitrogen levels on grain yield of Sorghum fodder. *Forage Research* 27(3): 199–202.

Bhattacharyya, P., and G. Chakraborty. 2005. Current status of organic farming in India and other countries. *Indian Journal of Fertilisers* 1(9): 111–123.

Bhattacharyya, T., D. K. Pal, C. Mandal, et al. 2013. Soils of India: Historical perspective, classification and recent advances. *Current Science* 104(10): 1308–1323.

Biswas, S. S., S. K. Singhal, D. R. Biswas, et al. 2017. Synchronization of nitrogen supply with demand by wheat using sewage sludge as organic amendment in an Inceptisol. *Journal of the Indian Society of Soil Science* 65(3): 264–273.

Brandon, C., and K. Hommann. 1995. The cost of inaction: Valuing the economy wide cost of environmental degradation in India. Mimeo. Asia Environment division, World Bank, Washington DC.

Castro, C., and T. J. Logan. 1991. Liming effects on the stability and erodibility of some Brazilian Oxisols. *Soil Science Society of America Journal* 55: 1407–1413.

Celik, I., I. Ortas, and S. Kilic. 2004. Effects of compost, mycorrhiza, manure and fertilizer on some physical properties of a Chromoxerert soil. *Soil and Tillage Research* 78: 59–67.

Chand, S., S. Kumar, B. P. Gautam, et al. 2015. Enhance the soil health, nutrient uptake and yield of crops under rice-wheat cropping system through green manuring. *Inter Journal of Tropical Agriculture* 33(3): 2075–2079.

Charreau, C. 1977. Some controversial technical aspects of farming systems in semi-arid West Africa. In: Cannell, G. H. (Ed.) *Proceedings of an International Symposium on Rainfed Agriculture in Semi-Arid Tropics*, April 17–22, Riverside, California. Cannell, University of California, Riverside, CA, pp. 313–360.

Chen, Y., M. Camps-Arbestain, Q. Shen, et al. 2018. The long-term role of organic amendments in building soil nutrient fertility: A meta-analysis and review. *Nutrient Cycling in Agroecosystems* 111: 103–125.

Chesti, M. H., and A. Tahir. 2012. Rhizospheric micro-flora, nutrient availability and yield of green gram (*Vigna radiata* L.) as influenced by organic manures, phosphate solubilizers and phosphorus levels in Alfisols. *Journal of the Indian Society of Soil Science* 60(1): 25–29.

Chivenge, P., B. Vanlauwe, and J. Six. 2011. Does the combined application of organic and mineral nutrient sources influence maize productivity? A meta-analysis. *Plant Soil* 342: 1–30.

Choudhari, J. S., and B. L. Pareek. 1976. Exchangeable reserve potassium in soils of Rajasthan. *Journal of the Indian Society of Soil Science* 24: 57–61.

Choudhari, J. S., S. V. Jain, and C. M. Mathur. 1979. Phosphorus fractions in some soil orders of Rajasthan and their sequence of weathering stage. *Annals of Arid Zone* 18: 260–268.

Clayden, B., B. K. Daly, R. Lee, et al. 1990. The nature, occurrence, and genesis of placic horizons. In: Kimble, J. M. (Ed.) *Proceedings of Fifth International Soil Correlation Meeting*. USDA-SCS, Lincoln, NE, pp. 88–104.

CRIDA-ORP. 1997. *Progress Report of Operation Research Projects of All India Coordinated Research Project on Dryland Agriculture*, CRIDA, Santoshnagar, Hyderabad, pp. 519.

De-Ruiter, P. C., A. M. Neutel, and J. C. Moore. 1998. Biodiversity in soil ecosystems: The role of energy flow and community stability. *Applied Soil Ecology* 10: 217–228.

De Vos, T. N. C., J. H., and Virgo K. J. 1969. Soil structure in Vertisols of the Blue Nile clay plains, Sudan. *European Journal of Soil Science* 20: 189–206.

Du, Z., S. Liu, K. Li, et al. 2009. Soil organic carbon and physical quality as influenced by long-term application of residue and mineral fertilizer in the North China Plain. *Australian Journal of Soil Research* 47: 585–591.

Dudal, R. 1965. A report on Dark clay soils of tropical and sub tropical regions. FAO, Rome, Agricultural Development Paper. pp. 1–148.

Dutta, B. K., and D. C. Joshi. 1993. Studies on potassium fixation and release in arid soils of Rajasthan. *Transactions of the Indian Society of Desert Technology* 18: 191–199.

Eldridge, S. M., K. Y. Chan, N. J. Donovan, et al. 2018. Agronomic and economic benefits of green-waste compost for peri-urban vegetable production – Implications for food security. *Nutrient Cycling in Agroecosystems* 111: 155–173.

FAO (Food and Agriculture Organization of the United Nations). 1998. *World Reference Base for Soil Resources.* World Soil Resources Reports 84. Food and Agriculture Organization of the United Nations, Rome, Italy.

Fertilizer Statistics 2017–2018. 63rd ed. Fertilizer Association of India, New Delhi, India.

Foss, J. E., F. R. Moormann, and S. Rieger. 1983. Inceptisols. In: Wilding, L. P., Smeck, N. E., and Hall, G. F. (Eds.) *Pedogenesis and Soil Taxonomy. II. The Soil Orders.* Elsevier, Amsterdam, the Netherlands, p. 410.

Franzluebbers, A. J. 2002. Water infiltration and soil structure related to organic matter and its stratification with depth. *Soil and Tillage Research* 66:97–105.

Gabhane, V. V., B. A. Sonune, and R. N. Katkar. 2014. Long term effect of tillage and integrated nutrient management on soil quality and productivity of rainfed cotton in Vertisols under semi-arid conditions of Maharashtra. *Indian Journal of Dryland Agricultural Research and Development* 29(2):71–77.

Goldberg, E. D. 1985. *Black Carbon in the Environment Properties and Distribution.* John Wiley & Sons, New York.

Gopinath, K. A., Ch. Srinivasarao, G. Venkatesh, et al., 2013. Climate change adaptation and mitigation potential of organic farming. In: *Adaptation and Mitigation Strategies for Climate Resilient Agriculture.* CRIDA, Hyderabad, India.

Gudadhe, N., M. B. Dhonde, and N. A. Hirwe. 2015. Effect of integrated nutrient management on soil properties under cotton chickpea cropping sequence in Vertisols of Deccan plateau of India. *Indian Journal of Agricultural Research* 49(3): 207–214.

Gupta, J. P. 1980. Effect of mulches on moisture and thermal regimes of soil and yield of pearl millet. *Annals of Arid Zone* 19: 132–138.

Gupta, J. P., and G. K. Gupta. 1986. Effect of tillage and mulching on soil environment and cowpea seedling growth under arid conditions. *Soil and Tillage Research* 7:233–240.

Hati, K. M., A. Swarup, A. K. Dwivedi, et al. 2007. Changes in soil physical properties and organic carbon status at the topsoil horizon of a Vertisol of central India after 28 years of continuous cropping, fertilization and manuring. *Agriculture Ecosystems & Environment* 119(1–2): 127–134.

Hati, K. M., A. Swarup, B. Mishra, et al. 2008. Impact of long-term application of fertilizer, manure and lime under intensive cropping on physical properties and organic carbon content of an Alfisol. *Geoderma* 148: 173–179.

Hazarika, M., P. K. Sharma, P. Borah, et al. 2016. Long term effect of integrated nutrient management on soil properties and productivity of toria (*Brassica campestris*) in an inceptisol under rainfed upland condition of North Bank Plains Zone of Assam. *In Indian Journal of Dryland Agricultural Research and Development* 31(1): 9–14.

Hegde, D. M. 1996. Long term sustainability of productivity in an irrigated sorghum-wheat system through integrated nutrient supply. *Field Crops Research* 48(2–3): 167–175.

Hegde, D. M. 1998. Integrated nutrient management for production sustainability of oilseeds – A review. *Journal of Oilseeds Research* 15(1): 1–17.

Hoyt, P. B. 1981. Improvement in soil tilth and rape seed emergence by lime application on acid soils in the Peace River region. *Canadian Journal of Soil Science* 61:91–98.

Indoria, A. K., K. L. Sharma, K. Sammi Reddy, et al. 2018. Alternative sources of soil organic amendments for sustaining soil health and crop productivity in India – impacts, potential availability, constraints and future strategies. *Current Science* 115(11): 2052–2062.

Kalibhavi, C. M., M. D. Kachapur, and R. H. Patil. 2002. Effect of organic and inorganic fertilizers on soil profile moisture and grain yield of rabi sorghum under rainfed situation. *Current Research – University of Agricultural Sciences* (Bangalore) 31(3/4): 51–53.

Kalyansundaram, N. K., R. G. Patil, G. N. Karelia, et al. 1993. Studies on soil and fertilizer potassium in north Gujarat region. In: *Potassium in Gujarat Agriculture.* Gujarat Agricultural University, Sardar Krushi Nagar, Gujarat, India, pp. 128–138.

Katkar, R. N., V. K. Kharche, B. A. Sonune, R. H. Wanjari, and M. Singh. 2012. Long term effect of nutrient management on soil quality and sustainable productivity under sorghum-wheat crop sequence in Vertisol of Akola, Maharashtra. *Agropedology* 22(2): 3–4.

Khambalkar, S. M., V. V. Gabhane, and S. V. Khambalkar. 2017. Studies on effect of integrated nutrient management on productivity of cotton in rainfed condition. *International Journal of Current Microbiology and Applied Sciences* 6(8):3639–3641. doi:10.20546/ijcmas.2017.608.439.

Kohn, W. 1975. The influence of long-term soil cultivation, fertilization and rotation on the chemical and physical properties and yielding ability of sandy loam soil. I. Soil chemical and physical properties. *Bayer Landwirtsh Jahrb* 52: 928–955.

Kundu, S., C. Srinivasarao, R. B. Mallick, T. Satyanarayana, R. P. Naik, A. Johnston, and B. Venkateswarlu. 2013. Conservation agriculture in maize (*Zea mays* L.) horsegram (*Macrotyloma uniflorum* L.) system in rainfed Alfisols for C sequestration, climate change mitigation. *Journal of Agrometeorology* 15: 144–149.

Kundu, S., M. Singh, J. K. Saha, et al. 2001. Relationship between C addition and storage in a Vertisol under soybean-wheat cropping system in sub-tropical central India. *Journal of Plant Nutrition and Soil Science* 164: 483–486.

Ladd, J. N., and M. Amato. 1985. Nitrogen cycling in legume-cereal rotations. In: *Proceedings of an International Symposium on Nitrogen Management in Farming Systems in Humid and Sub humid Tropics.* IITA, Ibadan, Nigeria, pp. 105–127.

Lal, J. K., B. Mishra, and A. K. Sarkar. 1996. Effect of fly ash on soil microbial and enzymatic activity. *Journal of the Indian Society of Soil Science* 44(1): 77–80.

Lal, R. 2004. Soil carbon sequestration impacts on global climate change and food security. *Science* 304: 1623–1627.

Maji, A. K. 2007. Assessment of degraded and wastelands of India. *Journal of Indian Society of Soil Science* 55(4): 427–435.

Majumdar, B., M. S. Venkatesh, and K. Kumar. 2002. Effect of nitrogen and farmyard manure on yield and nutrient uptake of turmeric (*Curcuma longa*) and different forms of inorganic N build-up in an acidic Alfisol of Meghalaya. *Indian Journal of Agricultural Sciences* 72(9): 528–31.

Malewar, G. U., and A. R. Hasnabade. 1995. Effects of long-term application of fertilizers and organic sources on some properties of Vertisol. *Journal of Maharashtra Agricultural Universities* 20(2): 285–286.

Mali, D. V., V. K. Kharche, S. D. Jadhao, et al. 2015. Soil biological health under long-term fertilization in sorghum-wheat sequence on swell-shrink soils of central India. *Journal of the Indian Society of Soil Science* 63(4): 423–428.

Manna, M. C., A. Swarup, R. H. Wanjari, et al. 2005. Long-term effect of fertilizer and manure application on soil organic carbon storage, soil quality and yield sustainability under sub-humid and semi-arid tropical India. *Field Crops Research* 93: 264–280.

Maruthi Sankar, G. R., K. L. Sharma, V. V. Gabhane, et al. 2014. Effects of long-term fertilizer application and rainfall distribution on cotton productivity, profitability, and soil fertility in a semi-arid Vertisol. *Communications in Soil Science and Plant Analysis* 45(3): 362–380.

Matamala, R., M. A. Gozaler-Meler, J. D. Jastrow, et al. 2003. Impacts of fine roots turnover on forest NPP and soil carbon sequestration potential. *Science* 302: 1385–1387.

Mathur, G. M., R. Deo, and B. S. Yadav. 2006. Status of zinc in irrigated north-west plain soils of Rajasthan. *Journal of Indian Society of Soil Science* 54(3): 359–361.

Mathur, S. K., N. K. Goswamy, and N. R. Talati. 1981. Fixation of potassium in north-west Rajasthan soils. *Transactions of the Indian Society of Desert Technology* 6: 29–32.

Meena, H. M., and R. P. Sharma. 2016. Long-term effect of fertilizers and amendments on different fractions of organic matter in an acid alfosols. *Communications in Soil Science and Plant Analysis* 47: 1430–1440.

Motsara, M. R. 2002. Available nitrogen, phosphorus and potassium status of Indian soils as depicted by soil fertilizer maps. *Fertilizer News* 47(8): 15–21.

Murthy, R. S. 1981. Distribution and properties of Vertisols and associated soils. In: *Improving the Management of India's Deep Black Soils: Proceedings of the Seminar on Management of Deep Black Soils for Increased Production of Cereals, Pulses, and Oilseeds, New Delhi, 21 May 1981.* ICRISAT, Hyderabad, India, pp. 9–16.

Naik, K. R., V. V. Gabhane, et al. 2018. Soil fertility and cotton productivity as influenced by potash management through *Gliricidia* green leaf manuring in vertisols. Special Issue ICAAASTSD-2018, *Multilogic in Science* 7: 207–209.

Parewa, H. P., J. Yadav, and A. Rakshit. 2014. Effect of fertilizer levels, FYM and bioinoculants on soil properties in inceptisol of Varanasi, Uttar Pradesh, India. *International Journal of Agriculture, Environment & Biotechnology* 7(3): 517–525.

Prasad, R., and M. Ram. 1985. Soils of North-eastern hills and their management. In: Biswas, B. C. and Yadav, D.S. (Eds.), *Soils of India and Their Management.* FAI Publication, New Delhi, pp. 266–283.

Ramteke, L. K., S. S. Sengar, and S. S. Porte. 2017. Effect of fly ash, organic manure and fertilizers on soil microbial activity in rice-wheat cropping system in Alfisols and Vertisols. *International Journal of Current Microbiology and Applied Sciences* 6(7): 1948–1952.

Rao, V. M. B., and S. P. Singh. 1993. Crop response to organic sources of nutrient in dryland conditions. In: Somani, L. L. (Ed.), *Advances in Dryland Agriculture.* Scientific Publishers, Rajasthan, India, pp. 287–304.

Rasool, R., S. S. Kukal, and G. S. Hira. 2007. Farmyard manure and inorganic fertilization effects on saturated hydraulic conductivity of soil and crop performance in *Oryza sativa-Triticum aestivum* and *Zea mays – Triticum aestivum* system. *Soil and Tillage Research* 77: 768–771.

Ravankar, H. N., K. T. Naphade, R. B. Puranik, et al. 1999. Productivity of sorghum-wheat sequence in Vertisol under long-term nutrient management and its impact on soil fertility. *PKV Research Journal* 23(1): 23–27.

Rose, D. A. 1991. The effect of long-continued organic manuring on some physical properties of soils. In: Wilson, W. S. (Ed.), *Advances in Soil Organic Matter Research*. Special Publication no. 90. Royal Society of Chemistry, Cambridge, UK, pp. 197–205.

Roth, C. H., and M. A. Pavan. 1991. Effects of lime and gypsum on clay dispersion and infiltration in samples of a Brazilian Oxisol. *Geoderma* 48: 351–361.

Sahrawat, K. L. 1980. Control of urea hydrolysis and nitrification in soil by chemicals prospects and problems. *Plant Soil* 57 (2–3): 335–352.

Sathish, A., B. K. Ramachandrappa, M. A. Shankar, et al. 2016. Long-term effects of organic manure and manufactured fertilizer additions on soil quality and sustainable productivity of finger millet under a finger millet–groundnut cropping system in southern India. *Soil Use and Management* 32: 311–321.

Shankaranarayana, H. S., and V. A. K. Sarma. 1982. Soils of India: B. Brief description of the benchmark soils. In Murthy, R. S., et al. (Ed.), *Benchmark Soils of India: Morphology, Characteristics and Classification for Resource Management*. National Bureau of Soil Survey and Land Use Planning, Maharashtra, India.

Sharma, B. D., P. S. Sidhu, Raj-Kumar, et al. 1997. Characterization classification and landscape relationship of Inceptisols in north-west India. *Journal of the Indian Society of Soil Science* 45: 167–74.

Sharma, K. L., G. J. Kusuma, U. K. Mandal, et al. 2008. Evaluation of long-term oil management practices using key indicators and soil quality indices in a semi-arid tropical Alfisol. *Australian Journal of Soil Research* 46: 368–377.

Sharma, K. L., J. Kusuma Grace, K. Srinivas, et al. 2009. Influence of tillage and nutrient sources on yield sustainability and soil quality under sorghum–mungbean system in rainfed semi-arid tropics. *Communications in Soil Science and Plant Analysis* 40: 2579–2602.

Sharma, K. L., U. K. Mandal, K. Srinivas, et al. 2004. Long-term soil management effects on crop yields and soil quality in a dryland Alfisol. *Soil and Tillage Research* 83(2): 246–259.

Shirale, S. T., and L. E. Khating. 2009. Effect of organic and inorganic nutrients on yield, nutrient uptake and balance in different cropping systems in Vertisol. *Annals of Plant Physiology* 23(1): 83–85.

Singh, G., R. Babu, and N. Bhushan. 1990. Soil erosion rates in India. *Journal of Soil and Water Conservation* 47(1): 97–99.

Singh, H. P., B. Venkateshwarlu, K. P. R. Vittal, et al. 2000. Management of rainfed agro-ecosystem. In: Yadav, J. S. P. and Singh, G. B. (Eds.), *Natural Resource Management for Agricultural Production in India*. International Conference on Managing Natural Resources for Sustainable Agricultural Production in the 21st Century, New Delhi, India, pp. 668–774.

Singh, J. P., and N. Ravisankar. 2015. Integrated farming systems for sustainable agricultural growth: Strategy and experience from research. In: Muralidharan, K., Prasad, M. V. R. and Siddiq, E. A. (Eds.), *Proceedings of National Seminar on "Integrated Farming Systems for Sustainable Agriculture and Enhancement of Rural Livelihoods," 13–14 December*. Retired ICAR employees Association (RICAREA), Hyderabad, India, pp. 9–22.

Singh, M., and K. S. Singh. 1981. Zinc and copper status of soils of Rajasthan. *Annals of Arid Zone* 20: 77–85.

Singh, O. P., and B. Datta. 1983. Characteristics of some hill soils of Mizoram in relation to altitude. *Journal of Indian Society of Soil Science* 31: 65–61.

Snapp, S. S., and S. N. Silim. 2002. Farmer preferences and legume intensification for low nutrient environments. *Plant Soil* 245: 181–192.

Soil Survey Staff. 1975. *Soil Taxonomy: A Basic System of Soil Classification for Making and Interpreting Soil Surveys*. Agriculture Handbook No. 436. Soil Conservation Service, U.S. Department of Agriculture, Washington, DC.

Sonune, B. A., K. B. Tayade, V. V. Gabhane, et al. 2003. Long-term effect of manuring and fertilization on fertility and crop productivity of Vertisols under sorghum-wheat sequence. *Crop Research* 25: 460–467.

Sreeramulu, K. R. and M. A. Shankar. 2016. Impact of green leaf manuring on changes in microbial counts and enzyme activity in soil grown with groundnut. *Mysore Journal of Agricultural Sciences* 50(3): 569–575.

Srinivasarao, Ch., A. N. Deshpande, B. Venkateswarlu, et al. 2012. Grain yield and carbon sequestration potential of post monsoon sorghum cultivation in Vertisols in the semi-arid tropics of central India. *Geoderma.* doi:10.1016/j.gederma.2012.01.023.

Srinivasarao, Ch., A. N. Ganeshamurthy, M. Ali, and R. N. Singh. 2006. Phosphorus and micronutrient nutrition of chickpea genotypes in a multi-nutrient deficient typic Ustochrept. *Journal of Plant Nutrition* 29(4): 747–763.

Srinivasarao, Ch., V. Girija-Veni, J. V. N. S. Prasad et al. 2017a. Improving carbon balance with climate-resilient management practices in tropical agro-ecosystems of Western India. *Carbon Management* 8. doi:10.1080/17583004.2017.1309202.

Srinivasarao, Ch., K. A. Gopinath. 2016a. Resilient rainfed technologies for drought mitigation and sustainable food security. *Mausam* 67(1): 169–182.

Srinivasarao, Ch., K. A. Gopinath, J. V. N. S Prasad, et al. 2016b. Climate resilient villages for sustainable food security in tropical India: Concept, process, technologies, institutions and impacts. *Advances in Agronomy* 140: 101–214.

Srinivasarao, Ch., R. Lal, J. V. N. S. Prasad, et al. 2015. Potential and challenges of rainfed farming in India. *Advances in Agronomy* 133: 113–181.

Srinivasarao, Ch., K. V. Rao, and M. C. Saxena. 2017b. Drought proofing through implementation of district agriculture contingency plans in India. Twelfth International Dryland Development Conference "Sustainable Development of Drylands in the Post 2015 World." 21–24 August 2016, Alexandria, Egypt. International Dryland Development Commission, pp. 17–18.

Srinivasarao, Ch., G. Ravindrachary, P. K. Mishra, et al. 2014a. *Rainfed farming – A Compendium of Doable Technologies.* All India Coordinated Research Project for Dryland Agriculture, Central Research Institute for Dryland Agriculture, Indian Council of Agricultural Research, Hyderabad, India.

Srinivasarao, Ch., S. B. Reddy, and S. Kundu. 2014b. Potassium nutrition and management in Indian agriculture: Issues and strategies. *Indian Journal of Fertilisers* 10(5): 51–80.

Srinivasarao, Ch., A. K. Shankar, and C. Shankar 2018. *Climate Resilient Agriculture – Strategies and Perspectives.* Intech Open.

Srinivasarao, Ch., B. Venkateswarlu, M. Dinesh Babu, et al. 2011a. *Soil Health Improvement with Gliricidia Green Leaf Manuring in Rainfed Agriculture.* Central Research Institute for Dryland Agriculture, Hyderabad, India.

Srinivasarao, Ch., B. Venkateswarlu, S. Dixit, et al. 2010. Productivity enhancement and improved livelihoods through participatory soil fertility management in tribal districts of Andhra Pradesh. *Indian Journal of Dryland Agricultural Research and Development* 25(2): 23–32.

Srinivasarao, Ch., B. Venkateswarlu, S. Dixit, et al. 2011b. *Livelihood Impacts of Soil Health Improvement in Backward and Tribal Districts of Andhra Pradesh.* Central Research Institute for Dryland Agriculture, Hyderabad, India, pp. 119.

Srinivasarao, Ch., S. P. Wani, K. L. Sahrawat, et al. 2009. Nutrient management strategies in participatory watersheds in semi-arid tropical India. *Indian Journal of Fertilisers* 5(12): 113–128.

Suryono, S., H. Widijanto, and E. M. Jannah. 2015. The balance of N, P, and manure fertilizer dosage on growth and yield of peanuts in alfisols dryland. *STJSSA* 12(1): 20–25.

Swindale, L. D. 1981. An overview of ICRISAT's research on the management of deep black soils. In: *Improving the Management of India's Deep Black Soils: Proceedings of the Seminar on Management of Deep Black Soils for Increased Production of Cereals, Pulses, and Oilseeds, New Delhi, 21 May 1981.* ICRISAT, Hyderabad, India, pp. 17–20.

Talati, N. R., G. S. Mathur, and S. C. Attri. 1975. Distribution of various forms of phosphorus in Northwest Rajasthan soils *Journal of the Indian Society of Soil Science* 23(2): 202–206.

Takkar, P. N. 1996. Micronutrient research and sustainable agricultural productivity in India. *Journal of the Indian Society of Soil Science* 44: 563–581.

Tama, K., and S. A. El-Swaify. 1978. Charge, colloidal and structural stability relationships in oxidic soils. In: Emerson, W. W., Bond, R. D., Dexter, A. R. (Eds.), *Modification of Soil Structure.* John Wiley, Chichester, UK, pp. 41–52.

Tarafdar, J. C., K. Bala, and A. V. Rao. 1989. Phosphatase activity and distribution of phosphorus in arid soil profiles under different land use patterns. *Journal of Arid Environments* 16: 29–34.

Terman, G. L. 1979. Volatilization of nitrogen as ammonium surface applied fertilizers, organic amendments, and crop residues. *Advances in Agronomy* 31: 189–223.

Thangavel, R., N. S. Bolan, M. B. Kirkham, et al. 2019. Soil organic carbon dynamics: Impact of land use changes and management practices: A review. *Advances in Agronomy.* doi:10.1016/bs.agron.

Venkateswarlu, B., A. V. Rao, and P. Raina. 1984. Evaluation of phosphorous solubilization by microorganisms from aridisols. *Journal of the Indian Society of Soil Science* 32: 273–277.

Venkateswarlu, J. 1984. Nutrient management in semi-arid red soils. Part II, Project Bulletin No. 8. AICRPDA, Hyderabad, India.

Venkateswarlu, J. 1987. Efficient resource management for drylands of India. *Advances in Soil Science* 7: 165–221.

Venkateswarlu, J., and B. R. Hegde. 1992. Nutrient management for sustainable production in coarse textured soils. In: *Nutrient Management for Sustained Productivity: Proceedings of the International Symposium, Punjab Agricultural University, 1992*. Ludhiana, India, pp. 188–194.

Verma, T. S., B. R. Tripathi, and S. D. Verma. 1987. A comparative study on the nature and properties of soils in relation to climate and parent material. *Journal of the Indian Society of Soil Science* 35: 85–91.

Vijayalakshmi, K. 1987. Soil Management for increasing productivity in semi-arid red soils: Physical aspects. Alfisols in the semi-arid tropics. In: *Proceedings of the Consultants' Workshop on the State of the Art and Management Alternatives for Optimizing the Productivity of SAT Alfisols and Related Soils, 1–3 December 1983*. ICRISAT Center, India.

Vineela, C., S. P. Wani, Ch. Srinivasarao, et al. 2008. Microbial properties of soils as affected by cropping and nutrient management practices in several long-term manurial experiments in the semi-arid tropics of India. *Applied Soil Ecology* 40: 165–173.

Virmani, S. M., K. C. Saharawat, and J. R. Burford. 1982. Physical and chemical properties of Vertisols and their management in Vertisols and rice soils of tropics. *Transactions of the Twelfth International Congress of Soil Science, Symposia Papers II, 8–16 February*, pp. 94–118.

Walia, C. S., and G. S. Chamuah. 1992. Soil profile development in relation to land use. *Journal of the Indian Society of Soil Science* 40: 220–222.

Walia, C. S., and G. S. Chamuah. 1996. Characterization of inceptisols of Arunachal Hills. *Journal of the Indian Society of Soil Science* 44: 179–182.

Wright, A. L., and F. M. Hons. 2005. Carbon and nitrogen sequestration and soil aggregation under sorghum cropping sequences. *Biology and Fertility of Soils* 41: 95–100.

Wu, S. P., and Z. S. Chen. 2005. Characteristics and genesis of Inceptisols with placic horizons in the subalpine forest soils of Taiwan.

Yadav, K. 2009. *Recent Advances in Dryland Agriculture*. GBPUAT, Pantnagar, India.

Zebarth, B. J., G. H. Nielsen, E. Hogue, et al. 1999. Influence of organic waste amendments on selected soil physical and chemical properties. *Canadian Journal of Soil Science* 79: 501–504.

7 Reducing Emission of Greenhouse Gases from Fertilizer Use in India

Sangeeta Lenka, Narendra Kumar Lenka,
and Himanshu Pathak

CONTENTS

7.1 INTRODUCTION

The increasing concentration of atmospheric greenhouse gases (GHGs) has made climate change inevitable. The agriculture sector is a potential contributor to total GHG emissions, with a share of about 24% (IPCC 2014b) of total anthropogenic emissions, and a growing global population means that agricultural production will remain high if food demands are to be met. Agriculture is a potential source and sink of three of these gases, carbon dioxide (CO_2), methane (CH_4), and nitrous oxide (N_2O), the so-called GHGs that contribute to global warming. Globally, agricultural soil emits nearly 60% of N_2O and nearly 50% of CH_4. At the national level, in India approximately 23% of total emissions from agricultural soils of this fertilizer are the largest source, contributing 77% to total direct N_2O emissions. The concentration in N_2O, which has a very high radiative forcing per unit mass or molecule (296 times higher than that of CO_2 on a 100-year period), have risen from a preindustrial value of 270 ppb to a 2011 value of 324 ppb (IPCC 2013) mainly due to human activities, primarily through agriculture and the increase in the use of industrial fertilizers. Beyond its greenhouse effect, N_2O is now also the major ozone depleting substance in the stratosphere (Ravishankara et al. 2009). Agricultural soils produce 3.25 (1.7–4.8) Tg N_2O-N $year^{-1}$ from the application of nitrogenous fertilizer and thus appear as the main anthropogenic source of this gas (IPCC 2013). Fertilizer use has been increasing linearly with food grain production to feed the growing population in India. Generally, high fertilizer application rates may cause higher rates of GHG emissions. Annual fertilizer use would need to increase by 50% in India to increase crop production by 20% from the current level to feed 1.7 billion people by 2050. While increasing crop production by 20% to feed the growing population remains a priority, the fertilizer industry has also prioritized climate-smart agriculture to minimize the impact on the environment. Without a reduction in emission intensity, increases in productivity cannot be sustainable over the long term. Reconciling the goals of increasing food demands with reducing GHG emissions from increased fertilizer use requires adoption of good practice guidelines for fertilizer use. It is thus important to consider which technologies and practices can be applied that meet not only adaptation needs, but also mitigation needs. Thus, management practices need to be identified to better utilize fertilizer, while at the same time safeguarding the environment. Therefore, this chapter aims to discuss the effect of natural and management factors that influence the biochemical and physical processes that determine GHG emissions from fertilizer use.

7.2 FERTILIZER USE AND EMISSION OF GHG FROM INDIAN AGRICULTURE

By sector, the largest sources of GHGs are the sectors of energy production (mainly CO_2 from fossil fuel combustion), and agriculture, forestry, and land use (mainly CH_4 and N_2O). The contribution of agriculture, forestry, and land use to total emissions decreased from 31% (2004) to 24% (2010). Identification of GHG sources and quantification of GHG emissions from the agriculture sector has passed through many phases of refinement. In agriculture, the non-CO_2 sources (CH_4 and N_2O) are reported as anthropogenic GHG emissions. The CO_2 emitted is considered neutral, being associated with annual cycles of carbon (C) fixation and oxidation through photosynthesis (IPCC 2013). Fertilizer and manure application form an integral part of agricultural soil management in agricultural production. According to a UN report, by 2050 India's population is likely to reach 1.7 billion (United Nations 2017), nearly equal to that of China and the United States combined. Food grain production must be about 333 million tons (MT) to feed the burgeoning population. Agriculture plays a vital role in India's economy. Over 58% of rural households depend on agriculture as their principal means of livelihood. Agriculture, along with fisheries and forestry, is one of the largest contributors to the gross domestic product (GDP). The share of agriculture in the GDP in 2015–2016 was 17.5% and in employment 50% (Deshpande 2014), and that tells a tale: agriculture is becoming less important to the economy while remaining critical to employment. Fertilizer is an important input, contributing 50% toward an increase in food grain production. Trends in fertilizer consumption and food grain production since 1950 in terms of total quantities in the country are presented in Figure 7.1. In India, increasing food grain production from 50.8 MT in 1950–1951 to 275.7 MT in 2016–2017 is accompanied by a considerable increase in fertilizer use, which increased from 0.1 MMT to 26.0 MMT (Fertiliser Association of India 2019). The compound annual growth rate (CAGR) in fertilizer consumption declined after 2000. While it increased at an annual growth rate of 10.63% between 1960–1961 and 2000–2001, it has grown at an annual rate of 2.63% between 2000–2001 and 2016–2017. Food grain production also registered a marginal reduction in growth rate from 2.21% to 2.00% during the same period. Therefore, fertilizer consumption significantly influences food grain production and the two complement each other. According to the FAO (FAOSTAT 2018), total GHG emissions (direct plus indirect emissions) from synthetic fertilizer in the year 2015 were 112 CO_2 eq. MMT when total consumption of fertilizer ($N + P_2O_5 + K_2O$) was 25.58 MMT (Fertiliser Association of India 2019), resulting in an emission factor of 4.38. The trend of GHG emissions (CO_2 equivalent) from fertilizer consumption since 1961 in India is shown in Figure 7.2. Therefore, GHG emissions increase with an increase in fertilizer use in agriculture. In India, the projected food grain demand by 2020 is 310.8 MT with fertilizer demand of 41.6 MMT (Kumar et al. 2016), and could have approximate GHG emissions of 184.79 CO_2 eq. MMT.

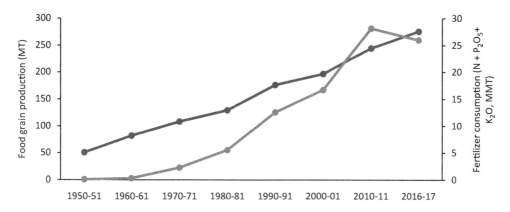

FIGURE 7.1   Trends in fertilizer consumption and food grain production since 1950 in terms of total quantities in India. (From *Fertilizer Statistics*. 62nd edn. The Fertilizer Association of India, New Delhi, 2016–2017. CIN: U85300 DL 1955NPL 002999.)

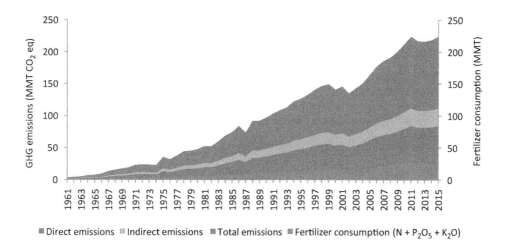

■ Direct emissions   ▨ Indirect emissions   ■ Total emissions   ■ Fertilizer consumption (N + P$_2$O$_5$ + K$_2$O)

FIGURE 7.2   Trends in GHGs emission (CO$_2$ equivalent) from fertilizer consumption since 1961 in India. (From *Fertilizer Statistics*. 62nd edn. The Fertilizer Association of India, New Delhi, 2016–2017. CIN: U85300 DL 1955NPL 002999; GHG emissions data taken from FAOSTAT, http://www.fao.org/faostat.)

7.3 MECHANISMS OF GHG EMISSIONS FROM SOIL

Understanding the mechanisms and biochemical processes of GHGs emissions from soil is important to identify the factors controlling production and mitigation of GHGs. The third most important anthropogenic GHG, N$_2$O, is produced during the following three processes in the soil as shown in Figure 7.3.

Nitrifier nitrification: Nitrification is the conversion of ammonium to nitrate, a microbially mediated process in the presence of oxygen in soil amended with manure or fertilizer, in soil following deposition of ammonia. N$_2$O is formed during ammonia (NH$_3$) oxidation to nitrite (NO$_2^-$) from the intermediate product hydroxylamine (NH$_2$OH) by chemodenitrification.

Denitrification: The conversion of nitrate (NO$_3^-$) back to atmospheric N$_2$, a microbially mediated process in the absence of oxygen in soil or manure, following manure or fertilizer application to the field (IPCC 2013).

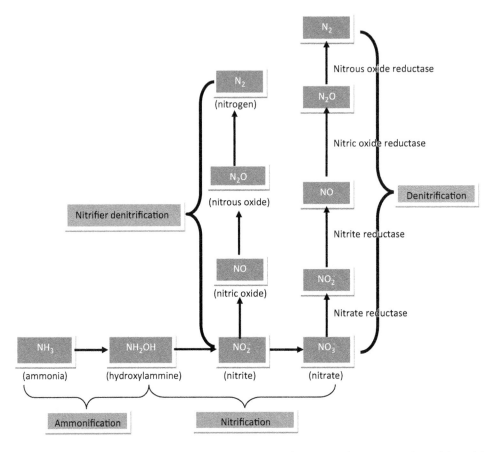

FIGURE 7.3 Pathways of nitrous oxide (N_2O) production in soil a schematic representation. (Adapted from Kool, D.M., et al., *Soil Biol. Biochem.*, 43, 174–178, 2011; Pilegaard, K., *Philos. Trans. Royal Soc. B Biol. Sci.*, 368, 20130126, 2013; Wrage, N., et al., *Soil Biol. Biochem.* 33, 1723–1732, 2001.)

Nitrifier denitrification: The oxidation of NH_3 to NO_2^- is followed by reduction of NO_2^- to NO, NO to N_2O, and N_2O to N_2. This is mainly carried out by autotrophic ammonia oxidizers.

CH_4 in soils is produced by methanogenesis under anaerobic conditions and is consumed by methanotrophic microorganisms that use O_2 and CH_4 for their metabolism under aerobic conditions. The three groups of organisms that may be involved in CH_4 flux (CH_4 oxidation and production) are methanotrophic bacteria, ammonia oxidizing bacteria, and methanogenic bacteria. Methanogenesis is frequently rate limited by the activities of the other members of the community, and particularly by how rapidly hydrogen or acetate is made available. A reduced and anaerobic condition is essential for methane production wetlands and paddy soils are major source of methane in agriculture.

$$CO_2 + 4H_2 \xrightarrow{\text{methanogens}} CH_4 + 2H_2O$$

CO_2 is produced in soil from respiration of plant roots and soil organisms, and from a negligible amount of chemical oxidation of C-containing minerals. The net ecosystem exchanges of CO_2 are balanced between photosynthesis and ecosystem respiration. Therefore, emission and mitigation of only non-CO_2 GHGs (N_2O and CH_4) are considered in the following sections.

7.4 FACTORS CONTROLLING GHG EMISSIONS FROM FERTILIZER

Fertilizers, both inorganic and organic, are the source of nutrients for soil microbes and plant respiratory processes in GHG emissions from soil. The application and management of fertilizers can contribute significant amounts of GHGs to the atmosphere. The GHG emissions vary with climate, soil properties and management, and fertilizer application and management. Understanding the major controlling factors and the combination of factors that minimize emissions is critical in the development of effective mitigation. The important factors controlling the GHG emission are discussed below.

7.4.1 CLIMATE

The climate of India varies from region to region due to its vast area and is home to diverse soil types. The tropical and subtropical climate of India results in a higher mean annual temperature (32°C–40°C) and precipitation (650 mm) than that of temperate countries. This higher mean annual temperature and precipitation increases the decomposition and mineralization of soil organic matter (SOM). The increased mineralization of SOM together with fertilizer application influences GHG emissions from soils. Further, the climate of a region directly affects soil temperature and moisture, which indirectly influences GHG emissions from fertilizer use. In general, GHG emissions are lower under drier climates than under wet, and also lower under cool than under warm temperatures.

7.4.2 OIL PROPERTIES

Soil physical and chemical properties include SOC content, mineral N content, water-filled pore space, pH, texture, drainage, temperature, moisture content, O_2 status, porosity, microbial abundance, and activity (Lenka and Lal 2013; Lenka et al. 2017). The N_2O fluxes correlate positively with soil nitrate and the nitrate/ammonium ratio and negatively with leaf litter and soil C/N ratio (Erickson et al. 2002). However, emission of CO_2 and CH_4 is positively correlated with C/N ratio. Decomposition rate and N mineralization, the processes that control N availability, influence N_2O fluxes from natural ecosystems (Werner et al. 2007). Low pH and soil moisture significantly suppress N_2O emission. Water-filled pore space (WFPS) dictates the amount of aeration and oxygen concentration in the soil, and thus nitrification is the main process of N_2O emission at <60% WFPS. Denitrification becomes more important at soil water contents >60% WFPS due to decreased oxygen supply (Zaman et al. 2012; Butterbach-Bahl et al. 2013; Huang and Gerber 2015). Availability of other elector donors like Fe_3^+, Mn_4^+, NO_3^-, and SO_4^{2-} in soil decrease emission of CH_4 from paddy fields. There is no association between pH and CH_4 production and consumption. CH_4 emission is suppressed at low temperature (<25°C) and high soil oxygen concentration. CH_4 and N_2O emission are more present in fine-textured soil than coarse-textured soil because soils with dominant fine pores support the formation of CH_4 and N_2O under anaerobic conditions (Zaman et al. 2012; Oertel et al. 2016). Fertile soil with higher organic C and N shows higher N_2O and CH_4 emission than less fertile soils (Butterbach-Bahl et al. 2013; Lenka et al. 2017).

7.4.3 FERTILIZER APPLICATION

Application of mineral N fertilizers into agricultural soils usually results in increasing N_2O emissions. At N rates not exceeding or equal to those required for maximum yields, N rates tend to create a linear response in N_2O emissions, with approximately 1% of applied mineral N lost as N_2O. Bhattacharyya et al. (2012) observed that N_2O emissions increased with application of inorganic fertilizer and rice straw compared to a control in four years of paddy cultivation. They further reported cumulative N_2O emissions were in the order of urea (1.0 kg ha^{-1}) > rice straw + urea (0.84 kg ha^{-1}) > rice straw + green manure (0.72 kg ha^{-1}) > control (0.23 kg ha^{-1}). In rice–wheat systems of Indo-Gangetic regions, urea application, either singly or in combination with rice straw, significantly increased the N_2O flux (Pathak et al. 2002; Bhatia et al. 2005). After nine years of experiment in soybean–wheat (*Glycine max-Triticum aestivum*), system application of inorganic fertilizer and organic manures increased the annual N_2O emission in Vertisols of central India (Lenka et al. 2017).

Organic or chemical fertilizer application enhances CH_4 production (Linquist et al. 2012b). Nitrogen fertilizers stimulate crop growth and provide more C substrates (via organic root exudates and sloughed-off cells) to methanogens for CH_4 production (Malla et al. 2005; Banger et al. 2012). In addition to crop growth, N fertilizers alter the activities of methanotrophs in the soils (Bodelier and Laanbroek 2004; Schimel 2004). However, there are contradictory reports on the effects of N fertilizers on methanotrophs in rice soils. For example, Bodelier et al. (2000) reported that application of urea at 200–400 kg N ha^{-1} stimulated the activities of methanotrophs, which resulted in greater CH_4 oxidation in the soil. In contrast, others have reported that N fertilizers suppress methanotrophs by attaching to the CH_4 oxidizing enzyme "methane monooxygenase" of methanotrophs (Alam and Jia 2012; Serrano-Silva et al. 2014). Furthermore, due to similar genetic structure, methanotrophs can switch substrates from CH_4 to ammonia when greater ammonium-N is available in soils (Alam and Jia 2012). These studies indicate the highly complex nature of the effects of N fertilizers on CH_4 production and oxidation processes in rice soils (Bodelier and Laanbroek 2004).

In a meta-analysis, Banger et al. (2012) reported that N fertilizers increased CH_4 emissions in 98 of 155 data pairs in rice soils. The response of CH_4 emissions per kgN fertilizer was significantly ($P < 0.05$) greater at < 140 kg N ha^{-1} than > 140 kg N ha^{-1}, indicating that substrate switch from CH_4 to ammonia by methanotrophs may not be a dominant mechanism for increased CH_4 emissions. On the contrary, decreased CH_4 emissions in intermittent drainage by N fertilizers indicate stimulation of methanotrophs in rice soils. The effects of N fertilizer–stimulated methanotrophs in reducing CH_4 emissions were modified by continuous flood irrigation due to the limitation of oxygen to methanotrophs. Greater response of CH_4 emissions per kg N fertilizer in urea than ammonia sulfate probably indicated the interference of sulfate in the CH_4 production process.

Seasonal flux of CH_4 increased by 94% following application of fertilizer-N (urea). Wide variation in CH_4 production and oxidation potentials was observed in rice soils tested. CH_4 oxidation decreased with soil depth, fertilizer-N and nitrification inhibitors, while organic amendment stimulated it (Adhya et al. 2000). Nitrification inhibitors also have a considerable impact on CH_4 emission through their inhibitory effect on CH_4 oxidation due to higher conservation of ammonium in soil (Adhya et al. 2000; Jain et al. 2000; Wang et al. 2014). Exposure of soil to NH_4 leads to an increase in the population of nitrifiers relative to methanotrophs and thus the overall CH_4 oxidation is reduced, as nitrifiers oxidize CH_4 less efficiently than methanotrophs (Chan and Parkin 2001; Serrano-Silva et al. 2014).

Apart from the quantity of fertilizer, the quality, mode, and time of fertilizer application also influence N_2O and CH_4 emissions from soil. N_2O emissions from agricultural land may increase by 35%–60% by 2030 from N fertilizer and manure production (Bruinsma 2003). However, the rate of growth could slow down because of efficient use of fertilizers (Bruinsma 2017) and management-related factors, including N application rate per fertilizer type, fertilizer application technique, application timing, tillage system, irrigation, incorporation of crop residues, and composition. Many of these factors also affect CH_4 and CO_2 emissions. N_2O production or consumption is generally controlled by nutrient availability in soil. It is, therefore, believed that fertilizer application increases the availability of nutrients, and increases global N_2O emissions by 10% (Bouwman et al. 1995). Butterbach-Bahl et al. (2013) mentioned the availability of reactive N as the major driver of N_2O soil emissions. Moreover, Palm et al. (2002) also attributed higher N_2O fluxes in cropping systems to N fertilization, and higher N_2O fluxes from tree-based systems to litter fall N, which are both components of nutrient availability.

7.4.4 Soil Management Practices

Soil management practices like tillage could have a large influence on the emission of N_2O and CH_4 from fertilizer applied in soil. Tillage is the major mechanism by which soil is exposed to oxidation and thus loss of soil C. Intensity of tillage before sowing a crop is known to profoundly affect soil aggregation, water-holding capacity, water-filled pore space, and the concentration of oxygen and CO_2 in soil air, which directly affect the quantity of GHG emissions from soil. No-till (NT) accumulates a mulch of crop residue on the soil surface, which can result in higher contents of soil water and labile SOC fractions. Wetter soil conditions with available substrate due to no-tillage management may increase emissions of N_2O (Ball et al. 1999; Venterea et al. 2005; Liu et al. 2006). However, some studies have shown lower N_2O emissions for NT soil or no difference between tillage systems, even with higher soil moisture content in NT. Contradictory findings may be a result of different cropping systems and soil types. Rochette (2008) recently summarized published experimental results showing that NT only increased N_2O emissions in poorly aerated soils.

   Tillage, residue, and fertilizer management also affect nitrate-N concentration, water content, aeration, and available C (Myrbeck 2014; Wang et al. 2015; Wiese et al. 2016), which, in turn, can impact N loss through denitrification and N_2O emissions (MacKenzie et al. 1997; Mangalassery et al. 2014; Krauss et al. 2017). However, Hao et al. (2001) found that removing residues decreased N_2O emissions from an irrigated soil in southern Alberta, particularly on plots tilled in autumn after harvest. N_2O flux from agricultural soils depends on a complex interaction between climatic factors, soil properties, and soil management.

7.5 MITIGATION STRATEGIES

A considerable effort has been made to develop new technology that minimizes nutrient and GHG emissions. It is possible that management and fertilizer practices conducted to reduce one GHG could favor conditions for production and emission of another. The Intergovernmental Panel on Climate Change (IPCC 2014a) assumes a default value of 1% of N content of the substrate, emitted as N_2O. As these emissions are the consequences of natural processes, they are difficult to control. The best possible approach is to increase N use efficiency. In addition, emission during fertilizer manufacturing can be reduced with new cleaning technology that can enable N_2O emission reduction by about 70%–90% (Kongshaug 1998; Pathak 2015). The only way to reduce N_2O emissions from agricultural soils is to affect the soil properties and climatic controls of GHGs emission through the following management strategies.

7.5.1 Tillage

Tillage operation is the physical manipulation of soil and its environment, which alters the biochemical processes that can significantly affect the production and consumption of N_2O. Tillage is the major mechanism by which soil is exposed to oxidation and thus loss of soil C. Intensity of tillage before sowing a crop is known to profoundly affect soil aggregation, water-filled pore space, and the concentration of oxygen and carbon dioxide in soil air, which directly affect the quantity of GHG emissions from soil (Buchkina et al. 2013; Butterbach-Bahl et al. 2013; Luckmann et al. 2014; Liu et al. 2015; Oertel et al. 2016; Krauss et al. 2017). Varying impacts of tillage on GHG mitigation has been reported by researchers globally (Lenka and Lenka 2014). This could be probably due to differences in soil, climate, crop, and management practices. Conservation tillage (CT) practices such as minimum tillage, NT, and reduced tillage increase climate change mitigation through reduced emission of CO_2 and increased methane uptake. On the contrary, positive (Skiba et al. 2002; Liu et al. 2011; Dendooven et al. 2012; Piva et al. 2012), negative (Malhi et al. 2006; Stockle et al. 2012; Smith et al. 2012), and no effect (Choudhary et al. 2002; Yamulki and Jarvis 2002) of CT have also been reported by researchers covering different ecosystems globally. Increased N_2O emissions have been related to increased denitrification under CT due to the formation of microaggregates within macroaggregates that create anaerobic micro sites with increased microbial activity, leading to greater competition for oxygen. Further, the residue retention under CT provides labile substrate to the denitrifying bacteria, which increase emission of N_2O by denitrification. Soil tillage has a profound influence on soil porosity, pore size distribution, and pore geometry, which affects the diffusion of gases between the soil and atmospheric. In general, CT such as NT reduces porosity and increases soil bulk density and soil penetration resistance, which decreases the efflux of GHGs (Mangalassery et al. 2014). Thus, the resulting retention of GHGs in soil such as CH_4 could increase the probability of CH_4 oxidation by methanotrophs. Further, tillage intensity is known to negatively affect methanotrophic population, which takes a long time to recover. This could be the reason for negative or no effect of CT on CH_4 and N_2O emissions. Therefore, the magnitude and direction of GHGs' efflux to or from soil depends on the resultant greater effect of soil physical or biological properties.

7.5.2 FERTILIZER MANAGEMENT

Fertilizer management that improves the efficiency of N fertilizers with the right combination of source, rate, placement, and timing of required N for growing relevant crops with high N uptake efficiency reduces the release of N_2O from soil (Chatskikh and Olesen 2007; Snyder et al. 2009; Buchkina et al. 2013). Soil nitrous oxide emission is reduced by reducing excess N inputs (inorganic and organic N fertilizer). High mineral N content (NH_4^+ or NO_3^-) decreases uptake of CH_4 from the atmosphere. This is because the CH_4 and NH_4 oxidizing bacteria can switch their substrate at a higher concentration of NH_3. The prevailing tropical climate in India is characterized by high temperature and moisture, which increases the loss of N_2O from fertilizer N applied at the time of sowing, compared to temperate climates. Therefore, strategies to curb this threat should also target the time and placement of N fertilizer application. The single-time application of N fertilizer at the time of sowing should be substituted by crop-demand-based or split application (Pathak and Wassmann 2007). Leaf color charts or site-specific models can be used to identify the crop stage and time of N fertilization in crops. Crop residue together with N fertilization also reduces the loss of N as N_2O and increases uptake of CH_4. The N fertilization should not be either at a too early or late crop growth stage where the crop requirement is less.

Organically managed soils increase the possibility of CH_4 uptake compared to inorganic and integrated uses of fertilizer (Pathak and Wassmann 2007; Linquist et al. 2012b). Integrated use of organic and inorganic fertilizer is widely advocated in both aerobic and anaerobic ecosystems to maintain soil health, sustain productivity, and sequester carbon. However, it increases GHG fluxes as well. Therefore, two important indices used to compare management practices for GHG mitigation are net global warming potential and GHG intensity (Piva et al. 2012; Valbuena et al. 2015). The net global warming potential (NGWP) is defined as the net fluxes for all three major biogenic GHGs (i.e., CO_2, N_2O, and CH_4). Greenhouse gas intensity (GHGI) is related to grain yield and is defined as the ratio of NGWP to grain yield expressed as Mg CO_2 eq Mg^{-1} grain yield. Though higher GWP has been reported by several workers in integrated use of inorganic and inorganic fertilizer, the NGWP and GHGI of these systems are relatively lower than that of inorganic fertilization (Lenka et al. 2017). Balanced fertilization, which increases crop yield and nutrient use efficiency, would significantly reduce GHG emissions. Application of plant nutrients like P, K, and S containing fertilizers distinctly inhibits CH_4 emission from flooded paddy fields (Babu et al. 2006). These nutrients probably influence plant growth, methanogenic microbial population, soil pH, and redox potential, which affect the emission of CH_4 from paddy fields. Water management plays a crucial role in CH_4 efflux from fertilized paddy fields. Alternate wetting and drying has significantly reduced emission of CH_4 compared to continuous flooding (Adhya et al. 2000).

Using the right method and instrument to apply fertilizers at the time of sowing also reduces the gaseous loss of N as N_2O. Use of a seed cum fertilizer drill properly calibrated to deliver the right amount of fertilizer at the desired depth would reduce the N_2O loss. The right placement of N fertilizer as an application technique will lower N_2O emissions. Deep placement of N deeper than 10 cm soil depth will increase N_2O emissions compared to that from surface broadcast and surface incorporation (Snyder et al. 2009). Further, deep placement of N fertilizer also influences CH_4 oxidation, albeit with variable reports of positive, negative, and no effects (Liu et al. 2006; Wang et al. 2012; Adviento-Borbe and Linquist 2016; Yao et al. 2017).

7.5.3 NITRIFICATION INHIBITORS

Application of nitrification inhibitors is advocated as a strategy to minimize fertilizer N losses and increase nitrogen use efficiency (Prasad and Power 1997). However, they can also have a significant influence on N_2O and CH_4 emissions from soil. Nitrification inhibitors slow down the breakdown of ammoniacal N (NH_4^+–N) to nitrite N (NO_2^-) and nitrate N (NO_3^-) pools, thus limiting the substrate pools available for N_2O production during the process of nitrification. During denitrification process when the soil NO_3^-–N is reduced to dinitrogen (N_2), N_2O is emitted. Thus nitrification inhibitors reduce the emission of N_2O, directly by reducing nitrification, and indirectly by reducing the availability of NO_3^- for denitrification (Adhya et al. 2000; Malla et al. 2005; Linquist et al. 2012a). In the rice–wheat systems of the Indo-Gangetic plain, Malla et al. (2005) reported that application of nitrification inhibitors like neem coated urea, coated calcium carbide, neem oil, dicyandiamide (DCD), hydroquinone, and thiosulphate decreased N_2O emission when compared to urea alone. Application of neem coated urea, coated calcium carbide, neem oil, and DCD reduced the emission of CH_4; hydroquinone and thiosulphate increased the emission compared to that of urea alone. Application of nitrification inhibitors increases the concentration of NH_4^+–N in soil. This inhibits the methane oxidation potential of methanotrophic bacteria, which has the ability to switch substrate between NH_4^+–N and CH_4 (Malla et al. 2005). Therefore, nitrification inhibitors have been reported to reduce N_2O emission by 63% to 9% (Adhya et al. 2000; Majumdar 2000; Malla et al. 2005; Prasad 2005; Sharma et al. 2008; Rees and Ball 2010; Linquist et al. 2012a). This wider range of N_2O reduction percentage shows that the efficacy of nitrification inhibitors depends on other controlling factors like soil type, climate, kind of fertilizer, and crop.

7.5.4 CROP

Fertilization together with the right selection of crop, especially cereals, also helps to mitigate GHG emissions. In a meta-analysis of 57 published studies, Linquist (2012a) showed that GHGI (i.e., yield-scaled GWP) is the highest for rice followed by that for maize and wheat. However, rice has been the staple food of many Indian states for ages. Therefore, when recommending mitigation strategies, cultural significance must also be considered. Site-specific legume-based crop rotations along with judicious and efficient irrigation and fertilizer management might reduce emission of GHGs. Selection of the appropriate high-yielding crop variety together with planting date and plant population could be the right choice to lower GHGI.

7.6 CONCLUSION AND WAY FORWARD

Increasing fertilizer use is inevitable for meeting the food demand of the growing population in India. GHG emissions, especially N_2O, are likely to increase with increasing fertilizer use. The strategy to reduce GHG emissions from fertilizer use in India must involve achieving the appropriate balance between increasing food production and crop management practices that increase nutrient use efficiency. Indicators like net global warming potential (net GWP) and GHGI may be used to compare management practices rather than single GHG flux. Generally, high-yielding crops have less GHG emissions per unit grain yield and higher soil carbon sequestration due to greater biomass C input. Therefore, the evaluation of mitigation strategies should be based on the tradeoff between soil N_2O and CH_4 emission vis-à-vis carbon sequestration potential and other ancillary benefits. Good practice guidelines for reducing GHG emission from fertilizer use include the following:

1. Selection of right crop variety, sowing/planting time, and optimum seed rate.
2. Right combination of rate, time, and method of fertilizer application to increase plant nutrient uptake and nutrient use efficiency.
3. Use of balanced fertilizer.
4. Integrated use of inorganic and organic fertilizer.
5. Crop residue retention and incorporation in field.
6. Use of nitrification inhibitors and biofertilizers.
7. Irrigation water management to increase nutrient and water use.

To feed the growing population while protecting the environment, future sustainable agriculture should explore systems with low GWP and GHGI at high crop productivity. Future research on evaluating the best management practices to reduce GHG emissions should observe the following:

1. Measure SOC content, bulk density, soil NO_3 leaching or content in drainage water, crop yield, and soil moisture content.
2. Launch an all-India coordinated research effort on GHG inventory, soil carbon storage, and crop yield across different cropping systems, soils, and climates of India. This effort should be initiated as soon as possible.
3. Account for residual soil N in choosing the dose of N to subsequent crops.
4. Use judiciously the N fertilizer on the basis of source, rate, time, placement, and method of application in conjunction with different residue management options for diverse crops.
5. Adopt precision agriculture by using sensors to trace the time and site of N deficiency in field and automatic N fertilization to enhance N use efficiency.

REFERENCES

Adhya, Tapan, Kollah Bharati, Santhosh R. Mohanty, Balasubramanian Ramakrishnan, V. Rajaramamohan Rao, Nambrattil Sethunathan, and Reiner Wassmann. 2000. "Methane Emission from Rice Fields at Cuttack, India." *Nutrient Cycling in Agroecosystems* 58 (1/3): 95–105. doi:10.1023/A:1009886317629.

Adviento-Borbe, M. Arlene, and Bruce Linquist. 2016. "N Fertilizer Placements and Greenhouse Gas Emissions from Continuously Flooded Rice Systems." *Geoderma* 266: 40–45.

Alam, M. Saiful, and Zhongjun Jia. 2012. "Inhibition of Methane Oxidation by Nitrogenous Fertilizers in a Paddy Soil." *Frontiers in Microbiology* 3: 246. doi:10.3389/fmicb.2012.00246.

Babu, Y. Jagadeesh, Dali Rani Nayak, and Tapan Adhya. 2006. "Potassium Application Reduces Methane Emission from a Flooded Field Planted to Rice." *Biology and Fertility of Soils* 42 (6): 532–41. doi:10.1007/s00374-005-0048-3.

Ball, Bruce C., Albert Scott, and John P. Parker. 1999. "Field N_2O, CO_2 and CH_4 Fluxes in Relation to Tillage, Compaction and Soil Quality in Scotland." *Soil and Tillage Research* 53 (1): 29–39. doi:10.1016/S0167-1987(99)00074-4.

Banger, Kamaljit, Hanqin Tian, and Chaoqun Lu. 2012. "Do Nitrogen Fertilizers Stimulate or Inhibit Methane Emissions from Rice Fields?" *Global Change Biology* 18 (10): 3259–67. doi:10.1111/j.1365-2486.2012.02762.x.

Bhatia, Artia, Surendra Pathak, Niveta Jain, Paramendra Singh, and Anil Kumar Singh. 2005. "Global Warming Potential of Manure Amended Soils under Rice-Wheat System in the Indo-Gangetic Plains." *Atmospheric Environment* 39 (37): 6976–84. doi:10.1016/j.atmosenv.2005.07.052.

Bhattacharyya, Pratap, Kumar Sekhar Roy, Subhasis Neogi, Tapan Adhya, Krs Sambasiva Rao, and Mihir Chandra Manna. 2012. "Effects of Rice Straw and Nitrogen Fertilization on Greenhouse Gas Emissions and Carbon Storage in Tropical Flooded Soil Planted with Rice." *Soil and Tillage Research* 124: 119–30. doi:10.1016/j.still.2012.05.015.

Bodelier, Paul L. E., and Hendrikus J. Laanbroek. 2004. "Nitrogen as a Regulatory Factor of Methane Oxidation in Soils and Sediments." *FEMS Microbiology Ecology* 47 (3): 265–77. doi:10.1016/S0168-6496(03)00304-0.

Bodelier, Paul L. E., Peter Roslev, Thilo Henckel, and Peter Frenzel. 2000. "Stimulation by Ammonium-Based Fertilizers of Methane Oxidation in Soil around Rice Roots." *Nature* 403 (6768): 421–24. doi:10.1038/35000193.

Bouwman, Alexandar F., Klaas W. Van der Hoek, and Jos Olivier. 1995. "Uncertainties in the Global Source Distribution of Nitrous Oxide." *Journal of Geophysical Research* 100 (D2): 2785. doi:10.1029/94JD02946.

Bruinsma, Jelle. 2003. "World Agriculture: Towards 2015/2030: An FAO Perspective." *Land Use Policy* 20 (4): 375. doi:10.1016/S0264-8377(03)00047-4.

Bruinsma, Jelle. 2017. *World Agriculture: Towards 2015/2030*. Routledge. doi:10.4324/9781315083858.

Buchkina, Natalya P., Elena Y. Rizhiya, Sergey V. Pavlik, and Eugene V. Balashov. 2013. "Soil Physical Properties and Nitrous Oxide Emission from Agricultural Soils." *Advances in Agrophysical Research*: 193–220. doi:10.5772/53061.

Butterbach-Bahl, Klaus, Elizabeth M. Baggs, Michael Dannenmann, Ralf Kiese, and Sophie Zechmeister-Boltenstern Butterbach-Bahl. 2013. "Nitrous Oxide Emissions from Soils: How Well Do We Understand the Processes and Their Controls?" *Philosophical Transactions of the Royal Society of London*. Series B, Biological Sciences 368 (1621): 20130122. doi:10.1098/rstb.2013.0122.

Chan, Alvarus S. K., and Timothy B. Parkin. 2001. "Methane Oxidation and Production Activity in Soils from Natural and Agricultural Ecosystems." *Journal of Environmental Quality* 30 (6): 1896–903. doi:10.2134/jeq2001.1896.

Chatskikh, Dmitri, and Jørgen E. Olesen. 2007. "Soil Tillage Enhanced CO_2 and N_2O Emissions from Loamy Sand Soil under Spring Barley." *Soil and Tillage Research* 97 (1): 5–18. doi:10.1016/j.still.2007.08.004.

Choudhary, M. A., Akmal Akramkhanov, and Surinder Saggar. 2002. "Nitrous Oxide Emissions from a New Zealand Cropped Soil: Tillage Effects, Spatial and Seasonal Variability." *Agriculture, Ecosystems and Environment* 93 (1–3): 33–43. doi:10.1016/S0167-8809(02)00005-1.

Dendooven, Luc, Leonardo Patiño-Zúñiga, Nele Verhulst, Marco Luna-Guido, Rodolfo Marsch, and Bram Govaerts. 2012. "Global Warming Potential of Agricultural Systems with Contrasting Tillage and Residue Management in the Central Highlands of Mexico." *Agriculture, Ecosystems and Environment* 152: 50–58. doi:10.1016/j.agee.2012.02.010.

Deshpande, Tanvi. 2014. "State of Agriculture in India." *PRS Legislative Research*, no. 113.

Erickson, Heather, Eric A. Davidson, and Michael Keller. 2002. "Former Land-Use and Tree Species Affect Nitrogen Oxide Emissions from a Tropical Dry Forest." *Oecologia* 130 (2): 297–308. doi:10.1007/s004420100801.

FAOSTAT. 2018. Accessed November 9. http://www.fao.org/faostat/en/#data/GY.

Fertilizer Statistics. 2016–2017. 62nd edn. The Fertilizer Association of India, New Delhi, CIN: U85300 DL 1955NPL 002999.

Fertiliser Association of India. 2019. Accessed 9 October. https://www.faidelhi.org/statistics/statistical-database.

Hao, Xiying, Chunhua Chang, Jack M. Carefoot, Hemry H. Janzen, and Benjamin H. Ellert. 2001. "Nitrous Oxide Emissions from an Irrigated Soil as Affected by Fertilizer and Straw Management." *Nutrient Cycling in Agroecosystems* 60 (1–3): 1–8. doi:10.1023/A:1012603732435.

Huang, Yuanyuan, and Stefen Gerber. 2015. "Global Soil Nitrous Oxide Emissions in a Dynamic Carbon-Nitrogen Model." *Biogeosciences* 12 (21): 6405–27. doi:10.5194/bg-12-6405-2015.

IPCC (Intergovernmental Panel on Climate Change). 2013. *Climate Change 2013: The Physical Science Basis*. Contribution of Working Group I to the Fifth Assessment Report of the Intergovernmental Panel on Climate Change.

IPCC (Intergovernmental Panel on Climate Change). 2014a. Climate Change 2014: Mitigation of Climate Change. doi:10.1017/CBO9781107415416.

IPCC (Intergovernmental Panel on Climate Change). 2014b. Climate Change 2014: Synthesis Report. Contribution of Working Groups I, II and III to the Fifth Assessment Report of the Intergovernmental Panel on Climate Change. Core Writing Team, R. K. Pachauri and L. A. Meyer. doi:10.1017/CBO9781107415324.004.

Jain, Manish, Shukal Kumar, Reiner Wassmann, Sudip Mitra, Shiva Dhar Singh, Jagdish P. Singh, Rajvir Singh, Ashok Kumar Yadav, and Shravan Gupta. 2000. "Methane Emissions from Irrigated Rice Fields in Northern India (New Delhi)." *Nutrient Cycling in Agroecosystems* 58 (1/3): 75–83. doi:10.1023/A:1009882216720.

Kongshaug, Gunnar. 1998. "Energy Consumption and Greenhouse Gas Emissions in Fertilizer Production." *IFA Technical Conference* 28: 18. doi:10.1590/S1415-43662013000600010.

Kool, Dorien M., Jan Dolfing, Nicole Wrage, and Jan Willem Van Groenigen. 2011. "Nitrifier Denitrification as a Distinct and Significant Source of Nitrous Oxide from Soil." *Soil Biology and Biochemistry* 43 (1): 174–78. doi:10.1016/j.soilbio.2010.09.030.

Krauss, Maike, Reiner Ruser, Torsten Müller, Sissel Hansen, Paul Mäder, and Andreas Gattinger. 2017. "Impact of Reduced Tillage on Greenhouse Gas Emissions and Soil Carbon Stocks in an Organic Grass-Clover Ley – Winter Wheat Cropping Sequence." *Agriculture, Ecosystems & Environment* 239: 324–33. doi:10.1016/J.AGEE.2017.01.029.

Kumar, Praduman, Pramod Kumar Joshi, and Surabhi Mittal. 2016. "Demand vs Supply of Food in India – Futuristic Projection." *Proceedings of the Indian National Science Academy* 82: 1579–86. doi:10.16943/ptinsa /2016/48889.

Lenka, Narendra Kumar, and Rattan Lal. 2013. "Soil Aggregation and Greenhouse Gas Flux after 15 Years of Wheat Straw and Fertilizer Management in a No-Till System." *Soil and Tillage Research.* doi:10.1016/j.still.2012.08.011.

Lenka, Sangeetha, Narendra Kumar Lenka, Amar Bahadur Singh, B. Singh, and Jyothi Raghuwanshi. 2017. "Global Warming Potential and Greenhouse Gas Emission under Different Soil Nutrient Management Practices in Soybean–Wheat System of Central India." *Environmental Science and Pollution Research* 24 (5). doi:10.1007/s11356-016-8189-5.

Lenka, Sangeeta, and Narendra Kumar Lenka. 2014. "Conservation Tillage for Climate Change Mitigation – The Reality." *Climate Change and Environmental Sustainability* 2 (1): 1. doi:10.5958/j.2320-642X.2.1.001.

Linquist, Bruce A., Maria Arlene Adviento-Borbe, Cameron M. Pittelkow, Chris van Kessel, and Kees Jan van Groenigen. 2012a. "Fertilizer Management Practices and Greenhouse Gas Emissions from Rice Systems: A Quantitative Review and Analysis." *Field Crops Research* 135: 10–21. doi:10.1016/j.fcr.2012.06.007.

Linquist, Bruce, Kees Jan Van Groenigen, and Maria Arlene Adviento-borbe. 2012b. "An Agronomic Assessment of Greenhouse Gas Emissions from Major Cereal Crops." *Global Change Biology* 18(1): 194–209. doi:10.1111/j.1365-2486.2011.02502.x.

Liu, Chunyan, Kai Wang, Shixie Meng, Xunhua Zheng, Zaixing Zhou, Shenghui Han, Deli Chen, and Zhiping Yang. 2011. "Effects of Irrigation, Fertilization and Crop Straw Management on Nitrous Oxide and Nitric Oxide Emissions from a Wheat-Maize Rotation Field in Northern China." *Agriculture, Ecosystems and Environment* 140 (1–2): 226–33. doi:10.1016/j.agee.2010.12.009.

Liu, Xuejun J., Arvin R. Mosier, Ardell D. Halvorson, and Fusuo S. Zhang. 2006. "The Impact of Nitrogen Placement and Tillage on NO, N_2O, CH_4 and CO_2 Fluxes from a Clay Loam Soil." 177–88. doi:10.1007/s11104-005-2950-8.

Liu, Yinglie, Ziqiang Zhou, Xiaoxu Zhang, Xin Xu, Hao Chen, and Zhengqin Xiong. 2015. "Net Global Warming Potential and Greenhouse Gas Intensity from the Double Rice System with Integrated Soil-Crop System Management: A Three-Year Field Study." *Atmospheric Environment* 116: 92–101. doi:10.1016/j.atmosenv.2015.06.018.

Luckmann, Monique, Daniel Mania, Melanie Kern, Lars R. Bakken, Åsa Frostegård, and Jörg Simon. 2014. "Production and Consumption of Nitrous Oxide in Nitrate-Ammonifying Wolinella Succinogenes Cells." *Microbiology (United Kingdom)* 160 (part 8): 1749–59. doi:10.1099/mic.0.079293-0.

MacKenzie, Angus F., Mingxian Fan, and François Cadrin. 1997. "Nitrous Oxide Emission as Affected by Tillage, Corn-Soybean-Alfalfa Rotations and Nitrogen Fertilization." *Canadian Journal of Soil Science* 77 (2): 145–52. doi:10.4141/S96-104.

Majumdar, Deepanjan. 2000. "Monitoring and Mitigation of Nitrous Oxide Emissions from Agricultural Fields of India: Relevance, Problems, Research and Policy Needs." *Current Science* 79 (10): 1435–39.

Malhi, Sukhdev S., Reynald Lemke, Zhaohui Wang, and Baldev S. Chhabra. 2006. "Tillage, Nitrogen and Crop Residue Effects on Crop Yield, Nutrient Uptake, Soil Quality, and Greenhouse Gas Emissions." *Soil and Tillage Research* 90 (1–2): 171–83. doi:10.1016/j.still.2005.09.001.

Malla, Ganesh, Arti Bhatia, Himanshu Pathak, Shiv Prasad, Niveta Jain, and Jayprakash Singh. 2005. "Mitigating Nitrous Oxide and Methane Emissions from Soil in Rice-Wheat System of the Indo-Gangetic Plain with Nitrification and Urease Inhibitors." *Chemosphere* 58 (2): 141–47. doi:10.1016/j.chemosphere.2004.09.003.

Mangalassery, Shamsudheen, Sofie Sjögersten, Debbie L. Sparkes, Craig J. Sturrock, Jim Craigon, and Sacha J. Mooney. 2014. "To What Extent Can Zero Tillage Lead to a Reduction in Greenhouse Gas Emissions from Temperate Soils?" *Scientific Reports* 4: 4586. doi:10.1038/srep04586.

Myrbeck, Åsa. 2014. Soil Tillage Influences on Soil Mineral Nitrogen and Nitrate Leaching in Swedish Arable Soils. PhD diss., Swedish University of Agricultural Sciences, Uppsala, 2014.

Oertel, Cornelius, Jörg Matschullat, Kamal Zurba, Frank Zimmermann, and Stefan Erasmi. 2016. "Greenhouse Gas Emissions from Soils—A Review." *Chemie der Erde – Geochemistry* 76 (3): 327–52. doi:10.1016/j.chemer.2016.04.002.

Palm, Cheryl A., Julio C. Alegre, Lamberto Arevalo, Patrick K. Mutuo, Arvin R. Mosier, and Richard Coe. 2002. "Nitrous Oxide and Methane Fluxes in Six Different Land Use Systems in the Peruvian Amazon." *Global Biogeochemical Cycles* 16 (4): 21-1–21-13. doi:10.1029/2001GB001855.

Pathak, Himanshu. 2015. "Greenhouse Gas Emission from Indian Agriculture: Trends, Drivers and Mitigation Strategies." *Proceedings of the Indian National Science Academy* 81: 1133–49. doi:10.16943/ptinsa/2015/v81i5/48333.

Pathak, Himanshu, Arti Bhatia, Shiv Prasad, Shalini Singh, Satendra Kumar, M. C. Jain, and Upendra Kumar. 2002. "Emission of Nitrous-Oxide from Rice-Wheat Systems of Indo-Gangetic Plains of India." *Environmental Monitoring and Assessment* 77 (2): 163–78. doi:10.1023/A:1015823919405.

Pathak, Himanshu, and Reiner Wassmann. 2007. "Introducing Greenhouse Gas Mitigation as a Development Objective in Rice-Based Agriculture: I. Generation of Technical Coefficients." *Agricultural Systems* 94 (3): 807–25. doi:10.1016/j.agsy.2006.11.015.

Pilegaard, Kim. 2013. "Processes Regulating Nitric Oxide Emissions from Soils." *Philosophical Transactions of the Royal Society B: Biological Sciences* 368 (1621): 20130126. doi:10.1098/rstb.2013.0126.

Piva, Jonatas Thiago, Jeferson Dieckow, Cimélio Bayer, Josiléia Acordi Zanatta, Anibal de Moraes, Volnei Pauletti, Michely Tomazi, and Maico Pergher. 2012. "No-Till Reduces Global Warming Potential in a Subtropical Ferralsol." *Plant and Soil* 361 (1–2): 359–73. doi:10.1007/s11104-012-1244-1.

Prasad, Rajendra. 2005. "R –w c S." *Advances* 86.

Prasad, Rajendra, and James F. Power. 1997. *Soil Fertility Management for Sustainable Agriculture*. Boca Raton, FL, CRC Press.

Ravishankara, Akkihebbal John, S. Daniel, and Robert W. Portmann. 2009. "Nitrous Oxide (N_2O): The Dominant Ozone-Depleting Substance Emitted in the 21st Century." *Science* 326 (5949): 123–25. doi:10.1126/science.1176985.

Rees, Robert M., and Bruce C. Ball. 2010. "Soils and Nitrous Oxide Research." *Soil Use and Management* 26 (2): 193–95. doi:10.1111/j.1475-2743.2010.00269.x.

Rochette, Philippe. 2008. "No-Till Only Increases N_2O Emissions in Poorly-Aerated Soils." *Plant and Soil Research* 101: 97–100.

Schimel, Joshua. 2004. "Playing Scales in the Methane Cycle: From Microbial Ecology to the Globe." *PNAS*. doi:10.1073pnas.0405075101.

Serrano-Silva, Nancy, Yohana Sarria-Guzmán, Luc Dendooven, and Marco Luna-Guido. 2014. "Methanogenesis and Methanotrophy in Soil: A Review." *Pedosphere* 24 (3): 291–307.

Sharma, Chhmendra, Manoj Kumar Tiwari, and Hemant Pathak. 2008. "Estimates of Emission and Deposition of Reactive Nitrogenous Species for India." *Current Science* 94 (11): 1439–46.

Skiba, Ute Maria, Sas Van Dijk, and Bruce C Ball. 2002. "The Influence of Tillage on NO and N_2O Fluxes under Spring and Winter Barley." *Soil Use and Management* 18: 340–45. doi:10.1079/SUM2002141.

Smith, Katy, Dexter Watts, Thomas Way, H. Allen Torbert, and Stepher Prior. 2012. "Impact of Tillage and Fertilizer Application Method on Gas Emissions in a Corn Cropping System." *Pedosphere* 22 (5): 604–15. doi:10.1016/S1002-0160(12)60045-9.

Snyder, Clifford S., Tom W. Bruulsema, Triniti L. Jensen, and Paul E. Fixen. 2009. "Review of Greenhouse Gas Emissions from Crop Production Systems and Fertilizer Management Effects." *Agriculture, Ecosystems and Environment* 133 (3–4): 247–66. doi:10.1016/j.agee.2009.04.021.

Stockle, Claudio, Stewart Higgins, Armen Kemanian, Roger Nelson, David Huggins, Jordi Marcos, and Harold Collins. 2012. "Carbon Storage and Nitrous Oxide Emissions of Cropping Systems in Eastern Washington: A Simulation Study." *Journal of Soil and Water Conservation* 67 (5): 365–77. doi:10.2489/jswc.67.5.365.

United Nations. 2017. "World Population 2017." *United Nations. Department of Economic and Social Affairs. Population Division*, 1–2. doi:10.1093/nar/gkl248.

Valbuena, Diego, Sabine Homann Kee Tui, Olaf Erenstein, Nils Teufel, Alan Duncan, Tahirou Abdoulaye, Braja Swain, Kindu Mekonnen, Ibro Germaine, and Bruno Gérard. 2015. "Identifying Determinants, Pressures and Trade-Offs of Crop Residue Use in Mixed Smallholder Farms in Sub-Saharan Africa and South Asia." *Agricultural Systems* 134: 107–18. doi:10.1016/j.agsy.2014.05.013.

Venterea, Rodney T., Dennis E. Rolston, and Zoe G. Cardon. 2005. "Effects of Soil Moisture, Physical, and Chemical Characteristics on Abiotic Nitric Oxide Production." *Nutrient Cycling in Agroecosystems* 72: 27–40. doi:10.1007/s10705-004-7351-5.

Wang, Hongguang, Zengjiang Guo, Yu Shi, Yongli Zhang, and Zhenwen Yu. 2015. "Impact of Tillage Practices on Nitrogen Accumulation and Translocation in Wheat and Soil Nitrate-Nitrogen Leaching in Drylands." *Soil and Tillage Research* 153: 20–27. doi:10.1016/j.still.2015.03.006.

Wang, Jun, Upendra M. Sainju, and Joy L. Barsotti. 2012. "Residue Placement and Rate, Crop Species, and Nitrogen Fertilization Effects on Soil Greenhouse Gas Emissions." *Journal of Environmental Protection* 3 (9): 1238–50.

Wang, Yuying, Chunsheng Hu, Hua Ming, Oene Oenema, Douglas A. Schaefer, Wenxu Dong, Yuming Zhang, and Xiaoxin Li. 2014. "Methane, Carbon Dioxide and Nitrous Oxide Fluxes in Soil Profile under a Winter Wheat-Summer Maize Rotation in the North China Plain." *PLoS One* 9 (6). doi:10.1371/journal.pone.0098445.

Werner, Christian, Klaus Butterbach-Bahl, Edwin Haas, Thomas Hickler, and Ralf Kiese. 2007. "A Global Inventory of N_2O Emissions from Tropical Rainforest Soils Using a Detailed Biogeochemical Model." *Global Biogeochemical Cycles* 21 (3). doi:10.1029/2006GB002909.

Wiese, Japie D., Johan Labuschagne, and Gert A. Agenbag. 2016. "Soil Water and Mineral Nitrogen Content as Influenced by Crop Rotation and Tillage Practice in the Swartland Subregion of the Western Cape." *South African Journal of Plant and Soil* 33 (1): 33–42. doi:10.1080/02571862.2015.1057772.

Wrage, Nicole, Gerard L. Velthof, Marinus L. Van Beusichem, and Oene Oenema. 2001. "Role of Nitrifier Denitrification in the Production of Nitrous Oxide." *Soil Biology and Biochemistry* 33 (12–13): 1723–32. doi:10.1016/S0038-0717(01)00096-7.

Yamulki, Sirwan, and Sal C. Jarvis. 2002. "Short-Term Effects of Tillage and Compaction on Nitrous Oxide, Nitric Oxide, Nitrogen Dioxide, Methane and Carbon Dioxide Fluxes from Grassland." *Biology and Fertility of Soils* 36 (3): 224–31. doi:10.1007/s00374-002-0530-0.

Yao, Zhisheng, Xunhua Zheng, Yanan Zhang, Chunyan Liu, Rui Wang, Shan Lin, Qiang Zuo, and Klaus Butterbach-Bahl. 2017. "Urea Deep Placement Reduces Yield-Scaled Greenhouse Gas (CH_4 and N_2O) and NO Emissions from a Ground Cover Rice Production System." *Scientific Reports*: 1–11. doi:10.1038/s41598-017-11772-2.

Zaman, Mohammed, Mathew L. Nguyen, Miloslav Simek, Shah Nawaz, Muhammed Jameel Khan, M. N. Babar, and S. Zaman. 2012. "Emissions of Nitrous Oxide (N_2O) and Di-nitrogen (N_2) from Agricultural Landscape, Sources, Sinks, and Factors Affecting N_2O and N_2 Ratios." *Greenhouse Gases – Emission, Measurement and Management*, edited by Guoxiang Liu, 1–32. doi:10.5772/32781.

8 Soil Health and Fertilizer Use in India

Bhagwati Prasad Bhatt, Surajit Mondal,
Kirti Saurabh, Sushanta Kumar Naik,
Karnena Koteswara Rao, and Akram Ahmed

CONTENTS

8.1   INTRODUCTION

Soil is one of the most complicated biological materials on the planet. Soils are the bedrock for the sustainable food system and for healthy lives. Soils produce 95% of the food for the growing population of the world. At the same time, soil provides living space for humans, as well as essential ecosystem services that are important for water regulation and supply, climate moderation, biodiversity conservation, carbon sequestration, and cultural services. Soil stores 10% of the total carbon dioxide emitted in the world (Patel 2016). Soil is the network of interacting living organisms within the earth's surface layer, which supports life above and below ground. Many soil processes like nutrient cycling, water infiltration, organic matter decomposition, etc. are affected by the type of living organisms of the soil. Fungi and bacteria help in decomposition of organic matter in the soil and earthworms digest organic matter, recycle nutrients, and make the surface soil richer. In a handful of fertile soil, there are more individual organisms than the total number of human beings that have ever existed (Salleh 2011). The soil has varying amounts of organic matter (living and dead organisms), minerals, and nutrients. Soil carbon is highly variable within the profile and controls many processes, such as the development of soil structure, water storage, and nutrient cycling. Soil carbon takes three distinct forms, namely, living carbon, labile carbon, and fixed carbon. Living carbon comprises microbes, fungi, plant roots, nematodes, earthworms, etc. Labile carbon (C) in the soil comprises decomposing (dead) plant and animal materials that are in a state of transition. Fixed C in soil consists of stable compounds as humates and glomalin. Sequestered C comprises fixed C plus the total living biomass. Soil high in organic carbon content is characterized by better rainfall infiltration and retention, and greater resilience to drought. Healthy soils are the basis for food, feed, fuel, fiber, water, and medicinal/herbal products important for human well-being. Soils are essential to the sustainability of natural and managed ecosystems, and play a key role in cycling C, storing and filtering water, and improving resilience to floods and droughts.

Soils are vulnerable to C loss through degradation, but regenerative land management practices can build soil and restore soil health. Soil erosion, loss of soil organic matter (SOM), and nutrient depletion are among the leading contributors to impaired soil health, reduced crop yields, and rural poverty. Soil health, an attribute of physical, chemical, and biological processes, is impaired by indiscriminate and intensive cultivation and by overmining of nutrients by crops with lesser replenishments through organic and inorganic sources of plant nutrients. Poorly managed soils are deficient in several major and micronutrients. These deficiencies not only influence crop yields but also the mineral content of seed and feed, and thus affect the health of crops, animals, and humans. The problem has been aggravated since the mid-1960s by intensive cropping but without supplementary input of micronutrients. Such a severe decline in soil health is one of the reasons for stagnating or declining agronomic yields (Kassam 2013). As described by Doran and Parkin (1994), soil health is "the capacity of a soil to function within the ecosystem and land-use boundaries, to sustain productivity, maintain environmental quality, and promote plant and animal health." Soil erosion, loss of SOM, and nutrient depletion are among the leading contributors to impaired soil health, reduced crop yields, and poverty. The decline in soil health is an important issue for sustaining crop productivity as well as human health. Moreover, intensive tillage and high water requirements mostly linked with the high use of chemical fertilizers and their overdependence have degraded soil health, resulting in a decline in soil C stocks (Kassam 2013). Maintaining soil health/quality is thus indispensable for sustaining agricultural productivity at the desired level.

India was facing the problem of food shortages during the period of independence and imported food grains from other countries. However, with focused efforts, the country attained self-sufficiency and agriculture production increased considerably. The use of fertilizers is an inevitable factor in improving soil fertility and productivity of crops regardless of the nature of the cropping sequence or environmental conditions. It has been unequivocally demonstrated that one-third of crop productivity is dictated by fertilizers, besides influencing use efficiencies of other agri-inputs (Indoria et al. 2018). In the past four decades, nutrient use efficiency (NUE) of crops has stagnated (FAI 2010). The nutrients that are left in the soil may enter into the aquatic environment, causing eutrophication. In addition to the low nutrient efficiencies, agriculture in developing countries, including India, is facing a problem of low SOM content, unbalanced fertilization, and low fertilizer response, which eventually leads to stagnation of crop yield (Biswas and Sharma 2008). The fertilizer response ratio in the irrigated areas of India has decreased drastically (NAAS 2009). The optimal NPK fertilizer ratio of 4:2:1 is ideal for crop productivity (NAAS 2009), while the current ratio in India is at 6.7:3.1:1 because of the excessive use of nitrogenous fertilizers. In order to achieve a target of 300 million tons (Mt) of food grains and to feed the burgeoning population of 1.4 billion in the year 2025, the country will require 45 Mt of nutrients as against a current consumption level of 23 Mt. The extent of multinutrient deficiencies is alarmingly increasing year by year, which is closely associated with a crop loss of nearly 25%–30%. There is a need for exploiting the potential of chemical fertilizers in Indian agriculture.

Enhancement and maintenance of soil productivity are essential to the sustainability of agriculture and for meeting basic needs of the rising population. India supports approximately 17% of the global population and 11% of the world's livestock population on merely 2.5% of the world's geographical area, adding immense pressure on agricultural land. The scope of increasing land area for producing food is rather limited. Bringing marginal lands under plough is risky and may pose a threat to fragile ecosystems. Hence, meeting food demand mostly depends upon the supply of plant nutrients and water as well as the capacity of soil to perform various functions related to nutrient cycling and rhizospheric environment. The availability of plant nutrients is becoming scarce due to rising costs, and that of water has always remained uncertain due to its dependence on climate. The situation is getting worse due to competition from other sectors and changing climatic patterns and also due to deteriorating soil quality because of increased anthropogenic activity. Numerous studies have documented a decline in soil productivity, produce quality, and groundwater quality, as well as a loss in soil biodiversity as a result of these factors (Bhandari et al. 2002; Manna et al. 2005; Panwar et al. 2010). It is a matter of great concern that degraded lands form more than 57% of the total reporting area against 17% in the whole world. Hence, a pressing need has arisen for managing our precious soil resources for improving and sustaining its various functions so that the demand for quality food is met and the quality of the environment is improved.

8.2 PRESENT SCENARIO OF SOIL HEALTH IN INDIA

Soil ecosystems are the basis of systems that sustain human existence. Soil erosion, loss of organic soil, and depletion of nutrients are among the major contributors to the impairment of soil health, decreased crop yields, and insufficiency. The United Nations proclaimed 2015 the Year of Soils, realizing the important role that soils play in food and environmental security and in mitigating climate change. Similarly, the International Soil Science Union (IUSS) proclaimed 2015–2024 to be the International Decade of Soil. The foundation for any harmonious and successful farming activity is to build upon and maintain soil quality. The connection between soil quality, agricultural practices, long-term soil productivity, sustainable soil management, agriculture, and quality of the environment is now commonly recognized (Figure 8.1), as it reflects the significance of soil conservation as a resource for future generations rather than "soil fertility," which has been generally connected only with crop yield (Gregorich et al. 2001).

Soil health is an attribute of physical, chemical, and biological processes and properties that are constantly decreasing and are often quoted as one of the factors for stagnating or decreasing crop yields and low input use efficiency. For sustainable crop development, understanding the soil environment

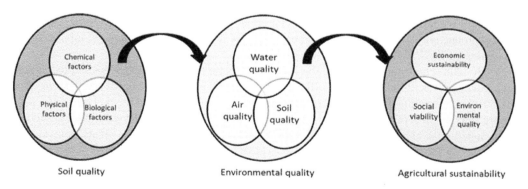

FIGURE 8.1 Relationship among soil quality, environmental quality, and agricultural sustainability. (From Gregorich, E.G., et al., *Can. J. Soil Sci.*, 81, 21–31, 2001.)

in which crops grow in totality, recognizing the constraints of that setting, and improving wherever possible without damaging the quality of the soil is crucial for all concerned, not just researchers.

Degradation of physical, chemical, and biological soil quality along with insufficient and imbalanced use of nutrients and neglect of organic manure is the cause of multinutrient deficiencies over time in many fields. This, combined with bad field water management, is the main cause of low crop productivity and decreased effectiveness in the use of nutrients and water. As such, it presents a major danger to the intrinsic ability of the soil to sustain productivity. Maintaining soil health/ quality is therefore essential to maintain the required level of agricultural productivity.

8.2.1 Soil Erosion

A decline in soil health status resulting in the ecosystem's reduced ability to provide its beneficiaries with products and services results in soil degradation. Degraded soils have a health status such that they do not provide their ecosystem with the normal goods and services of the specific soil. Of the complete geographic region of India (328.7 M ha), 304.9 M ha comprise the reporting region of 264.5 M ha used for agriculture, forestry, pasture, and other manufacturing of biomass. Many organizations have earlier evaluated the seriousness and magnitude of soil degradation in the nation (Figure 8.2). Approximately 146.8 M ha is degraded, according to the National Soil Survey and Land Use Planning Bureau (NBSS&LUP 2004). Water erosion is India's most severe issue of degradation, leading to loss and deformation of the topsoil and terrain. The average soil erosion rate was ~16.4 Mg ha^{-1} year^{-1} based on the first approximation assessment of current soil loss information, leading to an annual complete soil loss of 5.3 billion Mg across the nation (Dhruvanarayan and Ram 1983). Nearly 29% of the total eroded land is continuously lost to the ocean, while 61% is merely moved from one location to another, and the remaining 10% is deposited in reservoirs.

Globally, water erosion causes a loss of about 24 billion tons of fertile soil annually (FAO 2011). A recent domestic soil degradation database in India demonstrates that 120.7 M ha or 36.7% of the country's total arable and nonarable land surface is affected by multiple types of degradation, with soil erosion being its main contributor in 83 M ha (68.4%) (NAAS 2010a). A significant risk to soil health/quality and the quality of surface runoff is also water erosion. It results in a loss of soil organic carbon (SOC), imbalance in nutrients, soil compaction, and decline in soil biodiversity.

The annual soil loss rate in India is approximately 15.4 Mg ha^{-1}, leading to nutrient losses of 5.4 to 8.4 Gg, a decrease in crop productivity, floods and droughts, a decrease in reservoir capacity (1 to 2 percent annually), and biodiversity loss (Sharda and Ojasvi 2016). Loss of crop productivity, one of many negative impacts of soil erosion by water, has serious consequences for the country's food, livelihood, and environmental security. Major rainfed crops in India suffer an annual production loss of 13.4 Gg due to water erosion, which amounts to a loss of Rs 205.3 billion in monetary terms (Sharda and Dogra 2013).

In both rainfed and irrigated regions of India, soil degradation has become a serious issue. Economic loss due to soil degradation in India is high (Table 8.1). This loss includes a decrease in crop productivity and intensity of land use, changes in cropping patterns, high input use, and decreasing profit (Srinivasarao et al. 2013).

Reddy (2003) used the data set of the National Remote Sensing Agency (NRSA) to estimate the production loss in India at Rupees (Rs) 68 billion in 1988–1989. Additional salinization, alkalinization, and waterlogging losses were estimated at Rs 8 billion. Late in the process of extensive research on the effect of water erosion on crop productivity, it was found that water-related soil erosion resulted in an annual crop loss of 13.4 Mt in cereals, oilseeds, and pulse plants, equivalent to approximately $162 billion.

In order to maintain better soil quality under intensive farming technologies, emphasis should be placed on creating workable soil quality indicators and techniques for assessing and monitoring soil quality, assessing soil quality under specific soil management schemes (cropping, tillage, and water and nutrient use practices) and identifying the effects of aggrading/degrading management practices

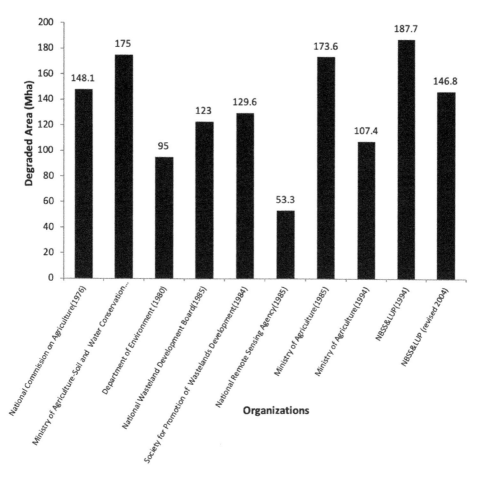

FIGURE 8.2 The extent of land degradation in India as evaluated by various organizations. (Modified from Bhattacharyya, R., et al., *Sustainability*, 7, 3528–3570, 2015.)

TABLE 8.1

Estimates of the Immediate Annual Price of Land Degradation in India

Parameters	NRSA (1990)	ARPU (1990)	Sehgal and Abrol (1994)
Area affected by soil erosion (M ha)	31.5	58.0	166.1
Area affected by salinization, alkalinization, and waterlogging (M ha)	3.2	—	21.7
Total area affected by land degradation (M ha)	34.7	58.0	187.7
Cost of soil erosion in lost nutrients (Rs billion)	18.0	33.3	98.3
Cost of soil erosion in lost production (Rs billion)	67.6	124.0	361.0
Cost of salinization, alkalinization, and waterlogging in lost production (Rs billion	7.6	—	87.6
Total direct cost of land degradation (Rs billion)	75.2	—	448.6

and sustaining soil quality. Also, strategies need to be formulated to decide the amount of organic matter to be added each year in the form of organic manures, organic waste, and organic residue to maintain or build up the SOM level in diverse soil types.

8.2.2 Physical Degradation

Soil physical degradation, caused by a decline in soil structure and tilth, leads to waterlogging, submergence, flooding, soil compaction, crusting, low infiltration, and impedance to root penetration. Physical deterioration of soil becomes a limiting factor in crop production. Some of these adverse changes are caused by inappropriate tillage, complete removal of crop residues from the soil, and the lack of sufficient energy for deep tillage or for adoption of agricultural practices for effective soil conservation.

Altogether, physical degradation contributes to the loss of agronomic productivity. Approximately 1.07 M ha of the land area is physically degraded, mostly by waterlogging owing to continuous ground flooding (0.88 M ha), and about 12.53 M ha of rainfed Vertisols stay fallow owing to temporary waterlogging during the summer (NAAS 2010a). Waterlogging alone leads to annual grain losses in India of 1.2 to 6.0 Gg (Bradon and Kishore 1995).

8.2.3 Chemical Degradation

Chemical soil degradation is attributed to (1) salinization/alkalinization, (2) acidification, (3) chemical soil toxification, and (4) depletion of nutrients and other nutrient-input-associated problems. There are presently around 6.74 M ha of salt-affected soils consisting of 3.79 M ha of high sodium (pH > 9.5) and about 3 M ha of salinity (including 1.25 M ha of coastal salinity) (NAAS 2010a).

Approximately 11 M ha of arable land has severe soil acidity (pH < 5.5) with very low productivity (<1 Mg/ha) owing to deficiencies and toxicity of certain nutrients. The poor soil health of acid soils is attributed to low fertility due to a mixture of toxicity of Al, Mn, and Fe, and deficiency of P, Ca, Mg, and K and some micronutrients such as B, Mo, and Zn, resulting in loss of crop productivity (NAAS 2010b).

With higher urbanization, the toxification of the soil through chemicals is growing. With heavy metals having carcinogenic impacts, more and more municipal and industrial waste is being dumped into the soil. In Bhopal, a study by the Indian Institute of Soil Science reported elevated concentrations of heavy metals (Cd, Cr, Cu, Pb, Ni, and Zn) in composts produced from blended municipalities in many towns of India. Groundwater contamination with arsenic (As) in the Bengal Delta basin, which lies between the Ganga and Padma Rivers of India and Bangladesh, is a major health concern. India, especially West Bengal, is one of the most affected areas, as 10 out of the 18 districts of the state are affected by As poisoning. Among these regions, Nadia District is one of the worst affected districts in terms of the level of As contamination and area coverage (Bhattacharya et al. 2009).

In addition, excessive use of fertilizers and pesticides for vegetables and other horticultural crops and decoration plants also leads to a buildup of toxic elements in the soil. Burning of fossil fuel and industrial emissions cause substantial air pollution that ultimately precipitates on soil and water.

When used for irrigation, polluted surface water and groundwater add a number of damaging chemicals to the soil. In some regions, there are reports of enhanced nitrate concentrations in the groundwater. It remains to be authenticated whether this is due to fertilizer N, animal production systems, or geogenic processes. Animal manures account for 5%–10% of N input in Indian agriculture but can be an important cause of nitrate pollution in some fields owing to leakage during storage and processing to groundwater. Depletion of nutrients, organic matter, and other nutrient-input-related issues aggravate soil degradation.

8.2.3.1 Poor Soil Fertility and Low and Imbalanced Nutrient Use

The soils of India are strongly depleted of their organic matter content and, therefore, have low fertility and poor structural attributes. Given low soil fertility and reduced soil and water resources, India must restore soil quality to boost productivity. It is also because of a low SOM content that soils covering approximately 59, 36, and 5 percent of the total agricultural area are characterized by low, medium, high stocks of N, respectively.

Similarly, soils of approximately 49%, 45%, and 6% of the land area are low, medium, and high in plant-available P, respectively; and soils of approximately 9%, 39%, and 52% of the land area are low, medium, and high in available K reserves, respectively (Chaudhari et al. 2015).

Not only are the intrinsic soil fertility and the nutrient input low, but there is also increasing evidence of rising P and K deficiencies, aggravated by the disproportionate or imbalanced implementation of greater P and K doses of N (Tewatia et al. 2017).

The N-based fertilizers make up a large proportion of the complete fertilizer material, almost 70%. There is growing proof that oilseeds, pulses, legumes, and high-yielding cereals are increasingly responding to S input. The S status of the soils of India is declining with each passing year. An increasing trend of S deficiency is reported by soil analysis and plant reaction information produced by the TSI-FAI-IFA project (1997–2006), which reinforced the finding of the ICAR system AICRP information. Of the more than 49,000 soil samples analyzed in 18 states, 46% were sulfur deficient, and another 30% were sulfur deficient samples that could be considered potentially sulfur deficient (Sulphur Institute 2019).

The micronutrient deficiency in crops is growing rapidly both in extent and intensity (see Figure 8.3), and as per the assessment made under the All India Coordinated Research Project on Micro and Secondary Nutrients and Pollutant Elements in Soils and Plants, nearly 49%, 15%, 6%, 8%, 11%, and 33% samples were found to be deficient in zinc, iron, manganese, copper, molybdenum, and boron, respectively, across the country, and hence all contribute toward poor soil health (Shukla et al. 2012).

It is expected that the micronutrient deficiency will increase in both quantity and magnitude with greater yields and more intensive agriculture. Limiting nutrients does not allow other nutrients to be fully exploited, and leads to reduced fertilizer reaction and reduced crop productivity. The present

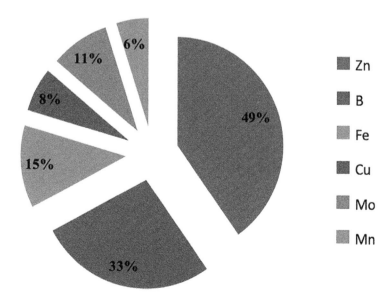

FIGURE 8.3 Micronutrient deficiency in Indian soils. (From Shukla, A.K., et al., *Indian J. Agron.*, 57, 123–130, 2012.)

gap between annual soil nutrient depletion and external source inputs is 10 Gg, which may be aggravated over time. This is one of the main causes of degradation of soil chemical properties leading to impaired soil health.

However, there is a big disparity in the consumption of fertilizers across Indian states on the input side. NPK's fertilizer input, 95.9 kg/ha, is much smaller in the country's western areas than the domestic average rate of 141.9 kg ha^{-1}. The average rate of fertilizer use is 140.9, 182.4, and 186.2 kg ha^{-1}, respectively, for the east, north, and south areas (FAI 2016). In addition, out of 565 districts surveyed for fertilizer consumption, 93 districts apply < 50 kg ha^{-1}, and another 124 districts fall into the NPK consumption category of 50–100 kg ha^{-1} (FAI 2016). Lack of guaranteed irrigation water supply is one of the main causes of low fertilizer input. Only 109 districts in the nation use NPK > 200 kg ha^{-1} of fertilizer. In Punjab, where over 98% of the region is irrigated, 19 out of 20 surveyed districts apply NPK > 200 kg ha^{-1} of fertilizer. As a result, the nutrient needs of crops and associated nutrient losses of Indian agriculture are very large (and growing each year). Therefore, it is important to assess the amount of nutrient added, removed, and remaining in agricultural soils (Figure 8.4). Practical information is needed on whether the nutrient status of soil (or a region) is being maintained, built up, or depleted. The net negative national budget is 9.7 million Mg for NPK, which is equivalent to an annual depletion rate of 19% N, 12% P, and 69% K. The current estimated net decline per ha from India's net sown area of 143 M ha amounts for 16 kg N, 11 kg P$_2$O$_5$, and 42 kg K$_2$O (69 kg N+P$_2$O$_5$ + K$_2$O). The large percentage for K is partially due to plants removing an average of 1.5 times more K than N, and K input via fertilizer is much smaller than that of N or P application (Tandon 2007). As a result, Indian farming operates at a net adverse plant nutrient equilibrium, resulting in chemical degradation and poor soil health.

8.2.3.2 Low Use of Organic Manures

Soil organic carbon (SOC) is an index of good soil health, and regular application of organic manure helps keep the soil high in SOC. The soils of India contain low levels of SOM, which is also decreasing over time (Figure 8.5). Therefore, without regular use of organic manures and recycling of crop residues, it is rather difficult to maintain good soil health, sustain productivity, and guarantee a high response to NPK fertilizers. Bulky organic waste is growing with fast urbanization, and its disposal and lucrative use in rural agriculture are hampered by transportability and high cost. The practice of green manuring is nearly forgotten.

SOC content is also the principal determinant of soil physical quality, chemical properties including soil nutrient status, and soil biological health. Long-term fertilizer experiments (LTFE)

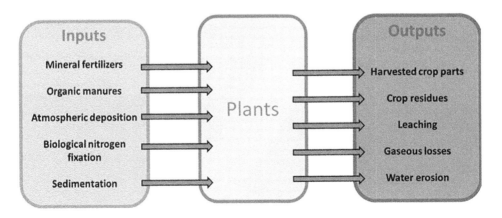

FIGURE 8.4 The nutrient balance (nutrient addition and removal) of agricultural soil. (Modified from Tandon, H.L.S., *Better Crops–India*, 15–19, 2007.)

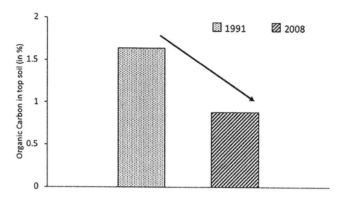

FIGURE 8.5   Change in soil organic carbon (in %) over time in India. (From FAOSTAT, http://www.fao.org/faostat/en/#data/RFN, accessed on 12 July 2019.)

TABLE 8.2
Status of Organic Soil Carbon (SOC) (g kg⁻1) in Long-Term Experiments with Fertilizers

Location	Control	N	NP	NPK	NPK + FYM
Akola	3.1	4.3	4.9	5.4	7.9
Bangalore	4.8	5.1	5.2	5.5	5.8
Barrackpore	5.5	6.8	7.2	7.2	9.0
Jabalpur	4.2	5.2	4.2	6.8	8.9
Ludhiana	2.8	3.9	3.9	4.2	5.3
New Delhi	3. 4	4.3	4.6	4.6	5.2
Palampur	7.9	8.8	9.3	10.2	13.7
Pantnagar	6.1	9.1	9.9	9.8	15.6
Parbhani	5.7	5.4	5.9	6.2	6.8
Ranchi	3.5	4.2	4.2	4.0	4.9

Source: NAAS, Soil health: New policy initiatives for farmers welfare, New Delhi, India, 2018.

have clearly demonstrated (Table 8.2) that balanced fertilization and input of farmyard manure (FYM) has maintained and/or enhanced SOC content in diverse soils (Singh and Wanjari 2017).

8.2.3.3   Skewed N:P:K Ratio

Degradation of soil health is also attributed to imbalanced use of fertilizer nutrients over a long time. The N:P:K use ratio has been skewed by suboptimal nutrient use against the ideal ratio of 4:2:1 ($N:P_2O_5:K_2O$) (Roy and Hammond 2004). There is an urgency to restore the balance of nutrients in states using high rates of urea. This nutrient imbalance has led to a widening gap between crop removal and application of fertilizer. Long-term coordinated fertilizer reaction studies in India have shown that soil P and K status in all soils have decreased when only N was applied, resulting in high degradation of soils, indicating bad soil health (Table 8.2). The years were specifically selected to indicate the effect on NPK use whenever there was any policy change in fertilizer prices as outlined in Table 8.3, i.e., 1991–1992 and 1992–1993, as well as the impact of the nutrient-based subsidy (NBS) scheme in 2010–2011 and 2011–2012.

Improper and imbalanced use of chemical fertilizer, as is apparent from broad NPK fertilizer usage ratios combined with less added organic manure, has led to degradation of soil health with

TABLE 8.3

Trends in NPK Use and Its Ratio in Indian Agriculture

Year	N (kg ha⁻¹)	P₂O₅ (kg ha⁻¹)	K₂O (kg ha⁻¹)	Total	NPK Use Ratio
1965–1966	3.70	0.85	0.50	5.05	7.4:1.7:1
1980–1981	21.31	7.03	3.61	31.95	5.9:1.9:1
1990–1991	43.06	17.34	7.15	67.55	6.0:2.4:1
2010–2011	83.76	40.72	17.78	142.26	4.7:2.3:1
2011–2012	88.36	40.42	13.15	141.93	6.7:3.1:1
2012–2013	86.60	34.25	10.61	131.46	8.2:3.2:1
2013–2014	83.35	28.03	10.44	121.83	8.0:2.7:1
2014–2015	85.45	30.75	12.77	128.96	6.7:2.4:1
2015–2016	87.58	35.18	12.11	134.87	7.2:2.9:1
2016–2017	84.37	33.80	12.65	130.82	6.7:2.7:1

Source: FAI, Fertiliser Statistics 2016–17, Fertiliser Association of India, New Delhi, India, 2017.

extensive multinutrient deficiencies, especially secondary and micronutrient deficiencies, namely sulfur (46%), zinc (49%), and boron (33%) (Shukla et al. 2012). Limiting nutrients, not enabling complete expression of other nutrients, significantly reduced the fertilizer reactions and stagnated the country's crop output. The reaction proportion of fertilizers (kg of grain per kg of nutrient) decreased almost four times in irrigated regions of India (from 13.4 in 1970 to around 3.2 in 2010) (NAAS 2018). The yield in irrigated areas stagnated at about 2.0 Mg ha⁻¹ over the years, with the contribution of fertilizers to total food grain production increasing from 39% (1970) to around 50% (1990) and thereafter it was almost stagnant at 40%. While only 54 kg fertilizer nutrients were required per ha during 1970 to maintain the yield level at around 2.0 Mg ha⁻¹, over five times as many fertilizer nutrients (280 kg ha⁻¹) are required presently to sustain the same yield level (Sharma 2008), indicating degradation of soil health owing to improper and imbalanced use of chemical fertilizer, and this is a matter of great concern.

The decline of soil health leads to low effectiveness of nutrient use. Real issues have been raised about the effectiveness of fertilizer use, in general, and N use effectiveness (NUE), in specific, for both financial and environmental reasons. Worldwide, NUE is as small as 33% for crop production (wheat, rice, corn, barley, sorghum, millet, oat, and rye). The annual loss of N fertilizer, worth up to Rs 72,000 crores in financial terms, is the uncounted 67% (NAAS 2005).

Many N retrieval studies undertaken on diverse crops in India have revealed unaccounted fertilizer N losses ranging from 20% to 50% based on soil health status and local circumstances. In India, though about 70% of the fertilizer used consists of nitrogen fertilizers and 80% of the fertilizer used is urea, its effectiveness of use is hardly 30%–50%. For other nutrient components, the effectiveness of use is 15%–20% (P), 60%–70% (K), 8%–10% (S), and 1%–5% (micronutrients), which is a significant problem. Low N effectiveness of use, owing to incorrect and imbalanced use of N resulting in bad soil health, eventually has global consequences. While the worldwide C cycle is disrupted by less than 10% owing to anthropogenic activity, the worldwide reactive N cycle is disrupted by more than 90%. Increasing agricultural use of N to satisfy continuous food production requirements, coupled with increasing discharge of N from dairies and industrial exhausts/effluents, makes it inevitable that N cycle disturbance will continue to be much quicker than that of carbon.

Further, the global warming potential of nitrous oxide is 298 times that of CO_2 as a prospective greenhouse gas. In addition, cow-dung cakes and crop residues are burned, which causes emissions of greenhouse gasses, plant nutrient losses, and organic carbon. Similarly, owing to low usage effectiveness, nitrogen fertilizers contribute about 77% of total direct nitrous oxide emissions from agricultural soils through the soil health-related denitrification method, being more from degraded soils.

Thus, India's challenge now is to identify methods to sustainably advance agricultural and industrial growth without adversely affecting nitrogen-related soil health, climate, and ecological processes.

8.2.4 Biological Degradation

Biological soil health degradation is caused by soil erosion and water runoff, leading to loss of fauna and flora and depletion of SOC content. These processes are exacerbated by extremes of acidity and alkalinity, addition of toxic substances, excessive use of chemicals, excessive tillage, extremes of climate, etc. Management procedures that decrease SOM or bypass biologically mediated nutrient cycling also tend to decrease soil microbial communities both in size and complexity. Both animals (fauna/microfauna) and plants (flora/microflora) are significant soil organisms in maintaining the general soil quality, fertility, and soil stability. They are closely connected with soil-borne biological and biochemical transformations. The soil health and the associated biological state provide an early indication of soil degradation that can assist in the timely adoption of extra prudent, viable soil-crop management methods. However, data on the desirable level of activity, numbers, and variety of soil organisms has yet to be established in order to preserve fertile and productive soil.

8.2.5 Crop Residue Menace

Residue management is essential, especially that of rice straw, which is poor feed for animals due to its high silica content but is often unsuccessfully managed due to the unavailability of proper processing facilities for its disposal. Traditionally, farmers graze cattle in their fields, sell residues as animal fodder and/or as biofuel to supplement their income, or burn the residues in situ, and the latter causes significant environmental and health damage. Residue removal leads to loss of a large amount of nutrients from the soil every year, and its retention is an obstacle to the sowing of the succeeding crop. The introduction of combined harvester (especially in the northwestern part of the Indo-Gangetic Plains [IGP] of India) leaves more loose straw on the ground, creating an obstacle to the sowing of the next wheat crop (Gupta et al. 2003). The shift in rice sowing coinciding with monsoon rains following a mandate of Punjab and Haryana states to stop groundwater overexploitation has actually left a small window of 15 days for rice harvest and sowing of wheat (Tallis et al. 2017). Farmers, therefore, burn the stubbles completely (chopped stubbles with stubble shaver and left to dry) or partially (only leftover loose stubble after combined harvester operation) for quick disposal of residues and to facilitate sowing of wheat. Burning releases several greenhouse gases and particulate matter or soot. Despite several adverse effects, residue burning is still rising in India (Figure 8.6). A very high level of air pollution (increase in black carbon particles and greenhouse gases) in Delhi and other north-western states in recent years is believed to be due to the burning of rice residues, which has become a serious health issue, and cause of the winter haze and smog (Sarkar et al. 2018). An estimated 26% of Delhi's air pollution in the winter months is due to the burning of rice residues (Sharma and Dikshit 2016), while deaths due to crop residues burning in India was estimated at 42,000 in 2010 (Lelieveld et al. 2015). Sustainable agricultural production in India is, therefore not possible with burning as an option for crop residue management, and we must opt for alternate practices (Tallis et al. 2017; Shyamsundar et al. 2019). While farmers continue to burn residues in fields despite a ban, the government of India released a national policy in 2014 (NPMCR 2014) and is preparing a roadmap to promote alternative uses of crop residue.

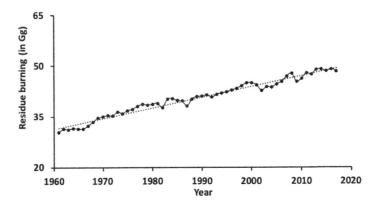

FIGURE 8.6 Amount of residue burning (in Gg) in India from 1961 to 2017. (From FAOSTAT, http://www. fao.org/faostat/en/#data/RFN, accessed on 12 July 2019.)

8.3 APPROACHES TO IMPROVE SOIL HEALTH

8.3.1 Conservation Agriculture

To arrest the widespread deterioration of soil health, conservation agriculture (CA) could be a viable option for restoring or maintaining soil health in a sustainable manner. CA covers a wide range of techniques based on three principles (Scopel et al. 2013): minimum soil disturbance during crop establishment; permanent soil cover by crop residues, cover crops, or growing crops; and crop rotation and diversification. CA is based on optimizing yields and profits to achieve a balance of agricultural, economic, and environmental benefits. It advocates that the combined social and economic benefits gained from combining production and protecting the environment, including reduced input and labor costs, are greater than those from production alone. Many researchers have already documented widespread benefits of CA in India. Excessive tillage in traditional agriculture destroys soil aggregates by mechanically breaking down aggregates, favors oxidation of SOM that was otherwise protected within aggregates, affects the population of soil organisms, especially macrofauna (earthworm, etc.), reduces macropores, and breaks down continuity of pore, resulting in more water runoff. Following the FAO's approach, adoption of CA can lead to (1) achievement of acceptable profits, (2) high and sustained production levels, and (3) conservation of the environment. It aims at reversing the process of degradation inherent to conventional agricultural practices like intensive agriculture and burning/removal of crop residues. Hence, it aims to conserve, improve, and make more efficient use of natural resources through integrated management of available soil, water, and biological resources combined with external inputs. It can also be referred to as resource-efficient or resource-effective agriculture.

CA practices pursued in many parts of the world are built on ecological principles of making land use more sustainable (Wassmann et al. 2009; Behera et al. 2010; Lal 2013, 2015, 2016). The adoption of CA for enhancing resource use efficiency (RUE) and crop productivity is the need of the hour as a powerful tool for the management of natural resources and to achieve sustainability in agriculture. Globally, CA is being practiced on about 180 M ha (Kassam et al. 2019). The leading countries adopting CA are the United States (43.2 M ha), Brazil (32.0 M ha), Argentina (31.0 M ha), Canada (19.9 M ha), and Australia (22.3 M ha). India is still lagging behind in adopting CA (Figure 8.7). Zero tillage (ZT), which is a component of CA, is recently gaining popularity in India. Over the past few years, adoption of CA in India has expanded to cover about 1.5 million hectares (Kassam et al. 2019). The major CA-based technology being adopted is ZT of wheat in the rice–wheat system of the IGP. In other cropping systems, the traditional agriculture-based

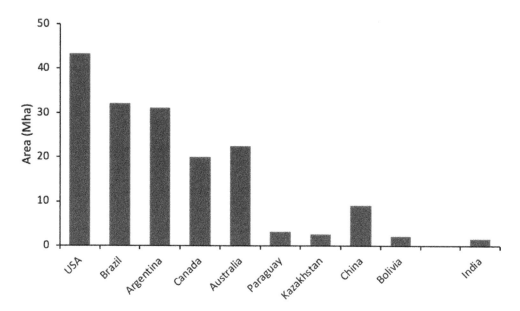

FIGURE 8.7 The extent of adoption of CA in leading countries and in India in 2015–2016. (Modified from Kassam, A., et al., *Int. J. Environ. Stud.*, 76, 29–51, 2019.)

systems are gradually undergoing a paradigm shift from intensive tillage to reduced/zero-tillage operations. There is need to adopt the other components of CA, namely, residue retention or cover crops and diversification of the cropping system, in the traditional system to maintain or increase the productivity in a sustainable manner and improve soil health. The much-needed diversification of rice–wheat system through crop intensification, introduction of pulse crops, relay cropping, etc. can also be achieved by the implementation of the CA. Naik et al. (2017) observed enrichment of total SOC by 17.2%, 12.6% and 11.0% in six-year old mango, guava and litchi orchards, respectively, in the eastern plateau and hill region over-orchard crops due to leaf litter decomposition." Mango and guava leaf litter constitute comparatively readily available sources of nutrients, and they could be suitable for short-term nutrient correction and sustainability of soil fertility. However, leaf litter from litchi caused a noticeable slow decay rate and is worthy to be used for organic matter buildup in a hot and dry subhumid climate under CA (Naik et al. 2018).

Increasing the adoption of CA in India has the following prospects:

1. Improvement in soil health – Limited soil disturbance and surface cover of soil by crop residues or growing of cover crops can improve soil health. Surface residue on decomposition increases the SOC, which in turn improves the physicochemical properties of soil (Mondal et al. 2019) and maintains a better environment for soil micro- and macrofauna.
2. Better residue management – Sowing of the crop with residue in the field under CA offers a better, environment-friendly option of residue management. Avoiding residue burning can lead to multiple benefits of better protection of soil from rain and wind, increased SOM content, better nutrient status, etc.
3. Decline in cost of production – Adoption of CA can reduce the cost of cultivation significantly due to savings of labor, diesel, and water. Many researchers have documented cost reductions ranging from Rs 2000 to Rs 3000 ($33 to $50) per hectare on account of CA adoption (Malik et al. 2005; RWC-CIMMYT 2005).
4. Less weed occurrence – *Phalaris minor*, the major weed in wheat in India, is reported to occur less under ZT, resulting in a reduction in the use of herbicides.

5. Diversification and intensification of cropping system – Adoption of CA with all its components gives farmers an opportunity for the cultivation of more crops, and when it is embraced with proper moisture conservation practices, cropping intensity could be enhanced in rice-fallow areas that otherwise remain fallow for lack of irrigation facilities.

6. Saving in water and nutrients – One of the most promising benefits of CA is soil water conservation. Residue retention, or live surface cover, facilitates rainwater infiltration and decreases the loss of water through evaporation. Furthermore, the advancement of sowing time in wheat also enables more efficient use of water. The residue retained under CA gets decomposed with time and supplies vital nutrients to the crop and thereby could save nutrient inputs in the form of fertilizers. Mondal et al. (2018) reported better soil moisture conservation under ZT wheat as compared to conventional practices, which further modifies the thermal regime of soil.

7. Yield benefits – If properly managed, ZT wheat gives more yield compared to traditionally cultivated wheat, and this yield improvement happens due to the associated effects of CA like better soil physical health, improved soil fertility, better soil moisture status, reduction in evaporation loss of water, better soil temperature regime, and crop rotational benefits. An increase of 200–500 kg ha^{-1} has been reported in ZT wheat as compared to farmers' practice in the rice–wheat system of the IGP (Hobbs and Gupta 2003).

8. Environmental benefits – Residue burning is a menace to Indian agriculture and is responsible for soil health deterioration and greenhouse gas emissions. Better residue management under CA results in better recycling of soil nutrients and less emission of CO_2, CH_4, and N_2O. It also enhances resource use efficiency and improves soil health.

8.3.2 DIRECT SEEDING FOR RICE CULTIVATION

Direct seeding on dry soils has been used by lowland rice farmers in Asian countries such as Indonesia and Philippines that grow two rice crops within a year in rainfed areas with long rainy season. The system is called *gogo-rancha* in Indonesia and *sabog-tanim* in the Philippines. In the United States, irrigated lowland rice farmers use direct seeding on dry soils. Direct seeding onto flooded and saturated soils has been used by farmers in South America and Europe. The adoption of direct seeding onto flooded and saturated soils has increased in Asia due to the labor shortage in rural areas. It is estimated to occupy 21% of the total rice area in Asia (Pandey and Velasco 2002; Bruinsma 2011) Direct seeding, however, requires large quantities of seed. Also, weed competition in direct seeded fields is high.

The agronomy prerequisites for direct-seeded rice (DSR) are machines, precise land leveling, optimal seed rate, and seeding depth for an initial good crop establishment, precise water management, efficient and economical weed management, and the ability to establish residue, whereas, at the variety front, anaerobic germination, seedling vigor, and lodging are the major factors. In the case of direct seeding in unpuddled soil, water and weed management and the ability of the seedlings to be established with residue are fundamentally important.

In dry DSR, the land is ploughed in dry conditions and leveled. Primed seeds (water soaked for 24 h and incubated) are broadcast, drilled, or sown in furrows, followed by flooding the field or keeping the soil moist for seven days to ensure germination. The optimization of seed rate is important under dry DSR. The optimum seed rate is 25 kg/ha, beyond which the yield return is diminished. For continuous drill-sowing, the seed rate is higher (50–80 kg/ha), but drill-sowing in rows with a spacing of 20 cm (row-to-row) and 0–15 cm (plant-to-plant) reduces the seed rate to only 15–25 kg/ha (Ladha et al. 2009).

Weed management imposes a great challenge in DSR. Basically, DSR seedlings are less competitive than seedlings from nurseries for transplanting. There will always be the initial flush of weeds, which is otherwise controlled by flooding. There could be increased reliance on limited herbicide chemistries, which leads to herbicide resistance and also shifts in weed communities. For integrated weed management in DSR, there are a few points to ponder like stale seedbed, mulching and ZT, use of suitable herbicides, crop rotation, intercropping of cover crops, hand weeding/mechanical weeding, competitive rice cultivars, water management, and herbicide-resistant rice.

8.3.3 CROP ROTATION/CROP DIVERSIFICATION

A permanent soil cover is important to protect the soil against the deleterious effects of exposure to rain and sun, to provide the micro- and macroorganisms in the soil with a constant supply of "food," and to alter the microclimate in the soil for optimal growth and development of soil organisms, including plant roots. In turn, it improves soil aggregation, soil biological activity, and soil biodiversity and carbon (C) sequestration (Ghosh et al. 2010).

The rotation of crops is not only necessary to offer a diverse "diet" to the soil microorganisms, but also for exploring different soil layers for nutrients that have been leached to deeper layers that can be "recycled" by the crops in rotation. Furthermore, a diversity of crops in rotation leads to diverse soil flora and fauna. Cropping sequence and rotations involving legumes help in minimal rates of the buildup of the population of pest species through life-cycle disruption, biological nitrogen fixation (BNF), control of off-site pollution, and enhancing biodiversity (Dumanski et al. 2006; Kassam and Friedrich 2009).

8.3.4 ADOPTION OF ZERO TILLAGE

ZT, which is recently gaining popularity due to its multiple benefits, particularly in wheat under a rice–wheat cropping system, is now practiced on about 3 M ha in the IGP of South Asia (Gupta and Seth 2007; Harrington and Hobbs 2009). Most positive impacts of ZT are on wheat productivity, profitability, resource-use efficiency, and soil health, especially in areas where wheat sowing is delayed due to late harvesting of rice (Erenstein and Laxmi 2008; Ladha et al. 2009). Increased yields of wheat (3%–12%, due to timely sowing) and reduction in the cost of cultivation due to absence of tillage (US$37–92 ha^{-1}) are the two most important drivers of wider adoption of ZT in the IGP (Hobbs and Gupta 2003; Gupta and Seth 2007; Erenstein and Laxmi 2008). However, cultivation of rice is still continuing with the traditional practices of puddling and transplanting, which are not only resource inefficient (water, labor, and fuel) but also delay timely sowing of the succeeding crop (wheat). To comprehend the full advantages of ZT, which otherwise are lost due to puddling in rice, serious efforts are being made to develop ZT rice followed by ZT wheat – commonly referred to as "double zero tillage."

In ZT, the crop is sown in single tractor operation by using the seed-cum-fertilizer drill in the presence of surface residue and thereby completely excluding primary tillage or field preparation. Sharma et al. (2005) have reported a reduction of Rs 2500 ($41.7) on land preparation and a saving of 50–60 liters of diesel per hectare by adopting ZT in the IGP. It also favors timely sowing of wheat, improves input use efficiency, saves precious water, and increases yields up to 20% (Sharma et al. 2005).

The adoption and spread of ZT wheat have been a success story in the northwestern parts of India due to the following reasons:

1. Enhancement in long-term C sequestration and buildup in SOM constitute a practical strategy to mitigate greenhouse gas emissions and impart greater resilience in production systems to climate change–related aberrations (Saharawat et al. 2010).
2. Enhancement of soil quality, i.e., soil physical, chemical, and biological conditions (Jat et al. 2009; Gathala et al. 2011).
3. Reduction in the cost of production by Rs 2000 to Rs 3000 ha^{-1} ($33 to $50) (Malik et al. 2005; RWC-CIMMYT 2005).
4. Reduction in the incidence of weeds, such as *Phalaris minor* in wheat (Malik et al. 2005).

5. Increased input use efficiency, particularly water and nutrients (Jat et al. 2012; Saharawat et al. 2012).
6. Increase in yield and productivity (4%–10%) (Gathala et al. 2011).
7. Advancement of the sowing date of wheat (Malik et al. 2005).
8. Improved environmental sustainability due to less greenhouse gas emissions (Pathak et al. 2011) and enhanced SOC by 14.6% (Samal et al. 2017)
9. Reducing environmental pollution, loss of nutrients, and avoiding serious health hazards by retaining crop residue instead of burning (Sidhu et al. 2007).
10. Offering better prospects for intensification and diversification of cropping systems (Jat et al. 2005).
11. Increasing resource use efficiency by conserving soil moisture and retaining crop residues, which with time get decomposed and provide vital nutrients to the growing crop (Jat et al. 2009).

Despite multiple benefits, the rate of adoption of ZT is slow due to lack of awareness among farmers regarding its usefulness, unavailability of suitable machinery (which could work under residue conditions), and lack of guidance at the field level. Government initiatives like the formation of custom hiring centers for farm machinery, incentive to farmers for adoption, and timely advice could significantly hasten the process of ZT adoption, which would not only reduce the cost of cultivation but also maintain soil health in a sustainable manner.

8.3.5 CROP RESIDUE MANAGEMENT

The rice–wheat system, the most important cropping system of India, is facing many problems, and the sustainability of the system is at stake. Low SOC, poor soil physical health, multiple nutrient deficiencies, and residue burning are some of the major reasons for declining system productivity (Ladha et al. 2000). Proper management of crop residues (CRs) is imperative for better soil health and recycling of plant nutrients.

In India, over 500 Gg of agricultural residues are produced every year (NICRA 2019). With the increased production of rice and wheat, residue production has also increased substantially. The RW system accounts for nearly one-fourth of the total CRs produced in India (Sarkar et al. 1999). The surplus CRs (i.e., total residues produced minus the amount used for various purposes) are typically burned on farm. The amount of surplus CRs available in India is estimated between 84 and 141 Mt year^{-1} where cereal crops contribute 58%. Of the 82 Mt of surplus CRs, nearly 70 MTs (44.5 Mt rice straws and 24.5 Mt wheat straws) are burned annually.

CRs are good sources of plant nutrients, the primary source of organic matter (as C constitutes about 40% of the total dry biomass) added to the soil, and are important components for the stability of agricultural ecosystems. In northwest India, the RW cropping system generates huge quantities of CRs. Emerging crop residue management (CRM) options in the IGP to avoid burning are to mulch with rice straw in ZT wheat and incorporate combine-harvested or even manually harvested (as in the middle and lower IGP) wheat straw (1–2 Mg ha^{-1}) in rice (Bijay-Singh et al. 2008; Sidhu et al. 2011). CRs retained on the soil surface provide soil and water conservation benefits, and increase subsequent crop yield. Conservation of soil and water resources is of paramount importance for sustaining cropland productivity in many semiarid environments. In these areas, as much as 50% of total evapotranspiration from a crop can be lost through evaporation from the soil surface. In fact, apart from adjusting the growing period of crops, as has been done for rice in Indian Punjab, mulching is the only practice that reduces the evapotranspiration by decreasing evaporation (Prihar et al. 2010). When all CRs are used as animal feed or removed for other purposes, the above benefits are lost. As a result, sustaining soil productivity becomes more difficult. However, it is possible to sustain crop production by using appropriate alternative practices, such as retaining partial residues, replacing the nutrients harvested in grain and CRs, growing forages that substitute for CRs, and adopting CA practices. Improvements in crop management, leading to greater CR production, may allow sufficient residues to be returned to fields and some to be removed without adversely affecting the soil environment.

8.3.5.1 Residue Management Strategies

- Incorporation of crop residue into soils through the adoption of CA practices to prevent soil erosion from wind and water and to augment soil moisture conservation (Bhatt and Khera 2006; Mondal et al. 2019).
- Promotion of diversified uses of crop residue for various purposes, namely, power generation, as industrial raw material for the production of bioethanol; packing material for fruits, vegetables, and glassware; utilization for the paper, board, and panel industry; biogas generation and composting; and mushroom cultivation in public-private partnership (PPP) mode (Raju et al. 2012; Singh et al. 2016; Trivedi et al. 2017).
- Promotion of the capacity building of various stakeholders, including farmers and extension functionaries, under crop development programs and organization of field-level demonstrations on the management of crop residues in all programs/schemes (Kumar et al. 2015).
- Provision of subsidies to the farmers for hiring resource conservation machinery from the Custom Hiring Centre (CHC)/Agriculture Service Centre (ASC) and promotion of the establishment of new CHCs/ASCs to ensure availability of such machines to farmers at the time of crop harvesting (Ranade et al. 2006; Chahal et al. 2014).
- Use of crop residue for the cultivation of mushrooms, particularly *Agaricus bisporus* (white button mushroom) and *Volvriella volvacea* (straw mushroom) (Choudhary et al. 2009; NPMCR 2014).
- Promotion of the use of crop residue for preparation of bioenriched compost and vermin compost and their utilization as farmyard manure (IARI 2012; Lohan et al. 2015).
- Promotion and encouragement of the use of crop residue and rice straw in paper, board, and panel and packing material (Kumar et al. 2015; Lohan et al. 2018).
- Promotion of collection of crop residue for feed, fuel, brick making, etc., and extending the subsidy for the transport of crop residue to fodder-deficient areas (Kumar et al. 2015; Trivedi et al. 2017; Lohan et al. 2018).
- Promotion of various interventions under ongoing schemes and programs for diversified use of crop residue as fuel for power plants, production of cellulosic ethanol, etc. in PPP mode (Lohan et al. 2018).
- Organization of training of farmers for creating awareness about the effects of crop residue burning, adoption of CA practices, and resource conservation technology through all ongoing state/center sector schemes (NPMCR 2014).
- Establishment of self-help groups and provision of subsidies to unemployed youth for the establishment of custom hiring centers to enhance the availability of resource conservation machinery (Kumar et al. 2015).
- Demonstrations of crop residue management technology on a large scale by the State Department of Agriculture and other government institutions by organizing on-farm demonstrations to create awareness and dissemination of various technologies and organizing field days under ongoing programs and schemes (NPMCR 2014).

8.3.6 Balanced Use of Fertilizer

Balanced fertilization refers to the application of plant nutrients in optimum quantities in the right proportion through appropriate methods at the time suited for a specific crop and agroclimatic situation. Balanced fertilization leads to restoration of soil health, while imbalanced fertilization leads to the mining of soil nutrients and SOC stocks, causing soil sickness and an uneconomic waste of scarce resources. It is only the restoration of soil health that leads to a sustainable land-use system where most food grain production continues to come from the land already appropriated for agriculture.

Nutrient balance is a key component in enhancing crop productivity. It ensures efficient use of all the nutrients since the deficiency of any one essential nutrient limits the efficient use of all other nutrients even when they are available in adequate quantities. Fertilizer promoters in India advocate the use of $N/P_2O_5/K_2O$ fertilizers in the ratio of 4:2:1 (Roy and Hammond 2004). This ratio is generally considered to be an ideal one for achieving the maximum use efficiency of each nutrient. If one nutrient is present in large amounts, it may depress the uptake of some other nutrient(s) and reduce crop yield. Thus, there must be a proper balance among the essential plant nutrients.

Balanced and judicious use of fertilizers is the key to efficient nutrient use and for maintaining soil productivity. A plant nutrient added to the soil will be efficiently used by the crop only if the other essential nutrients are present in adequate amounts. Balanced use of fertilizers should be aimed at the following considerations:

- Increasing crop yields
- Improving the quality of the produce
- Increasing profit margin
- Correcting inherent soil-nutrient deficiencies
- Maintaining or improving long-term soil fertility and productivity
- Advancing environmental safety
- Restoring fertility and productivity of the soil that has been degraded by wrong and exploitative activities in the past

8.3.7 Organic Agriculture (OA)

OA is one among the broad spectrum of production methods that are supportive of the environment. OA systems are based on specific standards precisely formulated for food production and aimed at achieving agroecosystems that are socially and ecologically sustainable (Trewavas 2001). OA is based on minimizing the use of external inputs through the use of on-farm resources efficiently compared to industrial agriculture. Thus, the use of synthetic fertilizers and pesticides is avoided (Lampkin 2002; Hole et al. 2005).

OA systems rely on the management of SOM to enhance the chemical, biological, and physical properties of the soil. One of the basic principles of soil fertility management in OA systems is that plant nutrition depends on "biologically derived nutrients" instead of using readily soluble forms of nutrients, and relatively less available forms of nutrients such as those in bulky organic materials are used (Trewavas 2001). This requires the release of nutrients to the plant via the activity of soil microbes and soil animals. Improved soil biological activity is also known to play a key role in suppressing weeds, pests, and diseases (Wall et al. 2015).

Animal dung, crop residues, green manure, biofertilizers, and biosolids from agroindustries and food processing wastes are some of the potential sources of nutrients of organic farming. While animal dung has competitive uses as fuel, it is extensively used in the form of farmyard manure (FYM). Development of several compost production technologies like vermicomposting, phosphor-composting, N-enriched phosphor-composting, etc. improve the quality of composts through enrichment with nutrient-bearing minerals and other additives (Mishra et al. 2013). These manures have the capacity to fulfill the nutrient demand of crops adequately and promote the activity of beneficial macro- and microflora in the soil (Watson et al. 2002).

There are several doubts in the minds of not only farmers but also scientists about whether it is possible to supply the minimum required nutrients to crops through organic sources alone. Even if it is possible, how are we going to mobilize the organic matter? At this juncture, it is neither advisable nor feasible to recommend the switchover from fertilizer use to organic manure under all agroecosystems. Presently, only 30% of India's total cultivable areas have irrigation facilities where the use of agrochemicals is more compared to that for the rain-fed zones. It is in rain-fed agriculture that ingenuity and efforts are required to increase crop productivity and farm production despite the recurrence of the environmental constraints of drought and water scarcity.

OA and food production systems are quite distinct from conventional farms in terms of nutrient management strategies. OA systems adopt management options with the primary aim of developing whole farms, like a living organism with balanced growth, in both crops and livestock holdings. Thus, the nutrient cycle is closed as far as possible. Only nutrients contained in the food and products harvested are exported out of the farm. Crop residue burning is prohibited; so also is the unscientific storage of animal wastes and its application in the fields. OA is, therefore, considered more environmentally friendly and sustainable than the conventional system. Farm conversion from a high-input, chemical-based system to an OA systems is designed after undertaking a constraint analysis for the farm with the primary aim of taking advantage of local conditions and their interactions with farm activities, climate, soil, and environment, so as to achieve (as far as possible) closed nutrient cycles with less dependence on off-farm inputs.

The so-called organic transition effect, in which a yield decline in the first one to four years of transition to OA, followed by a yield increase when soils have developed adequate biological activity (Liebhardt et al. 1989; Neera et al. 1999), has not been borne out in some reviews of yield comparison studies. Trials conducted on organic cotton at Nagpur indicated that after the third year, the organic plot, which did not receive fertilizers and insecticides, produced as much cotton as that cultivated with them (Rajendran et al. 2000). Similarly, studies conducted in Punjab clearly indicated that OA produced higher or equal yields of different cropping systems compared to those from chemical farming after an initial period of three years (Kler et al. 2002). A survey of 208 projects in developing tropical countries in which contemporary OA practices were introduced showed average yield increases of 5%–10% in irrigated crops and 50%–100% in rainfed crops (Pretty and Hine 2001).

8.3.8 Integrated Pest Management and Healthy Soil

Healthier soils produce crops that are less prone to pests and pathogens (Wall et al. 2015; Wall and Six 2015). Some soil-management practices boost plant-defense mechanisms, making plants more resistant and/or less susceptible to pests. Other practices or the favorable conditions they produce restrict the severity of pest damage by decreasing pest incidence or building beneficial organisms. Using multiple tactics rather than one major tactic like a single pesticide lessens pest damage through a third strategy: it diminishes the odds that a pest will adapt to the ecological pest management measures (Wall et al. 2015).

Practices that promote soil health constitute one of the fundamental pillars of ecological pest management (Altieri et al. 2005). When stress is alleviated, a plant can better express its inherent abilities to resist pests. Ecological pest management emphasizes preventive strategies that enhance the "immunity" of the agroecosystem. Farmers should be cautious of using reactive management practices that may hinder the crop's immunity. Healthier soils also harbor more diverse and active populations of the soil organisms that compete with, antagonize, and ultimately curb soil-borne pests (Ferris and Tuomisto 2015; Wall et al. 2015). Some of those organisms – such as springtails – serve as alternate food for beneficial organisms, thus maintaining viable populations of beneficial organisms in the field. Beneficial organisms can be favored by using crop rotations, cover crops, animal manures, and composts to supply them with additional food (de Vries et al. 2013; Tiemann et al. 2015).

Many researchers have suggested that increasing insect pest and disease pressure in agroecosystems is due to changes that have occurred in agricultural practices since World War II (Altieri and

Nicholls 2003). For example, the usage of fertilizers and pesticides has increased rapidly during this period, and evidence suggests that such excessive use of agrochemicals in conjunction with expanding monocultures has exacerbated pest problems (Conway and Pretty 1991). On the other hand, proponents of alternative agricultural methods contend that crop losses to insects and diseases are reduced with OA (Oelhaf 1978; Merrill 1983). Although this view is widespread, there have been surprisingly few attempts to test its validity. The few conducted studies suggest that lower pest pressure in OA systems could result from the greater use of crop rotation and/or preservation of beneficial insects in the absence of pesticides (Lampkin 1990). Alternatively, reduced susceptibility to pests may be a reflection of differences in plant health, as mediated by soil fertility management. Many researchers and also practicing farmers have observed that fertility practices that replenish and maintain high SOM and that enhance the level and diversity of soil macro- and microbiota provide an environment that through various processes enhances plant health (McGuiness 1993; Altieri and Nicholls 2003).

Despite the potential links between soil fertility and crop protection, the evolution of integrated pest management (IPM) and integrated soil fertility management (ISFM) have proceeded separately. The integrity of the agroecosystem relies on synergies of plant diversity and the continuing function of the soil microbial community, and its relationship with SOM. Most pest management methods used by farmers can be considered soil fertility management strategies and vice versa. There are positive interactions between soils and pests that once identified can provide guidelines for optimizing total agroecosystem function. Increasingly, new research is showing that the ability of a crop plant to resist or tolerate insect pests and diseases is tied to optimal physical, chemical, and mainly biological properties of soils. Soils with high SOM and active soil biology generally exhibit good soil fertility as well as complex food webs and beneficial organisms that prevent infection. On the other hand, farming practices that cause nutrition imbalances can lower pest resistance (Magdoff and van Es 2000).

8.4 CONCLUSION

It is difficult to maintain soil health and sustain productivity without regular application of organic amendments, recycling of available organic residues, and ensuring high responses to added fertilizers. Moreover, overdependence on only chemical fertilizers is posing a serious threat to the ecological balance. The enormous number of alternative sources of organic amendments available in the country for recycling and bioconversion should be explored to utilize their embedded nutrients and organic matter for sustainable soil health and crop growth. This will not only help meet the deficit of fertilizer nutrients but also conserve energy, minimize pollution, save foreign exchange, and improve fertilizer use efficiency. Recent scientific advancements need to be exploited for more effective, economical, and sustainable recycling of these alternative sources of organic amendments. Most importantly, as some of the studies have also revealed, the sole application of organic amendments cannot meet the nutrient requirements of the crops; hence, they should be used in conjunction with inorganic fertilizers as supplements for maintaining the desired crop productivity. In order to encourage the use of alternative sources of organic amendments in agriculture on a large scale, it is important to work at four levels: (1) focusing research for safe handling of the alternative sources of organics, using state-of-the-art technology; (2) increasing awareness among farmers and within rural and urban communities about the importance and potential of these organic amendments in improving soil health and crop productivity; (3) training and skill improvement of the communities for effective handling of the alternative sources of organic amendments; and (4) developing appropriate policies and by-laws for onsite safe processing of the alternative sources of organics by the industries and providing suitable incentives for encouraging the farmers to use them on a larger scale.

REFERENCES

Altieri, M.A., and C.I. Nicholls. 2003. Soil fertility management and insect pests: Harmonizing soil and plant health in agroecosystems. *Soil Tillage and Research* 72(2): 203–211.

Altieri, M.A., C.I. Nicholls, and M. Fritz. 2005. *Manage Insects on Your Farm: A Guide to Ecological Strategies*. Sustainable Agriculture Research and Education (SARE), College Park, MD.

ARPU. 1990. *Agro-climatic Regional Planning: An Overview*. Agro-Climatic Regional Planning Unit, New Delhi, India.

Behera, U.K., L.P. Amgain, and A.R. Sharma. 2010. Conservation agriculture: Principles, practices and environmental benefits. In *Conservation Agriculture*, ed. U.K. Behera, T.K. Das, and A.R. Sharma, pp. 28–41. Division of Agronomy, Indian Agricultural Research Institute, New Delhi, India.

Bhandari, A.L., J.K. Ladha, H. Pathak, A.T. Padre, D. Dawe, and R.K. Gupta. 2002. Yield and soil nutrient changes in a long-term rice-wheat rotation in India. *Science Society of America Journal* 66: 162–170.

Bhatt, R., and K.L. Khera. 2006. Effect of tillage and mode of straw mulch application on soil erosion in the submontaneous tract of Punjab, India. *Soil Tillage and Research* 88(1–2): 107–115.

Bhattacharya, P., A.C. Samal, J. Majumdar, and S.C. Santra. 2009. Transfer of arsenic from groundwater and paddy soil to rice plant (*Oryza sativa* L.): A micro level study in West Bengal, India. *World Journal of Agricultural Sciences* 5: 425–431.

Bhattacharyya, R., B. Ghosh, P. Mishra, et al. 2015. Soil degradation in India: Challenges and potential solutions. *Sustainability* 7(4): 3528–3570.

Bijay-Singh, Y.H. Shan, S.E. Johnson-Beebout, Yadvinder-Singh, and R.J. Buresh. 2008. Crop residue management for lowland rice-based cropping systems in Asia. *Advances in Agronomy* 98: 118–199.

Biswas, P.P., and S.P. Sharma. 2008. Nutrient management – Challenges and options. *Journal of the Indian Society of Soil Science* 56: 22–25.

Bradon, H., and N.M. Kishore. 1995. The cost of inaction valuing the economy wide cost of environmental degradation in India. In *Fifty Years of Natural Resource Management Research*, ed. G.B. Singh and B.R. Sharma. Indian Council of Agricultural Research, New Delhi, India.

Bruinsma, J. 2011. The resource outlook to 2050: By how much do land, water use and crop yields need to increase by 2050? In *Looking Ahead in World Food and Agriculture: Perspectives to 2050*. FAO. www.fao.org/docrep/014/i2280e/i2280e06.pdf.

Chahal, S.S., P. Kataria, S. Abbott, and B.S. Gill. 2014. Role of cooperatives in institutionalization of custom hiring services in Punjab. *Agricultural Economics Research Review* 27: 103–110.

Chaudhari, S.K., P.P. Biswas, I.P. Abrol, and C.L. Acharya. 2015. Soil and nutrient management policies. In *State of Indian Agriculture-Soil*, ed. H. Pathak, S.K. Sanyal, and P.N. Takkar, pp. 332–342. National Academy of Agricultural Sciences, New Delhi, India.

Choudhary, M., S. Dhanda, S. Kapoor, and G. Soni. 2009. Lignocellulolytic enzyme activities and substrate degradation by *Volvariella volvacea*, the paddy straw mushroom/Chinese mushroom. *Indian Journal of Agricultural Research* 43(3): 223–226.

Conway, G.R., and J.N. Pretty. 1991. *Unwelcome Harvest: Agriculture and Pollution*. Earthscan, London, UK.

de Vries, F.T., E. Thébault, M. Liiri, et al. 2013. Soil food web properties explain ecosystem services across European land use systems. *Proceedings of the National Academy of Sciences of the United States of America* 110(35): 14296–14301.

Dhruvanarayan, V.V.N., and B. Ram. 1983. Estimation of soil erosion in India. *Journal of Irrigation and Drainage Engineering* 109: 419–434.

Doran, J.W., and T.B. Parkin. 1994. Defining and assessing soil quality. In *Defining Soil Quality for a Sustainable Environment*, ed. J.W. Doran, D.C. Coleman, D.F. Bezdicek, and B.A.E. Stewart. Soil Science Society of America, Madison, WI.

Dumanski, J., R. Peiretti, J. Benetis, D. McGarry, and C. Pieri. 2006. The paradigm of conservation tillage. In *Proceedings of World Association of Soil and Water Conservation*, P1, pp. 58–64.

Erenstein, O., and V. Laxmi. 2008. Zero tillage impact in India's rice-wheat systems: A review. *Soil Tillage and Research* 100: 1–14.

FAI (Fertiliser Association of India). 2010. *Fertiliser Statistics 2009–2010*. Fertiliser Association of India, New Delhi, India.

FAI (Fertiliser Association of India). 2016. *Fertiliser Statistics 2015–16*. Fertiliser Association of India, New Delhi, India.

FAI (Fertilizer Association of India). 2017. *Fertiliser Statistics 2016–17*. Fertiliser Association of India, New Delhi, India.

FAO (Food and Agriculture Organization of the United Nations). 2011. *The State of the World's Land and Water Resources for Food and Agriculture (SOLAW): Managing Systems at Risk*. Food and Agriculture Organization of the United Nations, Rome, Italy.

FAOSTAT. 2019. http://www.fao.org/faostat/en/#data/RFN (accessed on 12 July 2019).

Ferris, H., and H. Tuomisto. 2015. Unearthing the role of biological diversity in soil health. *Soil Biology and Biochemistry* 85: 101–109.

Gathala, M.K., J.K. Ladha, V. Kumar, et al. 2011. Tillage and crop establishment affects sustainability of South Asian rice-wheat system. *Agronomy Journal* 103(4): 961–971.

Ghosh, P.K., A. Das, R. Saha, et al. 2010. Conservation agriculture towards achieving food security in north east India. *Current Science* 99(7): 915–921.

Gregorich, E.G., C.F. Drury, and J.A. Baldock. 2001. Changes in soil carbon under long-term maize in monoculture and legume-based rotation. *Canadian Journal of Soil Science* 81:21–31.

Gupta, R., and A. Seth. 2007. A review of resource conserving technologies for sustainable management of the rice–wheat cropping systems of the Indo-Gangetic Plains (IGP). *Crop Protection* 26: 436–447.

Gupta, R.K., R.K. Naresh, P.R. Hobbs, Z. Jiaguo, and J.K. Ladha. 2003. Sustainability of post-green revolution agriculture: The rice-wheat cropping systems of the Indo-Gangetic Plains and China. In *Improving the Productivity and Sustainability of Rice-Wheat Systems: Issues and Impacts*, ed. J.K. Ladha, et al., pp. 1–26. ASA Special Publication 65. American Society of Agronomy, Madison, WI.

Harrington, L.W., and P.R. Hobbs. 2009. The Rice-Wheat Consortium and the Asian Development Bank: A history. In *Integrated Crop and Resource Management in the Rice-Wheat System*, ed. J.K. Ladha, Yadvinder Singh, O. Erenstein, and B. Hardy, pp. 3–67. International Rice Research Institute, Los Baños, Philippines.

Hobbs, P.R., and R.K. Gupta. 2003. Resource-conserving technologies for wheat in the rice–wheat system. In *Improving the Productivity and Sustainability of Rice-Wheat Systems: Issues and Impacts*, ed. J.K. Ladha, J.E. Hill, J.M. Duxbury, R.K. Gupta, and R.J. Buresh, pp. 149–172. ASA Special Publication 65. American Society of Agronomy, Madison, WI.

Hole, D.G., A.J. Perkins, J.D. Wilson, I.H. Alexander, P.V. Grice, and A.D. Evans. 2005. Does organic farming benefit biodiversity? *Biological Conservation* 122(1): 113–130.

IARI (Indian Agricultural Research Institute). 2012. *Crop Residues Management with Conservation Agriculture: Potential, Constraints and Policy Needs*. Indian Agricultural Research Institute, New Delhi, India.

Indoria, A.K., K.L. Sharma, K. Sammi Reddy, et al. 2018. Alternative sources of soil organic amendments for sustaining soil health and crop productivity in India – Impacts, potential availability, constraints and future strategies. *Current Science* 115: 2052–2062.

Jat, M.L., S.K. Sharma, R.K. Gupta, K. Sirohi, and P. Chandana. 2005. Laser land leveling: The precursor technology for resource conservation in irrigated eco-system of India. In *Conservation Agriculture – Status and Prospects*, pp. 145–154. CASA, New Delhi, India.

Jat, M.L., M.K. Gathala, J.K. Ladha, et al. 2009. Evaluation of precision land levelling and double zero-till systems in the rice-wheat rotation: Water use, productivity, profitability and soil physical properties. *Soil Tillage and Research* 105(1): 112–121.

Jat, M.L., R.K. Malik, Y.S. Saharawat, R. Gupta, M. Bhag, and R. Paroda. 2012. *Regional Dialogue on Conservation Agriculture in South Asia – Proceedings & Recommendations*, APAARI, CIMMYT, ICAR, New Delhi, India, p. 32.

Kassam, A. 2013. Sustainable soil management is more than what and how crops are grown. In *Principles of Sustainable Soil Management in Agroecosystems*, ed. R. Lal and B.A. Stewart, pp. 337–399. CRC Press, Boca Raton, FL.

Kassam, A., T. Friedrich, and R. Derpsch. 2019. Global spread of conservation agriculture. *International Journal of Environmental Studies* 76(1): 29–51.

Kassam, A.H., and T. Friedrich. 2009. Perspectives on nutrient management in conservation Agriculture. Invited paper, IV World Congress on Conservation Agriculture, 4–7 February 2009, New Delhi, India.

Kler, D.S., S. Singh, and S.S. Walia. 2002. Studies on organic versus chemical farming. *Extended Summaries*. Vol. 1, *2nd International Agronomy Congress, 26–30 November 2002*, pp. 39–40. New Delhi, India.

Kumar, P., S. Kumar, and L. Joshi. 2015. *Socioeconomic and Environmental Implications of Agricultural Residue Burning: A Case Study of Punjab, India*. Springer Open.

Ladha, J.K., K.S. Fischer, M. Hossain, P.R. Hobbs, and B. Hardy. 2000. Improving the productivity and sustainability of rice-wheat systems of the Indo-Gangetic Plains: A synthesis of NARS-IRRI partnership research. Discussion paper No. 40. International Rice Research Institute, Los Baños, Philippines.

Ladha, J.K., V. Kumar, M.M. Alam, et al. 2009. Integrating crop and resource management technologies for enhanced productivity, profitability, and sustainability of the rice-wheat system in South Asia. In *Integrated Crop and Resource Management in the Rice-Wheat System of South Asia*, ed. J.K. Ladha, Y. Singh, O. Erenstein, and B. Hardy, pp. 69–108. International Rice Research Institute, Los Baños, Philippines.

Lal, R. 2013. Climate-resilient agriculture and soil organic carbon. *Indian Journal of Agronomy* 58(4): 440–450.

Lal, R. 2015. Restoring soil quality to mitigate soil degradation. *Sustainability* 7(5): 5875–5895.

Lal, R. 2016. Soil health and carbon management. *Food and Energy Security* 5(4): 212–222.

Lampkin, N. 1990. *Organic Farming*. Farming Press Books, Ipswich, UK.

Lampkin, N. 2002. *Organic Farming*. Old Pond, Ipswich, UK.

Lelieveld, J., J.S. Evans, M. Fnais, D. Giannadaki, and A. Pozzer. 2015. The contribution of outdoor air pollution sources to premature mortality on a global scale. *Nature* 525 (7569): 367–371.

Liebhardt, W.C., R.W. Andrews, M.N. Culik, et al. 1989. Crop production during conversion from conventional to low-input methods. *Agronomy Journal* 81: 150–159.

Lohan, S.K., J. Dixit, R. Kumar, et al. 2015. Biogas: A boon for sustainable energy development in India's cold climate. *Renewable and Sustainable Energy Reviews* 43: 95–101.

Lohan, S.K., H.S. Jat, A.K. Yadav, et al. 2018. Burning issues of paddy residue management in north-west states of India. *Renewable and Sustainable Energy Reviews* 81: 693–706.

Magdoff, F., and H. van Es. 2000. *Building Soils for Better Crops*. Sustainable Agriculture Research and Education (SARE), Washington, DC.

Malik, R.K., A. Yadav, and S. Singh. 2005. Resource conservation technologies in rice-wheat cropping system of Indo-Gangetic Plain. In *Conservation Agriculture: Status and Prospects*, ed. L.P. Abrol, R.K. Gupta, and R.K. Malik, pp. 13–22. Centre for Advancement of Sustainable Agriculture, New Delhi, India.

Manna, M.C., A. Swarup, R.H. Wanjari, et al. 2005. Long-term effect of fertilizer and manure application on soil organic carbon storage, soil quality and yield sustainability under sub-humid and semi-arid tropical India. *Field Crops Research* 93: 264–280.

McGuiness, H. 1993. *Living Soils: Sustainable Alternatives to Chemical Fertilizers or Developing Countries*. Consumers Policy Institute, New York.

Merrill, M.C. 1983. Eco-agriculture: A review of its history and philosophy. *Biological Agriculture & Horticulture* 1(3): 181–210.

Mishra, D., S. Rajvir, U. Mishra, and S.S. Kumar. 2013. Role of bio-fertilizer in organic agriculture: A review. *Research Journal of Recent Sciences* 2: 39–41.

Mondal, S., A. Das, S. Pradhan, et al. 2018. Impact of tillage and residue management on water and thermal regimes of a sandy loam soil under pigeonpea-wheat cropping system. *Journal of the Indian Society of Soil Science* 66(1): 40–52.

Mondal, S., T.K. Das, P. Thomas, et al. 2019. Effect of conservation agriculture on soil hydro-physical properties, total and particulate organic carbon and root morphology in wheat (*Triticum aestivum*) under rice (*Oryza sativa*)-wheat system. *Indian Journal of Agricultural Sciences* 89: 46–55.

NAAS (National Academy of Agricultural Sciences). 2005. Policy options for efficient nitrogen use. Policy paper No. 33, National Academy of Agricultural Sciences, New Delhi, India, p. 12.

NAAS (National Academy of Agricultural Sciences). 2009. Crop response and nutrient ratio. Policy paper no. 42. National Academy of Agricultural Sciences, New Delhi, India, pp. 1–16.

NAAS (National Academy of Agricultural Sciences). 2010a. Degraded and wastelands of India, status and spatial distribution. Indian Council of Agricultural Research, New Delhi, India, and National Academy of Agricultural Sciences, New Delhi, India.

NAAS (National Academy of Agricultural Sciences). 2010b. Exploring untapped potential of acid soils of India. Policy paper No. 48, National Academy of Agricultural Sciences, New Delhi, India, p. 8.

NAAS (National Academy of Agricultural Sciences). 2018. Soil health: New policy initiatives for farmers welfare. National Academy of Agricultural Sciences, New Delhi, India.

Naik, S.K., S. Maurya, and B.P. Bhatt. 2017. Soil organic carbon stocks and fractions in different orchards of eastern plateau and hill region of India. *Agroforestry Systems* 91: 541–552.

Naik, S.K., S. Maurya, D. Mukherjee, A.K. Singh, and B.P. Bhatt. 2018 Rates of decomposition and nutrient mineralization of leaf litter from different orchards under hot and dry sub-humid climate. *Archives of Agronomy and Soil Science* 64(4): 560–573.

NBSS&LUP. 2004. Soil Map (1:1 Million Scale), Nagpur, India.

Neera, P., M. Katano, and T. Hasegawa. 1999. Comparison of rice yield after various years of cultivation by natural farming. *Plant Production Science* 2: 58–64.

NICRA. 2019. Management of crop residues in the context of conservation agriculture. Draft base paper. New Delhi, India. www.nicra.iari.res.in/Data/FinalCRM.doc (last accessed 9 October 2019).

NPMCR. 2014. National policy for management of crop residues (NPMCR). Government of India, Ministry of Agriculture, Department of Agriculture & Cooperation, Natural Resource Management Division, Krishi Bhawan, New Delhi, pp. 1–7.

NRSA (National Remote Sensing Agency). 1990. IRS-Utilisation Programme: Soil Erosion Mapping. Project Report National Remote Sensing Agency, Hyderabad, India.

Oelhaf, R.C. 1978. *Organic Agriculture*. Halstead Press, New York.

Pandey, S. and L. Velasco. 2002. Economics of direct seeding in Asia: Patterns of adoption and research priorities. In *Direct Seeding: Research Strategies and Opportunities*, ed. S. Pandey, et al. pp. 3–14. IRRI, Philippines.

Panwar, N.R., J.K. Saha, T. Adhikari, et al. 2010. Soil and water pollution in India: Some case studies. IISS Technical Bulletin. Indian Institute of Soil Science, Bhopal. p. 40.

Patel, A. 2016. Addressing soil health management issues in India. *International Journal of Agriculture Innovations and Research* 5: 2319–1473.

Pathak, H., Y.S. Saharawat, M. Gathala, and J.K. Ladha. 2011. Impact of resource-conserving technologies on productivity and greenhouse gas emission in rice-wheat system. *Greenhouses Gases Science and Technology* 1: 261–277.

Pretty, J., and R. Hine. 2001. Reducing food poverty with sustainable agriculture: A summary of new evidence. SAFE Research Project, University of Essex, Wivenhoe Park, UK, p. 136.

Prihar, S.S., V.K. Arora, and S.K. Jalota. 2010. Enhancing crop water productivity to ameliorate groundwater decline. *Current Science* 99: 588–593.

Rajendran, T.P., M.V. Venugopalan, and P.P. Tarhalkar. 2000. Organic cotton farming in India. Technical bulletin No. 1/2000. Central Institute of Cotton Research, Nagpur, India, p. 39.

Raju, S.S., S. Parappurathu, R. Chand, P.K. Joshi, P. Kumar, and S. Msangi. 2012. Biofuels in India: Potential, policy and emerging paradigms. Policy paper. NCAP.

Ranade, D.H., M.C. Chourasia, N.K. Shrivastava, and D. Patidar. 2006. Improved tools and scope for their custom hiring in Malwa Region – A case study. *Agricultural Engineering Today* 30(1&2): 28–31.

Reddy, V.R. 2003. Land degradation in India: Extent, costs and determinants. *Economic and Political Weekly* 38: 4700–4713.

Roy, A.H., L.L. Hammond, A.R. Mosier, K.J. Syers, and J.R. Freney. 2004. Challenges and opportunities for the fertilizer industry. In *Agriculture and the Nitrogen Cycle*, ed. A.R. Mosier, pp. 233–243. Island Press, Washington.

RWC-CIMMYT. 2005. Agenda Notes. 13th Regional Technical Coordination Committee Meeting. RWC-CIMMYT, Dhaka, Bangladesh.

Saharawat, Y.S., J.K. Ladha, H. Pathak, M. Gathala, N. Chaudhary, and M.L. Jat. 2012. Simulation of resource-conserving technologies on productivity, income and greenhouse gas emission in rice-wheat system. *Journal of Soil Science and Environmental Management* 3(1): 9–22.

Saharawat, Y.S., B. Singh, R.K. Malik, et al. 2010. Evaluation of alternative tillage and crop establishment methods in a rice-wheat rotation in north western IGP. *Field Crops Research* 116: 260–267.

Salleh, A. 2011. Fungal "feng shui" designs healthy soil. News in Science, ABC Science. https://www.abc.net.au/science/articles/2011/12/07/3384792.htm?site=science/tricks&topic=latest.

Samal, S.K., K.K. Rao, S.P. Poonia, et al. 2017. Evaluation of long-term conservation agriculture and crop intensification in rice-wheat rotation of Indo-Gangetic Plains of South Asia: Carbon dynamics and productivity. *European Journal of Agronomy* 90: 198–208.

Sarkar, A., R.L. Yadav, B. Gangwar, and P.C. Bhatia. 1999. Crop residues in India. Technical bulletin. Directorate of Cropping System Research, Uttar Pradesh, India.

Sarkar, S., R.P. Singh, and A. Chauhan. 2018. Crop residue burning in northern India: Increasing threat to greater India. *Journal of Geophysical Research: Atmospheres* 123(13): 6920–6934.

Scopel, E., B. Triomphe, F. Affholder, et al. 2013. Conservation agriculture cropping systems in temperate and tropical conditions, performances and impacts. A review. *Agronomy for Sustainable Development* 33(1): 113–130.

Sehgal, J., and I.P. Abrol. 1994. *Soil Degradation in India: Status and Impact*. Oxford and IBH, New Delhi, India.

Sharda, V.N., and P. Dogra. 2013. Assessment of productivity and monetary losses due to water erosion in rainfed crops across different states of India for prioritization and conservation planning. *Agricultural Research* 2(4): 382–392.

Sharda, V.N., and P.R. Ojasvi. 2016. A revised soil erosion budget for India: Role of reservoir sedimentation and land-use protection measures. *Earth Surface Processes and Landforms* 41: 2007–2023.

Sharma, M., and O. Dikshit. 2016. Comprehensive study on air pollution and green house gases (GHGs) in Delhi. Report No. 208016. Indian Institute of Technology, Kanpur, India. http://delhi.gov.in/DoIT/Environment/PDFs/Final_Report.pdf.

Sharma, P., Tripathi, R.P., and S. Singh. 2005. Tillage effects on soil physical properties and performance of rice-wheat cropping system under shallow water table conditions of Tarai, Northern India. *European Journal of Agronomy* 23: 327–335.

Sharma, P.D. 2008. Nutrient management-challenges and options. *Journal of the Indian Society of Soil Science* 55: 395–403.

Shukla, A.K., S.K. Behera, Y.S. Shivay, P. Singh, and A.K. Singh. 2012. Micronutrient and field crop production in India: A review. IAC special issue, *Indian Journal of Agronomy* 57(3): 123–130.

Shyamsundar, P., N.P. Springer, H. Tallis, et al. 2019. Fields on fire: Alternatives to crop residue burning in India. *Science* 365(6453):536–538.

Sidhu, H.S., M. Singh, E. Humphreys, et al. 2007. The happy seeder enables direct drilling of wheat into rice straw. *Australian Journal of Experimental Agriculture* 47:844–854.

Sidhu, H.S., M. Singh, Yadvinder-Singh, J. Blackwell, V. Singh, and N. Gupta. 2011. Machinery development for crop residue management under direct drilling. In *Resilient Food Systems for a Changing World: Proceedings of the 5th World Congress on Conservation Agriculture. Incorporating 3rd Farming Systems Design Conference*, 25–29 September 2011, Brisbane, Australia, pp. 157–158. Australian Centre for International Agricultural Research, Canberra, Australia.

Singh, M., and R.H. Wanjari. 2017. *Annual Report 2016–17. All India Coordinated Research Project on Long-Term Fertilizer Experiments to Study Changes in Soil Quality, Crop Productivity and Sustainability.* AICRP (LTFE), ICAR–Indian Institute of Soil Science, Bhopal, India, pp. 1–118.

Singh, R., M. Srivastava, and A. Shukla. 2016. Environmental sustainability of bioethanol production from rice straw in India: A review. *Renewable and Sustainable Energy Reviews* 54: 202–216.

Srinivasarao, C.H., B. Venkateswarlu, R. Lal, A.K. Singh, and K. Sumanta. 2013. Sustainable management of soils of dry land ecosystems for enhancing agronomic productivity and sequestering carbon. *Advances in Agronomy* 121: 253–329.

Sulphur Institute. 2019. Learn more about sulphur: Status of Indian soils. https://www.sulphurinstitute.org/india/status.cfm (last accessed 9 October 2019).

Tallis, H., S. Polasky, P. Shyamsundar, et al. 2017. *The Evergreen Revolution: Six Ways to Empower India's No-Burn Agricultural Future.* University of Minnesota, St. Paul, MN; The Nature Conservancy, Washington, DC; International Maize and Wheat Improvement Center, Mexico; and Borlaug Institute for South Asia, Ludhiana, India.

Tandon, H.L.S. 2007. Soil nutrient balance sheets in India: Importance, status, issues, and concerns. *Better Crops–India*, 15–19.

Tewatia, R.K., R.K. Rattan, S. Bhende, and L. Kumar. 2017. Nutrient use and balance in India with special reference to phosphorus and potassium. *Indian Journal of Fertilisers* 13(4): 20–31.

Tiemann, L.K., A.S. Grandy, E.E. Atkinson, E. Marin-Spiotta, and M.D. McDaniel. 2015. Crop rotational diversity enhances belowground communities and functions in an agroecosystem. *Ecology Letters* 18(8): 761–771.

Trewavas, A. 2001. Urban myths of organic farming. *Nature* 410(6827): 409–410.

Trivedi, A., A.R. Verma, S. Kaur, et al. 2017. Sustainable bio-energy production models for eradicating open field burning of paddy straw in Punjab, India. *Energy* 127: 310–317.

Wall, D.H., and J. Six. 2015. Give soils their due. *Science* 347: 695–695.

Wall, D.H., U.N. Nielsen, and J. Six. 2015. Soil biodiversity and human health. *Nature* 528(7580): 69.

Wassmann, R., S.V.K. Jagadish, K. Sumfleth, et al. 2009. Regional vulnerability of climate change impacts on Asian rice production and scope for adaptation. *Advances in Agronomy* 102: 91–133.

Watson, C.A., D. Atkinson, P. Gosling, L.R. Jackson, and F.W. Rayns. 2002. Managing soil fertility in organic farming systems. *Soil Use and Management* 18: 239–247.

9 Applications of Isotopes in Fertilizer Research

S. M. Soliman and Y. G. M. Galal

CONTENTS

9.1 INTRODUCTION

Fertilizers are one of the essential inputs for maintaining or increasing the soil fertility level in intensive agricultural systems. The purpose of applying fertilizers is primarily to supply the crop with essential plant nutrients to ensure normal plant growth. The major plant nutrients (N, P, and K) have to be applied regularly to compensate for the amounts exported from the soil during harvest. Fertilizer use efficiency is a quantitative measure of the actual uptake of fertilizer nutrient by the plant in relation to the amount of nutrient added to the soil as fertilizer.

Radioisotopes are very valuable in guessing the quantity of nitrogen and phosphorus available in the soil. This guessing helps in determining the quantity of nitrogen and phosphorus fertilizers that should be applied to soil. Radioactive isotopes such as phosphorus-32 and nitrogen-15 have been labeled fertilizers and have been used to study the uptake, retention, and utilization of fertilizers. These isotopes provide a means to determine the quantity of fertilizer taken and lost to the environment by the plant (IAEA 2001; Hardarson 1990). In isotopic-aided fertilizer experiments, a labeled fertilizer is added to the soil and the amount of fertilizer nutrient that a plant has taken up is determined. In this way, different fertilizer practices (placement, timing, sources, etc.) can be studied. Recently, these means used for tracing the fate and distribution of nutrient elements in plant–soil interrelationships have been reviewed by Pourjafar (2017). The joint FAO/IAEA program on the application of nuclear techniques in food and agriculture has arranged and conducted several regional and international experimental works worldwide. Such tracer techniques were applied earlier by pioneers of the IAEA for tracing essential nutrients in plants, soil, and the surrounding environment, as well as estimating its efficient use by main strategic crops like rice, maize, and wheat (Hera 1995).

When comparing different nutrient sources, the term "available amount of a nutrient" has first to be defined. Only the plant can judge what is available since no chemical extraction can determine what is available to a plant. However, if the plant is used to measure which source of nitrogen or phosphorus is available and to what extent, one has to be able to discriminate between the sources, which is conveniently done by labeling one of the sources with an appropriate isotope. The nutrient supply from several fertilizer management practices, i.e., method of placement, timing of application, chemical and physical nature of sources, including symbiotic nitrogen fixation and interaction among them, and of these, cultural practices (irrigation, mulching, tillage, etc.), can be quantitatively evaluated using isotope techniques (Galal 2015).

The goal of nutrient management is to provide an adequate supply of all essential nutrients for a crop throughout the growing season. If the amount of any nutrient is limited at any time, there is a potential for loss in production. As crop yields increase and as increasing amounts of nutrients are exported from the fields where crops are grown, the nutrient supply in the soil can become depleted unless it is supplemented through application of fertilizers. Fertilizers need to be applied to all types of crop production systems in order to achieve yield levels that make the effort of cropping worthwhile. Modern fertilization practices, first introduced in the last half of the 1800s and based on the chemical concept of plant nutrition, have contributed very widely to the immense increase in agricultural production and have resulted in better-quality food and fodder. Furthermore, farmers' economic returns have increased substantially due to fertilizer use in crop production. In this respect, radioisotopes can be used to find out the effectiveness of different types of fertilizers; by labeling a fertilizer with a particular radioisotope, researchers can detect how much of it is taken up by the plant and how much is lost by any mechanism. It gives the chance to determine the most proper fertilization strategies, which leads to more efficient agriculture through a low-cost effective approach that minimize losses and make it friendlier to the environment (Canadian Nuclear Association 2016). In its report, the Canadian staff illustrated the worldwide role of nuclear technology and its applications in food production, where the FAO/IAEA division program listed the following compelling statistics:

- Thirty countries use nuclear science methods for improved irrigation and crop production.
- Forty-eight countries use nuclear tracer techniques and10 FAO/IAEA soil and water management guidelines to protect their farmlands.
- Ninety-five countries use isotopic and nuclear techniques to identify land and water management practices to improve nutrient and water use efficiency for crop productivity and environmental sustainability.

The application of tracer techniques in agriculture, in general, and specifically in fertilizer research, has many advantages, which are briefly described elsewhere (Alam et al. 2001). These advantages include its ability to easily recognize the presence of a single atom and molecule and their movement, as well as give the opportunity to follow up related nutritional processes. Also, it helps to define accurately the responsible mechanisms for the interaction between mineral nutrients and their impact on plant nutritional status.

9.1.1 DEMAND FOR FERTILIZER NUTRIENTS

The global demand for fertilizer nutrients (N, P_2O_5, and K_2O) for 2015 and the demand forecast estimates for 2016 to 2020 are summarized in Table 9.1. Total fertilizer nutrient demand was esti-mated to be 184.02 million tons in 2015 and is forecast to reach 186.67 million tons in 2016. With an average annual growth of 1.9% in the following years, it is expected to reach 201.66 million tons by the end of 2020. The demand for N, P_2O_5, and K_2O is forecast to grow annually by 1.5%, 2.2%, and 2.4%, respectively, for individual nutrients from 2015 to 2020. The balance of anticipated nutrients in 2020 is illustrated by Figure 9.1.

Fertilizer demand has historically been influenced by changing and often interrelated factors such as population and economic growth, agricultural production, prices, and government policies.

However, three developments distinguish the current state of agricultural markets from past fluc-tuations, namely, that the hike in world prices concerns nearly all major food and feed commodities, that record prices are being achieved at a time not of scarcity but of abundance, and that linkages between agricultural commodity markets and other markets are strengthening.

Such phenomena were already manifest in 2006 and strengthened in 2007, a year character-ized by persistent market uncertainty, record prices, and unprecedented volatility in grain markets.

TABLE 9.1

World Demand for Fertilizer Nutrient Use, 2015–2020 (thousand tons)

Year	2015	2016	2017	2018	2019	2020
Nitrogen (N)	110,027	11,575	113,607	115,376	117,116	118,763
Phosphate (P_2O_5)	41,151	41,945	43,195	44,120	45,013	45,858
Potash (K_2O_5)	32,838	33,149	34,048	34,894	35,978	37,042
Total (N+P_2O_5+K_2O)	**184,017**	**186,668**	**190,850**	**194,390**	**198,107**	**201,663**

Source: FAO, Summary report about world fertilizer trends and outlook to 2020, pp. 38, 2017; Food and Agriculture Organization of the United Nations, 2017, World fertilizer trends and outlook to 2020, http://www.fao.org/3/a-i6895e.pdf. Reproduced with permission.

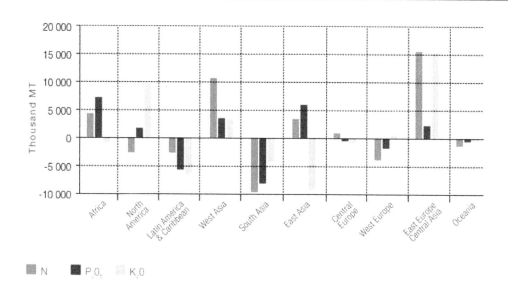

FIGURE 9.1 Anticipated nutrient balances in 2020. (From FAO, Summary report about World fertilizer trends and outlook to 2020, pp. 38, 2017.)

The magnitude and nature of these changes have led some observers to refer to a paradigm shift in agriculture away from decreasing real food prices over the past 30 years. Given the inextricable link between food production and fertilizer use, it is opportune to consider such changes when reviewing prospects for fertilizer demand and supply balances (Heffer 2007; Heffer and Prud'homme 2006). Wide areas of manufactured fertilizer were applicable under different cropping patterns.

9.2 WORLD NITROGEN FERTILIZER DEMAND

The forecast is for world nitrogen fertilizer demand to increase at an annual rate of about 1.4% until 2011/2012, which is an overall increase of 7.3 million tons. About 69% of this growth will take place in Asia. The world's largest consumers of nitrogen are East Asia, South Asia, North America, and West Europe. While their share of global consumption is modest, it is forecast that the relative contribution of Latin America, East Europe, and Central Asia (EECA) to the change in nitrogen use will be 10.4% and 5%, respectively. The relative contribution to the change in world nitrogen consumption by East Asia and South Asia is expected to be about 65% (Figure 9.2). North America is the largest importer, followed by South Asia, which still has a supply deficit of some 32%. East Asia will move from deficit to surplus during the outlook period (FAO 2009).

Increased requirements for food and other agricultural products will undoubtedly increase demand for N fertilizers. However, determining the magnitude of the increase is not straightforward. For example, Wood et al. (2004) pointed out that the 2.4% average annual growth in food consumption between 1961 and 2001 was accompanied by a 4.5% increase in fertilizer N use. They went on to explain that the increase in fertilizer use was largely due to a change in the structure of food demand, where consumption of meat products grew faster than cereals, increasing the demand for feed grains and for N. Projections of future fertilizer demand also involve assumptions about N use efficiency, measured as the amount of production resulting from each unit of fertilizer N used. Will Nitrogen use efficiency (NUE) decrease because higher application rates are used and the lower diminishing returns set in as farmers move up an unchanging N response curve? Or will it increase due to higher energy and input costs, improved management, better technology, and increased awareness of problems associated with inefficient use? Or will it be business as usual with no change from the past?

After exceptionally strong growth of world fertilizer demand in 2003/2004 and 2004/2005, Heffer and Prud'homme (2006) forecast that global N consumption would increase from 90.9 Mt N in 2005/2006 to 99 Mt in 2010/2011, corresponding to an average annual growth rate of 1.8%. All projections point to an increase in fertilizer consumption in the decades to come, but the magnitude of this increase depends greatly on the underlying assumptions. For instance, Wood et al. (2004) identified three different scenarios: (1) a scenario following trends since 1969 for both crop

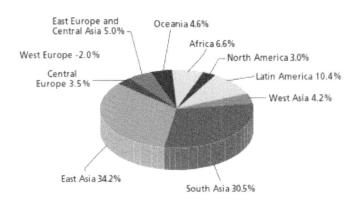

FIGURE 9.2 Regional and subregional contribution to change in world nitrogen consumption 2007/8–2011/12.

production and fertilizer use; (2) a scenario following the International Model for Policy Analysis of Agricultural Commodities and Trade (IMPACT) to project food production, and assuming constant NUE based on 1997 values; and (3) a scenario based on the IMPACT model for food production projections, and assuming relative NUE gains of 17% from 1997 levels by 2020, and of 30% by 2050. Wood et al. (2004) anticipated that fertilizer N use would grow around 1.8% annually in the short term. Average annual growth would then drop to 1.6% by 2020 and to 1.4% by 2050, unless NUE increases. With the NUE gains assumed in scenario 3, the average annual growth of fertilizer N use would drop to less than 0.5% after 2010.

Another recent analysis looking at crop-specific food production gave similar results but pointed out that such gains in NUE would require substantial additional investment in research and education (Dobermann and Cassman 2004). Forecasts to 2010/2011 by Heffer and Prud'homme (2006) tended to show that projections by Wood et al. (2004) under scenario 3 cannot be achieved, as these projections for 2020 would already be exceeded in 2010. Long-term projections are subject to great uncertainty and involve many critical assumptions about our ability to improve crop productivity as demand increases, while also improving NUE. Recent projections indicate that global demand for N fertilizers in 2050 could be between 107 and 171 Mt N. According to the four scenarios of the Millennium Ecosystem Assessment (2005), global fertilizer N consumption in 2050 is anticipated to be between 110 and 140 Mt N.

9.3 NITROGEN MANAGEMENT IN AGRICULTURE

Nitrogen (N) is a vital element for life. It is an essential component of all proteins and of deoxyribonucleic acid (DNA). On Earth, there are two pools of N, with relatively little exchange between them: the gaseous dinitrogen (N_2) of the atmosphere, which makes up about 99% of total N, and the 1% of N that is chemically bound to other elements such as carbon (C), hydrogen (H), or oxygen (O) and has been described as "reactive nitrogen" for its tendency to react with other elements (Galloway et al. 2004). Reactive N includes inorganic reduced forms (e.g., ammonia [NH_3] and ammonium [NH_4^+]), inorganic oxidized forms (e.g., nitrogen oxides [NO_x], nitric acid [HNO_3], nitrous oxide [N_2O], nitrate [NO_3^-], and nitrite [NO_2^-]) and organic compounds (e.g., urea, amines, proteins, and nucleic acids). Nitrogen in humus (decomposed organic matter found in soil) can be regarded as reactive in the long term only.

Gaseous N_2 cannot be used directly by plants, with the exception of some plant species (e.g., legumes) that have developed symbiotic systems with N_2-fixing bacteria. Owing to the strong bond between its two N atoms, N_2 is almost inert and thus nonreactive. It requires a high energy input to convert N_2 into plant-available, reactive N forms.

The variability in soil and climatic conditions associated with processes that affect nitrogen dynamics in the soil and their relationship with the plant may lead to changes in nitrogen availability and its requirement by the plant (Simili et al. 2008; Espindula et al. 2010). The soil-N cycle involves several N transformations, which essentially make soil organic N or fertilizer-N usable for plants. Scheffer and Schachtschabel (1998) demonstrated the processes that increase plant-available N are mineralization, nitrification, and biological N fixation, while processes such as ammonia (NH_3) volatilization, immobilization, denitrification, and leaching result in faster temporal or permanent N losses from the plant rooting zone (Figure 9.3).

Appropriate N inputs enhance soil fertility, sustainable agriculture, food security, (enough calories) and nutrition security (appropriate supply of all essential nutrients, including protein). On the other hand, when improperly managed, N inputs can be associated with a number of adverse effects on both the environment and human health. Lack of reactive N in the agroecosystem leads to soil fertility decline, low yields and crop protein content, depleted soil organic matter, soil erosion and, in extreme cases, desertification. Excess amounts of NO_3^- may move into groundwater and drinking water supplies, raising treatment costs faced by municipalities. Excess NO_3^- in drinking water wells also can be an issue in rural areas that are adjacent to farmland. In surface water, increased loading

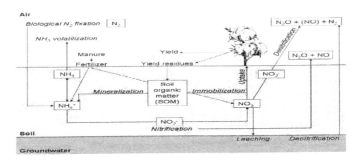

FIGURE 9.3 Simplified soil-N cycle. (Modified after Hofman and van Cleemput (2004) in Van Cleemput, O. and Boeckx, P., *Gayana Bot.*, 62, 98–109, 2005.)

of N-based nutrients can play a role in eutrophication, a process that contributes to ecological and resource degradation. In the atmosphere, NO_x and particulate matter can exacerbate several human health problems, from asthma to heart disease. Increasing the N_2O concentration in the atmosphere contributes to global warming. Adopting an integrated approach to nutrient management, maximizing the benefits and minimizing the risks associated with the use of N sources, contributes to raising crop productivity and N use efficiency.

9.4 NUTRIENTS AND CROP PERFORMANCE

Mineral elements have the same importance for all plants; however, the rate, quantity, and timing of uptake vary with crop, climate, variety, management, and soil characteristics. Nitrogen is the nutrient most limiting to crop production in most nonleguminous crops and the nutrient generally applied in the largest amount. These combined factors influence the nutritional need, nutrient content, and overall yield of a crop. Due to the prices of N-fertilizer, which represent a large portion of a producer's costs, it is very important to maximize fertilizer nutrient use efficiency. Timing fertilizer applications so that nutrients are available when plants need them should increase nutrient use efficiency and reduce potential adverse environmental effects. Knowing how nutrient needs change during the growing season is essential for matching nutrient supply with plant needs, especially for producers who can apply nutrients in season or for those considering controlled- and slow-release fertilizers.

9.5 NITROGEN USE EFFICIENCY

Intensive agriculture with large fertilizer input is one of the major sources of soil and groundwater contamination with nitrates (Schepers and Marter 1986). This trend is enhanced by mismanaged irrigation practices and induced groundwater recharge (Toussaint 2000).

Nitrogen use efficiency can be improved by other means as well, although practical manipulations of loss mechanisms can be difficult and expensive, and may increase one form of loss while reducing another. The use of conservation tillage practices can be expected to reduce erosion and runoff losses of N. Reducing water movement off a field, however, will likely increase the infiltration rate and thus NO_3 leaching and denitrification. Surface applications of wastes may also reduce soil-waste contact and accelerate waste drying, enhancing NH_3 volatilization but decreasing the rate of N mineralization. Other conservation practices that have the potential to reduce N losses include more efficient irrigation practices (e.g., versus flood), use of multiple cropping or winter cover crops to trap residual N from wastes, and controlled drainage systems or artificial wetlands to enhance denitrification in field border areas (Pierzynski et al. 1994). Plants only use about 50% of the applied N (Newbould 1989), which implies a large loss in money and energy.

Farmers use at least double the quantity required by the plants, and unused NO_3 is leached or denitrified N, causing environmental pollution (Byrnes 1990; Smith et al. 1990; Davies and Sylvester-Bradley 1995). N efficiency, the lowest and the grater applied N (Brown 1978), can be improved by maintain higher level of NH_4 as compared to those in the soil. This can be achieved NO_3 by nitrification inhibitors (Prasad and Power 1995), or slow release fertilizer (Shaviv and Mikkelsen 1993), or fertilization with ammonium fertilizers (Lips et al. 1990), or a large quantity of ammonium fertilizers (Shaviv 1988). On account of the above, studies carried out on nitrogen and ammonium nutrition in vegetables aimed to increase nitrogen use efficiency. A large part of such research work focuses on leaf vegetables that under high NO_3^- availability, N accumulate large amounts of nitrates (Santamaria et al. 1997; Santamaria and Elia 1997), a compound believed to be potentially toxic to human health (Walker 1990; Gangolli et al. 1994).

In general, fertilizer use efficiency (FUE) is a quantitative measure of the actual uptake of fertilizer nutrient by the plant in relation to the amount of nutrient added to the soil as fertilizer. A common form of expression of fertilizer use efficiency is plant recovery or "coefficient of utilization" of the added fertilizer. This is shown in the following equation:

$$\%\text{Utilization of added fertilizer} = \frac{\text{Amount of nutrient in the plant Dff}}{\text{Amount of nutrient applied as fertilizer}} \times 100$$

The concept of fertilizer use efficiency, however, is much broader. It implies not only the maximum uptake of the applied nutrient by the crop but also the availability of the applied nutrient under variable climatic and edaphic conditions. Environmental issues, such as pollution resulting from the fertilizer application, should also be considered. It is important to study the efficient use of fertilizers because we are interested in obtaining the highest possible yield with minimum fertilizer application. This is done in field trials by assessing the best fertilizer practices, such as sources, timing, and placement, and their interactions in different farming systems (FAO 1980, 1983a, 1985).

The measurement of fertilizer use efficiency can be established for each crop by carrying out, in practice, a series of carefully designed field experiments in several representative locations. These experiments were carried out over a period of time in order to estimate the effect of placement, timing, and source of fertilizer nutrient that will result in the most efficient fertilizer uptake by the crop. Yield, particularly economic yield, is generally the most important criterion for the farmer, but it is equally important that this yield is obtained with a minimum of fertilizer investment (minimum cost).

For example, Hamed (2013) and Hamed et al. (2019) gave an example about the efficient use of labeled ammonium sulfate enriched with 2% ^{15}N atom excess by wheat crop grown on two different-texture soils. They found that the water regime had a direct effect on NUE as a result of either enhancing the N uptake by wheat plants and/or reducing the amount of lost N. At the same time, both plant N uptake and N remained in the soil, and N losses varied significantly according to N application rate and mode in both soil textures (Figure 9.4a and b).

On the other hand, nitrogen fertilizer was efficiently used by wheat grain grown in clay soil to a higher extent, in general, than that grown in sand soil. Application of both 80% and 60% N rates reflected, to some extent, higher percent NUE than those applied at 100% N rate, especially in sand soil. It was obvious that the most efficient use of fertilizer N by grains occurred under W75, comparable to other water regimes. In this respect, there was no significant difference between the two modes of N application (Figure 9.5).

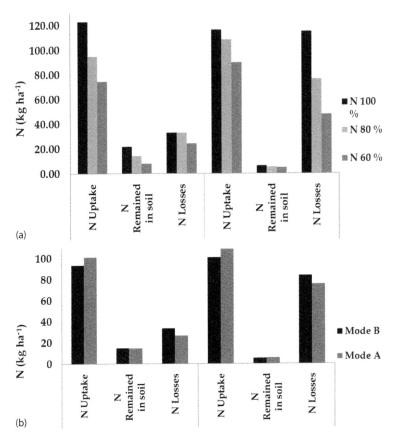

(a)

(b)

FIGURE 9.4 Effect of nitrogen application rate (a) and application mode (b) on nitrogen losses under different soil textures.

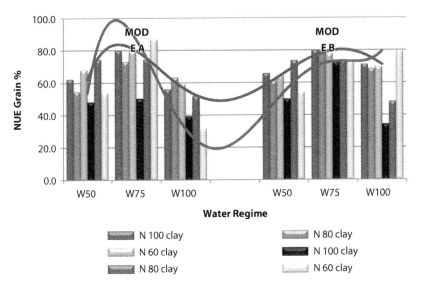

FIGURE 9.5 Nitrogen use efficiency (%) by wheat grain as affected by fertilizer N rate, mode of application, soil type, and water regime.

9.6   ISOTOPIC TECHNIQUES IN N FERTILIZER USE EFFICIENCY STUDIES

In isotopic-aided fertilizer experiments, a labeled fertilizer is added to the soil and the amount of fertilizer nutrient that a plant has taken up is determined. In this way, different fertilizer practices (placement, timing, sources, etc.) can be studied.

The first parameter to be determined when studying the fertilizer uptake by a crop by means of the isotope techniques is the fraction of the nutrient in the plant derived from the (labeled) fertilizer, i.e.: Ndff.

Often this fraction is expressed as a percentage, that is:

$$\%Ndff = Ndff \times 100.$$

The procedure followed in the calculation of this fraction and other derived parameters for nitrogen using ^{15}N labeled materials is given below.

9.6.1   MEASUREMENTS NEEDED FOR EXPERIMENTS WITH ^{15}N

In summary, for all field and greenhouse experiments with ^{15}N (or any other stable isotope)-labeled materials, the following basic primary data need to be recorded for each plot:

- Dry Matter (D.M.) yield for the whole plant or subdivided into plant parts.
- Total N concentration ($\%N$ in dry matter) of the whole plant or plant parts as in the previous bullet point. This is done by chemical methods, e.g., Kieldahl or combustion (Dumas).
- Plant $\%^{15}N$ abundance, which is analyzed by emission or mass spectrometry.
- Fertilizer $\%^{15}N$ abundance.
- ^{15}N labeled fertilizer(s) used and N rate(s) of application.

9.6.2   CALCULATIONS FOR EXPERIMENTS WITH ^{15}N

For these calculations, $\%^{15}N$ abundance is transformed into atom $\%^{15}N$ excess by subtracting the natural abundance (0.3663 atom $\%N$) from the $\%N$ abundance of the sample. Afterward, the following calculations can be made:

$$NUptake = N\% \times Dry\ yield$$

$$\%Ndff = \frac{\%N^{15}\ a.ex.\ in\ plant}{\%N^{15}\ a.ex.\ infertilizer} \times 100$$

$$Ndff\left(kg.ha^{-1}\right) = \%Ndff \times total\ N\ in\ Plant$$

9.6.3   QUANTIFICATION OF FERTILIZER N UPTAKE

The nitrogen isotope composition, i.e., the ^{15}N/total N ratio, of any material can be expressed as atom $\%^{15}N$ (a) or simply $\%^{15}N$ abundance. This isotopic ratio of a sample is measured directly in a single determination by optical emission or mass spectrometry. Since the $\%^{15}N$ natural abundance (a0) is 0.3663 atom $\%^{15}N$, this has to be subtracted from the $\%^{15}N$ abundance (a) of any enriched material to obtain the atom $\%^{15}N$ excess ($\%^{15}N$ a.e. = a') or ^{15}N enrichment. What then is Ndff? It is the fraction of N in the plant derived from the ^{15}N labeled fertilizer from simple isotope dilution principles.

Therefore, for the calculation of %Ndff it is necessary to determine the atom $\%^{15}N$ excess of the plant samples and of the fertilizer(s) used in the experiment. For instance, if Ndff = 0.25, this means that 1/4 of the nitrogen in the plant came from the fertilizer. If soil and fertilizer were the only sources of N available to the plant, then the remaining 3/4 of the nitrogen in the plant came from the soil. If these fractions are expressed in percentages, then %Ndff = 25% and %Ndfs = 75%, where %Ndfs is %N derived from soil

$$\%\,\mathrm{Ndfs} = 100 - \%\,\mathrm{Ndff}$$

$$\mathrm{Ndfs}\left(\mathrm{kg.ha}^{-1}\right) = \%\,\mathrm{Ndfs} \times \text{total N in (Grain OR Straw)}$$

$$\%\mathrm{NUE} = \frac{\mathrm{Ndff}\left(\mathrm{kg.ha}^{-1}\right)}{\text{Rate of fertilizer}\left(\mathrm{kg.ha}^{-1}\right)} \times 100$$

$$\% \text{ Fertilizer N remained in soil }\left(\mathrm{FNR}\right) = \frac{\%^{15}N \text{ a.e. in soil after harvest}}{\%^{15}N \text{ a.e. in fertilizer}}.$$

9.7 NITROGEN STATUS USING ISOTOPIC TECHNIQUE

9.7.1 PRINCIPLES AND APPLICATIONS OF ISOTOPES IN FERTILIZER EXPERIMENTS

The purpose of applying fertilizers is primarily to supply the crop with essential plant nutrients to ensure normal plant growth. The major plant nutrients (N, P, and K) have to be applied regularly to compensate for the amounts exported from the soil during harvest. Other plant nutrients such as Ca, Mg, and S, and the micronutrients, e.g., Zn, Mo, and B, may also need to be added to maintain adequate levels of these nutrients, or to correct deficiencies (FAO 1983b, 1984).

Fertilizers are applied to facilitate plant uptake of a particular nutrient. Increased uptake can lead to a yield response if the particular nutrient is a limiting factor. It is important, though, to note that the fertilizer is not applied to obtain a yield response but to feed the plant. The yield response is a consequence of the additional uptake of the nutrient when other production factors are adequate.

A combination of all the production factors and conditions in an agricultural system results in a given yield, and only if all factors are optimized (fertilizer, soil, plant, water, pest control, etc.) will yield be maximized. In fact, the contribution of fertilizer to increased yield is perhaps of the greatest importance among the purchased inputs. Fertilizer, when used in combination with the other adequate inputs such as high-yielding varieties and irrigation water, can result in a positive interaction, thereby further increasing its contribution to increased yield (Fried 1978).

In the 1980s, due to substantial increases in the cost of fertilizers and their limited supplies to resource-poor farmers, enhanced nutrient management was pursued through maximizing the efficiency of nutrient uptake from various inorganic and organic sources utilizing two complementary approaches: (1) identification and/or selection of plant genotypes efficient at low levels of soil-available nutrients and tolerant to predominant stress conditions, and (2) development of integrated plant nutrition systems to maximize yield responses and to reduce environmental contamination and degradation of natural resources (Zapata and Hera 1995).

In the case of the element N, which is another major plant nutrient, there is no suitable radioactive isotope that could be used by researchers to assess fertilizer use efficiency. In this case, they use the stable isotope of nitrogen or ^{15}N as the tracer. In nature, the element nitrogen consists of a mixture of two isotopes, $^{14}N + ^{15}N$, with abundances of 99.634% and 0.366%, respectively. There is consequently much more ^{14}N than ^{15}N in nature. However, researchers have been able to enrich the isotope ^{15}N in samples of natural nitrogen to produce nitrogen with a $^{15}N\%$ abundance greater than 0.366% so that this stable isotope may be used as a tracer in experiments such as fertilizer use efficiency studies. (See "Radioactivity Hall of Fame," Part V, L'Annunziata [2016], for a biographical sketch of Gustav Hertz, who was the first to devise a method of isotope enrichment.) The amount of enrichment of ^{15}N in excess of its natural abundance is defined as:

$$\%^{15}N \text{ atomic excess} = \%^{15}N \text{ in the sample} - 0.366\%.$$

Thus, as in the previous example taken in research with fertilizers labeled with a radioactive isotope, researchers can use a fertilizer with a known $\%^{15}N$ atomic excess to measure the plant utilization of the nitrogen from the fertilizer according to the following:

$$\%NDFF = \%^{15}N \text{ atomic excess in plant} / \%^{15}N \text{ atomic excess in fertilizer} \times 100$$

where %NDFF is the percent of nitrogen in the plant derived from the fertilizer. The $\%^{15}N$ atomic excess in the plant is always less than that of the fertilizer because the plant roots also absorb nitrogen from the soil, which contains ^{14}N at its natural abundance. Thus, the remaining nitrogen in the plant must be derived from the soil or

$$\%NDFS = 100\% - \%NDFF.$$

9.8 NONISOTOPIC CONCEPTS OF NUE ESTIMATION

Different definitions about NUE estimated with nonisotopic technique were used for calculating the different N use efficiency parameters. According to Abbasi et al. (2012, 2013), the different terms of NUE could be as follows:

9.8.1 AGRONOMIC EFFICIENCY OF APPLIED FERTILIZER N

NAE, (kg grain kg^{-1} N applied) = [grain yield (kg ha^{-1}) in N added plots − grain yield of control plots]/Total amount of N fertilizer applied

9.8.2 PHYSIOLOGICAL EFFICIENCY OF APPLIED N

(NPE, kg kg^{-1}) = [(dry matter (straw) yield (kg ha^{-1}) in N added treatment − dry matter (straw) yield of control plots)/(total N uptake by the fertilizer treatment − total N uptake in the control)]

9.8.3 NITROGEN USE EFFICIENCY

(NUE, %) = [(N uptake by the fertilized treatment − N uptake in the control)/total amount of N fertilizer applied] × 100.

The authors (we) believe that these items have disadvantages related to under- or above-estimated nitrogen content in plants, and intensive precautions are needed to establish perfectly controlled experimental work in sites. Therefore, the use of ^{15}N stable isotope may have an advantage in accurately tracing nitrogen nutrient gained by plants from fertilizers.

The current status of NUE for different crops and regions was reviewed by Fixen et al. (2015), who stated that an extensive review by Ladha et al. (2005) of 93 published studies revealed NUE measured in research plots (Table 9.2). This review provides widely varied estimates of the central tendency for NUE expressions for maize, wheat, and rice in different continents and regions. In addition, an excellent review of NUE measurements and calculations was summarized by Dobermann (2007).

TABLE 9.2
Common NUE values for N for Maize, Wheat, and Rice and for Various World Regions

Crop or Region	Number of Observations[a]	Average Rate of Fertilizer Use (kg ha^{-1})	PFP[b] (kg ha^{-1})	AE[b] (kg ha^{-1})	RE[b] (%)	PE[b] (kg ha^{-1})
Maize	35–62	123	72 (6)	24 (7)	65 (5)	37 (5)
Wheat	145–444	112	45 (3)	18 (4)	57 (4)	29 (4)
Rice	117–187	115	62 (3)	22 (3)	46 (2)	53 (3)
Africa	2–24	139	39 (11)	14 (6)	63 (5)	23 (6)
Europe	12–69	100	50 (6)	21 (9)	68 (6)	28 (6)
America	119–231	111	50 (5)	20 (7)	52 (6)	28 (8)
Asia		115	54 (3)	22 (2)	50 (2)	47 (3
Average/totals	**411**		**52 (2)**	**20 (2)**	**55 (2)**	**41 (3)**

Source: Ladha, J.K., et al., *Adv. Agron.*, 87, 85–156, 2005.

[a] Range in number of observations across NUE indices.

[b] PFP, partial factor productivity; AE, agronomic efficiency; RE, apparent recovery Efficiency by difference; PE, physiological efficiency value in parentheses is relative standard error of the mean (SEM/mean × 100).

9.9 ISOTOPIC METHODS

The recovery of N by crops may be determined using both unlabeled and labeled N-fertilizer (Harmsen and Moraghan 1988), but the measurement of the recovery of N-fertilizer in the soil and the subsequent calculation of N losses from the crop soil system can only be made using ^{15}N-labeled fertilizer (Powlson et al. 1992).

The use of ^{15}N isotopes for studying N-uptake by plants was reviewed by Hauck and Bremner (1976). The ^{15}N isotope labeling technique, involving the application of ^{15}N enriched fertilizers to soil, provides reliable integrated estimates of the proportions and amount of N$_2$-fixed by several grain and pasture legumes. The different isotopic methods used can be classed into four techniques, as follows.

9.9.1 ^{15}N NATURAL ABUNDANCE TECHNIQUE

The ^{15}N/^{14}N ratio of soil N has been observed to be usually slightly higher than that of atmospheric nitrogen. Hauck and Bremner (1976) analyzed clover (*Trifolium* spp.), soybean (*Glycine max* L. Merr), and grass for ^{15}N values. They found that the N$_2$-fixing species had lower ^{15}N values than the grass or the soil in which the tested species were grown. This method makes many assumptions, which are controversial, because at the natural abundance level, discrimination between ^{15}N and ^{14}N can cause serious errors.

9.9.2 THE ISOTOPE DILUTION TECHNIQUE

The isotope dilution principles and the equation involved have been outlined by Danso et al. (1986). Its use for estimating N$_2$-fixation requires the application of equal nitrogen rates and ^{15}N enrichment of a given fertilizer to both the legume and a suitable nonfixing reference control crop. In principle, since the legume and the reference crop are absorbing nitrogen from a similar zone, the ^{15}N/^{14}N ratio of soil-derived nitrogen is supported to be the same for both crops. However, in addition to the nitrogen absorbed from the soil, the N$_2$-fixing legume also assimilates atmospheric N$_2$ of a lower ^{15}N/^{14}N ratio than the soil nitrogen. This results in a dilution of the ^{15}N/^{14}N ratio of the soil nitrogen assessed by the reference crop. The extent to which this soil ^{15}N/^{14}N ratio is a dilution is an indication for the efficiency of N$_2$-fixation.

9.9.3 THE A-VALUE TECHNIQUE

On soils of low nitrogen status, nonfixing crops grow poorly, while N_2-fixing plants grow well. This could cause a mistake in growth and nitrogen uptake patterns between the reference and the fixing crop. But the addition of large amounts of fertilizer nitrogen to legumes has been shown to inhibit N_2-fixation (Allos and Bartholomew 1959; Wagner and Zapata 1982; Hardarson et al. 1984). The use of the A-value approach, however, allows the application of different nitrogen rates and ^{15}N enrichments of a given fertilizer to a legume and a reference crop to estimate the amount of the N_2-fixed (Fried and Broeshart 1975). The A-value for the nodulating legume represented soil and fixed nitrogen, while the A-value for the non-nodulating crop was reflected by only soil nitrogen.

The amount of symbiotically fixed nitrogen is calculated by multiplying the difference in A-value between the legume and the non-nodulating crop by percentage utilization of fertilizer nitrogen by the nodulating leguminous crop. Therefore, the usual calculation requires determination of an apparent A-value for both N_2-fixing and nonfixing crops.

In determining A-values, it is important to note the following:

• Since the available amount of nutrient in the soil is an inherent property of the soil, it will be constant for any set of experimental conditions.
• The "A-value" is a yield-independent parameter. It is only necessary to determine the respective proportions absorbed from each source, so as to determine the A-value of the soil. No yield data need be recorded. The absolute amounts of nutrient taken up from either source do not appear in the equation.
• The A-value for a particular soil remains constant even at different rates of application of the same fertilizer standard. In other words, the available amount of nutrient in the soil is independent of the rate of fertilizer applied. Thus, in soil fertility studies, it is sufficient to use only one rate of application to assess the nutrient supply of a soil and make relative comparisons of fertilizer treatments (Aleksic et al. 1968; Broeshart 1974).
• Any change in the set of experimental conditions (nature, source, placement, timing, etc.) will affect the magnitude of the A-value of the soil. Also, changes in harvesting times are important, since the plant samples reflect the nutrient isotopic composition of the soil from the seeding time until harvesting time. These changes of the A-value of a soil with time can be easily observed in a time-course study of nutrient uptake using labeled fertilizers (Rennie 1969; Smith and Legg 1971; Broeshart 1974; Zapata et al. 1987).
• The determination of the A-value of the soil has a number of practical applications, such as the quantitative evaluation of fertilizer practices, in particular fertilizer sources, and the design of further isotope-aided experiments (Rennie 1969; Broeshart 1974; Fried 1978; IAEA 1983).
• Extensive research work using A-values has been done for most plant nutrients, both macro- and micronutrients (Fried 1954; IAEA 1976, 1980, 1981; Vose 1980; Wagner and Zapata 1982).
• This method provides direct evidence for N_2-fixation, if the ^{15}N concentration in the plant exposed to $^{15}N_2$ is greater than 0.3663%. The extent to which ^{15}N is detected in the plant also provides an estimate of the proportion of the plant's nitrogen that was derived from fixation, and thus a direct method for quantifying N_2-fixed. Also, the use of gas-tight growth chambers for the assay of N_2-fixation in legumes has been described by Witty and Day (1978).

9.9.4 APPLICATION OF STABLE ISOTOPES IN FERTILIZATION PRACTICES

Application of ^{15}N tracer techniques gives us a chance to confirm some of the mechanisms responsible for enhancement of plant growth and nutrient acquisition. At the same time, it gives us the chance to recognize the accurate estimates of N derived from the different sources, i.e., soil, fertilizer, air

TABLE 9.3
Some Useful Stable Isotopes in Commercial Production

Atomic Number	Isotope	Natural Abundance %	Possible Enrichment %
6	^{13}C	1.07	>95
7	^{15}N	0.37	>99
8	^{18}O	0.21	>96
12	^{25}Mg	0.13	
14	^{28}Si	92.21	>99
16	^{33}S	0.76	>99
	^{36}S	0.02	>90
26	^{54}Fe	5.85	>99
30	^{68}Zn	18.75	>90
82	^{204}Pb	1.40	>85

Source: IAEA Bulletin no. 14, 2001.

and/or those derived and released from the organic originating materials. The stable isotopes ^{15}N and ^{34}S have been used mainly in these types of studies, reflecting their importance in crop nutrition. There are four stable or heavy isotopes of potential interest to researchers in soil and plant studies: ^{18}O, ^2H, ^{13}C, and ^{15}N. Some useful stable isotopes are commercially produced (Table 9.3).

Nitrogen is often the main factor limiting plant growth. There are twelve isotopes of nitrogen, many with extremely short half-lives. Of the radioactive isotopes, only ^{13}N with a half-life of 9.97 minutes has been used mainly in plant nitrogen translocation experiments (Caldwell et al. 1984). Two stable isotopes of N (Table 9.4) occur naturally in the atmospheric N_2.

Fertilizers are one of the essential inputs for maintaining or increasing the soil fertility level in intensive agricultural systems. The purpose of applying fertilizers is primarily to supply the crop with essential plant nutrients to ensure normal plant growth. The major plant nutrients (N, P, and K) have to be applied regularly to compensate for the amounts exported from the soil during harvest. Other plant nutrients such as Ca, Mg, S. and the micronutrients, e.g., Zn, Mo, and B, may also need to be added to maintain adequate levels of these nutrients, or to correct deficiencies (FAO 1983a, 1983b, 1984).

The role of applied nutrients in increasing crop yield is more obvious in deficient soils. Proper management of such nutrients could help in enhancement of nutrient uptake efficiency and maximizing benefits return. The yield response is a consequence of the additional uptake of the nutrient when other production factors are adequate.

TABLE 9.4
Several Isotopes of Nitrogen

Mass Number	Natural Abundance (%)	Half-Life (Time)
12	—	0.0125 sec
13	—	10.05 min
14	99.634 (light)	—
15	0.366 (heavy)	—
16	—	7.36 sec
17	—	4.14 sec

Soliman and Galal (2017) reviewed some contributions they and their colleagues in Egyptian Atomic Energy Authority had made using the ^{15}N tracer technique for differentiation between responses of different crops to fertilization practices and its effects on nitrogen use efficiency. Efficient use of mineral nitrogen either applied solely or in combination with organic composts was significantly affected by fertilization treatments and varied according to tested crop (Soliman et al. 2014). Nitrogen distribution within soil profile was varied according to crop, fertilizer rates, and types (Ghabour et al. 2015). They recorded that a half-dose of mineral fertilizer combined with a half-dose of organic compost resulted in accumulation of mineral-N in 0–20 cm soil depth. They concluded that portions of N remained in the soil and that the N uptake by plant and that lost from soil media were related to the rate of mineral fertilizer added. Other research conducted by Abdel-Salam et al. (2015) proved that combination of biofertilizers such as N-fixers or P-dissolvers along with the soluble fertilizer N would enhance the positive effect of N fertilization of sunflower plants, which reflected good fertilizer-^{15}N recovery. Rhizobium inoculation of pea crop has enhancement effects on portions of nitrogen derived from mineral fertilizer, and consequently its N recovery, either applied solely or in combination with organic compost (El-Sherbiny et al. 2014). Addition of organic residues has a positive effect on increasing the amount of fertilizer-N that remained in the soil after pea harvest.

9.10   EFFICIENT USE OF FERTILIZER BY CROPS

In his valuable textbook, L'Annunziata (2016) confirms the potential use of labeled fertilizer (radioactive or stable) and its vital role in the development and optimization of crop nutrition and water use efficiency. "He cites a paragraph from Menzel and Smith (1984), who noted that increasing fertilizer costs and the necessity for minimizing environmental pollution have given added impetus to fertilizer disposition studies…tracer fertilizers (i.e., isotope-labeled fertilizers) provide the only definitive means for determining both the behavior and fate of applied nutrients. A basic assumption is that the isotopically labeled element behaves in an identical manner to the non-labeled element both physically and chemically, and a plant cannot distinguish between [isotope] labels."

Also, L'Annunziata (2016) gave basic concepts and knowledge about the studies on fertilizer use efficiency, and we cited these basics as essential principals needed for researchers who have the ability to apply nuclear technique in fertilization practices. He noted that: radioactive isotopes are not applied in the open field; however, greenhouse experiments with radioactive isotopes of plant nutrients (e.g., ^{32}P, ^{33}P, ^{35}S, and ^{65}Zn) can provide scientists with the information needed to make recommendations to farmers to help achieve the optimum utilization of plant nutrient elements from soil and fertilizer. "For example, a research scientist may use a fertilizer labeled with the radioisotope of phosphorus, ^{32}P, and apply it to the soil to measure percentage use of the fertilizer nutrient by plant crops. The radioactive fertilizer phosphorus has a specific activity, defined as the intensity of the radioactivity per unit weight of the element or

$$^{32}P \text{ specific activity} = DPM / gP$$

where DPM is the radioactivity in units of disintegrations per minute, and gP represents the weight of the element phosphorus in grams." The weight of the phosphorus in the fertilizer is the sum of the weights of the radioactive isotope ^{32}P plus the natural stable isotope ^{31}P. However, the weight of the radioisotope tracers used in such studies is so small that the contribution of the radioisotope ^{32}P weight to the total weight of phosphorus can be neglected which demonstrate that 1 mCi of ^{32}P comprises only 3.5 ng or 3.5×10^{-9} g). When the researcher adds the radioactive fertilizer to the soil, the plant absorbs phosphorus from two sources, namely the phosphorus derived from the fertilizer (PDFF) and the phosphorus derived from the soil (PDFS). The specific activity of the radioactive tracer ^{32}P in the plant is lower than that of the radioactive fertilizer, because of the absorption of stable ^{31}P from the soil by the plant. The reduction in ^{32}P specific activity is referred to as isotope dilution, and the degree of isotope dilution serves to measure the proportions of phosphorus in the plant that came from the fertilizer and from the soil from the following:

$$\%PDFF = {}^{32}P \text{ specific activity in plant}/{}^{32}P \text{ specific activity in fertilizer} \times 100$$

where %PDFF is the percentage of phosphorus in the plant derived from the fertilizer. The remaining phosphorus in the plant is derived from the soil or

$$\%PDFS = 100 - \%PDFF.$$

Radioisotopes of phosphorus have relatively short half-lives (^{32}P, half-life = 14.3 days; ^{33}P, half-life = 25.3 days), which limits considerably the time available to carry out experiments with these isotopes as tracers. Long-term research on the dynamics of phosphorus in the soil is not possible with radioisotopes. Also, phosphorus has only one stable isotope, which eliminates any possibility of using stable isotopes of phosphorus as tracers. However, the IAEA Nuclear Technology Review (IAEA 2011) reports that a new approach to studying the dynamics of phosphorus in cropping systems has been investigated whereby researchers measure stable isotopes of oxygen in inorganic and organic soil phosphorus compounds. Tamburini et al. (2010) developed a method for accurately measuring the stable isotope oxygen-18 (^{18}O) in different soil phosphorus fractions. As reported by the IAEA (2011), "Soils under different farm management practices (e.g., fertilizer or manure applications) showed varying ^{18}O signatures in soil phosphorus indicating the potential of ^{18}O as an isotopic tracer for studying phosphorus cycling, tracing phosphorus sources and ultimately providing a better understanding of soil phosphorus dynamics in agro-ecosystems."

"Ryan (2008) has described the situation of fertilizers demand in the WANA region (Table 9.5). The greater requirement of nitrogen makes the proportions of P and K less than that of N. For example, ratios of applied NPK are 1.0, 0.30, and 0.30 in the United Kingdom and 1.0, 0.38, and 0.44 in the United States. Corresponding nutrient ratios are 1.0, 0.14, and 0.02 in Egypt; 1.0, 2.0, and 0.50 in Jordan; 1.0, 0.41, and 0.06 in Turkey; and 1.0, 0.83, and 0.06 in Tunisia. Theoretically, the nutrient needs of any crop are dependent on the crop species and the actual yields. The nutrients that do not come from the soil have to be supplied in fertilizer form, allowing for losses that inevitably occur. These discrepancies at least suggest that there is an imbalance with respect to the fertilizer nutrients applied in many countries of the WANA region."

More than 10 years ago, Roy et al. (2006) expanded on balanced fertilization, which, in turn, creates balanced plant nutrition. It is worth listing the following points that related to optimization of plant nutrition:

1. On many soils, application of N without addition of P and K made little sense.
2. Given the costs of crop production and the range of nutrients that can limit yields, fertilization with N, P, and K is counter-productive unless nutrients such as S, Zn, and B are also applied if they are deficient in the soil.
3. Balanced fertilization is the deliberate application of all nutrients that the soil cannot supply in adequate amounts for optimum crop yields.
4. There is no fixed recipe for balanced fertilization; it is soil and crop specific.
5. Any deficiency of one nutrient will severely limit the efficiency of others.
6. Imbalanced nutrition results in "mining" of soil nutrient reserves.
7. Luxury consumption is often a consequence of nutrients supplied in excess.
8. Imbalanced fertilization is inefficient, uneconomic, and wasteful.
9. Balanced fertilization depends on soil test values and crop removal.
10. If a soil is rich in one nutrient, fertilization should be directed towards the deficient nutrients or those in least supply.
11. Crop nutrient requirements are related to yield level.

TABLE 9.5
Fertilizer Nutrient Use (1000 Mg) and Ratios in Some Developed Countries and the West Asia–North Africa (WANA) Region

Consumption 2003/2004	Nitrogen	Phosphorus	Potassium		NPK Ratio	
China	24,745.0	9827.0	4663.0	1.0:	0.40:	0.19
Pakistan	2527.0	637.7	27.7	1.0:	0.25:	0.01
Cyprus	8.1	4.2	3.1	1.0:	0.52:	0.38
Turkey	1340.8	547.5	83.7	1.0:	0.41:	0.06
Egypt	1190.8	143.9	43.3	1.0:	0.12:	0.04
Iran	822.0	354.6	139.5	1.0:	0.43:	0.17
Morocco	237.0	124.0	55.0	1.0:	0.52:	0.23
Saudi Arabia	224.3	132.2	9.0	1.0:	0.59:	0.04
Syria	210.0	101.5	7.3	1.0:	0.48:	0.03
Iraq	83.2	106.8	0.7	1.0:	1.28:	0.01
Tunisia	56.0	41.0	5.0	1.0:	0.73:	0.09
Algeria	48.0	28.0	22.0	1.0:	0.58:	0.46
Libya	33.8	52.5	5.0	1.0:	1.55:	0.15
Sudan	41.9	1.8	2.7	1.0:	0.04:	0.06
Lebanon	22.1	10.0	8.8	1.0:	0.45:	0.4
Jordan	17.3	10.5	9.0	1.0:	0.61:	0.52
Uzbekistan	579.6	122.0	15.0	1.0:	0.21:	0.03
Turkmenistan	97.9	—	14.0	1.0:	0.00:	0.14
Kazakhstan	50.0	12.0	2.1	1.0:	0.24:	0.04
Tajikistan	26.0	—	—	1.0:	0.00:	0.00
Afghanistan	20.8	—	—	1.0:	0.00:	0.00
Azerbaijan	16.6	—	—	1.0:	0.00:	0.00
Kyrgyzstan	7.0	0.2	—	1.0:	0.03:	0.00

Source: Ryan, J., *Turk. J. Agric. For.*, 32, 79–89, 2008.

12. Fertilization with time can cause a build-up of P and K, thus reducing their fertilizer requirements.
13. The concept of balanced fertilization has expanded to integrated plant nutrition, embracing all sources of nutrients.
14. Integrated plant nutrition seeks to improve nutrient-use efficiency, build up nutrient stocks in the soil, and to limit losses to the environment.

9.11 PHOSPHORUS MANAGEMENT IN AGRICULTURAL SOILS

Phosphorus has one stable isotope (^{31}P) and several radioisotopes (from P and from ^{32}P to ^{38}P), but only two of them (^{32}P and ^{33}P) are suitable for agronomic studies (Zaharah and Zapata, 2003). The main characteristics of these radioactive P isotopes (i.e., half-life) and decay mode (i.e., radiation type and energy emitted) are shown in Table 9.6.

Taking the cost of the isotopes into consideration, ^{32}P is far cheaper than ^{33}P and also easier and faster to obtain from commercial suppliers. As beta emitters, ^{33}P and ^{32}P activities could be measured using liquid scintillation counting (LSC) methods. Technical steps and methodology have been briefly described elsewhere (Blair and Till 2003; L'Annunziata 2003). A historical review

TABLE 9.6

Summary of Main Characteristics of P Isotopes Used in Soil–Plant Studies

		Radiation Characteristics		
Isotope	**Half-Life (days)**	**Type**	**Energy (E_{max}) in MeV**	**Typical Applications**
^{32}P	14.3	β^-	1.71	Exchangeable P in soils
				P availability from P fertilizers
				Plant root distribution/activity
				Residual P fertilizer availability
^{33}P	24.4	β^-	0.248	Auto-radiography
				Diffusion in soils
				Double labeling with ^{32}P

Source: After Blair, G.J., et al., Phosphorus isotope tracer techniques: Procedures and safety issues, in *Use of Phosphorus Isotopes for Improving Phosphorus Management in Agricultural Systems*, IAEA TECDOC series no. 1805, Vienna, Austria, pp. 50–76, 2016. (International Atomic Energy Agency, Use of Isotope and Radiation Methods in Soil and Water Management and Crop Nutrition, Training Course Series No. 14, IAEA, Vienna (2001)).

about the beginnings of ^{32}P research was reported by Alam et al. (2001), who indicated that the application of ^{32}P radioisotope in short-term plant physiological studies began in 1939.

The principles of ^{32}P isotopic technique were perfectly reviewed by Goh and Adu-Gyamfi (2016). They revealed that radioisotopic techniques were first proposed by Dean et al. (1947) to study P uptake by crops from phosphate fertilizer, a technique still extensively used to study the P availability (A-value) to crops, P exchange (E-value), labile pool (L-value), and its dynamics in soil (McAuliffe et al. 1947; Larsen 1952; Fried 1964; Di et al. 1997). The A-value has been used widely not only to estimate the amount of plant-available P in soils but also to evaluate residual P from previous P fertilizer applications, P placement, and other factors. These values were found to be better and valuable for understanding and assessing phosphorus availability in the soil. Phosphate cycling in the soil-plant-environment system using tracers and definite terminology, principles, calculations, and methodology were previously described (Frossard et al. 2011; Blair et al. 2016).

The activity of a radioactive source or radionuclide sample is a measure of the total radiation emitted or representing the number of nuclei decaying per unit time. The SI unit of radioactivity is the Becquerel (symbol Bq). The Bacquerel is one nuclear disintegration per second (dps). Table 9.7 lists the radioactivity units of radioactive materials.

Determination of soil-available P using radioactive tracer was applied for determining the E-value or exchangeable P in the laboratory and the L-value or labile P using plants in the greenhouse, and the A-value or available P using plants grown in the field. Full characterization of the soil P status and its availability through the isotopic exchange kinetics (IEK) method provides better information for P management in agroecosystems. More details and description of the isotopic exchange kinetics method were reported earlier (Frossard and Sinaj 1997; Frossard et al. 2011). Recently, Nguyen et al. (2017) reviewed the basics and concepts used in the topic of the application of phosphorus radioisotopes in agricultural investigations. They summarized the principles and limitations involved in the use of phosphorus radioisotopes as tracers. They concluded that isotopic phosphorus can be used to measure P processes such as P dynamics, P kinetics, and mineralization of soil P_o, mycorrhizal acquisition of P_i, and to investigate the effects of farming practices such as P sources, green manure addition, and soil cultivation on soil Pi and Po. Since ^{32}P and ^{33}P are radioactive, their use in field experiments are limited in terms of quantifying on-farm P fluxes from different P sources as influenced by farm management practices. Although there exist limitations relating to the use of short-lived ^{32}P and ^{33}P in P cycling studies, these radioisotopes can be used in laboratory glasshouse experiments to investigate P transformation

TABLE 9.7

Activity Units of Radioisotopes

Curie (Ci)	Becquerel (Bq)	Disintegration per Second (dps)	Disintegration per Minute (dpm)
1 Ci	3.7×10^{10} Bq = 37 GBq	3.7×10^{10}	2.22×10^{12}
1 mCi	3.7×10^{7} Bq = 37 MBq	3.7×10^{7}	2.22×10^{9}
1 μCi	3.7×10^{4} Bq = 37 KBq	3.7×10^{4}	2.22×10^{6}
1 nCi	3.7×1 Bq	3.7×10	2.22×10^{3}
1 pCi	3.7×10^{-2} Bq	3.7×10^{-2}	2.22
	1 Bq	1	6×10
27,027 nCi	1 KBq	1×10^{3}	6×10^{4}
27,027 μCi	1 MBq	1×10^{6}	6×10^{7}
27,027 mCi	1 GBq	1×10^{9}	6×10^{10}

Source: International Atomic Energy Agency, Use of Isotope and Radiation Methods in Soil and Water Management and Crop Nutrition, Training Course Series no. 14, IAEA, Vienna, Austria (2001).

processes. Their chapter is very useful for young researchers and students who are involved in the application of radioisotopes in life sciences and particularly soil-plant-water relationships.

The principles of determination for quantification of P uptake by crops were outlined in IAEA (2016), which indicated that phosphorus isotopic composition, i.e., the ^{32}P/total P ratio, of any material is called specific radioactivity (SR). It was customary to express the specific activities of plant samples and fertilizer in Bq ^{32}P/g P at the time the samples were counted. It is important to note that the concept of specific radioactivity (ratio ^{32}P/total P) for radioisotopes is identical to that of ^{15}N atom excess (ratio ^{15}N/total N) for stable isotopes (Zapata 1990).

Specific radioactivity (SR) needs to determine the activity (Bq or dps) of the radioisotope by radio assay techniques using appropriate detectors, i.e., liquid scintillation counting or Cerenkov counting (for high-energy β emitters). In addition, total P concentration by any conventional chemical method should be quantified. These items lead to quantifying the P fraction derived from labeled P fertilizers (^{32}P or ^{33}P) following the isotope dilution principle:

$$Pdff = (SR\ plant\ sample / SR\ labeled\ fertilizer)$$

or as a percentage:

$$\%Pdff = (SR\ plant\ sample / SR\ labeled\ fertilizer) \times 100.$$

Standard equation and measurements required for P derived from solution or fertilizer applied to plants as well as P recovery in plants from different P sources like mineral fertilizers, organic residues, or compost and green manure were declared in the abovementioned document.

Previously, pot experiments using labeled P were conducted by Cabeza et al. (2011) to compare the effectiveness of recycled P products, triple superphosphate (TSP), and phosphate rock (PR) in increasing P uptake by maize under acid and neutral soil conditions. The isotopically exchangeable P (IEP) value can help assess the proportion of a P fertilizer that dissolves in soil water. They found that P fertilizer dissolves in the soil solution and P becomes adsorbed to soil particle surfaces or precipitates as a salt, but mostly remains in equilibrium with the soil solution. Addition of 60 mg P kg^{-1} soil added increased IEP (Figure 9.6) by 53 mg kg^{-1} in the acid soil and 42 mg kg^{-1} in the neutral soil, i.e., 88% and 70% of the TSP, remained in equilibrium with the soil solution even two years after its application.

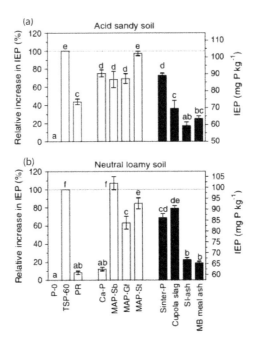

FIGURE 9.6 Absolute and relative increase in isotopically exchangeable P (IEP) in unplanted (a) acid sandy and (b) neutral loamy two years after fertilization. To calculate the relative increase in IEP, the increase of IEP from TSP was set to 100%. Data are means of three replicates, and error bars represent the standard error of the means. Different letters denote significant differences between treatments (Tukey, P\0.05).

In Europe, Tóth et al. (2014) reviewed high levels of soil P that were observed on areas with high input and high yields, like those in northwestern Europe, according to the Hungarian system, but high fertilizer doses on these areas were still needed to secure the required yield levels. Interestingly, we think the use of ^{32}P radiotracer may act perfectly to give a clear picture about the availability of P in such soils. Therefore, the system of the United Kingdom would not be recommended for additional fertilizer input on some of these areas of high P levels, e.g., in Belgium and the Netherlands, while in reality, they are constantly further fertilized (FAO 2013). Other sources beside chemical P fertilizers may add a considerable amount of P. For instance, organic manure adds considerable amounts of P in regions with high livestock densities (Figure 9.7). Inconsistency

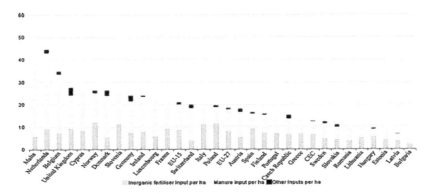

FIGURE 9.7 Phosphorus inputs on agricultural land by type of fertilizers (average 2005–2008, kg P ha^{-1}). (From Eurostat, Fertiliser Consumption and Nutrient Balance Statistics, http://epp.eurostat.ec.europa.eu/, last accessed September 2013.)

TABLE 9.8

Estimated vs. Reported P Fertilizer Amount in the United Kingdom and Hungary for the Reference Year 2009

	Estimated P Fertilizer Needed (ton P_2O_5)	Actual Fertilizer Use (ton P_2O_5)	
	Based on LUCAS topsoil data and the advisory systems of 1. DEFRA (2010) 2. Antal et al. (1979) 3. Combination of DEFRA (2010) and Antal et al. (1979)	Based on national statistics 1. DEFRA (2012b) 2. HCSO (2013)	Based on FAOSTAT (FAO 2013) Average of years 2008–2011
1. United Kingdom	145,896	109,267	173,250
2. Hungary	3,960,008	36,167	43,797
3. European Union (27 Member States)	3,849,873	—	2,365,502

between recommended and reported fertilizer applications (Table 9.8) proves differences in the farming practices in different regions of the European Union, while also reflecting the possible shortcomings of the fertilizer usage reporting systems. The observed differences certainly highlight the possibility of further optimizing P management within the European Union, as has been already advised in regional contexts by a number of authors (Valkama et al. 2009; Csathó and Radimszky 2011; Hejcmana et al. 2012). The arguments about the situation of P fertilization practices reveal the need for better statistical data on actual yield levels and P applications to optimize P management in the European Union.

A number of agronomic field experiments were conducted in different regions of sub-Saharan Africa (Chikowo et al. 2010), under a wide range of management and environmental conditions, to assess the associated variability in the efficiencies with which applied and available nutrients were taken up by tested crops. They considered N and P capture efficiencies (NCE and PCE, kg uptake kg^{-1} nutrient availability), and N and P recovery efficiencies (NRE and PRE, kg uptake kg^{-1} nutrient added). Both nitrogen and phosphorus nutrients were applied either solely or in combination in mineral or organic forms. They take the cropping systems into consideration. The response to nutrients was widely varied according to different climates spread out across regions from arid through wet tropics, coupled with an equally large array of management practices and interseason variability. They reported that NCE ranged from 0.05 to 0.98 kg kg^{-1} for the different systems (NP fertilizers, 0.16–0.98; fallow/cover crops, 0.05–0.75; animal manure, 0.10–0.74 kg kg^{-1}), while PCE ranged from 0.09 to 0.71 kg kg^{-1}, depending on soil conditions. The respective NREs averaged 0.38, 0.23, and 0.25 kg kg^{-1}. Cases were found where NREs were 1 for mineral fertilizers or negative when poor-quality manure immobilized soil N, while response to P was in many cases poor due to P fixation by soils. Other than good agronomy, it was apparent that flexible systems of fertilization that vary N input according to the current seasonal rainfall pattern offer opportunities for high resource capture and recovery efficiencies in semiarid areas. Finally, they suggested the use of cropping-systems modeling approaches to hasten the understanding of Africa's complex cropping systems.

Crop recovery of added P by labeling the P fertilizer with the radioisotope ^{32}P is the only direct method of estimating efficiency of a single application of P. However, this method is expensive and ^{32}P has a half-life of only 14.3 days, so most studies can only be short term. Crop recovery efficiency (RE) of added P by the difference method is a more common and widely used approach to estimate its efficiency. Phosphorus fertilizer is commonly thought to be very inefficient because its recovery by crops in the year of application is often only 10%–15%. The P not recovered by the crop is believed to be fixed in forms that are not plant available (Robertsa and Johnston 2015).

Using the radioactive isotope ^{32}P, Mattingly and Widdowson (1958) found that about 20% PUE was found in spring barley. Cases reviewed by Syers et al. (2008) noted P recovery using labeled ^{32}P fertilizer between 5% and 25%. Dhillon et al. (2017) showed that global PUE is generally low using the difference method and comparable to reported P effiencies on a smaller scale. They added that phosphorus use effiency estimated for the world using the difference method was 20%. P removed in the total cereal grain coming from the soil was 79.3%. This value was based on an average P fertilizer recovery of 20.7%.

Franzini et al. (2009) examined the use of ^{32}P as a tracer to quantitatively evaluate the effect of different ratios of TSP and PR ratios and P application rates on P recovery from PR by corn grown in a sand clay soil. They reported that the mixture of PR with TSP improved the P recovery from PR in the corn plant and that this effect increased proportionally to the TSP amounts in the mixture. When compared with the plant P recovery from TSP (10.52%), PR-P recovery (2.57%) was much lower even when mixed together in the ratio of 80% TSP: 20% PR. There was no difference in PR-P utilization by the corn plants with increasing P rates in the mixture (1:1 proportion). Therefore, PR-P availability is affected by the proportions of the mixtures with water soluble P, but not by P rates.

9.12 RADIOACTIVITY IN FERTILIZERS AND ITS EFFECT ON ENVIRONMENTAL QUALITY

Demand for phosphorus application in agricultural production is increasing quickly throughout the globe. The bioavailability of phosphorus is distinctively low due to its slow diffusion and high fixation in soils, which makes phosphorus a key limiting factor for crop production. Applications of phosphorus-based fertilizers improve soil fertility and agriculture yield, but at the same time concerns over a number of factors that lead to environmental damage need to be addressed properly (Gupta et al. 2014).

Phosphate, nitrogen, and potassium fertilizers, which are used predominantly in order to increase crops in agriculture, provide basic nutrients to plants. In NPK fertilizers, gamma activity shows a wide variation because of the difference in the factories of manufactured fertilizers and the difference in the places from which the raw minerals for manufacturing the fertilizers were taken (Hussain and Hussain 2011). Radionuclides in phosphate fertilizer belonging to ^{232}Th and ^{238}U from phosphate rocks series as well as radioisotope of potassium (^{40}K) are the major contributors of outdoor terrestrial natural radiation. The plants take some fractions of radioactivity, and radionuclides enter the food chain in this way (Bayrak et al. 2018). Phosphate rocks are largely used for the production of phosphoric acid, fertilizers, and gypsum (Jankovic et al. 2013; Sahu et al. 2014).

The natural radioactivity in phosphate rock depends on its origin. In sedimentary rock it is much higher than in volcanic rock. Granite rocks contain thorium in significant quantities. The main producers of phosphate rock are China, Morocco, Russia, and the United States. The radionuclide activity values differed among the districts, depending upon the geographic structures, rainfall amounts, and elevations of the districts. Some typical values of activity concentrations in phosphate rock are shown in Table 9.9. One reason for the increase in natural radiation involves the chemical fertilizers used in agriculture (Jankovic et al. 2013; Mir and Rather 2015; Durusoy and Yildirim 2017).

Phosphate ores typically contain about 1500 Bq/kg of uranium and radium, although some phosphates contain up to 20,000 Bq/kg of U_3O_8. In general, phosphate ores of sedimentary origin have higher concentrations of radionuclides of the uranium family. In 90% of cases, the ore is treated with sulfuric acid. The fertilizers become somewhat enriched in uranium (up to 150% relative to the ore), while 80% of the 226Ra, 30% of 232Th and 5% of uranium are left in phosphogypsum (Gaafar et al. 2016). Phosphoric acid is the starting material for TSP and ammonium phosphate fertilizers. Some typical values of activity concentrations in fertilizers are shown in Table 9.10.

TABLE 9.9
Some Typical Values of Activity Concentrations (Bq/kg) in Phosphate Rock

Location	^{226}Ra	^{238}U	^{232}Th	^{40}K
Morocco	1600	1700	10	20
Togo	1100	1300	30	4
Western Sahara	900	900	7	30
Syria	300	1000	2	—
USA	1600	150	20	—
Tunisia	800	1000	20	30
India	1290	1340	90	10

Source: Sahu, S.K., et al., *J. Radiat. Res. Appl. Sci.*, 7, 123–128, 2014.

TABLE 9.10
Some Typical Values of Activity Concentrations (Bq/kg) in Fertilizers in the World

Location	References	^{226}Ra	^{238}U	^{232}Th	^{40}K
Iraq	Hussain and Hussain (2011)		13–89	1-27	12–2276
Saudi Arabia	Alharbi 2013	64		17	2453
India	Shahul Hameed et al. (2014)		2–396	5–39	33–93
Egypt	Uosif et al. (2014)	12–244		3–99	109–670
Serbia	Jankovic et al. (2013)	87	220		4860
Croatia	Barisic et al. (1992)	75	120		
Egypt	Ghosh et al. (2008)	301		24	3
Egypt				125–239	446–882
Egypt		366		67	4

The radioactive content of the phosphatic fertilizers varies considerably and depends both on their concentrations in the parent mineral and on the fertilizer production process (Chaney 2012; Mortvedt and Beaton 2014). Uranium-238 concentrations can range from 7 to 100 pCi/g in phosphate rock (US EPA 2016) and from 1 to 67 pCi/g in phosphate fertilizers (Khater 2008; NCRP 1987; Hussein 1994). Where high annual rates of phosphorus fertilizer are used, this can result in uranium-238 concentrations in soils and drainage waters that are several times greater than are normally present (NCRP 1987; Barisic et. al. 1992). However, the impact of these increases on the risk to human health from radionuclide contamination of foods is very small (less than 0.05 mSv/y) (NCRP 1987; Hanlon 2012; Sharpley and Menzel 1987).

Most public and farmers were exposed to the natural radioactivity present in the sources used for fertilizer manufacturing or those applied through fertilization practices in the field. Some of these activities may be liberated to the underground water (Alcaraz Pelegrina and Martínez-Aguirre 2001). Phosphate is used in the production of some chemical fertilizers. Since phosphate contains some natural radionuclides like ^{238}U, ^{232}Th, and ^{40}K, fertilizers become the major contributor for outdoor terrestrial natural radiations. The radioactive content of phosphotic fertilizers varies considerably and depends both on their concentrations in the parent mineral and on the fertilizer production process (Chaney 2012; Mortvedt and Beaton 2014). Uranium-238 concentrations can range from 7 to 100 pCi/g in phosphate rock (US EPA 2016) and from 1 to 67 pCi/g in phosphate fertilizers

(Khater 2008; NCRP 1987; Hussein 1994). Where high annual rates of phosphorus fertilizer are used, this can result in uranium-238 concentrations in soils and drainage waters that are several times greater than are normally present (NCRP 1987; Barisic et al. 1992). However, the impact of these increases on the risk to human health from radionuclide contamination of foods is very small (less than 0.05 mSv/y) (NCRP 1987; Hanlon 2012; Sharpley and Menzel 1987). The phosphate material is very insoluble, and therefore in its original state is practically unavailable as a plant phosphors source (IAEA 1973). Among the constituents of agricultural phosphate fertilizers are potassium ores (potassium sulphate, potassium chloride) (Conceição and Bonotto 2006). Samples of granular and leafy types of NPK in addition to urea type collected from common markets in different regions in Iraq recorded high levels of radium-equivalent value in leafy-type NPK fertilizers, while urea types had no radionuclide. At the same time, the maximum specific activity and absorbed dose rate at 1m above the ground surface (nGy/h) after the agricultural application of NPK fertilizers was 0.15% of the world average outdoor exposure (Figure 9.8) due to terrestrial gamma radiation (Hussain and Hussain 2011).

The abovementioned naturally occurring radionuclides materials (NORM), including uranium and thorium series, are considered the largest contributor to radiation doses received by human beings. Components of ^{238}U, ^{235}U, and ^{232}Th series along with other nonseries radionuclides are given in Table 9.11.

Radioactivity in rock phosphate as a source of NORM was found to be varying from one place to another. This activity was transferred to phosphate fertilizer manufactured from rock-P source (Tufail et al. 2006). For example, a study was conducted to compare activity concentration in Egyptian and Japanese phosphate fertilizers (Hassan et al. 2017). Table 9.12 shows the variation in activities between the two countries.

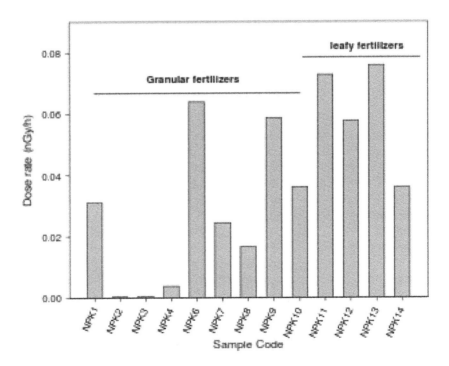

FIGURE 9.8 Absorbed dose rate 1 m above the ground surface (nGy/h) originated from the agricultural application of NPK fertilizers. (After Hussain, R.O. and Hussain, H.H., *Braz Arch Biol Technol.*, 54, 777–782, 2011.)

TABLE 9.11

Principal Natural Radionuclides Decay Series

Nuclide	Half-Life	Major Radiation	Nuclide	Half-Life	Major Radiation
^{238}U	4.47 BY	α, X	^{219}Ra	4.0 sec	α
^{234}Th	24.1 D	β, γ, X	^{215}Po	1.78 m sec	α
^{234}Pa	1.17 min	β, γ	^{211}Pb	36.1 min	β, γ
^{234}U	245,000 Y	α, X	^{211}Bi	2.13 min	α
^{230}Th	77,000 Y	β, α, γ	^{207}Tr	4.77 min	β, γ
^{226}Ra	1600 Y	α, γ	^{287}Pb	Stable	—
^{222}Rn	3.83 D	α	^{232}Th	14.1 BY	α, X
^{218}Po	3.05 min	α	^{228}Ra	5.75 Y	β
^{214}Bi	19.7 min	β, γ	^{228}Th	1.91 Y	α, X
^{214}Pb	26.8 min	β, γ, X	^{228}Ac	6.13 hr	β, γ, X
^{214}Po	164 μ sec	α	^{224}Ra	3.66 D	α, γ
^{210}Pb	22.3 Y	β, α, γ	^{220}Rn	55.6 sec	α
^{210}Bi	5.01 D	β	^{216}Po	0.15 sec	α
^{210}Po	138 D	α	^{212}Pb	10.64 hr	β, γ, X
^{206}Pb	Stable	—	^{212}Bi	60.6 min	α, β, γ
^{235}U	7.1×10^9 Y	α	^{212}Po	0.305 μ sec	α
^{231}Th	25.5 hrs	β, γ	^{208}Ta	3.07 min	β, γ
^{231}Pa	3.25×10^4 Y	α	^{208}Pb	Stable	—
^{227}Ac	21.8 Y	β, γ	Non-Series Radionuclides		
^{227}Th	18.5 sec	α	^{40}K	1.28 BY	β, γ

TABLE 9.12

Activity Concentrations of ^{226}Ra, ^{232}Th, and ^{40}K in Egyptian and Japanese Fertilizers

	Code		Composition		^{226}Ra		^{232}Th		^{40}K	
Sample	Egypt	Japan	Egypt	Japan	Egypt	Japan	Egypt	Japan	Egypt	Japan
1	EF1	JF1	SSP (15% P$_2$O$_5$)	P. acid 10%	761 ± 31	25 ± 1	67 ± 13	5 ± 2	251 ± 94	3909 ± 21
2	EF2	JF2	SSP (12% P$_2$O$_5$)	P. acid 8%	557 ± 19	62 ± 1	15 ± 6	15 ± 1	175 ± 27	3280 ± 17
3	EF3	JF3	SSP (16% P$_2$O$_5$)	P. acid 20%	782 ± 24	74 ± 1	14 ± 8	6 ± 1	222 ± 24	48 ± 2
4	EF4	JF4	NPK	P. acid 20%	443 ± 11	200 ± 2	ND	12 ± 1	88 ± 22	231 ± 4
5	EF5	JF5	Triple	P. acid 17%	312 ± 14	1264 ± 5	ND±	8 ± 2	175 ± 31	31 ± 5

Source: Hassan, N.M., et al., *J. Chem.*, Volume 2017, Article ID 9182768, 8 pages, 2017.

Their data from Hassan et al. (2017) indicated that the radionuclide concentrations in Japanese fertilizer were less than those of Egyptian fertilizers except for potassium, as seen in Figure 9.9. The radionuclide concentration of ^{40}K is much higher in Japanese fertilizer samples and especially sample JF-1. Both Egyptian and Japanese fertilizers maintain radionuclide concentrations less than the recommended limits by UNSCEAR (2008).

Comparison of the estimated radiological indexes values during their work (Hassan et al. 2017) and other values in previous studies in literature indicated that the radium equivalent in the Egyptian and Japanese fertilizers was less than its value for fertilizer used in Algeria and Brazil but was greater than its value for fertilizer used in Saudi Arabia and Bangladesh. The gamma indexes had the same trend as radium equivalent (Table 9.13).

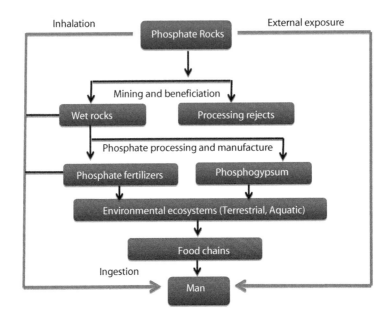

FIGURE 9.9 Environmental pathways of natural radionuclides from phosphate rocks. (From Khater, A.E.M., et al., *J Environ. Radioact.*, 55, 255–267, 2001.)

TABLE 9.13

Comparison of Radiological Indexes in Egyptian and Japanese Fertilizers and Their Values in Literature

Country	Sample	Radium Equivalent (Bq/kg)	Gamma Index ($I\gamma$)	References
Egypt	Fertilizer	613 ± 33	2.06 ± 0.11	Hassan et al. (2017)
Japan	Fertilizer	454 ± 5	1.63 ± 0.08	Hassan et al. (2017)
Algeria	NPK	1168	9.6	Boukhenfouf and Boucenna (2011)
Brazil	NPK	1772	12.3	Becegato et al. (2008)
Saudi Arabia	NPK	275	—	Alharbi (2013)
Egypt (Qena)	Phosphate fertilizer	462	3.1	Ahmed and El-Arabi (2005)
Bangladesh	Triple superphosphate	374	—	Alam et al. (1997)

Phosphate ores in Egypt reflected a radium concentration in the same range as UNSCEAR's typical value (Hassan et al., 2016). The radionuclide concentrations of ^{226}Ra, ^{232}Th, and ^{40}K maintained in phosphate ores of several countries is consistent with those of Egypt (Table 9.14).

Another contribution is by Sahu et al. (2014), who reported the radioactivity in rock phosphate and phosphogypsum in India compared to other countries, as listed in Table 9.15. They indicated higher levels of natural radionuclides in the gypsum ponds and rock silo than other locations in the plant premises. The ^{238}U levels in the phosphogypsum sample were higher than those compared worldwide. Values of the activities due to ^{226}Ra, ^{232}Th, and ^{40}K were varied according to phosphatic fertilizer types from the Pakistani market as shown in Table 9.16, after Khan et al. (2004). They found that the concentration of ^{226}Ra in all phosphatic fertilizers as well as phosphate rock samples is much higher than ^{232}Th and ^{40}K. All the analyzed phosphatic fertilizers and phosphate

TABLE 9.14

Specific Activities of ^{226}Ra, ^{232}Th, and ^{40}K in Phosphate Ores Used in Several Countries All over the World

Country	Activity Concentration (Bq kg^{-1})			References
Phosphate ore	^{226}Ra	^{232}Th	^{40}K	
Saudi Arabia	513	39	242	Al-Zahrani et al. (2011)
Egypt	840	395	398	El-Taher and Makhluf (2010)
South Korea	—	4.0	49	Chang et al. (2008)
Brazil	256	3238	1202	Conceição and Bonotto (2006)
Nigeria/Sokoto	—	16	40	Ogunleye et al. (2002)
Tanzania (Arusha)	—	350	280	Banzi et al. (2000)
Sudan (Uro)	4131	7.5	62.3	Sam et al. (1999)
Jordan	1044	2	8	Olszewska-Wasiolek (1995)
Tunisia	821	29	32	Olszewska-Wasiolek (1995)
Tanzania (Arusha)	5022	717	286	Makweba and Holm (1993)
Morocco	1600	20	10	Guimond (1990)
Egypt	871 ± 92	19 ± 2	176 ± 18	Hassan et al. (2016)

Source: Hassan et al., *J. Taibah Univ. Sci.*, 10, 296–306, 2016.

TABLE 9.15

Radioactivity (Bq/g) in Phosphate Rocks and Phosphogypsum

		Phosphate Rock Activity (Bq/g)			
Location	References	^{226}Ra	^{238}U	^{232}Th	^{40}K
Morocco	Guimond and Hardin (1989)	1.6	1.7	0.01	0.02
Taiba-Togo (Western Sahara)		1.1	1.3	0.03	0.004
		0.9	0.9	0.007	0.03
Syria	Attar et al. (2011)	0.3	1.0	0.002	—
Florida	Guimond (1990)	1.6	1.5	0.02	—
Tunisia	Olszewska-Wasiolek (1995)	0.8	1.0	0.02	0.03
India	Sahu et al. (2014)	1.29	1.34	0.09	0.01
	Phosphogypsum Activity				
Florida	Olszewska-Wasiolek(1995)	0.9	0.069	0.01	—
Brazil	Mazzilli et al. (2000)	0.6	0.04	0.1	0.02
Brazil		0.2	0.04	0.1	0.01
Syria	Attar et al. (2011)	0.3	0.03	0.002	—
Egypt	Ahmed (2005)	0.1	—	0.04	0.5
Spain	José et al. (2009)	0.8	0.08	—	—
India	Sahu et al. (2014)	0.3	0.03	0.01	0.005

rock samples have shown a higher amount of ^{226}Ra due to deposits of uranium in the rock phosphate. It could result in significant radiation exposure if these fertilizers are handled in places with poor ventilation that could lead to radon accumulation.

The pathways of natural radionuclides from phosphate fertilizers to the environment and finally to the public (Figure 9.9) demonstrated the need to recognize and calculate the exposure rate comparing to Egypt and the world average (Table 9.17). Khater et al. (2001) listed the calculated

TABLE 9.16

Specific Gamma-Ray Activities Due to ⁴⁰K, ²²⁶Ra, and ²³²Th in Different Brands of Phosphatic Fertilizers Available in Pakistan

Fertilizer	Specific Gamma-Activities (Bq kg⁻¹)		
	^{40}K	^{226}Ra	^{232}Th
Single Superphosphate	221.2	556.3	49.7
Triple Superphosphate	142.5	558.6	84.8
Nitrophos	205.7	389.4	79.9
Mono Ammonium Phosphate	137.7	560.9	85.1
Di Ammonium Phosphate	237.5	545.3	65.5
Phosphate Rock	207.3	439.5	50.4

Source: Khan, K., et al., *Geol Bull.*, 37, 59–64, 2004.

TABLE 9.17

Calculated Exposure Rate (nGy/h) at 1 m above the Ground Due to Natural Radionuclides in Wet Phosphate Rock and Soil

	^{226}Ra	^{232}Th	^{40}K	Total
nGy/h per Bq/kg	0.461	0.623	0.041	—
Wet rock	132	14.8	0.89	148
Soil	11.8 (8.8–15.1)[a]	18.1 (10.5–27.5)	5.34 (3.1–6.8)	35.3 (22.3–49.4)
Egypt	—	—	—	32 (8–93)
World	—	—	—	55

Source: Khater, A.E.M., et al., *J Environ. Radioact.*, 55, 255–267, 2001.

[a] Mean (range).

exposure rates (nGy/h) at 1 m above the ground due to natural radionuclides in wet phosphate rock and soil and recorded that for soil the exposure rate (35 nGy/h) is comparable to the Egyptian average (32 nGy/h) and less than the world average (55 nGy/h) (UNSCEAR 1993).

It is of interest to follow up the phosphate manufacture processes and its by-products. In this respect, the phosphate processing operations comprise the mining and milling of phosphate ore and then the manufacture of phosphate products by either the wet or the thermal process. More than 70% of the ore being beneficiated in several process steps to increase the P_2O_5 concentration before delivery are wet processes. The main route for more than 90% is then acidulation with sulfuric acid, besides nitric and hydrochloric acid in minor extent with the main by-product of gypsum sulfate (phosphogypsum), of which 4–5 tons are received when 3 tons of ore are turned into 1 ton of P_2O_5 (Figure 9.10).

The activities of radionuclides may have an impact on soil and plant upon fertilization with such phosphatic fertilizers. In this regard, Alsaffar et al. (2016) found that the radioactivity produced by the addition of fertilizers to pots planted with rice was apparently insignificant compared with that of soil alone. They also indicated that ²²⁶Ra concentrations in rice grains were increased with increasing urea. Additionally, ²²⁶Ra concentrations in grains also slightly increased with increasing both NPK and NPK+Mg rates, but ²²⁶Ra concentration was found to be lower than that of urea. Therefore, risk assessment due to application of such fertilizers should be taken into consideration in comparison with the average annual ingestion dose worldwide.

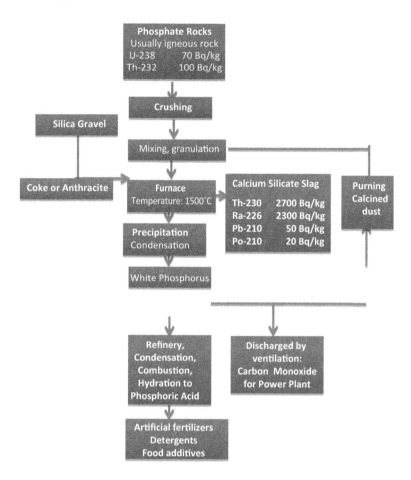

FIGURE 9.10 Flow of radionuclides in thermal process of phosphorus production. (From Scholten, L.C., Approaches for regulating management of large volumes of waste containing natural radionuclides in enhanced concentrations, Official Publication of the European Communities, 1996; Penfold, J.S.S., et al., Establishment of reference levels for regulatory control of workplaces where materials are processed which contain enhanced levels of naturally occurring radionuclides, NRPB report on contract number 95-ET-009, 1997.)

REFERENCES

Abbasi, M.K., Tahir, M.M. and Rahim, N. (2013). Effect of N fertilizer source and timing on yield and N use efficiency of rainfed maize (*Zea mays* L.) in Kashmir–Pakistan. *Geoderma*, 195–196, 87–93.

Abbasi, M.K., Tahir, M.M., Sadiq, A., Iqbal, M. and Zafar, M. (2012). Yield and nitrogen use efficiency of rainfed maize response to splitting and nitrogen rates in Kashmir, Pakistan. *Agronomy Journal*, 104, 48–57.

Abdel-Salam, A.A., Soliman, S.M., Galal, Y.G.M., Zahra, W.R., Moursy, A.A. and Hekal, M.A. (2015). Response of sunflower (*Helianthus annuus* L.) to N-application and biofertilization with assessment of fertilizer N recovery by N versus subtraction method. *Journal of Nuclear Technology in Applied Science*, 3(3), 157–169.

Ahmed, N.K. (2005). Measurement of natural radioactivity in building materials in Qena City, Upper Egypt. *Journal of Environmental Radioactivity*, 83, 91–99.

Ahmed, N.K. and El-Arabi, A.G.M. (2005). Natural radioactivity in farm soil and phosphate fertilizer and its environmental implications in Qena governorate, Upper Egypt. *Journal of Environmental Radioactivity*, 84(1), 51–64.

Alam, M.N., Chowdhury, M.I., Kamal, M., Chose, S., Banu, H. and Chakraborty, D. (1997). Radioactivity in chemical fertilizers used in Bangladesh. *Applied Radiation and Isotopes*, 48(8), 1165–1168.

Alam, S.M., Ansari, R. and Khan, M.A. (2001). Application of radioisotopes and radiation in the field of agriculture: Review. *On Line Journal of Biological Sciences*, 1(3), 82–86.

Alcaraz Pelegrina, J.M. and Martínez-Aguirre, A. (2001). Natural radioactivity in groundwaters around a fertilizer factory complex in south of Spain. *Applied Radiation and Isotopes*, 55, 419–423.

Aleksic, Z., Broeshart, H. and Middelboe, V. (1968). The effect of nitrogen fertilization on the release of soil nitrogen. *Plant and Soil*, 29, 474–478.

Alharbi, W.R. (2013). Natural radioactivity and dose assessment for brands of chemical and organic fertilizers used in Saudi Arabia. *Journal of Modern Physics*, 4(3), 344–348.

Alley, M.M. and Vanlauwe, B. (2009). *The Role of Fertilizers in Integrated Plant Nutrient Management*. IFA, Paris, France; TSBF-CIAT, Nairobi, Kenya.

Allos, H.F. and Bartholomew, W.V. (1959). Replacement of symbiotic fixation with available nitrogen. *Soil Science*, 87, 61–67.

Alsaffar, M.S., Suhaimi Jaafar, M., Ahmad Kabir, N. and Nisar, A. (2016). Impact of fertilizers on the uptake of ^{226}Ra, ^{232}Th, and ^{40}K by pot grown rice plants. *Pollution*, 2(1), 1–10.

Al-Zahrani, J.H., Alharbi, W.R. and Adel Abbady, G.E. (2011). Radiological impacts of natural radioactivity and heat generation by radioactive decay of phosphorite deposits from Northwestern Saudi Arabia. *Australian Journal of Basic and Applied Sciences*, 5, 683–690.

Antal, J. et al. (1979). Fertilization guidelines. N, P, K műtrágyázási irányelvek. In: *Műtrágyázási irányelvek és üzemi számítási módszer. I.rész*, edited by I. Buzás, et al. MEM Növényvédelmi és Agrokémiai Központ, Budapest P 95, pp. 1–47 (in Hungarian).

Attar, L.A., Al-Oudat, M., Kanakri, S., Budeir, Y., Khalily, H. and Hamwi, A.A. (2011). Radiological impacts of phosphogypsum. *Journal of Environmental Management*, 92, 2151–2158.

Banzi, F.P., Kifanga, L.D. and Bundala, F.M. (2000). Natural radioactivity and radiation exposure at the Minjingu phosphate mine in Tanzania. *Journal of Radiological Protection*, 20, 41–51.

Barisic, D., Lulic, S. and Miletic, P. (1992). Radium and uranium in phosphate fertilizers and their impact on the radioactivity of waters. *Water Research*, 26(5), 607–611. doi:10.1016/0043-1354(92)90234-U.

Bayrak, G., Keleş, E. and Atik, D. (2018). Estimation of radioactivity caused by chemical fertilizers on trakya sub-region soils and its potential risk on ecosystem. *European Journal of Sustainable Development*, 7(3), 413–424.

Becegato, V.A., Ferreira, F.J.F. and Machado, W.C.P. (2008). Concentration of radioactive elements (U, Th and K) derived from phosphatic fertilizers in cultivated soils. *Brazilian Archives of Biology and Technology*, 51(6), 1255–1266.

Blair, G.J., Adu-Gyamfi, J.J. and Zapata, F. (2016). Phosphorus isotope tracer techniques: Procedures and safety issues. In: *Use of Phosphorus Isotopes for Improving Phosphorus Management in Agricultural Systems*. IAEA TECDOC Series no. 1805. Vienna, Austria, pp. 50–76.

Blair, G.J. and Till, A.R. (2003). *Guidelines for the Use of Isotopes of Sulfur in Soil-Plant Studies*. IAEA Training Manual Series no. 20. IAEA, Vienna, Austria.

Boukhenfouf, W. and Boucenna, A. (2011). The radioactivity measurements in soils and fertilizers using gamma spectrometry technique. *Journal of Environmental Radioactivity*, 102(4), 336–339.

Broeshart, H. (1974). Quantitative measurement of fertilizer uptake by crops. *Netherlands Journal of Agricultural Science*, 22, 245–254.

Brown, R.H. (1978). A difference in N use efficiency in C3 and C4 plants and its implications in adaptation and evolution. *Crop Science*, 18, 93–98.

Byrnes, B.H. (1990). Environmental effects of N fertilizer use – An overview. *Fertilizer Research*, 26, 209–215.

Cabeza, R., Steingrobe, B., Römer, W. and Claassen, N. (2011). Effectiveness of recycled P products as P fertilizers, as evaluated in pot experiments. *Nutrient Cycling in Agroecosystems*, 91, 173–184.

Caldwell, C.D., Fensom, D.S., Bordeleau, K., Thompson, R.G., Drouin, R. and Didsbury, R. (1984). Translocation of 13N and 11C between nodulated roots and leaves in alfalfa seedlings. *Journal of Experimental Botany*, 35, 431–444.

Canadian Nuclear Association. (2016). *The Role of the Nuclear Industry in the World*. Final report, pp.18. https://cna.ca/wp.../The-Role-of-the-Nuclear-Industry-FINAL.pdf.

Chaney, R.L. (2012). Food safety issues for mineral and organic fertilizers. *Advances in Agronomy*, 117, 51–99.

Chang, B.U., Koh, S.M., Seo, J.S., Yoon, Y.Y., Row, J.W. and Lee, D.M. (2008). Nationwide survey on the natural radionuclides in industrial raw minerals in South Korea. *Journal of Environmental Radioactivity*, 99, 455–460.

Chikowo, R., Corbeels, M., Mapfumo, P., Tittonell, P., Vanlauwe, B. and Giller, K.E. (2010). Nitrogen and phosphorus capture and recovery efficiencies, and crop responses to a range of soil fertility management strategies in sub-Saharan Africa. *Nutrient Cycling in Agroecosystems*, 88, 59–77. doi:10.1007/s10705-009-9303-6.

Conceição, F.T. and Bonotto, D.M. (2006). Radionuclides, heavy metals and fluorine incidence at Tapira phosphate rocks, Brazil, and their industrial (by) products. *Environmental Pollution*, 139, 232–243.

Csathó, P. and Radimszky, L. (2011). Towards sustainable agricultural NP turnover in the EU 27 countries. A review. In: *Land Quality and Land Use Information in the European Union*, edited by T. Nemeth. Publication Office of the European Union, Luxembourg, pp. 69–85.

Danso, S. K. A., Hardarson, G. and Zapata, F. (1986). Assessment of dinitrogen fixation potentials of forage legumes with ^{15}N technique. In: *Potentials of Forage Legumes in Farming Systems of Sub-Saharan Africa*, edited by I. Haque, M. Jutzi and P.J.H. Neate, pp. 26–57. ILCA, Ethiopia.

Davies, D.B. and Sylvester-Bradley, R. (1995). The contribution of fertiliser nitrogen to leachable nitrogen in the UK: A review. *Journal of the Science of Food and Agriculture*, 68, 399–406.

Dean, L.A., Nelson, W.L., MacKenzie, A.J., Armiger, W.H. and Hill, W.L. (1947). Application of radioactive tracer technique to studies of phosphatic fertilizer by crops. *Soil Science Society of America Proceedings*, 12, 107–112.

DEFRA. (2010). *Fertiliser Manual (RB209)*. 8th ed. Department for Environment, Food & Rural Affairs (DEFRA).

DEFRA. (2012). British Survey of Fertilizer Practice, 2012. https://www.gov.uk.

Dhillon, J., Torres, G.E., Driver, E., Figueiredo, B. and Raun, W.R. (2017). World phosphorus use efficiency in cereal crops. *Agronomy Journal*, 109(4).

Di, H.J., Condron, L.M. and Frossard, E. (1997). Isotope techniques to study phosphorus cycling in agricultural and forest soils: A review. *Biology and Fertility of Soils*, 24, 1–12.

Dobermann, A. (2007). Nutrient use efficiency – Measurement and management. In: *IFA International Workshop on Fertilizer Best Management Practices*. Brussels, Belgium, pp. 1–28.

Dobermann, A. and Cassman, K.G. (2004). Environmental dimensions of fertilizer nitrogen; What can be done to increase nitrogen use efficiency and ensure global food security? In: *SCOPE 65; Agriculture and the Nitrogen Cycle; Assessing the Impacts of Fertilizer Use on Food Production and the Environment*, edited by A.R. Mosier, J.K. Syers and J.R. Freney. Island Press, Washington, DC. doi:10.21013/jte.v8.n3.p1.

Durusoy, A. and Yildirim, M. (2017). Determination of radioactivity concentrations in soil samples and dose assessment for Rize Province, Turkey. *Journal of Radiation Research and Applied Sciences*, 10(4), 348–352.

El-Sherbiny, A.E., Galal, Y.G.M., Soliman, S.M., Dahdouh, S.M., Ismail, M.M. and Fathy, A. (2014). Fertilizer nitrogen balance in soil cultivated with pea (*Pisum sativum* L.) under bio and organic fertilization system using ^{15}N stable isotope. *Proceedings of 4th International Conference of Radiation Research and Applied Sciences*, Taba, Egypt, pp. 85–96.

El-Taher, A. and Makhluf, S. (2010). Natural radioactivity levels in phosphate fertilizer and its environmental implications in Assuit governorate, Upper Egypt. *Indian Journal of Pure & Applied Physics*, 48, 697–702.

Espindula, M.C., Rocha, V.S., Souza, M.A., Grossi, J.A.S. and Souza, L.T. (2010). Doses e formas de aplicação de nitrogênio no desenvolvimento e produção da cultura do trigo. *Ciência e Agrotecnologia*, 34(6), 1404–1411.

Eurostat. (2013). Fertiliser Consumption and Nutrient Balance Statistics. http://epp.eurostat.ec.europa.eu/ (last accessed September 2013).

FAO. (1980). *Maximizing the Efficiency of Fertilizer Use by Grain Crops*. FAO Fertilizer and Plant Nutrition Bulletin no. 3. FAO, Rome, Italy.

FAO. (1983a). *Fertilizer Use under Multiple Cropping Systems*. FAO Fertilizer and Plant Nutrition Bulletin no. 6. FAO, Rome, Italy.

FAO. (1983b). *Micronutrients*. FAO Fertilizer and Plant Nutrition Bulletin no. 7. FAO, Rome, Italy.

FAO. (1984). *Fertilizer and Plant Nutrition Guide*. FAO Fertilizer and Plant Nutrition Bulletin no. 9. Rome, Italy.

FAO. (1985). *Efficient Fertilizer Use in Acid Upland Soils of the Humid Tropics*. FAO Fertilizer and Plant Nutrition Bulletin no. 10. Rome, Italy.

FAO. (2009). FAOSTAT. FAO Statistics Division. http://faostat3.fao.org.

FAO. (2013). FAOSTAT. Food and Agriculture Organization of the United Nations. http://faostat.fao.org.

FAO. (2017). *Summary Report about World Fertilizer Trends and Outlook to 2020*. FAO, Rome, Italy, pp. 38.

Fixen, P., Brentrup, F., Bruulsema, T.W., Garcia, F., Norton, R. and Zingore, Sh. (2015) Nutrient/fertilizer use efficiency: Measurement, current situation and trends. In: *Managing Water and Fertilizer for Sustainable Agricultural Intensification. International Fertilizer Industry Association (IFA), International Water Management Institute (IWMI), International Plant Nutrition Institute (IPNI), and International Potash Institute (IPI)*, edited by P. Drechsel, P. Heffer, H. Magen, R. Mikkelsen, and D. Wichelns. Paris, France, pp. 8–38.

Franzini, V.I., Muraoka, T. and Mendes, F.L. (2009). Ratio and rate effects of ^{32}p-triple superphosphate and phosphate rock mixtures on corn growth. *Scientia Agricola (Piracicaba, Braz.)*, 66(1), 71–76.

Fried, M. (1954). Quantitative evaluation of processed and natural phosphates. *Agriculture Food Chemistry*, 2, 241–244.

Fried, M. (1964). 'E", 'L" and "A" value. In: *Transactions of the 8th International Congress Soil Science*, IV, The Academy of the Socialist Republic of Romania ASIN: B00MNS3WLQ, Bucharest, pp. 29–39.

Fried, M. (1978). Direct quantitative assessment in the field of fertilizer management practices. In: *Proceedings of the 11th International Congress Soil Science*, vol. 3, International Soil Science Society, Edmonton, Canada, pp. 103–129.

Fried, M. and Broeshart, H. (1975). An independent measurement of the amount of nitrogen fixed by a legume crop. *Plant and Soil*, 43, 707–711.

Frossard, E. et al. (2011). The use of tracers to investigate phosphate cycling in soil-plant systems. In: *Phosphorus in Action, Soil Biology*, edited by E.K. Bunemann et al. Springer-Verlag, Berlin, Germany, 26: 59–91.

Frossard, E. and Sinaj, S. (1997). The isotopic exchange technique; A method to describe the availability of inorganic nutrients. Applications to K, PO_4, SO_4, and Zn. *Isotopes in Environmental and Health Studies*, 33, 61–77.

Gaafar, I., El-Shershaby, A., Zeidan, I. and Sayed El-Ahll, L. (2016). Natural radioactivity and radiation hazard assessment of phosphate mining, Quseir-Safaga area, Central Eastern Desert, Egypt. NRIAG *Journal of Astronomy and Geophysics*, 5, 160–172.

Galal, Y.G.M. (2015). *Application of Stable Isotopes in Fertilization Practices.*

Galloway, J.N., Dentener, F.J., Capone, D.G., Boyer, E.W., Howarth, R.W., Seitzinger, S.P., Asner, G.P. et al. (2004). Nitrogen cycles: Past, present, and future. *Biogeochemistry*, 70, 153–226.

Gangolli, S.D., Van Den Brandt, P.A., Feron, V.J., Jan-Zowsky, C., Koeman, J.H., Speijers, G.J.A., Spiegelhalder, B., Walker, R. and Winshnok, J.S. (1994). Assessment nitrate, nitrate and N nitroso compounds. *European Journal of Pharmacology: Environmental Toxicology and Pharmacology*, 292, 1–38.

Ghabour, S., Galal, Y.G.M., Soliman, S.M., El-Sofi, D.M., Morsy, A.A. and El-Sofi, M.M. (2015). Nitrogen distribution in soil profile under sesame, maize and wheat crops as affected by organic and inorganic nitrogen fertilizers using N technique. *International Journal of Plant & Soil Science*, 6(5), 294–302.

Ghosh, D., Deb, A., Bera, S., Sengupta, R. and Patra, K.K. (2008). Measurement of natural radioactivity in chemical fertilizer and agricultural soil: Evidence of high alpha activity. *Environmental Geochemistry and Health*, 30 (1), 79–86.

Goh, K.M. and Adu-Gyamfi, J.J. (2016). Chemical, nuclear and radioisotopic tracer techniques for studying soil phosphorus forms and estimating phosphorus availability, fluxes and balances. In: *Use of Phosphorus Isotopes for Improving Phosphorus Management in Agricultural Systems*. IAEA TECDOC series no. 1805, Vienna, Austria, pp. 21–49.

Guimond, R.J. (1990). Radium in fertilizers. In: *Environmental Behavior of Radium*. Technical Report no. 310. International Atomic Energy Agency (IAEA), Vienna, Austria, pp. 113–128.

Guimond, R.J. and Hardin, J.M. (1989). Radioactivity released from phosphate-containing fertilizers and from gypsum. *Radiation Physics and Chemistry*, 34(2), 309e315.

Gupta, D., Chatterjee, S., Datta, S., Veer, V. and Walther, C. (2014). Role of phosphate fertilizers in heavy metal uptake and detoxification of toxic metals. *Chemosphere*, 108, 134–144.

Hamed, L.M.M. (2013). Nitrogen balance in soil profile as affected by different soil types, soil water regimes, nitrogen rates and application methods using ^{15}N tracer technique. PhD thesis, Department of Agricultural and Environmental Sciences, Production, Landscape, Agro-Energy (DiSAA), Milan, Italy, pp. 132.

Hamed, L., Galal, Y.G.M., Emara, E.I.R. and Soliman, M.A.E. (2019). Optimum applications of nitrogen fertilizer and water regime for wheat (*Triticum aestivum* L.) using ^{15}N tracer technique under Mediterranean environment. *Egyptian Journal of Soil Science*, 59(1), 41–52.

Hanlon, E.A. (2012). Naturally occurring radionuclides in agricultural products. University of Florida. edis. ifas.ufl.edu. Retrieved 17 July 2014.

Hardarson, C., Zapata, F. and Danso, S.K.A. (1984). Field evaluation of symbiotic nitrogen fixation by rhizobial strains using ^{15}N methodology. *Plant and Soil*, 82, 369–375.

Hardarson, G. (1990). *Use of Nuclear Techniques in Soil Plant Relationships*. Training Course Series no. 2, IAEA, Vienna, Austria.

Harmsen, K. and Moraghan, J.T. (1988). A comparison of the isotope recovery and difference methods for determining nitrogen fertilizer efficiency. *Plant and Soil*, 105, 55–67.

Hassan, N.M., Mansour, N.A., Fayez-Hassan, M. and Sedqy, E. (2016). Assessment of natural radioactivity in fertilizers and phosphate ores in Egypt. *Journal of Taibah University for Science*, 10, 296–306.

Hassan, N.M., Chang, B.-U. and Tokonami, S. (2017). Comparison of natural radioactivity of commonly used fertilizer materials in Egypt and Japan. *Journal of Chemistry*, 2017, article ID 9182768. doi:10.1155/2017/9182768.

Hauck, R.D. and Bremner, J.M. (1976). Use of tracers for soil and fertiliser nitrogen research. *Advances in Agronomy*, 28, 219–266.

HCSO. (2013). Hungarian Central Statistical Office. Központi Statisztikai Hivatal, KSH. http://www.ksh.hu.

Heffer, P. (2007). *Medium-Term Outlook for World Agriculture and Fertilizer Demand 2006/07–2011/12*. International Fertilizer Industry Association, Paris, France.

Heffer, P. and Prud'homme, M. (2006). *Medium-Term Outlook for Global Fertilizer Demand, Supply and Trade – 2006–2010*. Summary Report. IFA, Paris, France.

Hejcmana, M., Kunzová, E. and Srek, P. (2012). Sustainability of winter wheat production over 50 years of crop rotation and N, P and K fertilizer application on illimerized luvisol in the Czech Republic. *Field Crops Research*, 139, 30–38.

Hera, C. (1995). Contribution of nuclear techniques to the assessment of nutrient availability for crops. In: *Integrated Plant Nutrition Systems*. FAO Fertilizer and Plant Nutrition Bulletin no. 12. FAO, Rome, Italy, pp. 307–331.

Hussain, R.O. and Hussain, H.H. (2011). Investigation the natural radioactivity in local and imported chemical fertilizers. *Brazilian Archives of Biology and Technology*, 54(4), 777–782.

Hussein, E.M. (1994). Radioactivity of phosphate ore, superphosphate, and phosphogypsum in Abu-zaabal phosphate. *Health Physics*, 67(3), 280–282.

IAEA (International Atomic Energy Agency). (1973). *Safe Handling of Radionuclides*. Safety Series no. 1. International Atomic Energy Agency, Vienna, Austria.

IAEA (International Atomic Energy Agency). (1976). *Tracer Manual on Crops and Soils*. Technical Report Series no. 171. International Atomic Energy Agency, Vienna, Austria.

IAEA (International Atomic Energy Agency). (1980). *Soil Nitrogen as Fertilizer or Pollutant*. Panel Proceedings Series. International Atomic Energy Agency, Vienna, Austria.

IAEA (International Atomic Energy Agency). (1981). *Isotopes in Day to Day Life*. International Atomic Energy Agency, Vienna, Austria, 64.

IAEA (International Atomic Energy Agency). (1983). *A Guide to the Use of Nitrogen-15 and Radioisotopes in Studies of Plant Nutrition; Calculations and Interpretation of Data*. IAEA TECDOC-288. International Atomic Energy Agency, Vienna, Austria.

IAEA (International Atomic Energy Agency). (2001). *Use of Isotope and Radiation Methods in Soil and Water Management and Crop Nutrition*. Manual Training Course Series no. 14. International Atomic Energy Agency, Vienna, Austria.

IAEA (International Atomic Energy Agency). (2011). *Nuclear Technology Review 2011*. International Atomic Energy Agency, Vienna, Austria, pp. 40–41.

IAEA (International Atomic Energy Agency). (2016). *Use of Phosphorus Isotopes for Improving Phosphorus Management in Agricultural Systems/International Atomic Energy Agency*. IAEA TECDOC series no. 1805. International Atomic Energy Agency, Vienna, Austria, p. 96.

Jankovic, M., Nikolic, J., Pantelic, G., Rajacic, M., Sarap, N. and Todorovic, D. (2013). Radioactivity in chemical fertilizers. In: *Proceedings of 9th Symposium of the Croatian Radiation Protection Association*, p. 578. Krk (Croatia). 10–12 April 2013. ISBN 978-953-96133-8-7.

José, M.A., Rafael, G.T. and Guillermo, M. (2009). Extensive radioactive characterization of a phosphogypsum stack in SW Spain: ^{226}Ra, ^{238}U, ^{210}Po concentrations and ^{222}Rn exhalation rate. *Journal of Hazardous Materials*, 164, 790–797.

Khan, K., Orfi, S. and Khan, H.M. (2004). Distribution of primordial radionuclides in phosphate fertilizers and rock deposits of Pakistan. *Geological Bulletin, University of Peshawar*, 37, 59–64.

Khater, A.E.M. (2008). Uranium and heavy metals in phosphate fertilizers. www.radioecology.info. Archived from the original (PDF) on 24 July 2014. Retrieved 17 July 2014.

Khater, A.E.M., Higgy, R.H. and Pimpl, M. (2001). Radiological impacts of natural radioactivity in Abu-Tartor phosphate deposits. *Journal of Environmental Radioactivity*, 55, 255–267.

L'Annunziata, M.F. (Ed.) (2003). *Handbook of Radioactivity Analysis*. 2nd ed. Academic Press, San Diego, CA, pp. 1273.

L'Annunziata, M.F. (Ed.) (2016). *Radioactivity: Introduction and History, from the Quantum to Quarks*. 2nd ed. Elsevier, Amsterdam, the Netherlands.

Ladha, J.K., Pathak, H., Krupnick, T.J., Six, J. and van Kessel, C. (2005). Efficiency of fertilizer nitrogen in cereal production: Retrospects and prospects. *Advances in Agronomy*, 87, 85–156.

Larsen, S. (1952). The use of ^{32}P in studies on the uptake of phosphorus by plants. *Plant and Soil*, 4, 1–10.

Lips, S.H., Leidi, E.O., Siberbush, M., Soares, M.I.M. and Lewis, O.E.M. (1990). Physiological aspects of ammonium and nitrate fertilization. *Journal of Plant Nutrition*, 13, 1271–1289.

Makweba, M.M. and Holm, E. (1993). The natural radioactivity of the rock phosphates, phosphatic products and their environmental implications. *Science of the Total Environment*, 133, 99–110.

Mattingly, G.E.G., and Widdowson, F.V. (1958). Uptake f phosphorus from ^{32}P labelled superphosphate by field crops. Part 1. Effects of simultaneous application f non-radioactive phosphorus fertilizers. *Plant Soil*, 9, 286–304.

Mazzilli, B., Palmiro, V., Saueia, C. and Nisti, M. B. (2000). Radiochemical characterization of Brazilian phosphogypsum. *Journal of Environmental Radioactivity*, 49, 113–122.

McAuliffe, C.D., Hall, N.S. and Dean, L.A. (1947). Exchange reactions between phosphates and soils: Hydroxylic surfaces of soil minerals. *Soil Science Society of America, Proceedings*, 12, 119–123.

Menzel, R.G. and Smith, S.J. (1984). Soil fertility and plant nutrition. In: *Isotopes and Radiation in Agricultural Sciences, Soil-Plant-Water Relationships*, vol. 1, edited by M.F. L'Annunziata, J.O. Legg. Academic Press, London, UK, pp. 1–34.

Millennium Ecosystem Assessment. (2005). *Ecosystems and Human Well-Being: Synthesis*. Island Press, Washington DC.

Mir, F. and Rather, S. (2015). Measurement of radioactive nuclides present in soil samples of district Ganderbal of Kashmir Province for radiation safety purposes. *Journal of Radiation Research and Applied Sciences*, 8, 155–159.

Mortvedt, J.J. and Beaton, J.D. (2014). Heavy metal and radionuclide contaminants in phosphate fertilizers. Archived from the original on 26 July 2014. Retrieved 16 July 2014.

NCRP. (1987). *Radiation Exposure of the U.S. Population from Consumer Products and Miscellaneous Sources*. Supersedes NCRP Report No. 56, National Council on Radiation Protection and Measurements, pp. 29–32. Retrieved 17 July 2014. ISBN 0-913392-94-4.

Newbould, P. (1989). The use of nitrogen fertilizer in agriculture. Where do we go practically and ecologically? *Plant Soil*, 115, 297–311.

Nguyen, L., Zapata, F. and Adu-Gyamfi, J. (2017). The use of phosphorus radioisotopes to investigate soil phosphorus dynamics and cycling in soil–plant systems. In: *Soil Phosphorus*, edited by R. Lal and B.A. Stewart. Advances in Soil Science. CRC Press, Taylor & Francis Group, Boca Raton, FL, pp. 285–312.

Ogunleye, P.O., Mayaki, M.C. and Amapu, I.Y. (2002). Radioactivity and heavy metal composition of Nigerian phosphate rocks: Possible environmental implications. *Journal of Environmental Radioactivity*, 62, 39–48.

Olszewska-Wasiolek, M. (1995). Estimates of the occupational radiological hazards in phosphate fertilizers industry in Poland. *Radiation Protection Dosimetry*, 58, 269–276.

Penfold, J., Degrange, J.P., Mobbs, S. and Schneider, T. (1997). *Establishment of Reference Levels for Regulatory Control of Workplaces Where Materials Are Processed Which Contain Enhanced Levels of Naturally Occurring Radionuclides*. NRPB report on contract number 95-ET-009. Chilton, Didcot, UK.

Pierzynski, G.M., Sims, J.T. and Vance, G.F. (1994). *Soils and Environmental Quality*, Lewis Publications, Boca Raton, FL, 313.

Pourjafar, M. (2017). Radioisotopes in agricultural industry. *IRA International Journal of Technology & Engineering*, 8(3), 39–43.

Powlson, D.S., Hart, P.B.B.B., Poulton, P.R., Johnston, A.E. and Jenkinson, D.S. (1992). Influence of soil type crop management and weather on the recovery of ^{15}N-labelled fertilizer applied to winter wheat in spring. *Journal of Agricultural Science*, 118–100.

Prasad, R. and Power, J.F. (1995). Nitrification inhibitors for agriculture, health and the environment. *Advances in Agriculture*, 54, 234–281.

Rennie, D.A. (1969). The significance of the A value concept in field fertilizer studies. In: *Monitoring of Radioactive Contamination on Surfaces*, IAEA (ed.) Technical Report Series no. 120. Vienna, Austria, pp. 132–145.

Robertsa, L. and Johnston, A.E. (2015). Phosphorus use efficiency and management in agriculture. *Resources, Conservation and Recycling*, 105, 275–281.

Roy, R.N., Finck, A., Blair, G.J. and Tandon, H.L.S. (2006). *Plant Nutrition for Food Security: A Guide to Integrated Nutrient Management*. FAO Fertilizer and Plant Nutrition Bulletin no. 16. FAO, Rome, Italy.

Ryan, J. (2008) A perspective on balanced fertilization in the Mediterranean region. *Turkish Journal of Agriculture and Forestry*, 32, 79–89.

Sahu, S.K., Ajmal, P.Y., Bhangare, R.C., Tiwari, M. and Pandit, G.G. (2014). Natural radioactivity assessment of a phosphate fertilizer plant area. *Journal of Radiation Research and Applied Sciences*, 7, 123–128.

Sam, A.K., Ahmad, M.M.O., El-Khngi, F.A., El-Nigumi, Y.O. and Holm, E. (1999). Radiological and assessment of Uro and Kurun rock phosphates. *Journal of Environmental Radioactivity*, 24, 65–75.

Santamaria, P. and Elia, A. (1997). Producing nitrate-free endive heads; effect of nitrogen form on growth, yield and ion composition of endive. *Journal of the American Society for Horticultural Science*, 122, 140–145.

Santamaria, P., Elia, A., Gonnella, M. and Serio, F. (1997). Ways of reducing nitrate content in hydroponically grown leafy vegetables. In: *Proceedings of the 9th International Congress on Soilless Culture*, pp. 437–451.

Scheffer, F. and Schachtschabel, P. (1998). *Lehrbuch der Bodenkunde*. 14th ed. Enke Verlag, Stuttgart, Germany.

Schepers, J.S. and Marter, D.L. (1986). Public perception of ground-water quality and producers dilemma. In: *Agricultural Impacts on Groundwater*. National Water Well Association, Westerville, OH; Water Well Journal Publishing Company, Dublin, OH, pp. 399–411.

Scholten, L.C. (1996). *Approaches for Regulating Management of Large Volumes of Waste Containing Natural Radionuclides in Enhanced Concentrations*. Publications Office of the European Union, Luxembourg.

Shahul Hameed, P., Sankaran Pillai, G., and Mathiyarasu, R. (2014). A study on the impact of phosphate fertilizers on the radioactivity profile of cultivated soils in Srirangam (Tamil Nadu, India). *Journal of Radiation Research and Applied Sciences*, 7, 463–471.

Sharpley, A.N. and Menzel, R.G. (1987). The impact of soil and fertilizer phosphorus on the environment. *Advances in Agronomy*, 41, 297–324. doi:10.1016/s0065–2113(08)60807-x.

Shaviv, A. (1988). Control of nitrification rate by increasing ammonium concentration. *Fertilizer Research*, 17, 177–188.

Shaviv, A. and Mikkelsen, R.L. (1993). Controlled-release fertilizers to increase efficiency of nutrient use and minimize environmental degradation review. *Fertilizer Research*, 35, 1–12.

Simili, F.F., Reis, R.A., Furlan, B.N., Paz, C.C.P., Lima, M.L.P. and Bellingieri, P.A. (2008). Resposta do híbrido de sorgo-sudão à adubação nitrogenada e potássica; composição química e digestibilidade in vitro da matéria orgânica. *Ciência e Agrotecnologia*, 32(2), 474–480.

Smith, S.J. and Legg, J.O. (1971). Reflections on the A value concept of soil nutrient availability. *Soil Science*, 112, 373–375.

Smith, S.J., Schepers, J.S. and Porter, L.K. (1990). Assessing and managing agricultural nitrogen losses to the environment. *Advances in Soil Science*, 14, 1–43.

Soliman, S.M. and Galal, Y.G.M. (2017). Contribution of nuclear science in agriculture sustainability. *Arab Journal of Nuclear Science and Applications*, 50(2), 37–48.

Soliman, S.M., Ghabour, S., Galal, Y.G.M., El-Sofi, D.M., Moursy, A.A. and El-Sofi, M.M. (2014). Alternative strategies for improving nitrogen nutrition of some economical crops using N stable isotope. *International Journal of Current Microbiology and Applied Science*, 3(7), 970–983.

Tamburini, F. et al. (2010). A method for the analysis of the $\delta^{18}O$ of inorganic phosphate extracted from soils with HCl. *European Journal of Soil Science*, 61(6), 1025–1032.

Tóth, G., Guicharnaud, R.-A., Tóth, B. and Hermann, T. (2014). Phosphorus levels in croplands of the European Union with implications for P fertilizer use. *European Journal of Agronomy*, 55, 42–52.

Toussaint, B. (2000). Groundwater monitoring in Germany – An overview. IHP/OHP-Jahrbuch der BR Deutschland, 1991–*1995*. Koblenz, Teil I Sammelband, pp. 73–77.

Tufail, M., Akhtar, N. and Waqas, M. (2006). Rock phosphate the feed stock of phosphate fertilizers used in Pakistan. *Health Physics*, 90(4), 361–367.

UNSCEAR. (1993). *Ionizing Radiation: Sources and Biological Effects*. United Nations Scientific Committee on the Effects of Atomic Radiation Report. UNSCEAR, New York.

UNSCEAR. (2008). *Sources and Effects of Ionizing Radiation*. Report to the General Assembly, with Scientific Annexes. UNSCEAR, New York.

Uosif, M., Mostafa, A., Elsaman, R. and Moustafa, E. (2014). Natural radioactivity levels and radiological hazards indices of chemical fertilizers commonly used in Upper Egypt. *Journal of Radiation Research and Applied Sciences*, 7, 430–437.

US EPA. (2016). TENORM: Fertilizer and fertilizer production wastes. Retrieved 30 August 2017.

Valkama, E., Uusitalo, R., Ylivainio, H., Virkajarvi, P. and Turtol, E. (2009). Phosphorus fertilization: A meta-analysis of 80 years of research in Finland. *Agriculture, Ecosystems and Environment*, 130, 75–85.

Van Cleemput, O. and Boeckx, P. (2005). Alteration of nitrogen cycling by agricultural activities, and its environmental and health consequences. *Gayana Botanica*, 62(2), 98–109.

Vose, P.B. (1980). *Introduction to Nuclear Techniques in Agronomy and Plant Biology*. Pergamon Press, Oxford.

Wagner, G.H. and Zapata, F. (1982). Field evaluation of reference crops in the study of nitrogen fixation by legumes using isotope techniques. *Agronomy Journal*, 74, 607–612.

Walker, R. (1990). Nitrates, nitrites and N-nitrosocompounds: A review of the occurrence in food and diet and the toxicological implications. *Food Additives & Contaminants*, 6, 717–768.

Witty, J.F. and Day, J.M. (1978). Use of $^{15}N_2$ in the evaluating a symbiotic N_2 fixation. In: *Isotopes in Biological Dinitrogen Fixation*. IAEA, Vienna, Austria.

Wood, S., Henao, J. and Rosegrant, M. (2004). The role of nitrogen in sustaining food production and estimating future nitrogen fertilizer needs to meet food demand. In: *Agriculture and the Nitrogen Cycle; Assessing the Impacts of Fertilizer Use on Food Production and the Environment*, edited by A.R. Mosier, J.K. Syers, and J.R. Freney. SCOPE 65, Paris, France, pp. 245–260.

Zaharah, A.R. and Zapata, F. (2003). The use of P isotope techniques to study soil P dynamics and to evaluate the agronomic effectiveness of phosphate fertilizers. In: *Proceedings of the International Symposium Direct Application of Phosphate Rock and Related Appropriate Technology*, edited by S.S.S. Rajan and S.H. Chien. IFDC Publication G-1, Muscle Shoals, AL, pp. 225–235.

Zapata, F. (1990). Isotope techniques in soil fertility and plant nutrition studies. In: *Use of Nuclear Techniques in Studies of Soil-Plant Relationships*, edited by G. Hardarson. IAEA Training Course Series no. 2. IAEA, Vienna, Austria, pp. 109–127.

Zapata, F., Danso, S.K.A., Hardarson, G. and Fried, M. (1987). Time course of nitrogen fixation in field grown soybean using N-15 methodology. *Agronomy Journal*, 79, 172–176.

Zapata, F. and Hera, C. (1995). Enhancing nutrient management through use of isotopic techniques. In: *Proceedings of the International Symposium Nuclear and Related Techniques in Soil/Plant Studies for Sustainable Agriculture and Environmental Preservation*, October 1994, Vienna. IAEA bSTI/PUB/947, Vienna, Austria, pp. 83–105.

10 Managing Fertilizers in Soils of Paddy Rice

Pauline Chivenge, Michelle Anne Bunquin,
Kazuki Saito, and Sheetal Sharma

CONTENTS

10.1 INTRODUCTION

Rice (*Oryza* spp.) is a grass that evolved from a semiaquatic ancestor and is native to Asia and Africa, whose cultivation has been in existence since the beginning of organized agriculture, more than 3500 years ago (Sweeney and McCouch 2007). Rice, wheat (*Triticum aestivum*), and maize (*Zea mays*) constitute the main cereal crops grown and consumed globally. However, rice is the staple food in many low- and middle-income countries, constituting about half the global population, providing about 25% of the global human energy consumption per capita and about 16% of protein (GRiSP 2013). Global paddy rice production has been increasing over time from the 1960s with a worldwide production of about 770 million tons in 2017 (Figure 10.1) (FAOSTAT 2019). The global rice production trends are closely influenced by the production trends in Asia, where about 90% of the rice is produced, with the Americas and Africa producing 4.5% each on average. Since most of the rice is produced in Asia, it follows that most of it is grown on smallholder farms. The greater proportion of the rice produced in Asia is also consumed in Asia, with the remaining being exported to other regions. China is the leading rice producer globally, followed by India, accounting for 28% and 22%, respectively, of the total world production in 2017 (FAOSTAT 2019). However, India has the largest harvested area, contributing 26% of the global rice area compared to 18% in China.

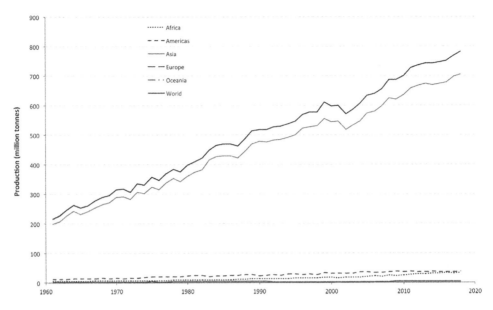

FIGURE 10.1   Global rice production trends. (From FAOSTAT, http://www.fao.org/faostat/en/-data, 2019.)

The global demand for rice has been increasing with increasing population, but with greater proportional demand increasing in Africa, where around 50% is imported from other regions.

Rice is grown on 167 million hectares under varying soil and climatic conditions from temperate to tropical, across latitudes, and in a wide range of production ecosystems across 118 countries (GRiSP 2013). Rice yields vary widely across geographical regions, countries, seasons, and ecosystems with global average paddy yields of 4.6 t ha^{-1}, while in Asia it is 4.8 t ha^{-1}, but Africa lags behind at 2.4 t ha^{-1} (FAOSTAT 2019). These yields are below the genetic and climatic yield potentials, representing about 65% of potential yields for Asia (Dobermann et al. 2002; Peng et al. 2010), and thus there is a huge exploitable yield gap, which is even larger for Africa (van Ittersum et al. 2016; Van Oort et al. 2017). Nutrient management has a huge role for closing the rice yield gap in Asia and Africa (Tsujimoto et al. 2019), where nutrient use efficiency has remained low (Haefele et al. 2013; Saito et al. 2013). While total rice production is lowest in Oceania and Europe (Figure 10.1), rice yields are highest in these regions, with average yields of 9.4 and 6.3 t ha^{-1}, respectively (FAOSTAT 2019). Efficient management of nutrients and other resources have contributed to the higher yields in these regions. Consequently, any significant advancements in closing the rice yield gap will require concerted efforts in improving nutrient management, particularly in Asia, where modern high-yielding varieties that are responsive to nutrient addition have been widely adopted (Saito et al. 2018).

This chapter starts with a description of rice production systems and soil characteristics in lowland rice in comparison with upland soils. Next, innovations to increase rice productivity through external nutrient inputs are provided and historical and current use of fertilizer in lowland are described, including site-specific nutrient management (SSNM) as an essential approach for improving productivity and nutrient use efficiency for rice. Decision support tools and the challenges and opportunities for managing soil fertility and adapting crop management in paddy soils are discussed.

10.2   VARIABILITY OF RICE CROPPING ECOSYSTEMS AND THEIR SOILS

There are three main rice-growing ecosystems, classified as irrigated lowland, rainfed lowland, and rainfed upland. About 75% of the world's rice is produced from irrigated ecosystems, while 20% is produced on rainfed lowland ecosystems (GRiSP 2013). Rainfed upland rice ecosystems are highly heterogeneous and prone to multiple stresses such as weeds, pests, and diseases, in addition to drought. Since most of the rice, about 95%, is produced in irrigated and rainfed lowland environments, this chapter largely focuses on these two ecosystems, which are generally regarded as paddy rice ecosystems (Kögel-Knabner et al. 2010). Together with improved rice varieties and irrigation, fertilizer inputs have contributed to great yield increases in lowland rice throughout the world (Saito et al. 2018). Furthermore, most of future additional production will come from lowland rice to meet increasing demand. Paddy rice is grown on bunded fields, creating semiaquatic environments where irrigation ensures production of one, two, or three crops per year, while rainfed lowland environments are flooded with rainwater kept within bunds. Paddy rice is primarily puddled and transplanted, although wet- and dry-direct seeding is on the increase.

In lowland rice, which includes rainfed lowland, irrigated lowland, deep-water, and mangrove swamp (Saito et al. 2013), fields are usually flooded during part or all of the growing season. Upland rice, on the other hand, is generally grown on level or sloping, unbunded fields and without flooding. Upland rice is grown under crop rotation systems with other crops or under slash-and-burn systems. Rice yields are generally higher in irrigated ecosystems than in rainfed lowland and upland ecosystems where rice plants could be affected by drought and/or excessive water (Tanaka et al. 2017). Low yields and yield growth in Africa are partially due to the fact that the share of lowland rice, especially irrigated lowland, is smaller than in the other regions, although there is huge potential for irrigation expansion (Saito et al. 2013). While continuous rice monocropping systems in irrigated lowland systems with appropriate fertilizer management practices could maintain rice yields (Dobermann et al. 2000; Bado et al. 2010), upland rice yields in continuous rice monocropping are rapidly declining (Peng et al. 2006a; Saito et al. 2006).

Within irrigated lowlands, higher rice yields are obtained in high-latitude areas that have long day length and where intensive farming techniques are practiced (e.g., southwestern Australia and northern California), or in low-latitude desert areas that have very high solar energy (e.g., Egypt). In the tropics, paddy rice is grown in the river deltas where water saturation can be maintained between bunds, typically in areas that have distinct dry and wet seasons. However, rainfed lowland ecosystems vary widely, with some being more favorable than others, depending on rainfall distribution, soil texture, landscape position, among other factors. In most rainfed lowland ecosystems, one rice crop is grown in a year, in the wet season, followed by an upland crop, usually a pulse, or is left fallow. In irrigated ecosystems, two or three rice crops can be produced in a year, as continuous rice monocropping or in rotation with upland crops, typically wheat or maize, which are usually grown during the dry season. In lowland, rice fields are primarily puddled and rice plants are transplanted, although wet- and dry-direct seeding is on the increase. For irrigated rice in the tropics, rice yields are generally higher in the dry season than in the wet seasons. This is due to cloud cover in the wet season, which reduces solar radiation and temperature (Peng et al. 2004; van Oort et al. 2015), with increased occurrence of pests and diseases (Flinn and De Datta 1984).

As rice is cultivated in a wide range of conditions, it is grown on a wide range of soils with huge differences in soil quality. Numerous studies have characterized paddy soils in Asia and Africa (Kawaguchi and Kyuma 1977; Hirose and Wakatsuki 2002; Saito et al. 2013). Recently, Haefele et al. (2014) assessed soil quality in global rice production areas using the fertility capability soil classification system. They computed the distribution of 20 soil constraints, and used these to categorize soils as "good," "poor," "very poor," or "problem soil" for rice production. Rice growth on "problem soils" is likely to be limited by salinity; very low pH; P deficiency; Fe/S/Al toxicity; nutrient deficiencies of N, Zn, K, P, Cu, and Mo; or high pH causing P, Fe, and Zn deficiency. "Very poor" soils are highly weathered with very limited indigenous nutrient supplies, low nutrient retention capacity, frequent and often severe P deficiency, acidic to very acidic soil reaction, and Fe/Al toxicity. "Poor" soils include those with only minor constraints and/or limited soil fertility, whereas "good" soils are much less weathered, have considerably higher natural soil fertility than poor and very poor soils, and soil constraints are minor or absent. The results showed that rice production on "good," "poor," "very poor," and "problem" soils accounted for 44%, 18%, 33%, and 5% of the total rice area, respectively. Asia had the largest percentage of rice on good soils (47%), whereas rice production on good soils accounted for only 18% in Africa. Also, at the global level, irrigated soils had higher and lower share of good and very poor soils, respectively, than rainfed lowland soils, whereas there was no large difference in the share of poor and problem soils between the two systems. The most common soil chemical problems in rice fields are very low inherent nutrient status, very low pH, high P fixation, and widespread soil physical problems. Especially severe in rainfed production systems are very shallow soils and low water-holding capacity. Furthermore, Van Oort (2018) mapped abiotic stresses relevant for rice in Africa, including two soil constraints (iron toxicity and salinity/sodicity), and reported the countries most affected and total potentially affected. The results showed that 12% and 2% of rice area were potentially affected by iron toxicity and salinity/sodicity, respectively.

10.3   PADDY RICE SOILS – HOW THEY DIFFER FROM UPLAND SOILS

Paddy rice cultivation is highly dependent on soil moisture conditions, with most of the paddy soils being wetlands, inundated by water part of the major growing season or continuously. Paddy soils are considered as manmade wetlands found in low-lying areas (Yoon 2009; Kögel-Knabner et al. 2010), and are usually heavy textured with a reduced soil horizon showing signs of permanent or intermittent waterlogging, such as segregations of Fe and Mn in the soil matrix (Kirk 2004). These soils were created by years of manipulation of the soil conditions through the creation of artificial submergence through puddling, i.e., tillage of wet soil after flooding for a few days, followed by levelling. Traditionally, puddling was done using ploughs pulled by buffalos, with wet tillage breaking down the soil aggregates, removing air from soil pores, and creating a thick mud layer that reduced water percolation into deeper soil layers. This, together with the mechanical pressure exerted during ploughing by buffalos or tractors, which are now used more widely, created a plough pan at about 15–25 cm depth that had a high bulk density and further limited water percolation, promoting water saturation, and thus maintaining submerged conditions. Repeated puddling over time led to the creation of these manmade wetlands, which are classified Hydragic Anthrosols (Figure 10.2) (IUSS Working Group 2006). They form from a wide range of parent material where the plough pan is part of an anthraquic horizon (e.g., Apg and Ardp in Figure 10.2) that is reduced and has stagnic conditions with distinct mottling (Kögel-Knabner et al. 2010), masking the original character of the soil (Kirk 2004). These hydrological characteristics in paddy rice ecosystems make them vastly different from upland agroecosystems. The chemical species of most nutrients are in their reduced condition in lowland soils, while they are oxidized in upland soils.

Land preparation in lowland rice soils through puddling collapses soil pores and decreases the volume of soil air as water replaces oxygen, decreasing the soil strength (Sharma and De Datta 1985). Flooding predominantly produces anaerobic soil conditions, with fluctuating reduced and oxidized environments depending on the water level, presence of facultative, and obligate microorganisms and organic matter content (Fageria et al. 2011). The saturated conditions deplete the supply of atmospheric oxygen into the soil, enabling various facultative and obligate anaerobes to use oxidized compounds as electron acceptors for respiration, thus converting them to reduced forms (Pezeshki and DeLaune 2012). Within a few days of submergence, a thin yellowish or reddish brown soil layer develops on the surface, forming the aerobic or oxidized soil layer as oxygen diffuses slowly through the floodwater (Reddy et al. 1984). Figure 10.2 shows an anaerobic or reduced layer that can be found underneath the aerobic layer with limited oxygen penetration, unlike in upland soils where the entire soil profile is largely aerobic.

Oxidation-reduction (redox) reactions commonly occur in lowland soils since there are components that serve as electron donors and acceptors, creating a series of changes that have important implications for fertility and its management in lowland rice (Sahrawat 2015). The redox reactions occurring in paddy soils regulate the availability of nutrients for uptake by rice, depending on soil pH, the source of nutrients, and the timing of their application. The major electron donors in upland soils are decomposing plant litter

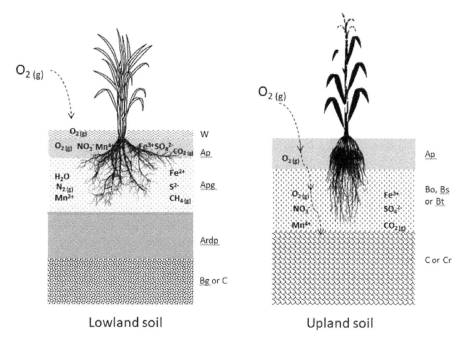

FIGURE 10.2 Typical soil profiles for lowland and upland soils, showing the nutrient species found in each of the soils and their oxidation states as influenced by redox. In lowland soils the characteristic horizons are: W: standing water; Ap: the aerobic puddled layer; Apg: the upper part of an anthraquic puddled horizon, showing a reduced matrix and some oxidized root channels; Ardp: the lower part of the anthraquic horizon characterized with high bulk density and a plough pan; and Bg or C: a hydragic horizon that usually depicts redoximorphic features, in some instances with iron and manganese concretions (Jahn et al. 2006). Upland soils are more variable and horizons can be as follows: Ap, plough layer; Bo, Bs or Bt, the accumulation layer of residual and illuvial iron and clays; and C or Cr, the parent material or unconsolidated bedrock.

and soil organic matter, releasing carbon dioxide, water, and energy for microorganisms in the presence of oxygen as the primary electron acceptor (Ponnamperuma 1972; Sahrawat 2015). In submerged soils, where dissolved oxygen is depleted upon flooding, resulting in low or negative redox potentials, secondary electron acceptors such as sulfates, nitrates, Fe^{3+}, and Mn^{4+} are utilized instead of oxygen (Liesack et al. 2000). Several studies have adopted a microcosm study protocol by Patrick (1966) to evaluate redox-related transformations of N and other micronutrients in flooded soils at controlled conditions in microcosms (Bunquin et al. 2017; Yu and Rinklebe 2011). As oxygen is depleted from the soil, NO_3^- is utilized by facultative anaerobes and is rapidly reduced to N_2. Patrick and Jugsujinda (1992) describe the approximate redox potentials at which oxidized species become unstable and get reduced.

10.3.1 Soil Nutrient Supplying Capacity and Nutrient Dynamics

Rice yield obtained from a particular field depends on the quantities of nutrients that are taken up by the plant during the growth cycle, either from the soil's indigenous nutrients (natural reserves) or from external inputs (Haefele et al. 2013). Indigenous or natural nutrient supply traditionally played a key role in securing the sustainability of irrigated rice systems, but their contribution has decreased since the beginning of the Green Revolution (Kirk and Olk 2000). Since the Green Revolution, there has been decreased sedimentation in many areas, removal or burning of straw, reduced use of green manure in intensive rice cropping systems, and short fallow periods (Cassman et al. 1996), affecting the overall nutrient cycling and soil organic matter content. In aerobic soil, microbial decomposition of organic matter is facilitated by a wide range of microorganisms including fungi, heterotrophic bacteria, and actinomycetes, because there is enough oxygen available from the atmosphere to supply their needs, releasing CO_2 and a high amount of energy. This is substantially different in a submerged soil, where the breakdown of soil organic matter is much slower and is only restricted to bacterial microflora, due to a slow oxygen renewal rate through the floodwater, and is usually depleted within a day or so after flooding, leading to lower gross N mineralization and SOM buildup (Patrick and Mahapatra 1968; Buresh et al. 2008).

Nitrogen is generally the most limiting nutrient in most rice soils (Wade et al. 1999; Dobermann et al. 2003a; Saito et al. 2019). On-farm experiments conducted on irrigated rice production systems in five Asian countries to assess indigenous macronutrient (N, P, and K) supply showed that on average, N was the most limiting nutrient (3.9 t ha^{-1}), followed by K (5.1 t ha^{-1}) and P (5.2 t ha^{-1}) (Dobermann et al. 2003a). Wade et al. (1999) conducted fertilizer trials on rainfed lowland rice production systems in five Asian countries and observed a 0.4 t ha^{-1} difference in rice yield on average between unfertilized and PK-fertilized plots, whereas yields for NPK-fertilized plots were 1.5 t ha^{-1} greater than PK-fertilized plots. In Africa, diagnostic trials were conducted on 30 sites in 17 countries across three major rice production systems. Rice yields without N, P, and K were 68%, 84%, and 89%, respectively, of yields in the NPK treatment, respectively (Saito et al. 2019). Site mean yields were significantly lower without N, P, and K in 93%, 60%, and 50%, respectively, of sites as compared to the NPK treatment. These three studies clearly suggest that N is the most limiting of the three macronutrients.

10.3.1.1 Nitrogen

The indigenous N supply lowland rice soil is usually insufficient to support high yields, yet it is the most abundant element in the atmosphere (Dobermann and Fairhurst 2000). Indigenous N supply includes that coming from the soil, mainly soil organic matter, atmospheric deposition as dust and rainfall, irrigation water, floodwater and sediments, and biological N fixation. Oxygen depleted paddy soils provide favorable environments for biological N fixation, which can sustain rice yields at low levels even without the addition of N fertilizers. Roger and Ladha (1992) estimated that biological N fixation contributes on average 30 kg N ha^{-1} season^{-1} when no fertilizer N is added to the soil. The fixed N contributes to the organic N pool in the soil.

Soil organic matter, which is derived from plant residues and manures, plays an important role in the supply of N to rice. Soils with high organic matter content tend to have greater N supply, requiring less additions from external sources to meet the demands of a rice crop compared to soils with low organic matter content. At an intermediate level, redox potentials are referred to as "healthy reducing conditions" that can be maintained by Fe and Mn dynamics in paddy soils (Patrick and Mahapatra 1968), and N mineralization is promoted. As organic matter decomposes, largely mediated by microbes with close interactions between C and N cycles, organic N is mineralized into NH_4^+, which is the dominant N form in anaerobic soils because in the absence of oxygen it is not nitrified to form nitrates, as happens in aerobic soils (Figure 10.3) (Buresh et al. 2008). The NH_4^+ is taken up by rice plants, fixed on clay minerals, immobilized by soil microbes, or subject to volatilization, depending on soil conditions.

While soil organic matter mineralization is lower in paddy soils than in aerobic soils, the rate of immobilization is also slower, with a higher net mineralization occurring in paddy soils than in aerobic soils. Early reviews have shown that N mineralization in paddy soils was considerably higher than mineralization under well-drained conditions (Patrick and Mahapatra 1968) and incubation experiments suggested that yield and N uptake were greater for lowland than in upland conditions (Yoneyama et al. 1977). Being a critical process in the soil–plant system, the extent of N mineralization is determined by several factors, such as quality and management of crop residue, environmental factors, soil type, and soil and fertilizer N (Singh et al. 2005; Sahrawat 2010). The rate of decomposition of organic material and the mineralization–immobilization behavior is influenced by the biochemical composition. High-quality crop residues with high N concentrations, low cellulose and lignin content, and low C:N and lignin:N ratios, such as legumes, tend to have

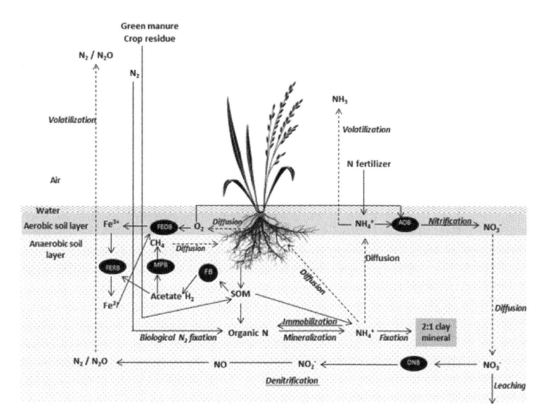

FIGURE 10.3 Schematic presentation of the interactions between the oxygen-releasing root and aerobic and anaerobic microbial processes involved in C- and N-cycling as well as in methane emission from flooded soils and sediments. Solid arrows indicate microbial processes, while dotted arrows depict diffusion or volatilization processes. Organisms are shown as circles, namely: FB: fermenting bacteria; MPB: methane-producing bacteria; FERB: iron-reducing bacteria; MOB: methane-oxidizing bacteria; FEOB: iron-oxidizing bacteria; AOB: ammonium-oxidizing bacteria; DNB: denitrifying bacteria.

high N mineralization (Chen et al. 2014). In addition, the placement of crop residues depends on tillage systems where surface application works under zero tillage practice as dominated by fungal-based food webs with slower decomposition and higher nutrient retention, while incorporation supports conventional tillage as characterized by bacterial-based food webs with faster decomposition and mineralization (Singh et al. 2005). Temperature and moisture conditions also affect the rate of mineralization since mesophyllic microorganisms dominating in the soil are optimally active at 25°C–35°C, while optimum moisture conditions are 80% of the field capacity (Guntiñas et al. 2012). Clay-textured soils have been observed to release about twice as much residue N than in sandy soils (Becker et al. 1994), while mineralization was inhibited at soil pH below 5, which is generally the pH of a soil when it is well drained (Patrick and Mahapatra 1968; Fu et al. 1987).

Rice plants under anaerobic conditions exchange gases through the aerenchyma, enabling some oxygen to be transported to the rhizosphere, which can facilitate nitrification of small amounts of ammonium to nitrates. These nitrates, together with that leached from the aerobic layers, can be leached below the root zone due to its negative charge-repelling soil colloid surfaces. In a study by Singh et al. (1991), leaching losses of N as urea, NH_4^+, and NO_3^- were about 6% of the total urea-N and 3% of the total ammonium sulfate-N applied in three equal split doses, and urea applied basally in a single dose resulted in 13% N leaching losses, about three times lower when potassium nitrate was applied in three split doses. Nitrates in paddy soils can also be denitrified, to form N_2O, a greenhouse gas and ultimately, N_2 gas, which are lost from the soil. Loss of applied N through denitrification in lowland rice cultivation has been reported to be higher than that from wheat culti-vation, but can vary from negligible to 46% depending on urea application and crop establishment methods (Buresh and De Datta 1990), whereas earlier studies showed 5%–10% in continuously flooded rice-cropped soils and 40% in fallow soil (Fillery and Vlek 1982). However, the major N loss pathway in paddy soils is through ammonia volatilization, which can result in up to 50% loss of applied fertilizer N (Mikkelsen et al. 1978; De Datta and Buresh 1989; De Datta et al. 1989). This is because NO_3^--based fertilizers are generally not used in lowland rice, as the NO_3^- would readily be lost from the soil, with consequences on climate change since N_2O is a potent greenhouse gas.

Due to the abundance of NH_4^+ over NO_3^- in paddy soils, rice plants take up N mainly as NH_4-N, which is loosely bound to the water as NH_4OH and converted to NH_3 gas and water molecules. Loss of N due to NH_3 volatilization is influenced by pH, temperature of the floodwater, algal growth, crop growth, and other soil properties (De Datta 1987). N loss through NH_3 volatilization ranges from negligible to almost 60% of the applied N in flooded soils (Choudhury and Kennedy 2005). The pH of standing water in a submerged field increases during the day, related to photosynthetic activity of algae using up CO_2 (i.e., decrease in H_2CO_3 concentration). Consequently, N fertilizer application should be done at the start of the day when the pH is still low to prevent formation and volatilization of ammonia.

10.3.1.2 Phosphorus

Phosphorus is mainly in the inorganic phosphate form, is an essential soil constituent in flooded soils, and is particularly important during early growth stages, but the requirement is less pro-nounced than that for N. Deficiency of available P is the second-most important nutritional disorder of lowland rice, especially in the highly weathered acidic soils of the tropics that contain large quantities of Al and Fe oxides, which cause P-fixation. The different forms of inorganic phosphate as outlined by Chang and Jackson (1957) include aluminum phosphates, iron phosphates, calcium phosphates, reductant soluble or the occluded phosphates, and the soil solution orthophosphates as PO_4^{3-}, HPO_4^{2-}, or $H_2PO_4^-$, where the predominant form depends on soil pH, and the latter is the more preferred by plants for uptake. These forms and their distribution vary among soil types (Patrick and Mahapatra 1968). Initially, P availability increases after submergence, so it is not considered as a yield-limiting nutrient in rice growth, but P solubility may not last long due to the increasing sorption capacity of ferrous hydroxides favoring the formation of insoluble Fe-P fraction. In anaerobic soil and chang-ing soil pH, P becomes more available as ferric phosphates are reduced to ferrous phosphates, liberating the sorbed-P. Interestingly, flooded soils release more P than well-drained soil if the soil solution content is low to begin with, though reduced soils have a higher sorption capacity (Fageria et al. 2003). Several studies indicated the relationship between the extent of P sorption and the amount oxalate extractable iron in flooded soils, and it was observed that 84% of P is sorbed when the oxalate extractable Fe is 3000 mg kg^{-1} because oxalate can extract the most reactive, poorly crystalline, small-size, and high-surface-area Fe oxides (Willett and Higgins 1978; Khalid et al. 1979; Shahandeh et al. 1994). Diffusion, which is the main mechanism of P supply to roots, is enhanced by flooding, and a study by Turner and Gilliam (1976) confirmed that P uptake and shoot growth increased with increasing moisture level, in fact at the highest when saturated with water. The requirement for optimum growth is 0.12–0.20 mg P L^{-1} in the soil solution, and about 90% rice yield can be produced at 0.10 mg P L^{-1} soil solution (Hossner and Baker 1988). This preview provides a summary, and a comprehensive discussion of P dynamics in flooded soils was described in Ponnamperuma (1972) and Hossner and Baker (1988).

10.3.1.3 Potassium

Changes in the redox potential do not seem to affect soil potassium directly but somehow decrease its availability. Upon submergence, there is a cation exchange between the surface of negatively charged 1:1 clays and the soil solution, increasing the solution K and enhancing K uptake by the roots, but this soil type is prone to leaching with a high percolation rate, resulting in K removal

(Dobermann and Fairhurst 2000). Low oxygen diffusion environments in lowland soils differ from aerated soils. In addition to the several reasons mentioned above, there is a crescent diffusion of K located in the interlayers of illite, a 2:1 type of clay mineral (da Silva et al. 2015), which is considered a nonexchangeable K. Flooding of lowland soils with this type of clay mineral that causes K fixation reduces the solution K concentration at first, thus utilizing the nonexchangeable K, as the cation over anion intake is balanced, while Fe is oxidized by rice roots, releasing H^+ and O_2 (Dobermann and Fairhurst 2000), resulting in displacement of K^+ ions by H^+ in the interlayer positions, and thus eventually increasing the amount of K^+ in solution.

In a previous study by Dobermann et al. (1996), micaceous and vermicullitic soils containing high amounts of Ca and Mg induced K deficiency because of the preferential adsorption of K to cation exchange sites despite the lyotropic series, which reduced K activity in the soil.

The soil chemical reactions involving K are more simpler than those for N and P. Potassium is not complexed into soil organic matter and do not volatilize or form insoluble precipitates, but rather, soil K is greatly influenced by the associated soil mineral fraction. Thus its availability is associated with properties such as soil K concentration, soil texture, soil pH, cation exchange capacity, temperature, soil moisture, soil aeration, yield level, and root growth patterns (Fageria et al. 2003). This has resulted in limited research on efficient K fertilization practices and diagnosis and prevention of K deficiency, and it has only been during the review of de Datta and Mikkelsen (1985) that the research on K nutrition of rice has begun to increase our knowledge of the K nutritional requirements of rice.

Crop residue management practices were observed to have a positive impact on K uptake in rice with about 179 kg kg^{-1} when crop residue is incorporated, which is 8.5% higher than when crop residue is removed (Dotaniya 2013). A rice crop contains 200–300 kg ha K, where the majority of K from the rice straw that is returned to the field the field is recycled back into the soil, while K fertilization is required if straw is removed to prevent from depleting the soil K (de Datta and Mikkelsen 1985).

Unlike N, P, and Zn, K has not been a major yield-limiting nutrient in many rice-producing areas, and fewer studies on K nutrition are available in the literature, but as grain yield increases the demand for K also increases. Rice was characterized as having a high capacity to absorb plant available soil K because of its fibrous root system followed by depleting K in the growth medium, since high amounts of K were taken up during the vegetative and early reproductive growth stages.

Taking into account the chemistry of soil organic matter and monitoring the indigenous nutrient supply as a measure of soil quality are essential in understanding the soil nutrient supplying capacity in lowland soils. Interactions among nutrients have a great impact on the physiological and agronomic efficiency that results from nutrient applications.

10.3.2 MANAGEMENT PRACTICES

Rice is commonly grown by transplanting seedlings into puddled soil. Puddling, which is essential for rice cultivation, impoverishes the soil physical condition, increases bulk density, and reduces hydraulic conductivity. Puddling benefits rice by reducing water percolation losses, controlling weeds, facilitating easy seedling establishment, and creating anaerobic conditions to enhance nutrient availability (Aggarwal et al. 1995). But repeated puddling adversely affects soil physical properties by destroying soil aggregates, reducing permeability in subsurface layers, and forming hardpans at shallow depths (Sharma and De Datta 1985; Aggarwal et al. 1995), all of which can negatively affect the following non–rice upland crops in rotation (Hobbs and Gupta 2000; Tripathi et al. 2005).

Puddling is energy consuming and deteriorates the soil health for growing succeeding crops. Repeated cultivation of rice leads to the formation of hardpan below the plow layer, deteriorates the soil structure, inhibits root elongation, and delays the planting of a succeeding crop. Continuous rice cultivation for longer periods with poor crop management practices has often resulted in loss of soil fertility, and in turn has led to multiple nutrient deficiencies (Ladha et al. 2004). Under puddled conditions, rice undergoes several changes, namely, going from an aerobic to an anaerobic environment, resulting in several physical and electrochemical transformations. Puddling operation consumes water and energy, breaks the capillary pores, destroys the soil aggregates, disperses the fine clay particles, and lowers soil strength in the puddle layer (Sharma and De Datta 1985). Imbalanced use of N-fertilizer in rice may increase the leaching of nitrates beyond the root zone, leading to groundwater pollution in rural areas (Singh et al. 2002a).

10.4 EXTERNAL SOURCES OF NUTRIENTS

Among external inputs, mineral fertilizers are the most common source of nutrients in rice cropping systems, although organic inputs are also used. Different compound and straight fertilizers for the major elements, N, P, K, and S, and micronutrients, mostly Zn, are available on the market. While N is generally the most limiting nutrient in paddy soils and is applied in large quantities, its application is often associated with losses resulting in low recovery efficiencies (Buresh et al. 2008). In addition, there is a lack of synchrony between the supply and demand of nutrients, particularly N, which is often applied early before peak demands by the crop. Given the huge losses that are associated with N fertilization and the poor synchrony, slow- or controlled-release fertilizers such as prilled urea or coated urea granules are increasingly becoming available on the market (Mi et al. 2019). These have been evaluated in the field and shown to improve yield and to enhance N utilization efficiency in rice cropping systems, where the availability of N is slow and likely match the demand by the crop and thus improve synchrony (Kondo et al. 2005; Chalk et al. 2015). However, these are still expensive and out of the reach of many smallholder farmers in rice-growing countries. P and K fertilizers can be added in paddy soils based on input-output budgets, taking into consideration different inputs such as manures, crop residues, irrigation water, and atmospheric deposition and dust. The nutrient input-output budget in a typical irrigated rice field has been described by Dobermann and Fairhurst (2000).

There are different schools of thought on the importance of organic nutrient sources in rice cropping systems, with some suggesting that rice yield gains from organic resources can be realized in the long term. However, an analysis of 25 long-term experiments with double and triple cropping rice or rice–cropping systems across six countries in Asia showed no differences in yield where farmyard manure was where only NPK fertilizer was added (Dawe et al. 2003). Further, they observed that continuous addition of organic resources did not improve yield trends compared to where only NPK fertilizers were added. While the organic resources in the 25 studies were not added as a complement to the NPK fertilizers, the findings suggest that organic resources may not play an important role in improving yields in intensive rice-rice and rice-wheat systems. Similarly, in a three-year study of P sources across three sites in Madagascar under irrigation, there were no rice yield responses to the application of farmyard manure application but to triple superphosphate (Andriamananjara et al. 2016). In addition, farmyard manure did not improve P utilization efficiency of the triple superphosphate when applied in combination. This suggests that farmyard manure has little or no influence on P availability in rice soils. In contrast, however, Ding et al. (2018) observed 7.8% greater yields when organic resources were added compared to where only NPK fertilizers were applied in a meta-analysis in China. Thus, there is still debate on the role of organic resources in rice production.

Rice straw is the main organic material available for most rice farmers and serves as an important source of K since 80%–85% of the K taken up by the plant remains in the rice straw. At harvest, rice straw contains 0.5%–0.8% N, 0.07%–0.12% P_2O_5, 1.16%–1.66% K_2O, 0.05%–0.1% S, and 4%–7% Si (Ponnamperuma 1984; Dobermann and Fairhurst 2000). Since N content in straw is low, large quantities of straw would be needed to supply adequate amounts of N. However, the straw has to decompose before the nutrients can become available for uptake, and this depends on soil type and season. Additionally, only a proportion of the nutrients become available in the season of application. For example, in a study on an alluvial soil in Vietnam, about 67%–69% of the rice straw had decomposed by the time rice had reached physiological maturity (Thuan and Long 2010). Given the low use of organic inputs, this chapter will focus on mineral fertilizers and their management. It is acknowledged, however, that some farmers integrate the use of mineral fertilizers with organic fertilizers in one form or another.

10.5 HISTORICAL AND CURRENT USE OF MINERAL FERTILIZER

Introduced in the mid-1960s, the Green Revolution represents the most dramatic shift in agricultural practices in human history. Rice production significantly increased following the Green Revolution, with most of the increases coming from increased yields, mostly in Asia. This was largely due to the use of improved rice varieties, which were mostly semi-dwarf and short duration in combination with the use of fertilizers, pesticides including herbicides, and increased access to irrigation (Yoshida 1972). The release of semi-dwarf, short-duration, and high-yielding varieties, starting with IR8 in 1966, triggered the onset of the Green Revolution. Due to the availability of short-duration rice varieties and increased access to irrigation, rice cropping was intensified to double or triple rice cropping being common in irrigated ecosystems. Fertilizer use in Asia increased from 13 Mt in 1950 to 110 Mt in 1984 due to the striking yield increases and the immediate economic gains to the farmers with the use of fertilizers with nutrient-responsive varieties. However, the application of N fertilizers outpaced the application of other macronutrients, particularly K, which generally has low responses to K fertilization, leading to imbalances in plant nutrition and negative K balances (Dobermann et al. 1998). Following the yield increases, the yield growth rate declined in the mid-1980s in double and triple rice cropping systems despite breeding research efforts to develop varieties with greater yield potentials (Dawe and Dobermann 1999). In addition, there was a rapid decline in the efficiency of fertilizer uptake by plants, indicating that increased fertilizer use outpaced yield improvements (Cassman and Pingali 1995; Tilman et al. 2002). At modern fertilizer application rates, only 30%–50% of N is recovered by the crop, indicating that 50%–70% of applied N remains unused or is lost from the soil (Dobermann 2000; Ladha et al. 2005). Furthermore, soil-related environmental problems have arisen in different areas. For example, excessive and inappropriate use of fertilizers pollute waterways.

Fertilizer use has been increasing since the beginning of the Green Revolution, but the increase has been variable across regions and also varies by crop (FAO 2006; Lu and Tian 2017). China, the United States, and India account for 50% of global fertilizer use, whereas more than 50% of the global N fertilizer is used in Asia, while less than 5% is used in Africa (FAOSTAT 2019). Much of this is because of skewed fertilizer prices and the fertilizers not being available on a timely basis when needed for crop production, mostly in Africa, although global quantities are adequate for use in all the countries (FAO 2006). China is the largest consumer of world fertilizer in agriculture, consuming about 27% of the total global N fertilizer used in 2017 (FAOSTAT 2019). On average, N fertilizer use in China was 193 kg N ha^{-1} in 2006 (Heffer 2009), with an average grain yield of 6.27 t ha^{-1} (FAOSTAT 2019). These rates are excessive, as less than 40% of the applied N is converted into harvestable N (Lassaletta et al. 2014) and has resulted in surface and groundwater pollution, causing ecological problems and increased greenhouse gas emissions. While fertilizer use is important for improving crop yields and the attainment of food security in the world, the overuse in some regions and for some crops such as rice brings the sustainability issue into question.

10.5.1 SUSTAINABILITY

Increased use of fertilizers, particularly N, significantly improved crop yields globally, including rice (Cassman et al. 2003; Lu and Tian 2017). Farmers in Asia often apply large quantities of N fertilizers in order to increase rice yields, but the quantities are often in excess of plant requirements for N, which often results in losses of N to the environment. However, grain yield response diminishes as N fertilizer increases, and in some instances overapplication of N fertilizer may cause lodging and susceptibility to pest and disease damage, in both cases causing yield losses (Balasubramanian et al. 1998; Duy et al. 2004). This inefficient use of fertilizers, coupled with increasing water and labor scarcity and the changing climate, threaten the productivity and sustainability of rice-based systems. The emerging energy crisis and rising fuel prices, the rising cost of cultivation, and emerging socioeconomic changes such as urbanization, migration of labor, preference for nonagricultural work, and concerns about farm-related pollution further threaten the productivity and sustainability of rice-based cropping systems (Ladha et al. 2011). Agronomic management and technological innovations are needed to address these concerns in Asia. Some of the potential strategies for sustaining the productivity of rice systems are: (1) reduction of the rice monoculture and diversification of the cropping system, and (2) enhancing the input use efficiency in existing double and triple rice-based cropping systems through improved technology and management practices.

Nutrient management remains an important component for sustainable rice production in Asia in order to meet the challenges of rising demand for the commodity driven by population growth, while protecting the environment. This is also essential for ensuring that rice farming is profitable. Challenges with declining yields and nutrient use efficiency require sophisticated nutrient management in order to deal with the multiple requirements of productivity, profitability, and environmental protection, and this led to the development of the site-specific nutrient management (SSNM) approach in the 1990s (Dobermann et al. 2002). This approach is a form of precision nutrient management based on scientific principles and aims to improve yield within the same fields without expanding to new areas while enhancing nutrient use efficiency and protecting the environment and thus promoting sustainable intensification of rice cropping systems.

10.6 PRECISION NUTRIENT MANAGEMENT – SSNM

10.6.1 THE NEED

Rice fields in Asia and Africa tend to be small – often about one hectare or less, with high spatial variability with respect to nutrient status, use of varieties, and crop management by farmers. Consequently, nutrient management in smallholder rice production systems needs to be tailored to suit the conditions of different farm conditions and seasons. Opportunities exist to increase the effectiveness of nutrient use and thus income for rice farmers through precision nutrient management. The management of fertilizers can be particularly critical for profitable rice farming in Asia because fertilizers are typically the second-largest input cost after labor (Pampolino et al. 2007). A study in seven major irrigated rice areas in six Asian countries showed fertilizers represented 11%–28% of the annual costs of farmers for producing rice (Moya et al. 2004). Fertilizers must be applied at appropriate times and rates for efficiently increasing yield per unit of nutrient applied. This is because substantial portions of added fertilizer N can be lost from rice soils as gases through ammonia volatilization and nitrification denitrification (Buresh et al. 2008).

Climate change introduces new dynamics and uncertainties into agricultural production systems. It affects agriculture through different means that include changes in average temperature, rainfall, and climate extremes; changes in atmospheric carbon dioxide; changes in ozone concentration; changes in pest and diseases; and deviations in nutritional quality. Rice fields in Asia and Africa are small and highly heterogenous in soil nutrient status; yet rice farmers are often provided with blanket N, P, and K fertilizer recommendations for all fields within a geographical or administrative area. Such recommendations ignore variations across fields in the supply of essential nutrients from the soil (Dobermann et al. 2003a, 2003b) and variations in crop management practices, which can influence the needs of the rice crop for fertilizer and the yields obtained in farmers' fields (Witt et al. 2007). The use of blanket nutrient recommendations across a rice production area can lead to low nutrient use efficiencies and leakages of nutrients to the environment (Singh and Singh 2008). The management of fertilizers for high yields and higher efficiency of nutrient use in rice production could consequently benefit from a cost-effective, rapid, and easy-to-use approach to handle the field-specific needs of a rice crop.

10.6.2 Concept

Site-specific nutrient management (SSNM) is a dynamic, field-specific, and season-specific nutrient management approach that optimizes nutrient supply and demand according to differences in indigenous nutrient supply, nutrient cycling, and target yield (Dobermann et al. 2002, 2004). It is based on a set of scientific principles with the aim of increasing yields and profit for the farmer by increasing nutrient use efficiency while reducing losses and hence ensuring environmental sustainability. Nutrients are applications distributed through the season as and when needed by the crop, matching demand and supply. It was developed and refined through years of research by the International Rice Research Institute (IRRI) and partners across Asia beginning in the 1990s (Dobermann et al. 2004). It provides scientific principles for calculating field- and season-specific N, P, and K requirements for cereal crops, including rice, before the start of the season (Dobermann et al. 2002; Buresh et al. 2010). SSNM was initially conceptualized for rice using principles from the Quantitative Evaluation of the Fertility of Tropical Soils (QUEFTS) model (Janssen et al. 1990) to estimate the requirement for a fertilizer nutrient from the gap between the total amount of nutrient required by the crop to achieve a specific target yield and the indigenous supply of the nutrient (Witt and Dobermann 2004). Estimating the indigenous nutrient supply enables effective utilization of nutrients existing in the soil. The indigenous supply of nutrients for the SSNM approach is determined based on plant response using nutrient omission plot technique trials (Dobermann et al. 2003a, 2003b). The SSNM approach was initially developed for rice but has also been applied to wheat (Khurana et al. 2008) and maize (Witt et al. 2009).

The optimal amount of N, P, and K fertilizer for a particular field is dependent on the indigenous supply of the nutrients for the particular field, and the target yield, which is dependent on variety, climate, water management or regime, and crop management. The SSNM approach adjusts inputs of fertilizers based on a supply of indigenous nutrients originating from soil, plant residues, manures, and irrigation water. Indigenous N is influenced by soil organic matter, biological N fixation, irrigation water, and organic fertilizer. The amount of N fertilizer requirement is calculated based on the anticipated gain in yield and an attainable agronomic efficiency of N using equation (10.1):

$$FN = \left(GY - GY_{0N}\right)/AE_N \qquad (10.1)$$

where FN is the fertilizer N requirement (kg N ha^{-1}), GY is the attainable target yield (kg ha^{-1}), GY_{0N} is the grain yield without N fertilizer (kg ha^{-1}), and AE_N is the attainable agronomic efficiency of N fertilizer (kg grain yield increase kg^{-1} fertilizer N applied).

For N management, the SSNM approach includes both the determination of the optimal amount of N fertilizer to be applied in a particular field and its distribution through the season so as to meet the peak demands for N by the crop, depending on varietal needs (Dobermann et al. 2002; Witt et al. 2007; Peng et al. 2010). While farmers apply a greater proportion of the N fertilizer in the early stages of the crop, this has resulted in greater early growth vigor, which has, however, not translated into greater grain yield benefits at maturity.

The determination of fertilizer K and P requirements using the SSNM approach ensures enough quantities are applied to avoid yield loss due to deficiencies while ensuring profitability. Field-specific P and K management combines the yield gain approach and nutrient balances to ensure efficient utilization of the nutrients and avoid mining the soil (Buresh et al. 2010). The approach uses the internal nutrient efficiency combined with estimates of attainable yield, nutrient balances, and probable yield gains from added nutrient within specific fields (Witt et al. 2007; Buresh et al. 2010). The amount of K fertilizer is influenced by irrigation water, crop residue management, and supply for the soil. Considering that more than 80% of the K taken up by a rice crop is in the straw, it follows that return of rice residues to the field contributes to the soil K and hence fertilizer K needs are reduced. However, the incorporation of rice straw can significantly increase greenhouse gas emissions, particularly methane under continuous flooding, which is typical for paddy soils (Wassmann et al. 2007; Sander et al. 2014).

10.6.3 Evaluation of SSNM

Fertilization using the SSNM approach has been shown to increase yield and income in rice pro-duction (Dobermann et al. 2004; Witt et al. 2007) and provide environmental benefits by reducing N losses and greenhouse gas emissions (Pampolino et al. 2007). Researcher-managed trials were conducted across rice-production areas in Asia and Africa to evaluate use of SSNM to manage fer-tilizer N, P, and K and was shown to effectively increase yield, agronomic nutrient use efficiency, and net income, as shown in Table 10.1. The table presents a summary of 32 studies where SSNM was evaluated in comparison to the farmer practice (FFP) in seven countries in Asia and one coun-try in Africa, Senegal. On average, rice yields are 8.6% greater where SSNM is used compared to the FFP. However, the yield differences vary under different conditions. The SSNM approach was developed mostly under irrigated lowland environments. While most of the studies were conducted in irrigated ecosystems, rice yields were also greater with SSNM than FFP in rainfed lowland eco-systems (Biradar et al. 2006; Haefele and Konboon 2009; Banayo et al. 2018), although the yield differences were smaller compared to irrigated ecosystems.

In a recent study in the Philippines, Banayo et al. (2018) observed a 6% yield advantage where the SSNM approach was used compared to the farmer practice under lowland rainfed conditions. In that study, they observed that farmers tended to use greater quantities of N and P fertilizers than with SSNM, and this resulted in a net increased income on average of \$154 ha^{-1} season^{-1}. The yield increases with SSNM fertilization, despite lower quantities than the farmer practice, was attributed to the distribution of N fertilizer. On average, farmers apply 54% of the total N in the first 15 days after transplanting, whereas with SSNM 29% of the N was applied within the first 15 days after transplanting. Rice yields were about 50% greater when fertilized using the SSNM approach com-pared to the farmer practice study in Northern Karnataka in India under rainfed conditions (Biradar et al. 2006). In the same study, rice yields with the state blanket recommendation were higher than the farmer practice, but were still more than 25% lower than SSNM, confirming that blanket recom-mendations are not efficient in improving farmer yields. These studies show that there are positive benefits with SSNM under rainfed conditions, but that the extent of the benefits varies with locations and conditions under the rainfed systems.

Most of the SSNM evaluation studies for rice were established using the transplanting method, and yield differences were greater than where rice was direct seeded (Table 10.1). In most of the rice-growing regions farmers are increasingly shifting from transplanted to direct-seeded rice sys-tems, largely due to increasing water scarcity associated with climate change, and labor shortages and the associated economic gains (Kumar et al. 2018). Transplanting is labor intensive, given that labor constitutes the largest cost in rice production (>60%) (Pampolino et al. 2007). However, weed challenges with direct seeding can affect the responses to fertilizer management (Rao et al. 2007; Farooq et al. 2011). A recent study showed that ensuring late season application of N fertilizer at the heading enhanced rice yields under direct-seeded rice (Liu et al. 2019). This is in line with the SSNM recommendations that generally recommend distribution of N fertilizer to meet the crop N demands, while avoiding applying large quantities early in the season before the plants establish an effective root network to take up the nutrients. In a study in Thailand, Satawathananont et al. (2004) observed both positive and negative yield benefits with SSNM compared to farmer practice under direct-seeded rice in the wet and dry seasons, respectively. In contrast, in Vietnam, yield benefits were positive in both the wet and dry seasons under direct-seeded rice (Tan et al. 2004). These results suggest that SSNM is beneficial in terms of yield increases compared to the farmer practice under direct-seeded rice, but the benefits are likely dependent on water and weed management.

Agronomic efficiency of N is generally greater with SSNM than FFP, with an average 15.9 and 11.8 kg grain kg^{-1} N applied, respectively (Table 10.1). This suggests reduced N losses and hence greater environmental sustainability. In a study across more than 120 sites in China over several seasons, Peng et al. (2010) observed greater agronomic efficiency of N with SSNM fertilization compared to the farmer practice. In these trials, N fertilizer was reduced by about 40% using the

TABLE 10.1

Rice Yield, Agronomic Use Efficiency of N (AEN), and Partial Factor Productivity of N Fertilizer (PFPN) under Site-Specific Nutrient Management (SSNM) Compared with Farmer Field Practice (FFP) across Eight Countries from 32 Different Studies

Country (# of Sites)	Cropping System β	Season/# of Seasons §	Crop Estab.Ψ	Grain Yield kg ha⁻¹			AEN kg Grain kg⁻¹ N		PFPN kg Grain kg⁻¹ N		Source
				SSNM	FFP	% Change	SSNM	FFP	SSNM	FFP	
Irrigated											
China (1)	R-R	DS/2	TPR	5800	5400	6.9	11.4	6.3	48.5	36.5	Wang et al. (2001)
	R-R	WS/2	TPR	6900	6400	7.2	11.4	6.3	48.5	36.5	
Philippines (1)	R-R	DS/2	TPR, DSR	5725	5100	10.9	16.3	13.0	44.8	42.4	Gines et al. (2004)
	R-R	WS/2	TPR, DSR	4675	4300	8.0	13.0	11.5	58.0	62.8	
Thailand (1)	R-R-R	DS/2	DSR	4600	4700	−2.2	8.9	10.5	38.0	40.5	Satawathananont et al. (2004)
	R-R-R	WS/2	DSR	4990	4750	4.8	8.4	6.5	47.5	52.5	
Vietnam (1)	R-R-R	DS/2	DSR	5700	5300	7.0	22.0	17.0	58.5	51.8	Tan et al. (2004)
	R-R-R	WS/2	DSR	3625	3480	4.0	17.0	12.5	39.3	31.0	
Vietnam (1)	R-R-M	DS/2	TPR	6200	6025	2.8	17.0	14.7	64.3	61.0	Son et al. (2004)
	R-R-M	WS/2	TPR	6150	6000	2.4	18.0	13.5	74.0	60.0	
Indonesia (1)	R-R	DS/2	TPR	3900	3625	7.1	10.3	8.4	23.5	19.3	Abdulrachman et al. (2004)
	R-R	WS/2	TPR	5100	4925	3.4	15.0	9.9	48.5	37.5	
India (2)	R-R-P	DS/2	TPR	6550	5788	11.6	15.2	13.6			Nagarajan et al. (2004)
	R-R-P	WS/2	TPR	5493	4913	10.6	15.2	13.6			
China (1)	R-R	DS/3	TPR	5867	5833	0.6	11.2	6.3	47.0	34.3	Guanghuo et al. (2004)
	R-R	WS/3	TPR	7083	6550	7.5	13.6	7.0	56.5	38.2	
India (2)	R-R	DS/1	TPR	12850	12000	6.6			51.0	48.0	Pampolino et al. (2007)
Vietnam (3)	R-R	DS/1	DSR	9833	9167	6.8			55.0	47.0	
Philippines (2)	R-R	DS/1	TPR and DSR	10400	9700	6.7			47.0	39.0	
India (6)	R-W	DS/3	TPR	6000	5117	14.7	16.6	9.4	44.3	35.4	Khurana et al. (2007)
China (4)	R-R	–	TPR	5900	5700	3.4			44.0	31.0	Hu et al. (2007)
China (4)	R-R	f2	TPR	7425	7163	3.5	14.5	3.6	123.4	34.8	Peng et al. (2006b)
Philippines (1)		DS/2	TPR	6800	6150	9.6	19.2	23.1	48.0	68.5	
India (1)	R-R-R	DS/3	TPR	4833	4300	11.0			38.3	37.7	Sharma et al. (2019)
India (2)		DS/2	TPR	6400	5900	7.8					Rajendran et al. (2010)
		WS/2	TPR	6050	5750	5.0					
Senegal		DS/1	DSR	7467	5967	20.1					Saito et al. (2015)

(Continued)

TABLE 10.1 (Continued)

Rice Yield, Agronomic Use Efficiency of N (AEN), and Partial Factor Productivity of N Fertilizer (PFPN) under Site-Specific Nutrient Management (SSNM) Compared with Farmer Field Practice (FFP) across Eight Countries from 32 Different Studies

Country (# of Sites)	Cropping System β	Season/# of Seasons §	Crop Estab. Ψ	Grain Yield kg ha⁻¹		% Change	AEN kg Grain kg⁻¹ N		PFPN kg Grain kg⁻¹ N		Source
				SSNM	FFP		SSNM	FFP	SSNM	FFP	
Bangladesh (6)	R-W	WS/2	TPR	4400	4000	9.1	15.0	10.3			Alam et al. (2006)
Bangladesh (5)	R-W	WS/2	TPR	4450	3863	13.2	16.8	11.1	36.8	31.5	Alam et al. (2005)
		DS/2	TPR	5963	5513	7.5	21.5	15.9	44.6	34.5	
India (5)	R-W	–	TPR	5220	4400	15.7					Singh et al. (2013)
India (1)	R-W	–		9110	5640	38.1					Singh et al. (2015)
Vietnam (2)	R-R-R	DS/		6870	6363	7.4					van Hach and Tan (2007)
	R-R-R	WS/		4743	4530	4.5					
China (3)	R-R	WS/1	TPR	6538	6550	-0.2					Xu et al. (2010)
China (3)		/3	TPR	9167	8933	2.5	21.5	21.0	54.0	54.5	Xu et al. (2017)
		/3	TPR	8967	8367	6.7	16.0	9.5	59.0	44.0	
		DS/3	TPR	7567	7000	7.5	16.0	12.5	52.0	49.5	
		WS/3	TPR	7667	7133	7.0	13.0	10.0	50.0	48.0	
India (1)		WS/1		5784	4627	20.0					Mandal et al. (2015)
Thailand (8)				4029	4478	-11.2			175.0	134.0	Attanandana et al. (2010)
India	R-W	DS/2	TPR	6000	5050	15.8	16.0	9.5			Khurana et al. (2009)
Nepal		DS		5460	4430	18.9					Gupta et al (2016)
Nepal (1)		WS/1	TPR	6350	4620	27.2					Marahatta (2017)
India (1)	R-W	WS and DS/2	TPR	6675	6064	9.2	26.6	19.4	56.6	46.6	Qureshi et al. (2018)
Vietnam (3)		DS/1		6090	6060	0.5	15.1	13.7			Khuong et al. (2007)
		WS/1		5150	4990	3.1	15.2	12.3			
Rainfed											
Philippines (4)	R	WS/4	TPR	4538	4228	6.8					Banayo et al. (2018)
India	R		TPR	5520	3686	33.2					Biradar et al. (2006)
Thailand (2)	R	–	TPR	3050	3125	-2.5					Haefele and Konboon (2009)

Cropping system β: R-R is rice-rice; R-R-R is rice-rice-rice; R-R-M is rice-rice-maize; R-R-P is rice-rice-pulse, R-W is rice-wheat; R is rice only – for rainfed systems.

Season/# of seasons §: Represents season; dry season (DS) and wet season (WS); number of seasons for which the study was conducted.

Crop estab. Ψ: Crop establishment method – TPR is transplanted and DSR is direct-seeded rice.

SSNM approach as compared to the farmer practice, and the yields were greater with the SSNM approach than with the farmer practice. Similarly, Wang et al. (2001) observed greater agronomic efficiency of N with SSNM than with farmer practice on 21 sites in China. These observations confirm that farmers in Asia, particularly in China, use excessive amounts of fertilizers, especially N. A study conducted on 179 sites in eight regions in six countries showed greater agronomic efficiency of N with the SSNM approach compared with farmer practice, both when N fertilizer quantity applied with SSNM fertilization was greater or smaller than the farmer practice. These observations suggest that the SSNM approach adjusts the N fertilizer upward or downward, depending on whether the farmers are applying less N or in excess, depending on the target yield and indigenous supply of N. As a result, it seems that, based on the SSNM evaluation trials conducted, SSNM optimizes nutrient application in terms of quantity and distribution, resulting in greater synchrony between demand and supply of nutrients.

The examples cited here show that SSNM is effective in improving rice yields, agronomic efficiency of N fertilizer, and profit based on partial factor of productivity. The SSNM approach was shown to improve crop yields versus the farmer practice, which is often based on blanket recommendations (Wang et al. 2001; Dobermann et al. 2002; Peng et al. 2010), while reducing fertilizer application (Peng et al. 2010). The increase in grain yield while lower amounts of fertilizer are applied in some instances has been associated with increased timing of application, particularly for N. In recent years, SSNM has been identified as one of the options for sustainable intensification of rice production in Asia and as a climate-smart technology based on increased resource use efficiency while reducing greenhouse gas emissions (FAO 2014). Given that the SSNM approach is knowledge intensive and calculation of fertilizer recommendations for different farmers would take time and probably become impractical to reach many farmers, decision-support tools have become necessary in translating the SSNM principle into practice.

10.7   DECISION-SUPPORT TOOLS

10.7.1   The Evolution of Decision-Support Tools

The SSNM approach has evolved over 20 years along a research-to-impact pathway beginning in 1996 in a study on Reversing Trends in Declining Productivity (RTDP) followed by a study on Reaching toward Optimal Productivity (RTOP) in 2001–2004 at the IRRI. During that period, research partners linked with extension workers to extend technology by implementing activities through multidisciplinary teams at pilot sites. The technology was refined over time, including the development of leaf color charts for N fertilizer management (Singh et al. 2002b; Witt et al. 2005), which were used to facilitate uptake of SSNM by farmers through extension. However, there was little diffusion from pilot sites to other locations, and researchers had initiated the activities rather than extension workers, so there was little uptake by other organizations. In 2005 the Irrigated Rice Research Consortium (IRRC) was formulated across several countries in Asia, where the first two years focused on establishing guidelines for regions and countries and developing country-specific literature on the SSNM approach in local languages. While this resulted to expanded acceptance by research organizations, it was difficult to facilitate interactive decisions with printed materials, limiting the uptake of SSNM by farmers and extension. The challenges of rapid dissemination and uptake of SSNM were emphasized by the required intensive knowledge of the specific field conditions and the different recommendations received by extension workers from various organizations.

Witt and Dobermann (2004) proposed the development of decision-support tools that use computerized information systems to translate SSNM extension guidelines in addition to the traditional printed extension material. Using the principles of SSNM, an information and communication technology (ICT) decision-support tool, the Nutrient Manager for Rice (Buresh et al. 2014; Sharma et al. 2019), was developed in 2008 to give field-specific fertilizer recommendations to individual fields for smallholder farmers. The Nutrient Manager for Rice was targeted for irrigated and rainfed lowland rice farmers with the aim to increase productivity and net income by $100 ha^{-1} season^{-1} at the farm level. The tool was tailored for countries and regions where agricultural technicians and farmers could get access to SSNM recommendations. Initially distributed as an offline version on compact discs, this had great scientific impact. However, the limited availability of computers became a constraint, limiting the ease of use for extension workers and farmers. The tool was later upgraded as a web-based tool which could be accessed using a mobile phone, computer, laptop, or any other ICT device with internet. With high demand from farmers and extension workers alike for additional information on crop management, the nutrient manager for rice was upgraded into Rice Crop Manager (RCM; http://cropmanager.irri.org), which gives climate-informed agroadvisory services to farmers, including the selection of suitable varieties, weed management options, and nursery management (Buresh et al. 2019). This has been released for the Philippines, India, Bangladesh, Indonesia, and Vietnam. Rice Crop Manager is a web-based decision-support tool that can be accessed using computers, tablets, or smartphones connected to the internet. The tool is designed to be operated by extension workers or farmer technicians who ask farmers 25–30 easy-to-answer questions about farm conditions and practices. At the end of the questionnaire, Rice Crop Manager automatically calculates and gives easy-to-follow nutrient and crop management recommendations that can be printed on one page before the start of the season. To date, more than 1.85 million recommendations on integrated crop management have been generated and given to rice farmers in the Philippines, and those who followed the recommendations increased their yield by an average of 640 kg ha^{-1} season^{-1} (http://news.irri.org/). The large-scale dissemination of RCM in 2018 also showed an average added net benefit for the farmers was US$ 215 ha^{-1} season^{-1}.

During the same time that IRRI was developing Nutrient Manager for Rice and RCM, AfricaRice was developing RiceAdvice and the International Plant Nutrition Institute (IPNI) was developing Nutrient Expert (NE), both based on SSNM principles. Beta versions of Nutrient Expert for maize were developed in 2011 for South Asia, China, Kenya, and Zimbabwe, while similar versions of NE for wheat were developed for South Asia as well as China, where the field-validated versions were publicly released in 2013 (Pampolino et al. 2012). In 2009, Nutrient Expert for Hybrid Maize was developed for favorable tropical environments (e.g., Southeast Asia) and underwent field evaluation in Indonesia and the Philippines. Nutrient Expert was also developed for rice for China and South Asia for rice where nutrient management recommendations are based on SSNM principles using the expected yield response, target agronomic efficiency of applied N, and the QUEFTS model (Xu et al. 2017). The complexities of nutrient management principles were developed into simplified easy-to-use software suited for farm advisers, extension agents, and industry agronomists. The Nutrient Expert obtains information from farmers based on their own experience, knowledge, and practices and gives fertilizer management guidelines that suit specific field characteristics and locally available fertilizer sources. Nutrient Expert fertilizer recommendations are developed following the "4R" principles of nutrient management (applying the right source of nutrient at the right rate and the right time in the right place). Currently, Nutrient Expert has been applicable for specific crops and geographies.

AfricaRice initiated the development of RiceAdvice (https://www.riceadvice.info/en/) in 2013, and a first version was released and tested in 2014. Since then, various improvements have been made to the tool, and it was released in Google Play in 2016 (Saito and Sharma 2018). The "RiceAdvice" is a free app for Android™ that can work offline, but an internet connection is needed when data needs to be sent to a server and when updates are required. Up to early 2019, more than 50,000 RiceAdvice guidelines have been generated in West Africa (Nigeria, Mali, Senegal, and Burkina Faso). Farmers using RiceAdvice report yield gains between 0.6 and 1.8 t ha^{-1} and income gains between US$100 and 200 ha^{-1}. RiceAdvice is operated by extension workers and lead farmers who enter information such as rice-growing condition, variety, typical practices, expected sowing date, fertilizer availability, and market price into the application. Once the details are filled in, farmers can set yield targets based on their available budget or desired or recommended production level. In case farmers do not have access to the internet or do not have a smartphone or tablet, the offline version of RiceAdvice can be used or other rice value-chain actors can use the decision-support tool to provide farmers with field-specific guidelines for crop and nutrient management practices. The tool provides guidelines on crop and nutrient management practices at the start of the season and for in-season practices, e.g., fertilizer application rate and timing.

10.7.2 CHALLENGES

Leaf color charts provide a simple, inexpensive, and easy-to-use decision-support tool for N fertilizer management. Farmers in Asia adopted leaf color charts (Islam et al. 2007), but only used them for a few years. Other organizations in the region did not take up leaf color charts, limiting their promotion. There is still ongoing distribution of leaf color charts in Asia, with possible expansion to Africa, where they can be used for rice and other cereal crops. However, knowledge and management of P and K is needed to avoid nutrient imbalances.

The introduction of ICT-based decision support tools has been seen as an option for greater spread of SSNM recommendations to farmers. However, internet availability and stability in rural areas, where most smallholder farmers are located, is generally poor, limiting the spread of the information. There is therefore a need to develop applications that can be accessible and give recommendations in the offline mode. In addition, many farmers do not have their own smartphone or tablet to use the decision-support tools by themselves. Furthermore, farmers have limited access to extension services, as the number of public extension workers is limited, especially in Africa. The challenge is to identify suitable and local-specific business models on the use of the tools, and scale out such models and the tools to both farmers and service providers. Once the business model is developed, farmers may have to pay for the recommendations. For example, in Nigeria and Mali, farmers are willing to pay up to US$5 for each recommendation generated by RiceAdvice to service providers. Service providers do not have to pay for the tools at this moment, as these are freely available for download. Different types of rice value-chain actors could become service providers and include this service in their businesses. They can provide the service to the farmers who have no access to a smartphone, tablet, or internet or who are illiterate. The introduction of service providers could overcome these two constraints. Another challenges is farmers' low level of literacy and lack of ICT skills, limiting their access and use smartphones, tablets, and internet, which are often common challenges for almost all ICT tools.

10.8   FINAL THOUGHTS AND RESEARCH GAPS, CONCLUSIONS

The SSNM approach has been shown to improve rice yields, agronomic efficiency of N, and partial factor of productivity. This means that the SSNM approach offers an alternative to the traditional blanket recommendations that are associated with the overapplication and loss of nutrients in some regions, resulting in environmental degradation and reduced profit for rice farmers. Because the SSNM approach is knowledge intensive, the development of ICT tools has offered an easier alternative for the dissemination of SSNM recommendations.

While soil sampling and testing have traditionally formed the basis of fertilizer recommendations in farming systems, work on SSNM has shown no relationship between soil test results and the yield performance of lowland rice. This is likely because the soil being tested is dried and has changed into an aerobic state and thus is not in its natural field condition, altering the concentrations. The fertilizer recommendations using the SSNM approach have been dependent on nutrient omission plot trial data to determine indigenous nutrient supply. However, that requires a season of testing with nutrient omission plots before the tool can effectively give reliable fertilizer recommendations. The ICT tools developed based on SSNM – Rice Crop Manager, Nutrient

Expert, and RiceAdvice – use the existing nutrient omission plot technique trial data and refine it to suit local conditions using farmer knowledge of farm conditions to develop the fertilizer recommendations. To improve the robustness of the fertilizer recommendations using the ICT tools, soil testing may be essential. There is a need to calibrate soil tests for lowland rice systems that can closely be linked with rice productivity.

Technological approaches and algorithms for developing fertilizer requirements tailored to field-specific needs of crops in irrigated rice-based cropping systems must be based on robust scientific principles applicable across the field-level variability and diversity of crop-growing conditions. In the Philippines, where the greater proportion of the activities using Rice Crop Manager have been conducted, there is little diversification. Thus there is a need to extend the research and perform evaluation of the tool in diversified rice cropping systems such as rice-wheat, rice-maize, or rice-legume systems to allow for the provision of fertilizer recommendations to suit such systems. However, Rice Crop Manager in India has been developed and evaluated in diversified rice cropping systems. At the same time, approaches must be relatively simple with minimal characterization or interviewing of farmers for each field in order to ensure rapid, cost-effective delivery of field-specific guidelines to millions of small-scale farmers. Technological approaches and algorithms should strive to draw upon existing research information in order to avoid delays in reaching farmers with practical solutions based on scientific principles.

Seasonal and annual climatic variations among regions are still lacking in the tools, which must be taken into account because temperature and moisture regimes are some of the major constraints to achieving high productivity levels, especially in rainfed crops grown by low-resource farmers. Nonetheless, the digital tools have the potential to revolutionize agriculture in the developing world given the rapid spread of mobile technology. These tools enable widespread scaling of relevant and contextual fertilizer management recommendations unique to farmers' conditions.

REFERENCES

Abdulrachman, S., Susanti, Z., Pahim, A. D., Dobermann, A., and Witt, C. 2004. Site-specific nutrient management in intensive irrigated rice systems of West Java, Indonesia. In: Dobermann, A., Witt, C., and Dawe, D. (Eds.), *Increasing Productivity of Intensive Rice Systems through Site-Specific Nutrient Management*. Science Publishers, Enfield, NH and International Rice Research Institute, Los Baños, Philippines, pp. 171–192.

Aggarwal, G., Sidhu, A., Sekhon, N., Sandhu, K., and Sur, H. 1995. Puddling and N management effects on crop response in a rice-wheat cropping system. *Soil and Tillage Research* **36**: 129–139.

Alam, M. M., Ladha, J., Khan, S. R., Khan, A., and Buresh, R. 2005. Leaf color chart for managing nitrogen fertilizer in lowland rice in Bangladesh. *Agronomy Journal* **97**: 949–959.

Alam, M. M., Ladha, J., Rahman, Z., Khan, S. R., Khan, A., and Buresh, R. 2006. Nutrient management for increased productivity of rice–wheat cropping system in Bangladesh. *Field Crops Research* **96**: 374–386.

Andriamananjara, A., Rakotoson, T., Razanakoto, O., Razafimanantsoa, M.-P., Rabeharisoa, L., and Smolders, E. 2016. Farmyard manure application has little effect on yield or phosphorus supply to irrigated rice growing on highly weathered soils. *Field Crops Research* **198**: 61–69.

Attanandana, T., Kongton, S., Boonsompopphan, B., Polwatana, A., Verapatananirund, P., and Yost, R. 2010. Site-specific nutrient management of irrigated rice in the Central Plain of Thailand. *Journal of Sustainable Agriculture* **34**: 258–269.

Bado, B. V., Aw, A., and Ndiaye, M. 2010. Long-term effect of continuous cropping of irrigated rice on soil and yield trends in the Sahel of West Africa. *Nutrient Cycling in Agroecosystems* **88**: 133–141.

Balasubramanian, V., Morales, A., Cruz, R., and Abdulrachman, S. 1998. On-farm adaptation of knowledge-intensive nitrogen management technologies for rice systems. *Nutrient Cycling in Agroecosystems* **53**: 59–69.

Banayo, N. P. M. C., Haefele, S., Desamero, N. V., and Kato, Y. 2018. On-farm assessment of site-specific nutrient management for rainfed lowland rice in the Philippines. *Field Crops Research* **220**: 88–96.

Becker, M., Ladha, J. K., Simpson, I. C., and Ottow, J. C. G. 1994. Parameters affecting residue nitrogen mineralization in flooded soils. *Soil Science Society of America Journal* **58**: 1666–1671.

Biradar, D. P., Aladakatti, Y. R., Rao, T. N., and Tiwari, K. N. 2006. Site-Specific Nutrient Management for Maximization of Crop Yields in Northern Karnataka.. *Better Crops* **90**: 33–35.

Bunquin, M. A. B., Tandy, S., Beebout, S. J., and Schulin, R. 2017. Influence of soil properties on zinc solubility dynamics under different redox conditions in non–calcareous soils. *Pedosphere* **27**: 96–105.

Buresh, R., and De Datta, S. 1990. Denitrification losses from puddled rice soils in the tropics. *Biology and Fertility of Soils* **9**: 1–13.

Buresh, R. J., Castillo, R., van den Berg, M., and Gabinete, G. 2014. *Nutrient Management Decision Tool for Small-Scale Rice and Maize Farmers.* Technical Bulletin 190. Food and Fertilizer Technology Center, Taipei, Taiwan.

Buresh, R. J., Laureles, E., Dela Torre, J. C., Samson, M., Guerra, M., Sinohin, P. J., and Castillo, R. 2019. Site-specific nutrient management for rice in the Philippines: Calculation of field-specific fertilizer requirements by rice crop manager. *Field Crops Research* **239**: 35–64.

Buresh, R. J., Pampolino, M. F., and Witt, C. 2010. Field-specific potassium and phosphorus balances and fertilizer requirements for irrigated rice-based cropping systems. *Plant and Soil* **335**: 35–64.

Buresh, R. J., Reddy, K. R., and van Kessel, C. 2008. Nitrogen transformations in submerged soils. In: Schepers, J. S., Raun, W. R., Follett, R. F., Fox, R. H. and Randall, G. W. (Eds.). *Nitrogen in Agricultural Systems Agronomy Monograph* **49**: 401–436.

Cassman, K., Dobermann, A., Cruz, P. S., Gines, G., Samson, M., Descalsota, J., Alcantara, J., Dizon, M., and Olk, D. 1996. Soil organic matter and the indigenous nitrogen supply of intensive irrigated rice systems in the tropics. *Plant and Soil* **182**: 267–278.

Cassman, K. G., Dobermann, A., Walters, D. T., and Yang, H. 2003. Meeting cereal demand while protecting natural resources and improving environmental quality. *Annual Review of Environment and Resources* **28**: 315–358.

Cassman, K. G., and Pingali, P. L. 1995. Extrapolating trends from long-term experiments to farmers' fields: The case of irrigated rice in Asia. In: Barnett, V., Payne, R., and Steiner, R. (Eds.), *Agricultural Sustainability in Economic, Environmental, and Statistical Terms.* John Wiley & Sons, London, UK, pp. 64–84.

Chalk, P. M., Craswell, E. T., Polidoro, J. C., and Chen, D. 2015. Fate and efficiency of 15N-labelled slow- and controlled-release fertilizers. *Nutrient Cycling in Agroecosystems* **102**: 167–178.

Chang, S. C., and Jackson, M. L. 1957. Fractionation of soil phosphorus. *Soil Science* **84**: 133–144.

Chen, B., Liu, E., Tian, Q., Yan, C., and Zhang, Y. 2014. Soil nitrogen dynamics and crop residues. A review. *Agronomy for Sustainable Development* **34**: 429–442.

Choudhury, A., and Kennedy, I. 2005. Nitrogen fertilizer losses from rice soils and control of environmental pollution problems. *Communications in Soil Science and Plant Analysis* **36**: 1625–1639.

da Silva, L., Pocojeski, E., Britzke, D., Kaefer, S., Griebeler, G., and dos Santos, D. 2015. Potassium availability to flooded rice in lowland soils. *Pesquisa Agropecuária Tropical* **45**: 379–387.

Dawe, D., and Dobermann, A. 1999. Defining productivity and yield. IRRI Discussion Papers Series no. 33. International Rice Research Institute, Makati City, Philippines.

Dawe, D., Dobermann, A., Ladha, J. K., Yadav, R. L., Bao, L., Gupta, R. K., Lal, P., et al. 2003. Do organic amendments improve yield trends and profitability in intensive rice systems? *Field Crops Research* **83**: 191–213.

De Datta, S. 1987. Advances in soil fertility research and nitrogen fertilizer management for lowland rice. In: Banta, S.J. (Ed.), *Efficiency of Nitrogen Fertilizers for Rice.* International Rice Research Institute, Los Baños, Philippines, pp. 27–41.

De Datta, S., and Buresh, R. 1989. Integrated nitrogen management in irrigated rice. In: Stewart, B. A. (Ed.), *Advances in Soil Science.* Springer, New York, pp. 143–169.

De Datta, S., Obcemea, W., Real, J., Trevitt, A., Freney, J., and Simpson, J. 1989. Measuring nitrogen losses from lowland rice using bulk aerodynamic and nitrogen-15 balance methods. *Soil Science Society of America Journal* **53**: 1275–1281.

de Datta, S. K., and Mikkelsen, D. S. 1985. Potassium nutrition of rice. In Munson, R. D. (Ed.), *Potassium in Agriculture.* American Society of Agronomy, Madison, WI, pp. 665–699.

Ding, W., Xu, X., He, P., Ullah, S., Zhang, J., Cui, Z., and Zhou, W. 2018. Improving yield and nitrogen use efficiency through alternative fertilization options for rice in China: A meta-analysis. *Field Crops Research* **227**: 11–18.

Dobermann, A. 2000. Future intensification of irrigated rice systems. In: Sheehy, J. E., Mitchell, P. L., and Hardy, B. (Eds.), *Redesigning Rice Photosynthesis to Increase Yield.* International Rice Research Institute/Elsevier, Makati City, Philippines, pp. 229–247.

Dobermann, A., Cassman, K., Mamaril, C., and Sheehy, J. 1998. Management of phosphorus, potassium, and sulfur in intensive, irrigated lowland rice. *Field Crops Research* **56**: 113–138.

Dobermann, A., Cruz, P. C. S., and Cassman, K. G. 1996. Fertilizer inputs, nutrient balance, and soil nutrient-supplying power in intensive, irrigated rice systems. I. Potassium uptake and K balance. *Nutrient Cycling in Agroecosystems* **46**: 1–10.

Dobermann, A., Dawe, D., Roetter, R. P., and Cassman, K. G. 2000. Reversal of rice yield decline in a long-term continuous cropping experiment. *Agronomy Journal* **92**: 633–643.

Dobermann, A., and Fairhurst, T. H. 2000. *Rice: Nutrient Disorders & Nutrient Management*. International Rice Research Institute, Los Baños, Philippines.

Dobermann, A., Witt, C., Abdulrachman, S., Gines, H., Nagarajan, R., Son, T., Tan, P., Wang, G., Chien, N., and Thoa, V. 2003a. Estimating indigenous nutrient supplies for site-specific nutrient management in irrigated rice. *Agronomy Journal* **95**: 924–935.

Dobermann, A., Witt, C., Abdulrachman, S., Gines, H., Nagarajan, R., Son, T., Tan, P., Wang, G., Chien, N., and Thoa, V. 2003b. Soil fertility and indigenous nutrient supply in irrigated rice domains of Asia. *Agronomy Journal* **95**: 913–923.

Dobermann, A., Witt, C., and Dawe, D. 2004. *Increasing the Productivity of Intensive Rice Systems through Site-Specific Nutrient Management*. Science Publishers, Enfield, NH, and International Rice Research Institute, Los Baños, Philippines.

Dobermann, A., Witt, C., Dawe, D., Abdulrachman, S., Gines, H., Nagarajan, R., Satawathananont, S., Son, T., Tan, P., and Wang, G. 2002. Site-specific nutrient management for intensive rice cropping systems in Asia. *Field Crops Research* **74**: 37–66.

Dotaniya, M. L. 2013. Impact of crop residue management practices on yield and nutrient uptake in rice-wheat system. *Current Advances in Agricultural Sciences* **5**: 269–271.

Duy, P. Q., Abe, A., Hirano, M., Sagawa, S., and Kuroda, E. 2004. Analysis of lodging-resistant characteristics of different rice genotypes grown under the standard and nitrogen-free basal dressing accompanied with sparse planting density practices. *Plant Production Science* **7**: 243–251.

Fageria, N. K., Carvalho, G. D., Santos, A. B., Ferreira, E. P. B., and Knupp, A. M. 2011. Chemistry of lowland rice soils and nutrient availability. *Communications in Soil Science and Plant Analysis* **42**: 1913–1933.

Fageria, N. K., Slaton, N., and Baligar, V. 2003. Nutrient management for improving lowland rice productivity and sustainability. *Advances in Agronomy* **80**: 63–152.

FAO. 2006. *Fertilizer Use by Crop*. FAO Fertilizer and Plant Nutrition Bulletin no. 17, FAO, Rome, Italy: 109 pp.

FAO. 2014. *A Regional Rice Strategy for Sustainable Food Security in Asia and the Pacific*. RAP Publication 2014/05. FAO, Regional Office for Asia and the Pacific, Bangkok, Thailand: 52 pp.

FAOSTAT. 2019. Food and Agriculture Organization crop statistics. http://www.fao.org/faostat/en/#data.

Farooq, M., Siddique, K. H. M., Rehman, H., Aziza, T., Lee, D.-J., and Wahide, A. 2011. Rice direct seeding: Experiences, challenges and opportunities. *Soil and Tillage Research* **111**: 87–98.

Fillery, I., and Vlek, P. 1982. The significance of denitrification of applied nitrogen in fallow and cropped rice soils under different flooding regimes I. Greenhouse experiments. *Plant and Soil* **65**: 153–169.

Flinn, J. C., and De Datta, S. 1984. Trends in irrigated-rice yields under intensive cropping at Philippine research stations. *Field Crops Research* **9**: 1–15.

Fu, M. H., Xu, X. C., and Tabatabai, M. A. 1987. Effect of pH on nitrogen mineralization in crop-residue-treated soils. *Biology and Fertility of Soils* **5**: 115–119.

Gines, H., Redondo, G., Estigoy, A., and Dobermann, A. 2004. Site-specific nutrient management in irrigated rice systems of Central Luzon, Philippines. In: Dobermann, A., Witt, C., and Dawe, D. (Eds.), Increasing Productivity of Intensive Rice Systems through Site-Specific Nutrient Management. Science Publishers, Enfield, NH and International Rice Research Institute, Los Baños, Philippines, pp. 145–169.

GRiSP (Global Rice Science Partnership). 2013. *Rice Almanac*. 4th ed. International Rice Research Institute, Los Baños,Philippines, 283 p.

Guanghuo, W., Sun, Q., Fu, R., Huang, X., Ding, X., Wu, J., He, Y., Dobermann, A., and Witt, C. 2004. Site-specific nutrient management in irrigated rice systems of Zhejiang Province, China. In: Dobermann, A., Witt, C., and Dawe, D. (Eds.), *Increasing Productivity of Intensive Rice Systems through Site-specific Nutrient Management*. Science Publishers, Enfield, NH and International Rice Research Institute, Los Baños, Philippines, pp. 243–263.

Guntiñas, M. E., Leirós, M. C., Trasar-Cepeda, C., and Gil-Sotres, F. 2012. Effects of moisture and temperature on net soil nitrogen mineralization: A laboratory study. *European Journal of Soil Biology* **48**: 73–80.

Gupta, G., Shrestha, A., Shrestha, A., and Amgain, L. P. 2016. Evaluation of different nutrient management practice in yield and growth in rice in Morang district. *Advances in Plants and Agriculture Research* **6**: 187–191.

Haefele, S., and Konboon, Y. 2009. Nutrient management for rainfed lowland rice in northeast Thailand. *Field Crops Research* 114: 374–385.

Haefele, S., Nelson, A., and Hijmans, R. J. 2014. Soil quality and constraints in global rice production. *Geoderma* 235: 250–259.

Haefele, S. M., Saito, K., N'Diaye, K. M., Mussgnug, F., Nelson, A., and Wopereis, M. C. 2013. Increasing rice productivity through improved nutrient use in Africa. In: Wopereis, M. C., Johnson, D. E., Ahmadi, N., Tollens, E., and Jalloh, A. (Eds.) *Realizing Africa's Rice Promise*, CABI: 250–264.

Heffer, P. 2009. *Assessment of Fertilizer Use by Crop at the Global Level*. International Fertilizer Industry Association, Paris. www.fertilizer.org/ifa/Home-Page/LIBRARY/Publication-database.html/Assessment-of-Fertilizer-Use-by-Crop-at-the-Global-Level-2006-07-2007-08.html2.

Hirose, S., and Wakatsuki, T. 2002. Restoration of inland valley ecosystems in West Africa. Association of Agriculture & Forestry Statistics, pp. 574.

Hobbs, P., and Gupta, R. 2000. Sustainable resource management in intensively cultivated irrigated rice–wheat cropping systems of the Indo-Gangetic Plains of South Asia: Strategies and options. In: *Proceedings of the International Conference on Managing Natural Resources for Sustainable Production in 21st Century*. Indian Society of Soil Science, New Delhi, India, pp. 14–18.

Hossner, L. R., and Baker, W. H. 1988. Phosphorus transformations in flooded soils. In: Hook, D. D. et al. (Eds.), *The Ecology and Management of Wetlands*. Donal D. Hook Publishers, pp. 293–306.

Hu, R., Cao, J., Huang, J., Peng, S., Huang, J., Zhong, X., Zou, Y., Yang, J., and Buresh, R. J. 2007. Farmer participatory testing of standard and modified site-specific nitrogen management for irrigated rice in China. *Agricultural Systems* 94: 331–340.

Islam, Z., Bagchi, B., and Hossain, M. 2007. Adoption of leaf color chart for nitrogen use efficiency in rice: Impact assessment of a farmer-participatory experiment in West Bengal, India. *Field Crops Research* 103: 70–75.

IUSS Working Group. 2006. World reference base for soil resources 2006: A framework for international classification, correlation and communication. *World Soil Resources Reports*, No. 103. FAO, Rome.

Janssen, B. H., Guiking, F., van der Eijk, D., Smaling, E., Wolf, J., and van Reuler, H. 1990. A system for quantitative evaluation of the fertility of tropical soils (QUEFTS). *Geoderma* 46: 299–318.

Kawaguchi, K., and Kyuma, K. 1977. Paddy soils in tropical Asia. Their material nature and fertility. In: *Paddy Soils in Tropical Asia: Their Material Nature and Fertility*. University of Hawaii Press, Honolulu, HI.

Khalid, R. A., Patrick, W. H., and Peterson, F. J. 1979. Relationship between rice yield and soil phosphorus evaluated under aerobic and anaerobic conditions. *Soil Science and Plant Nutrition* 25: 155–164.

Khuong, T. Q., Thi, T., Huan, N., Tan, P., and Buresh, R. 2007. Effect of site specific nutrient management on grain yield, nutrient use efficiency and rice production profit in the Mekong Delta. *Omonrice* 158: 153–158.

Khurana, H. S., Phillips, S. B., Alley, M. M., Dobermann, A., Sidhu, A. S., and Peng, S. 2008. Agronomic and economic evaluation of site-specific nutrient management for irrigated wheat in northwest India. *Nutrient Cycling in Agroecosystems* 82: 15–31.

Khurana, H. S., Phillips, S. B., Dobermann, A., Sidhu, A. S., and Peng, S. 2007. Performance of site-specific nutrient management for irrigated, transplanted rice in northwest India. *Agronomy Journal* 99: 1436–1447.

Khurana, H. S., Sidhu, A., Singh, B., and Singh, Y. 2009. Performance of site-specific nutrient management in a rice-wheat cropping system. *The Proceedings of the International Plant Nutrition Colloquium XVI*, UC Davis, California.

Kirk, G. 2004. *The Biogeochemistry of Submerged Soils*. Wiley, Chichester, UK.

Kirk, G., and Olk, D. 2000. Carbon and nitrogen dynamics in flooded soils. *Proceedings of the Workshop on Carbon and Nitrogen Dynamics in Flooded Soils*, 19–22 April 1999. International Rice Research Institute (IRRI), Los Baños, Philippines.

Kögel-Knabner, I., Amelung, W., Cao, Z., Fiedler, S., Frenzel, P., Jahn, R., Kalbitz, K., Kölbl, A., and Schloter, M. 2010. Biogeochemistry of paddy soils. *Geoderma* 157: 1–14.

Kondo, M., Singh, C. V., Agbisit, R., and Murthy, M. V. R. 2005. Yield response to urea and controlled-release urea as affected by water supply in tropical upland rice. *Journal of Plant Nutrition* 28: 201–219.

Kumar, V., Jat, H. S., Sharma, P. C., Balwinder-Singh, Gathala, M. K., Malik, R. K., Kamboj, B. R., et al. 2018. Can productivity and profitability be enhanced in intensively managed cereal systems while reducing the environmental footprint of production? Assessing sustainable intensification options in the bread-basket of India. *Agriculture, Ecosystems & Environment* 252: 132–147.

Ladha, J., Khind, C., Gupta, R., Meelu, O., and Pasuquin, E. 2004. Long-term effects of organic inputs on yield and soil fertility in the rice–wheat rotation. *Soil Science Society of America Journal* 68: 845–853.

Ladha, J. K., Pathak, H., Krupnik, T. J., Six, J., and van Kessel, C. 2005. Efficiency of fertilizer nitrogen in cereal production: Retrospects and prospects. *Advances in Agronomy* 87: 85–156.

Ladha, J. K., Reddy, C. K., Padre, A. T., and van Kessel, C. 2011. Role of nitrogen fertilization in sustaining organic matter in cultivated soils. *Journal of Environmental Quality* **40**: 1756–1766.

Lassaletta, L., Billen, G., Grizzetti, B., Anglade, J., and Garnier, J. 2014. 50 year trends in nitrogen use efficiency of world cropping systems: The relationship between yield and nitrogen input to cropland. *Environmental Research Letters* **9**: 1–9.

Liesack, W., Schnell, S., and Revsbech, N. P. 2000. Microbiology of flooded rice paddies. *FEMS Microbiology Reviews* **24**: 625–645.

Liu, H., Won, P. L. P., Banayo, N. P. M. C., Nie, L., Peng, S. B., and Kato, Y. 2019. Late-season nitrogen applications improve grain yield and fertilizer-use efficiency of dry direct-seeded rice in the tropics. *Field Crops Research* **233**: 114–120.

Lu, C., and Tian, H. 2017. Global nitrogen and phosphorus fertilizer use for agriculture production in the past half century: Shifted hot spots and nutrient imbalance. *Earth System Science Data* **9**: 181–192.

Mandal, M., Dutta, S., Majumdar, K., Satyanarayana, T., Pampolino, M., Govil, V., Johnston, A., and Shrotriya, G. 2015. Enhancing rice yield, profitability, and phosphorus use efficiency in West Bengal using the Nutrient Expert® fertilizer decision support tool. *Better Crops – South Asia*: 12–14.

Marahatta, S. 2017. Increasing productivity of an intensive rice based system through site specific nutrient management in Western Terai of Nepal. *Journal of Agriculture and Environment* **18**: 140–150.

Mi, W., Gao, Q., Xia, S., Zhao, H., Wu, L., Mao, W., Hu, Z., and Liu, Y. 2019. Medium-term effects of different types of N fertilizer on yield, apparent N recovery, and soil chemical properties of a double rice cropping system. *Field Crops Research* **234**: 87–94.

Mikkelsen, D., De Datta, S., and Obcemea, W. 1978. Ammonia volatilization losses from flooded rice soils 1. *Soil Science Society of America Journal* **42**: 725–730.

Moya, P., Dawe, D., Pabale, D., Tiongco, M., Chien, N., Devarajan, S., Djatiharti, A., Lai, N., Niyomvit, L., and Ping, H. 2004. The economics of intensively irrigated rice in Asia. In: Dobermann, A., Witt, C., and Dawe, D. (Eds.), *Increasing Productivity of Intensive Rice Systems through Site-Specific Nutrient Management*. Science Publishers, Enfield, NH and International Rice Research Institute, Los Baños, Philippines, pp. 29–58.

Nagarajan, R., Ramanathan, S., Muthukrishnan, P., Stalin, P., Ravi, V., Babu, M., Selvam, S., Sivanatham, M., Dobermann, A., and Witt, C. 2004. Site-specific nutrient management in irrigated rice systems of Tamil Nadu, India. In: Dobermann, A., Witt, C., and Dawe, D. (Eds.), *Increasing Productivity of Intensive Rice Systems through Site-Specific Nutrient Management*. Science Publishers, Enfield, NH and International Rice Research Institute, Los Baños, Philippines, pp. 101–123.

Pampolino, M., C., W., Pasuquin, E., Johnston, A., and Fisher, M. J. 2012. Development approach and evaluation of the Nutrient Expert software for nutrient management in cereal crops. *Computers and Electronics in Agriculture* **88**: 103–110.

Pampolino, M. F., Manguiat, I. J., Ramanathan, S., Gines, H., Tan, P., Chi, T., Rajendran, R., and Buresh, R. 2007. Environmental impact and economic benefits of site-specific nutrient management SSNM in irrigated rice systems. *Agricultural Systems* **93**: 1–24.

Patrick, W. H. 1966. Apparatus for controlling the oxidation-reduction potential of waterlogged soils. *Nature* **212**: 1278.

Patrick, W. H., and Jugsujinda, A. 1992. Sequential reduction and oxidation of inorganic nitrogen, manganese, and iron in flooded soil. *Soil Science Society of America Journal* **56**: 1071–1073.

Patrick, W. H., and Mahapatra, I. 1968. Transformation and availability to rice of nitrogen and phosphorus in waterlogged soils. *Advances in Agronomy* **20**: 323–359.

Peng, S., Bouman, B., Visperas, R. M., Castañeda, A., Nie, L., and Park, H.-K. 2006a. Comparison between aerobic and flooded rice in the tropics: Agronomic performance in an eight-season experiment. *Field Crops Research* **96**: 252–259.

Peng, S., Buresh, R. J., Huang, J., Yang, J., Zou, Y., Zhong, X., Wang, G., and Zhang, F. 2006b. Strategies for overcoming low agronomic nitrogen use efficiency in irrigated rice systems in China. *Field Crops Research* **96**: 37–47.

Peng, S., Buresh, R. J., Huang, J., Zhong, X., Zou, Y., Yang, J., Wang, G., Liu, Y., Hu, R., and Tang, Q. 2010. Improving nitrogen fertilization in rice by site-specific N management: A review. *Agronomy for Sustainable Development* **30**: 649–656.

Peng, S., Huang, J., Sheehy, J. E., Laza, R. C., Visperas, R. M., Zhong, X., Centeno, G. S., Khush, G. S., and Cassman, K. G. 2004. Rice yields decline with higher night temperature from global warming. *Proceedings of the National Academy of Sciences of the United States of America* **101**: 9971–9975.

Pezeshki, S. R., and DeLaune, R. D. 2012. Soil oxidation-reduction in wetlands and its impact on plant functioning. *Biology* **1**: 196–221.

Ponnamperuma, F. 1972. The chemistry of submerged soils. *Advances in Agronomy* **24**: 29–96.

Ponnamperuma, F. 1984. Straw as a source of nutrients for wetland rice. *Organic Matter and Rice* **117**: 136.

Qureshi, A., Singh, D. K., and Kumar, A. 2018. Climate smart nutrient management (CSNM) for enhanced use efficiency and productivity in rice and wheat under rice-wheat cropping system. *International Journal of Current Microbiology and Applied Sciences* **7**: 4166–4176.

Rajendran, R., Stalin, P., Ramanathan, S., and Buresh, R. 2010. Site-specific nitrogen and potassium management for irrigated rice in the Cauvery Delta. *Better Crops* 7.

Rao, A. N., Johnson, D. E., Sivaprasad, B., Ladha, J. K., and Mortimer, A. M. 2007. Weed management in direct-seeded rice. *Advances in Agronomy* **93**: 153–255.

Reddy, K. R., Patrick, W. H., and Broadbent, F. E. 1984. Nitrogen transformations and loss in flooded soils and sediments. *CRC Critical Reviews in Environmental Control* **13**: 273–309.

Roger, P.-A., and Ladha, J. 1992. Biological N 2 fixation in wetland rice fields: Estimation and contribution to nitrogen balance. In: Bohlool, J. K. L. G. B. (Ed.), *Biological Nitrogen Fixation for Sustainable Agriculture*. Springer, pp. 41–55.

Sahrawat, K. L. 2010. Nitrogen mineralization in lowland rice soils: The role of organic matter quantity and quality. *Archives of Agronomy and Soil Science* **56**: 337–353.

Sahrawat, K. L. 2015. Redox potential and pH as major drivers of fertility in submerged rice soils: A conceptual framework for management. *Communications in Soil Science and Plant Analysis* **46**: 1597–1606.

Saito, K., Asai, H., Zhao, D., Laborte, A. G., and Grenier, C. 2018. Progress in varietal improvement for increasing upland rice productivity in the tropics. *Plant Production Science* **21**: 145–158.

Saito, K., Diack, S., Dieng, I., and N'Diaye, M. K. 2015. On-farm testing of a nutrient management decision-support tool for rice in the Senegal River Valley. *Computers and Electronics in Agriculture* **116**: 36–44.

Saito, K., Linquist, B., Keobualapha, B., Phanthaboon, K., Shiraiwa, T., and Horie, T. 2006. Cropping intensity and rainfall effects on upland rice yields in northern Laos. *Plant and Soil* **284**: 175–185.

Saito, K., Nelson, A., Zwart, S. J., Niang, A., Sow, A., Yoshida, H., and Wopereis, M. 2013. Towards a better understanding of biophysical determinants of yield gaps and the potential for expansion of the rice area in Africa. *Realizing Africa's Rice Promise* 188–203.

Saito, K., and Sharma, S. 2018. *e-Agriculture Promising Practice: Rice Crop Manager and RiceAdvice: Decision Tools for Rice Crop Management*. http://www.fao.org/publications/card/en/c/I9039EN.

Saito, K., Vandamme, E., Johnson, J.-M., Tanaka, A., Senthilkumar, K., Dieng, I., Akakpo, C., Gbaguidi, F., Segda, Z., and Bassoro, I. 2019. Yield-limiting macronutrients for rice in sub-Saharan Africa. *Geoderma* **338**: 546–554.

Sander, B. O., Samson, M., and Buresh, R. J. 2014. Methane and nitrous oxide emissions from flooded rice fields as affected by water and straw management between rice crops. *Geoderma* **235**: 355–362.

Satawathananont, S., Chatuporn, S., Niyomvit, L., Kongchum, M., Sookthongsa, J., and Dobermann, A. 2004. Site-specific nutrient management in irrigated rice systems of Central Thailand. In: Dobermann, A., Witt, C., and Dawe, D. (Eds.), *Increasing Productivity of Intensive Rice Systems through Site-specific Nutrient Management*. Science Publishers, Enfield, NH and International Rice Research Institute, Los Baños, Philippines, pp. 125–143.

Shahandeh, H., Hossner, L. R., and Turner, F. T. 1994. Phosphorus relationships in flooded rice soils with low extractable phosphorus. *Soil Science Society of America Journal* **58**: 1184–1189.

Sharma, P. K., and De Datta, S. 1985. Effects of puddling on soil physical properties and processes. In: *Soil Physics and Rice*. International Rice Research Institute, Los Baños, Philippines, pp. 217–234.

Sharma, S., Panneerselvam, P., Castillo, R., Manohar, S., Raj, R., Ravi, V., and Buresh, R. J. 2019. Web-based tool for calculating field-specific nutrient management for rice in India. *Nutrient Cycling in Agroecosystems* **113**: 21–33.

Singh, A., Choudhury, B., and Bouman, B. 2002a. Effects of rice establishment methods on crop performance, water use, and mineral nitrogen. In: *Water-Wise Rice Production*. International Rice Research Institute, Los Baños, Philippines, pp. 237–246.

Singh, B., and Singh, Y. 2008. Reactive nitrogen in Indian agriculture: Inputs, use efficiency and leakages. *Current Science* **94**: 1382–1393.

Singh, B., Singh, Y., Ladha, J. K., Bronson, K. F., Balasubramanian, V., Singh, J., and Khind, C. S. 2002b. Chlorophyll meter– and leaf color chart–based nitrogen management for rice and wheat in Northwestern India. *Agronomy Journal* **94**: 821–829.

Singh, V., Shukla, A., Singh, M., Majumdar, K., Mishra, R., Rani, M., and Singh, S. 2015. Effect of site-specific nutrient management on yield, profit and apparent nutrient balance under pre-dominant cropping systems of Upper Gangetic Plains. *Indian Journal of Agricultural Sciences* **85**: 335–343.

Singh, V. K., Dwivedi, B. S., Buresh, R. J., Jat, M., Majumdar, K., Gangwar, B., Govil, V., and Singh, S. K. 2013. Potassium fertilization in rice–wheat system across Northern India: Crop performance and soil nutrients. *Agronomy Journal* **105**: 471–481.

Singh, Y., Khind, C. S., and Singh, B. 1991. Efficient management of leguminous green manures in wetland rice. *Advances in Agronomy* **45**: 135–189.

Singh, Y., Singh, B., and Timsina, J. 2005. Crop residue management for nutrient cycling and improving soil productivity in rice-based cropping systems in the tropics. *Advances in Agronomy* **85**: 269–407.

Son, T. T., van Chien, N., Thoa, V. T. K., Dobermann, A., and Witt, C. 2004. Site-specific nutrient management in irrigated rice systems of the Red River Delta of Vietnam. In: Dobermann, A., Witt, C., and Dawe, D. (Eds.), *Increasing Productivity of Intensive Rice Systems through Site-Specific Nutrient Management*. Science Publishers, Enfield, NH and International Rice Research Institute, Los Baños, Philippines, pp. 217–242.

Sweeney, M., and McCouch, S. 2007. The complex history of the domestication of rice. *Annals of Botany* **100**: 951–957.

Tan, P. S., Tuyen, T. Q., Huan, T. T. N., Khuong, T. Q., Hoai, N. T., Diep, L. N., Dung, H. T., Phung, C., Lai, N. X., and Dobermann, A. 2004. Site-specific nutrient management in irrigated rice systems of the Mekong Delta of Vietnam. In: Dobermann, A., Witt, C., and Dawe, D. (Eds.), *Increasing Productivity of Intensive Rice Systems through Site-specific Nutrient Management*. Science Publishers, Enfield, NH and International Rice Research Institute, Los Baños, Philippines, pp. 193–215.

Tanaka, A., Johnson, J.-M., Senthilkumar, K., Akakpo, C., Segda, Z., Yameogo, L. P., Bassoro, I., Lamare, D. M., Allarangaye, M. D., and Gbakatchetche, H. 2017. On-farm rice yield and its association with biophysical factors in sub-Saharan Africa. *European Journal of Agronomy* **85**: 1–11.

Thuan, H. N., and Long, D. T. 2010. Use rice straw from previous season for the following season on degraded soil, Bac Giang Province, Vietnam. *Journal of Science and Development* **8**: 843–849.

Tilman, D., Cassman, K. G., Matson, P. A., Naylor, R., and Polasky, S. 2002. Agricultural sustainability and intensive production practices. *Nature* **418**: 671.

Tripathi, R., Sharma, P., and Singh, S. 2005. Tilth index: An approach to optimize tillage in rice–wheat system. *Soil and Tillage Research* **80**: 125–137.

Tsujimoto, Y., Rakotoson, T., Tanaka, A., and Saito, K. 2019. Challenges and opportunities for improving N use efficiency for rice production in sub-Saharan Africa. *Plant Production Science*. doi:10.1080/13439 43X.2019.1617638.

Turner, F. T., and Gilliam, J. W. 1976. Increased P processes in flooded soils. *Plant and Soil* **45**: 365–377.

van Hach, C., and Tan, P. S. 2007. Study on site-specific nutrient management (SSNM) for high-yielding rice in the Mekong Delta. *Omonrice* **15**: 144–152.

van Ittersum, M. K., van Bussel, L. G., Wolf, J., Grassini, P., van Wart, J., Guilpart, N., Claessens, L., de Groot, H., Wiebe, K., Mason-D'Croz, D., Yang, H., Boogaard, H., van Oort, P. A. J., van Loon, M. P., Saito, K., Adimo, O., Adjei-Nsiah, S., Agali, A., Bala, A., Chikowo, R., Kaizzi, K., Kouressy, M., Makoi, J. H. J. R., Ouattara, K., Tesfaye, K., and Cassman, K.G. 2016. Can sub-Saharan Africa feed itself? *Proceedings of the National Academy of Sciences* **113**: 14964–14969.

van Oort, P. 2018. Mapping abiotic stresses for rice in Africa: Drought, cold, iron toxicity, salinity and sodicity. *Field Crops Research* **219**: 55–75.

van Oort, P., Saito, K., Dieng, I., Grassini, P., Cassman, K. G., and van Ittersum, M. K. 2017. Can yield gap analysis be used to inform R&D prioritisation? *Global Food Security* **12**: 109–118.

van Oort, P., Saito, K., Tanaka, A., Amovin-Assagba, E., van Bussel, L., van Wart, J., De Groot, H., van Ittersum, M., Cassman, K., and Wopereis, M. 2015. Assessment of rice self-sufficiency in 2025 in eight African countries. *Global Food Security* **5**: 39–49.

Wade, L., Amarante, S., Olea, A., Harnpichitvitaya, D., Naklang, K., Wihardjaka, A., Sengar, S., Mazid, M., Singh, G., and McLaren, C. 1999. Nutrient requirements in rainfed lowland rice. *Field Crops Research* **64**: 91–107.

Wang, G., Dobermann, A., Witt, C., Sun, Q., and Fu, R. 2001. Performance of site-specific nutrient management for irrigated rice in southeast China. *Agronomy Journal* **93**: 869–878.

Wassmann, R., Butterbach-Bahl, K., and Dobermann, A. 2007. Irrigated rice production systems and greenhouse gas emissions: Crop and residue management trends, climate change impacts and mitigation strategies. *CAB Reviews: Perspectives in Agriculture, Veterinary Science, Nutrition and Natural Resources* **2**.

Willett, I., and Higgins, M. 1978. Phosphate sorption by reduced and reoxidized rice soils. *Soil Research* **16**: 319–326.

Witt, C., Buresh, R. J., Peng, S. B., Balasubramanian, V. A. D. 2007. Nutrient management. In: Fairhurst, T., Witt, C., Buresh, R., and Dobermann, A. (Eds.), *Rice: A practical guide to nutrient management*. International Rice Research Institute (IRRI), Los Baños, Philippines and International Plant Nutrition Institute (IPNI) and International Potash Institute (IPI), Singapore, 1–45.

Witt, C., and Dobermann, A. 2004. Toward a decision support system for site-specific nutrient management. In: Dobermann, A., Witt, C., and Dawe, D. (Eds.), *Increasing Productivity of Intensive Rice Systems through Site-Specific Nutrient Management*. Science Publishers, Enfield, NH and International Rice Research Institute, Los Baños, Philippines, 359–395.

Witt, C., Pasuquin, E., Mutters, R., and Buresh, R. 2005. New leaf color chart for effective nitrogen management in rice. *Better Crops* **89**: 36–39.

Witt, C., Pasuquin, J., Pampolino, M., Buresh, R., and Dobermann, A. 2009. *A Manual for the Development and Participatory Evaluation of Site-Specific Nutrient Management for Maize in Tropical, Favorable Environments*. International Plant Nutrition Institute, Penang, Malaysia.

Xu, X., He, P., Yang, F., Ma, J., Pampolino, M. F., Johnston, A. M., and Zhou, W. 2017. Methodology of fertilizer recommendation based on yield response and agronomic efficiency for rice in China. *Field Crops Research* **206**: 33–42.

Xu, Y., Nie, L., Buresh, R. J., Huang, J., Cui, K., Xu, B., Gong, W., and Peng, S. 2010. Agronomic performance of late-season rice under different tillage, straw, and nitrogen management. *Field Crops Research* **115**: 79–84.

Yoneyama, T., Lee, K., and Yoshida, T. 1977. Decomposition of rice residues in tropical soils. *Soil Science and Plant Nutrition* **23**: 287–295.

Yoon, C. G. 2009. Wise use of paddy rice fields to partially compensate for the loss of natural wetlands. *Paddy and Water Environment* **7**: 357.

Yoshida, S. 1972. Physiological aspects of grain yield. *Annual Review of Plant Physiology* **23**: 437–464.

Yu, K., and Rinklebe, J. 2011. Advancement in soil microcosm apparatus for biogeochemical research. *Ecological Engineering* **37**: 2071–2075.

11 Nuclear Powered Agriculture
Fertilizers and Amendments

Darryl D. Siemer

CONTENTS

11.1 INTRODUCTION

Justus von Liebig developed the "Law of the Minimum" to explain the basics of improving soils. His principle states that plant growth is determined by the scarcest, "limiting" nutrient – if even one of the required nutrients is deficient, plants will not thrive and produce at their optimum levels. This chapter describes how a properly implemented "nuclear renaissance" could address three issues key to assuring genuinely sustainable agriculture – sufficient water, sufficient nutrients, and enough power/energy to properly plant, harvest, transport, and distribute the resulting crops. I shall be assuming that by circa 2100 AD, the people responsible for developing such a renaissance will have done their job, which means that many of the constraints facing farmers today will have disappeared, thereby allowing "unconventional" solutions to their descendants' technical issues. Those solutions will include irrigation with desalinated seawater, fertilization with powdered plutonic rock (basalt) plus ammonia and/or nitric acid, and the use of equipment powered one way or another with clean (no greenhouse gas emissions) nuclear energy.

11.2 THE FUTURE'S ENERGY REQUIREMENTS

Since this chapter is about how a properly implemented "nuclear renaissance" (not just more of the same) could solve most of the future's agricultural problems, let's begin by coming up with an estimate of how much energy/power that would require.

The exceptionally "rich" lifestyle of the United States' ~320 million people is nominally supported by about 99.5 EJ (98 quads) of raw/primary energy per annum, which figure has remained roughly constant for over two decades (LLNL 2018). It's "nominal" because its consumer-driven economy consumes energy and other resources from an area outside of and greater than that of the United States, which isn't counted in such compilations.

Approximately 80% of the United States' primary/raw energy is provided by fossil fuels, which translates to a mean per capita raw/primary energy consumption rate (power) of 9860 watts [99.5E +18J/3.15E+7/320E+6], or about eight [99.5/3.2e+8/570/7.5E+9] times that of the world's average person today. Since one joule's worth of raw/primary (heat) energy currently provides about 0.4 joules worth of useful "energy services" (the efficiency of most of fossil fuel's applications is Carnot-limited) and Europeans apparently live almost as well, consuming one-half that much raw/primary energy per capita, let's assume that supporting the lifestyles of each of the future's equitably EU-rich people would require ~2 kW's [9860 × 0.5 × 0.4 = 1972 ≈ 2000] worth of energy services (electricity). Consequently, a world with 11.2 billion such individuals (the United Nations' current best guess) must possess power plants able to supply an average power of about 22.4 [11.2E+9 × 2000 × 3.15E+7/1E+ 12/3.17e+7 = 22.4] TWe (terawatt electrical – that's about 3.5 times as much useful energy as today's people consume). Finally, assuming that each individual region's peak power demand is about 40% higher than its average and that no magic worldwide, zero-loss, "super grid" exists, our descendants would need ~30,000 [22.4 × 1.4 × 1012/109] one GWe power plants to live that well.

That power could not be generated with fossil fuels because even if there were enough of them (there isn't), burning it would have catastrophic consequences. For example, the raw/primary (heat) energy represented by the world's remaining 1139 billion Mg (tons) of coal reserves, 187 trillion m^3 of natural gas, and 1.707 trillion barrels of petroleum (BP 2017) is about 5.0E+22 J's, which, if consumed by 40% Carnot (heat to electricity) efficient power plants, could generate 22 TWe for 29 years – ~35% of a current first-world human life span. Additionally, those reserves collectively contain about 1200 Gt (10 + 9 Mg) of carbon, which, if converted to CO_2 and dumped into the atmosphere, would push global warming well past any of the "tipping points" (irreversible change in the earth's climate) suggested by the world's climate modeling experts (Hansen 2008).

That much energy could not be produced with the same sorts of "burner-type" reactors utilized in today's nuclear power plants, either. The reason for this is that they are extremely inefficient in terms of fuel (uranium) use – only about one of 160 uranium atoms mined/processed is actually "burned"; the rest go to waste. This means that enough such reactors to generate 22.4 GWe of power would run through 100% of the world's "affordable" uranium reserves within about five years. To become "renewable," nuclear power must be generated with reactors that consume nearly 100% of the uranium (or thorium) mined/processed, i.e., with breeder-type reactors coupled with fuel recycling/ reprocessing systems. That's certainly possible, but civilian nuclear power remains in "technological lock-in" because its first movers succeeded in establishing a profitable business model that didn't emphasize efficiency or long-term sustainability. It worked for quite a long time, but its drawbacks eventually rendered nuclear power much less attractive than it should/could be (Cowan 1990).

In order to better understand what these facts, figures, and trends mean, it's useful to consider them on a longer time scale than that which we customarily employ. Figure 11.1 was excerpted from a paper written/delivered by the one of the petroleum industry's most influential geologists (and, eventually, most influential gadfly), Professor M. King Hubbert, 63 years ago (Hubbert 1956). It depicts mankind's total energy consumption extending from the dawn of recorded history 5000 years ago to 5000 years in the future based upon two assumptions: that human population eventually stabilizes and that we choose to replace finite fossil fuels with a sustainable (breeder reactor based) nuclear fuel cycle before civilization collapses. While the fossil fuel industries' champions and apologists have repeatedly proven Professor Hubbert "wrong" by cherry-picking assumptions, time scales, and data sets, the fact remains that his figure's message, i.e., that "on such a time scale, the discovery, exploitation, and exhaustion of the earth's fossil fuels will constitute an ephemeral event," is correct.

Another relevant observation was made by English-born American economist and philosopher Kenneth Boulding in a *Science* article almost two decades later (Abelson 1975): "Anyone who believes in indefinite growth in anything physical, on a physically finite planet, is either mad or an economist." (That's why economics is still deemed to be a "dismal" science.)

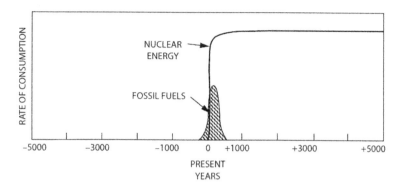

FIGURE 11.1 Mankind's long-term energy consumption. (From Hubbert, M.K., *Nuclear Energy and the Fossil Fuels,* Publication no. 95, Shell Development Company, Exploration and Production Research Division, Houston, TX, June 1956, http://www.hubbertpeak.com/hubbert/1956/1956.)

11.3 AGRICULTURAL ENERGY REQUIREMENTS

11.3.1 DESALINATION

Water is a renewable resource with both highly variable and limited availability. Rainfall, temperature, evaporation, and runoff determine its total availability, and human decisions determine who gets what. Nearly every country in the world experiences water shortages during part of the year, and more than 80 of them suffer from serious water shortages. Clean water resources per capita are declining rapidly as human population increases. Population growth reduces water availability per person and stresses the entire environmental system. Pollution, erosion, runoff, and salinization associated with irrigation, plus the overall inefficient use of water, contribute to the decline in water resources. Allocation of increasingly scarce fresh water generates conflicts between and within countries, industries, and individual communities, with the majority of it everywhere consumed by agriculture. Water shortages are also severely reducing biodiversity in both aquatic and terrestrial ecosystems (Pimentel and Burgess 2013).

The world's 11.2 billion people circa 2100 won't be able to feed themselves with its remaining arable land unless they become able to irrigate it. Because irrigated land almost always produces higher yields than do rain-fed farms and permits double and sometimes even triple cropping in warmer regions, such lands currently provide around 40 percent of global cereal supply. Pumping water onto approximately 10% of the world's total currently arable land (around 300 M ha) consumes around 0.225 EJ/yr. Another 0.05 EJ/yr of indirect energy is devoted to the manufacture and delivery of irrigation equipment (Smil 2008). Around two-thirds of the water currently used for irrigation is drawn from underground aquifers. Energy-intensive electricity-powered deep well pumping accounts for about two-thirds of that, and projections suggest that it will become ~90% by 2050 when shallow reserves are almost depleted (by 2100 they will be totally depleted unless radical changes take place). Current aquifer water extraction rates exceed recharge rates – grossly so in many places. Additionally, global warming is simultaneously exacerbating droughts and melting the glaciers feeding the rivers that provide much of the world's cheap-to-deliver irrigation water. Global warming has caused Mount Kilimanjaro's "snows" to disappear along with most of those within the United States' Glacier National Park. Building more dams won't solve the problem because dams do not create water. Additionally, a comprehensive review of Nigeria's outside-funded dam projects (Tomlinson 2018) concluded that while they do make money for local promoters and the outsiders that fund/support them, they decrease net agricultural productivity by turning fertile downstream floodplains into deserts. In addition to killing wetland-dependent wildlife, those dams have served to lower, not raise, the incomes of far more people than have benefited. Most such dams also don't generate nearly as much electrical power as "promised" due to inadequate maintenance and low (water-limited) capacity factors. Finally, at best, dams represent a temporary fix for the problems that they are built to address because their reservoirs will eventually fill with mud.

Water shortages plus the high cost of desalination – primarily due to high electricity costs – have led some countries rich enough to do so (e.g., China) to reduce their own crop production and rely more heavily upon imported grains.

Because today's situation is unsustainable, I'll next assume that most of the water irrigating the future's farmlands would be generated by desalinating seawater – the earth's only truly inexhaustible/sustainable water source. Israel's solution to its water issues (Israel 2018), demonstrates how a properly managed future technological society could address the whole world's water-related woes. About 20% of Israel's area – about 4500 km^2 – is devoted to agriculture. Its 8.5 million people are fed by ~1.045E+9 m^3 of fresh water applied to its mostly irrigated farmland (Israel 2018). The World Bank reports that about 37.7% of the world's total land area is considered agricultural land, of which approximately 10.6% or about 14 million km^2 is deemed to be arable. Extrapolating Israel's agricultural water demand to that of the whole world suggests that our descendants will need something like 3.2E+12 m^3 (1.045E+9 × 14E+6/4500) of irrigation water. Assuming the ~3 kWh/m^3 energy requirement of today's most popular approach to desalination, reverse osmosis (RO), providing it would require an energy input of 3.51 + 19 joules/a, which corresponds to the full-time output of ~1113 one-GWe ("full sized") nuclear reactors.

The contractual cost of the world's biggest (~1 million m^3/day) RO-based desalination plant (located in Saudi Arabia) is $1.89 billion (Desalination 2020). Collectively, these numbers suggest that building enough RO plants to irrigate the future's farmlands to the same degree that Israel currently does would require a one-time capital expenditure of $16.6 trillion [3.2E+12 × 0.00189/ (365 × 1E+6) – about 75% of the United States' current annual GDP. Similarly addressing California's Central Valley's chronic irrigation water problems should cost only about $40 billion.

In principle, at least, Siemen's electrodialysis-based desalination technology would require only about one-half that number of reactors and is also less apt to become fouled by seawater's other-than-salt impurities (Hussain and Abolaban 2014).

To continue, the world's farmlands' average elevation is about 600 m (2000 feet) above sea level. Assuming that all of the world's desalinated irrigation water would have to be pumped uphill that far, the energy needed to do so would be 1.88E+19 joules [3.2E+12 m^3 × 1000 kg/m^3 × 600 m × 9.8 m/s], which would require another 596 full-sized power plants.

It's useful to note here that the Western world's recent Middle East military incursions will probably end up costing its citizens ~30 times more ($4–$6 trillion) than it would to have built enough nuclear-powered desalination plants to provide sufficient fresh water for everyone living there and thereby address a root cause of that region's almost perpetual turmoil. For example, a recent paper in the *Proceedings of the (US) National Academy of Sciences* (Kelley et al. 2015) points out that the chief driver for today's Syrian conflict is the unrest/poverty generated by relentlessly worsening droughts, not a desire for "regime change."

Another plus for desalination is that its product does not add additional salts to soil and is also better at remediating already oversalinized soils than is groundwater. A final advantage is that because deionized water doesn't already contain near-equilibrium levels of calcium, magnesium, carbonate/bicarbonate, silica, etc., it is a better rock solvent (more corrosive) than is groundwater. The next section will reveal why this is important.

11.3.2 FERTILIZERS

11.3.2.1 Nitrogen

We'll start with an assumption that the future will be fed with today's four most popular food crops (maize or Zea mays, wheat, rice, and soybeans) in the same proportion that they are raised today. We'll also assume that 100% of the world's nominally arable land will be devoted to their cultivation and that

TABLE 11.1

Current Basis Crop Characteristics

Crop	ha*1E+6[a]	Mg/ha[a]	g N/m²/a[b]	% P[c]	%K[c]	kcal/g[c]
Soy	124	2.8	6	70	1.80E+00	4.46
Rice	162	4.5	18	0.34	0.29	3.63
Wheat	218	3.5	8	0.32	0.39	3.32
Maize	191	5.7	15	0.21	0.29	3.65

[a] https://apps.fas.usda.gov/psdonline/circulars/production.pdf.
[b] Cao, P. et al., *Earth Syst. Sci. Data*, 10, 969–984, 2017.
[c] US Agricultural Research Service National Nutrient Database for Standard Reference Legacy Release. https://ndb.nal.usda.gov/ndb/foods/show/305217.

TABLE 11.2

Scale up to 14 Million Km² (2.01 times larger acreage)

Crop	Scaled ha	t N Scaled	Tons Crop	Tot kcal
Soy	2.50E+08	1.50E+07	6.99E+08	3.12E+15
Rice	3.26E+08	5.87E+07	1.47E+09	5.33E+15
Wheat	4.39E+08	3.51E+07	1.54E+09	5.10E+15
Maize	3.85E+08	5.77E+07	2.19E+09	8.00E+15
Sum	1.4e+7	1.67E+08	5.90E+09	2.16E+16
tot kcal at 2500/person/day=			1.022E+16	kcal
excess food energy =			4.75036E+19	Joules

the same yields/ha, composition, and fertilizer needs will not change. Table 11.1 contains the necessary data. The sum of the acreage devoted to their production is currently 6.95 million ha (sum of column 2).

Table 11.2 summarizes the results of extrapolating acreage up to 14 million ha, calculating the recommended amount of nitrogen, and adding up the amounts of P and K required to replace that removed by harvesting the crops. At the bottom, the table also compares the total "food value" (calories) of that food relative to that required by 11.2 billion people consuming 2500 kcal/day.

Assuming the total nitrogen application rate given in Table 11.2 (1.67E+08 Mg), fertilizing the world's cropland would require 2.01E+8 Mg of ammonia each year (one kg N ≈ 1.21 kg of ammonia). An up-to-date estimate (Thyssenkrupps 2019) of ammonia synthesis's energy requirements concludes that each Mg of ammonia made with electrochemically generated hydrogen, pressure swing-generated atmospheric nitrogen, and conventional Haber Bosch processing equipment (in other words, "conventionally") would require about 10 MWh's [10 × 3.6E+9 J] worth of electricity. That, in turn, suggests that satisfying my scenario's nitrogenous fertilizer requirement would require the full-time output of another ~230 one-GWe power plants.

11.3.3 EVERYTHING ELSE THAT PLANTS REQUIRE

Because…

1. Much of the world's farmland has already lost a great deal of its topsoil via erosion,
2. Much of its remaining topsoil is mineral depleted,
3. Basaltic rocks are both intrinsically basic (contain a good deal of magnesium and calcium) and rapidly weathered by natural phenomena and thereby constitute the mechanism

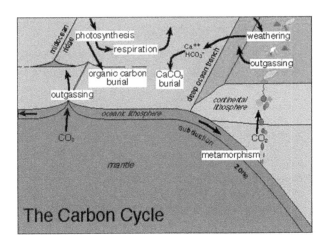

FIGURE 11.2   The earth's carbon cycle. (From http://www.columbia.edu/~vjd1/carbon.htm.)

limiting the earth's atmospheric CO_2 concentration (Figure 11.2) via "mineralization" extant in cultivated soils (Moulton et al. 2000),

4. Most of the earth's crust consists of basalt, a good deal of which is either on or close to the surface of its continents,

5. Such rock contains relatively high concentrations of potassium and phosphorous along with all of the other biologically important elements, which explains why soils comprised of weathered volcanic ash are exceptionally productive (Hensel 1894; Beerling et al. 2018),

...I'll next assume that the phosphorous and potassium required to produce the future's food circa 2100 AD will be provided by amending its farmland with powdered basalt. In order to be effective, any such amendment must weather rapidly enough to release sufficient potassium and phosphorous to support high-yield agriculture, which, in turn, means that the raw silicate rock surfaces must be "fresh" (not already equilibrated with the atmosphere and thereby covered/blocked with secondary phases), ground to a considerably smaller particle size than is the quarry waste–type soil amendment rock powder currently being marketed to hobby farmers, and also mixed with root-zone topsoil, not just dumped upon the surface of the ground (Priyono and Gilkes 2004; Campbell 2009). Based upon the rather limited amount of scientifically planned/supervised experimentation described in the open-access (not paywalled) technical literature, I'm next going to assume that this would require grinding it so that the particles comprising >80% of it possess diameters <10 microns. Since rock grinding is highly energy intensive – much more so than is simply recovering it from a quarry's rock outcrop or waste dump – the cost estimate for this part of my scenario will be based upon that step's energy demand plus the resulting powder's transport and distribution costs.

First, how much powdered rock must be made? The crops assumed above contain phosphorous and potassium that unless recycled back to those fields in some fashion (nightsoils?) must be replaced each year. The compositions of flood basalts vary considerably, but since all originate from the earth's fairly well-mixed underlying magma, for the following estimate I'm going to assume a composition with which I'm familiar (Leeman 1982; Siemer 2019b, see Table 11.3) – that of the basalt comprising Idaho's "Craters of the Moon" National Monument and covering much of the rest of Idaho's Snake River Plain. It contains an average of 0.61 wt% K_2O and 0.55 wt% P_2O_5, which translates to 0.00506 tons of K and 0.00231 Mg of P per Mg of rock dust. Since for this combination of crops, phosphorus comprises Liebig's limiting nutrient, the amount of powdered basalt required to replace that removed with the grain comes to 1.26E+10 Mg [2.91E+7 Mg P/0.00231 Mg P/Mg basalt] – an average of about 9 Mg/ha over 14 million km² of arable land.

TABLE 11.3

Weight % Composition of Major Basalt Types in the Northwest United States

	1	2	3	4	5	6	7	8	9	10	11
SiO_2	49.70	48.64	49.50	50.36	54.37	51.19	47.45	48.73	47.50	47.04	47.84
TiO_2	1.54	1.29	2.31	1.56	1.17	3.13	3.62	3.30	3.79	2.70	1.07
Al_2O_3	17.93	17.54	16.61	15.54	15.28	14.07	13.84	13.88	12.50	15.11	16.35
FeO	9.79	10.12	12.14	11.25	9.46	13.91	15.22	14.41	17.53	13.51	10.18
MnO	0.14	0.18	0.17	0.20	0.16	0.23	0.23	0.20	0.27	0.20	0.18
MgO	6.99	8.09	4.95	6.68	5.91	4.39	5.99	5.88	4.41	7.65	9.66
CaO	10.02	10.61	9.02	10.67	9.79	8.48	9.71	9.72	8.80	10.06	12.34
Na_2O	2.91	2.83	3.57	2.95	2.80	2.72	2.31	2.42	2.44	2.54	2.09
K_2O	0.73	0.46	1.26	0.57	0.77	1.22	0.72	0.73	1.23	0.61	0.17
P_2O_5	0.26	0.23	0.47	0.22	0.29	0.67	0.91	0.73	1.54	0.58	0.18

Source: Leeman, W.P., Olivine tholeiitic basalts of the Snake River Plain, Idaho, in B. Bonnichsen and R. M. Breckenridge, Eds., *Cenozoic Geology of Idaho, Idaho Bureau of Mines and Geology Bulletin*, 26, pp. 181–191, 1982.

A review of rock-grinding technologies (Jankovic 2003) suggests that producing one Mg (ton) of <10 micron basalt powder would require about 100 kWh's worth of electricity. If so, making 1.26E+10 Mg of it would require 4.16E+18 J, which if spread out uniformly throughout one year would require the full-time output of 131 one-GWe power generators.

If that powder were to be transported an average of 1930 km (1200 miles) from mine to farm via an electrified rail system as energy efficient as that currently used to move US coal (185 km/L diesel fuel/short ton), its energy cost would be about 2.43E+17 J (assumes 1.1 Mg/short ton, 33% heat-to-mechanical engine efficiency, and 44.5 MJ/kg diesel fuel with an SpG of 0.85). Trebling that figure to account for fuel consumed by trucks and tractors at the rail heads brings the total to 7.28E+17 joules/a, which corresponds to an annual transportation/distribution energy demand requiring another 23 one-GWe power supplies to satisfy.

Note that the rock powder's transportation cost is much lower than its grinding cost and both are much lower than the cost of the desalinated water and/or nitrogenous fertilizer.

A rock powder application rate of 9 Mg/ha/year is not really very large because it represents only about 0.5% of the mineral matter already within a six-inch deep (typical annual crop root zone) layer of normal density/composition topsoil and is also similar to what current farming practices often lose via wind/water erosion (5–30 Mg/ha/a; see Pimentel and Burgess 2013). Consequently, since my scenario's rock grinding/distribution costs are a fraction of its irrigation water and nitrogenous fertilizer costs, it might be a good idea to at least start out with considerably larger powder application rates, perhaps 40–50 Mg/ha. Doing so would also reduce the chance of crop "starving" due to slower-than-I've assumed weathering rates.

11.3.4 OTHER REASONS FOR POWDERED ROCK FERTILIZATION

11.3.4.1 Addressing Already Oversalinized Soils

Here's a statement cut and pasted from one of the best-written soil science papers I've seen to this day (Said Hassan and El Toney 2014):

With unique 92+ essential elements, consisting of a broad range of trace minerals along with small amounts of macro elements, the soil nutrients which have been slowly lost through the ages by erosion, leaching and farming are gently regenerated. Rock dust not only improves soil vitality, but also increases plants' overall health while strengthening immunity and pest resistance. Resulting in a completely natural and disease-free final product, unique in flavor and mineral composition.

The paper goes on to show that powdered basalt would address another of the issues that is becoming more prevalent throughout much of the world, the oversalination of irrigated soils. The reason for that is that freshly powdered basalt exhibits a huge ion exchange capacity that "pulls" excess ions (e.g., Na$^+$) out of soil solutions.

11.3.4.2 "Climate Change"

Finally, another virtue of such "natural" fertilization is that the weathering of powdered basalt would retain (sequester) soil-generated CO_2 that would otherwise "respire" back to the atmosphere. The weathering of 1.26E+10 Mg of "Snake River Plain Basalt" [10.06 wt% CaO and 7.65 wt% MgO='s 7.35 milliequivalents' [0.1006 × 2/(40 + 16)+0.0765 × 2/(24.32 + 16)] worth of base per gram] would therefore in effect remove about 4.1 GT of carbon dioxide from the atmosphere – that's about 11% of today's anthropogenic CO_2 emission rate. A good deal of theorizing has been done about how the "enhanced weathering" of mafic rocks could address global warming (see Schuiling and Krijgsman 2006; Hartmann et al. 2013), but very little realistic experimental work has been done to date. Thankfully, that subject is beginning to receive more attention (Taylor et al. 2017), and some hopefully realistic experimental studies have apparently begun (Beerling et al. 2018).

11.3.5 OTHER AGRICULTURAL ENERGY REQUIREMENTS

I'll next assume that the rest of my scenario's on-farm agricultural energy can be approximated as follows:

- Assumptions
    - 200 horsepower-to-the-ground tractors drag 5-meter-wide machinery at 8 km/hour over 14 million km^2 of land three times per year.
    - For simplicity's sake I'll next assume that those tractors, possess 72% efficient drive trains powered with 95% efficient electric motors and 50 kWh "TESLA 3" battery pack modules …
    - …and that the overall energy source (nuclear reactor) to battery to motor efficiency is 90%
- Calculations
    - Kilometers traveled by tractors per year = 3 × 14E+6 km^2 × 1000 m/km/5 m = 8.3E+9 km.
    - Hours so-traveled per year = 8.3E+8/8 = 1.06E+9.
    - 200 hp = 5.36E+8 J/hour = 149 kWh.
    - At 75% discharge/cycle (greater discharge seriously reduces battery life), each battery module would provide 50 × 0.75 = 37.5 kWh of energy per discharge cycle.
    - Which, if 200 hp is to be delivered to the load, would last 0.72 × 0.9 × 37.5 = 0.172 hours (10.3 minutes).
    - Total from source energy required per year = 1.05E+9 hrs × 5.36E+8 J/hour/0.9/ .72/0.95 = 9.14E+17 J/year.
    - That's 29 full-sized 1GWe power plants required to power the world's tractors
    - At 3 cents/kWh, that's $7.6 billion worth of electricity.
    - However, since farm tractors run only, let's say, 10% of the time and for, let's say, ~6 hours each time, the number of battery modules required would be ~4.19E+7 [1.08E+9 h/yr/ (.1 × 24 × 365) h/yr × 6 hr/.172 hr/L (that's 39 battery modules per tractor), which at $7000 each represents a whole world up-front cost of $293 billion).
    - Since Mr. Musk (2019) currently claims that those modules could sell for $7000 apiece (far less/kWh than do his/TESLA's "Power Walls") and will last 1500 cycles, their steady state replacement cost would be 28.5 $billion per year.

These figures suggest (to me anyway) that battery-powered tractors are unlikely to replace internal combustion engine powered tractors.

11.4   PUTTING EVERYTHING INTO PERSPECTIVE

My ballpark calculations up to here were based upon an assumption that 100% of the world's arable land would be devoted to producing high-yield crop food for humans as efficiently as is currently done in the United States' "corn belt." That's unrealistic in the sense that the amount of food so produced would provide each of the future's 11.2 billion people about 5259 kcal's worth of food per day – twice as much as a typical, mostly sedentary, civilized person is apt to need.

Let's utilize what we've come up with to test another much-mentioned "what if." The difference between 5259 and 2500 kcal/day for 11.2 billion people amounts to a total of 4.73E+19 J's worth of carbohydrate/fat "fuel," which if it were to be burned by 40% Carnot-efficient "biofuel" power plants would generate the same amount of power as would 1500 one-GWe nuclear reactors. That's *almost* enough to power the future's agricultural activities but under 7% of our descendants' total energy demand. An appropriate conclusion would be that while biofuels might have some useful niche applications (e.g., "bio oil"–fueled farm machinery; see Cataluña et al. 2013), they certainly could not power everything.

The 1886 [1113 + 360 + 230 + 131 + 23 + 29] GWe's worth of power plants required to implement the agricultural aspects of my scenario's clean/green utopian future is about five times that currently generated by all of the world's nuclear reactors. However, it represents only about 8% of the total energy services required by 11.2 billion people as rich as are the European Union's now.

All of the necessarily huge machinery and manufacturing facilities required to implement this or any other technological fix capable of "saving the world" would be much cheaper to build and operate much more efficiently with reliable power than with that provided by intermittent sources such as windmills and solar panels. The worldwide average "capacity factor" (energy actually generated/nameplate rating = "CF") for solar panels is about 15% and about 30% for wind turbines. While it would indeed be "possible" to run desalination/ammonia plants, rock crushers, farm tractors, locomotives, etc., with windmills and/or solar panels, doing so would be terribly expensive, dangerous, and frustratingly unproductive to both such machinery's owner-operators and their customers. It would also require that 1/CF times as much machinery be built/maintained to do the job at the same rate that a CF ≈ 1.0, breeder reactor–powered world would require – typically ~3 times as much machinery for wind and about 7 times as much if solar powered.

It would require somewhat over 37 million 30% CF, 2 MW-rated windmills to generate an average of 22.4 TWe. Assuming 265 watt-rated, 15% CF rooftop-type solar panels were to be used instead, the number required would be 564 billion [22.4E+12/(265 × 0.15)]. The current retail cost (as purchased, not installed) of such panels is about $300 each, which adds up to an up-front cost of $169 trillion, which figure is even greater than what the batteries required to render such a power system somewhat "reliable" would cost.

While it's true that, in *principle*, batteries could render wind/solar power "reliable," it is unrealistic to assert that they actually could. For instance, let's assume that the future's political leaders decided that their constituents could get by with just one day's worth of battery-type energy storage and that it would be supplied with huge banks of Tesla's state-of-the-art "Tesla 3 battery modules."

One day's worth of 22.4 TWe power adds up to 1.935E+19 Joules or 5.38E+11 kWr. A state-of-the-art (after two decades' worth of development) "Tesla 3 battery module" possesses 50 kWh worth of storage capacity and is now supposed to cost only about $7000. Consequently, the up-front cost of that fine-sounding "what if" is $75.3 trillion, and, since it's unlikely that such batteries would last much more than a decade, its battery "maintenance cost" would be about $7.5 trillion per year.

For this sort of application, lithium ion batteries don't scale for other reasons, one of which has to do with their most expensive component, cobalt. When a lithium ion battery charges/discharges, the cobalt within its cathode shifts back and forth between its tri- and quadrivalent oxidation states.

Since …

- Lithium ion battery's voltage is about 3.5,
- One equivalent's worth of charge is 965,000 coulombs (one Faraday), and
- Equivalent weight for this reaction is 58.9 grams,

…one kWh's worth of storage would require an absolute minimum of 621 grams [3.6E+6/96500 × 58.9/3.5] of cobalt. One day's worth of 22.4 TW power storage via that sort of battery translates to a cobalt requirement of 3.36E+11 kg. Last year's (2018) total world cobalt production was about 1.5E+8 kg or just 0.04% of this particular "what if's" requirement for it.

As is the case with today's unsustainable light water–type nuclear reactors (LWRs), lithium ion batteries represent a prematurely locked-in technology poorly suited to this "new" problem that it is supposed to address. However, there are several somewhat more promising alternatives. For example, the liquid metal batteries developed by MIT Professor Sadoway and colleagues can't fail the way that lithium ion batteries do (no dendrites) and are also apt to be cheaper because they feature liquid (molten) – not solid, cheap – not expensive, metallic electrodes separated by a molten salt electrolyte (Yin et al. 2018). Their operation relies upon density/SPG differences: a high-density (e.g., antimony) metal electrode lies on the bottom of the cell. Immediately above it is an intermediate-density molten electrolyte (e.g., a low-melting mix of lithium and potassium chloride salts), and a low-density liquid metal (e.g., lithium) electrode floats on top. Like oil and vinegar, those fluids naturally self-segregate, and charging/discharging them doesn't generate the sorts of irreversible physical changes that eventually cause any solid electrode-based battery system to fail (lithium ion batteries are unlikely to last more than ten years in such applications). Professor Sadoway's group at MIT has looked into a plurality of cheap metal (e.g., sodium, calcium, magnesium) battery chemistries. What makes this sort of battery especially promising is that unlike any of the competing solid electrode–based systems, examples have retained >99 percent of their initial capacity after over 5000, 100% charge/discharge cycles – a characteristic absolutely necessary for a genuinely practical, grid-relevant sized (>>1 MWh) electricity storage system. Professor Sadoway estimates that his batteries will eventually cost about $75/KWh, which, if they last for 20 years, at one charge/discharge cycle per day, corresponds to an LCOS of just $10.3/MWh [$75/20/(365 cycles/year × 0.001 MWh/cycle)].

Unfortunately, enough of even such cheap batteries to store one day's worth of 22 TWe power (approximate world demand circa 2100 AD) would cost ~39.6 trillion of today's US dollars.

Thomas Edison's ~120-year-old nickel-iron (Ni Fe) battery (Wikipedia 2019) has proven to be the most durable (by far) solid electrode–based storage battery. It's extremely robust, tolerant of abuse (overcharge, overdischarge, and short-circuiting), and often exhibits very long life even if so treated. However, due to its low specific energy/power, poor charge retention, and high cost of manufacture (retail cost per kWh is currently about 60% higher than TESLA's "Power Wall"), other types of rechargeable batteries have largely displaced them. However, Edison's battery is experiencing resurgence in some quarters (as the "Battolyser") because overcharging them generates hydrogen that can be stored and used for many purposes including energy regeneration via either combustion or fuel cells. In that manner, they could accomplish both short-term and long-term energy storage. However, even if they were to last for 30 years, their relatively high up-front cost (currently about $83) translates to a "levelized cost of storage" (LCOS) of $76/MWh.

11.5 CONCLUSIONS

This chapter's "technical" arguments should (but probably won't) convince most of the people reading them that it would make far more sense to utilize wind and solar power for the niche-type applications for which they are actually suited than to try to force our demand-driven technological civilization to adapt an energy system based upon them (Jacobson and Delucchi 2009;

Jacobson et al. 2017). The one thing that's absolutely certain now is that what's now being done to address the future's energy-related issues isn't working (Jackson et al. 2018).

At this point in time, my scenario's clean/green utopian future is just an updated rehash of the same "Age of Substitutability" envisioned by Alvin Weinberg (Goeller and Weinberg 1974) and M. King Hubbert over 60 years ago. The reasons why mankind hasn't already developed the technologies required to realize their vision has more to do with political drivers and humans', not Mother Nature's, rules and propensities. While it is almost 100% certain that nuclear power could indeed become simultaneously cheap, clean, and genuinely renewable, it seems unlikely that such a "renaissance" is apt to happen in the Western world for the reasons that I've identified in a book (Siemer 2019a). However, it is rather likely that China will do what must be done to render such a "Nuclear Green New Deal" possible.

REFERENCES

Abelson, P. H. 1975. *Energy for Tomorrow.* Seattle, Washington: University of Washington Press.

Beerling, D. J., Leake, J. R., Long, S. P., Scholes, J. D., Ton J., Nelson P. N., et al. 2018. Farming with crops and rocks to address global climate, food and soil security. *Nature Plants*, 4, 138–147.

BP. 2017. *BP Energy Outlook.* 2017 ed. https://www.bp.com/content/dam/bp/business-sites/en/global/corporate/pdfs/energy-economics/energy-outlook/bp-energy-outlook-2017.pdf.

Brown, T. 2018. ThyssenKrupp's "green hydrogen and renewable ammonia value chain." *Ammonia Industry.* https://ammoniaindustry.com/thyssenkrupps-green-hydrogen-and-renewable-ammonia-value-chain/.

Campbell, N. S. 2009. The use of rockdust and composted materials as soil fertility amendments. PhD thesis. University of Glasgow, Glasgow, Scotland. http://theses.gla.ac.uk/617/.

Cao, P., et al. 2017. Historical nitrogen fertilizer use in agricultural ecosystem. *Earth System Science Data*, 10, 969–984. doi:10.5194/essd-10-969-2018.

Cataluña, R., Kuamoto, P. M., Petzhold, C. L., Caramão, E. B., Machado, M. E., and da Silva, R. 2013. Using bio-oil produced by biomass pyrolysis as diesel fuel. *Energy Fuels*, 27(11), 6831–6838. doi:10.1021/ef401644v.

Cowan, R. 1990. Nuclear power reactors: A study in technological lock-in. *Journal of Economic History*, 50(3), 541–567.

Desalination. 2020. Water supply and sanitation in Saudi Arabia. https://en.wikipedia.org/wiki/Water_supply_and_sanitation_in_Saudi_Arabia.

Goeller, H. E., and Weinberg, A. M. 1974. The age of substitutability. *Science*, 191(4228), 683.

Hansen, J., 2008. Tipping point: Perspective of a climatologist. In E. Fearn, Ed., *State of the Wild 2008–2009: A Global Portrait of Wildlife, Wildlands, and Oceans.* Wildlife Conservation Society/Island Press, University of Chicago Press, pp. 6–15.

Hartmann, J., West, A. J., Renforth, P., Köhler, P., De La Rocha, C. L., et al. 2013. Enhanced chemical weathering as a geoengineering strategy to reduce atmospheric carbon, supply nutrients, and mitigate ocean acidification. *Reviews of Geophysics*, 51(2), 113–149.

Hensel, J. 1894. *Bread from Stones: A new and rational system of land fertilization*, A. J.Tafel, Philadelphia.

Hubbert, M. K. 1956. *Nuclear Energy and the Fossil Fuels.* Publication no. 95. Shell Development Company, Exploration and Production Research Division, Houston, TX. http://www.hubbertpeak.com/hubbert/1956/1956.

Hussain, A., and Abolaban, F. 2014. Nuclear desalination: A viable option for producing fresh water-feasibility and techno-economic studies. *Life Science Journal*, 11(1), 301–307.

Israel. 2018. Water Supply and Sanitation in Israel. https://en.wikipedia.org/wiki/Water_supply_and_sanitation_in_Israel.

Jackson, R. B., Le Quéré, C., Andrew, R. M., Canadell, J. G., Korsbakken, J. I., Liu, Z., Peters, G. P., and Zheng, B. 2018. Global energy growth is outpacing decarbonization. *Environmental Research Letters.* doi:10.1088/1748-9326/af303.

Jacobson, M. Z., and Delucchi, M. A. 2009. A plan to power 100 percent of the planet with renewables. *Scientific American*, 26.

Jacobson, M. Z., Delucchi, M. A., Zack, A. F. Bauer, Z. A. F., Goodman, S. C., Chapman, W. E., et al. 2017. 100% Clean and renewable wind, water, and sunlight (WWS) all sector energy roadmaps for 139 countries of the world. 27 January 2017. Stanford University, Stanford, California. http://web.stanford.edu/group/ef.

Jankovic, A. 2003. Variables affecting the fine grinding of minerals using stirred mills. *Minerals Engineering*, 16(4), 337–345.

Kelley, C. P., Mohtadi, S., Cane, M. A., Seager, R., and Kushnir, Y. 2015. Climate change in the fertile crescent and implications of the recent Syrian drought. *PNAS*, 112(11), 3241–3246.

Leeman, W. P. 1982. Olivine tholeiitic basalts of the Snake River Plain, Idaho. In B. Bonnichsen and R. M. Breckenridge, Eds., *Cenozoic Geology of Idaho, Idaho Bureau of Mines and Geology Bulletin*, 26, pp. 181–191.

LLNL. 2018. Lawrence Livermore National Laboratory's annual energy flow charts (its Sanky-diagrams, see https://flowcharts.llnl.gov/commodities/energy/) are based upon DOE/EIA figures which consider both data and assumptions regarding the efficiency with which relevant "sectors" (residential, commercial, industrial, and transportation) convert primary energy to useful "energy services." Recently, its assumptions have varied more than the data.

Moulton, K. L., West, J., and Berner, R. A. 2000. Solute flux and mineral mass balance approaches to the quantification of plant effects on silicate weathering. *American Journal of Science*, 300(7), 539–570.

Musk, E. 2019. Elon Musk makes incredible claims about Tesla Model 3 longevity, will offer battery module replacement. https://electrek.co/2019/04/13/tesla-model-3-longevity-claims-elon-musk/.

Pimentel, D., and Burgess, M. 2013. Soil erosion threatens food production. *Agriculture*, 3(3), 443–463.

Priyono, J., and Gilkes R. J. 2004. Dissolution of milled-silicate rock fertilizers in the soil. *Australian Journal of Soil Research*, 42(4). doi:10.1071/SR03138.

Said Hassan, M., and El Toney, M. 2014. Increasing productivity of bean by healing salinity of soils. *Asian Journal of Agriculture and Food Science*, 2(1).

Schuiling, R. D., and Krijgsman, P. 2006. Enhanced weathering: An effective and cheap tool to sequester CO_2. *Climate Change*, 74(1–3), 349–354.

Siemer, D. 2019a. *Nuclear Power: Policies, Practices, and the Future*. John Wiley & Sons and Scrivener Publishing LLC, Hoboken, NJ.

Siemer, D. 2019b. Silicate weathering to mitigate climate change. In R. Lal and B. A. Stewart, Eds., *Advances in Soil Science Soil and Climate*. Taylor & Francis, Boca Raton, FL, pp. 249–265.

Smil, V. 2008. *Energy in Nature and Society – General Energetics of Complex Systems*. MIT Press, Cambridge, MA.

Taylor, L. L., Beerling, D. J., Quegan, S., and Banwart, S. A. 2017. Simulating carbon capture by enhanced weathering with croplands: An overview of key processes highlighting areas of future model development. *Biology Letters*, 13(4), 20160868.

Tomlinson, J., 2018. Nigerian briefing: How engineers can help secure a sustainable economy. *Proceedings of the Institution of Civil Engineers – Energy*. doi:10.1680/jener.17.00024.

Wikipedia. 2019. Nickel iron battery. https://en.wikipedia.org/wiki/Nickel%E2%80%93iron_battery.

Yin, H., Chung, B., Chen, F., Ouchi, T., Zhao, J., Tanaka, N., and Sadoway, D. R. 2018. Faradaically selective membrane for liquid metal displacement batteries. *Nature Energy*. doi:10.1038/s41560-017-0072-1.

12 Nitrogen Dynamics and Management in Rainfed Drylands
Issues and Challenges

Rachid Bouabid, Brahim Soudi, and Mohamed Badraoui

CONTENTS

12.1 DRYLAND CHARACTERISTICS

Dryland agriculture occurs in many parts of the world. It is quite difficult to give a precise definition of dryland agriculture. For the purpose of the present chapter, dryland agriculture or dryland farming is referred to as that occurring in areas where precipitation is insufficient so that water is the key limiting factor for crop production. Dryland agriculture, dryland farming, and rainfed farming are often used interchangeably. The term "dry" is used to describe areas characterized by low amounts of rainfall and where shortage of water affects vegetation growth, productivity, and other related agricultural activities. Limitation of water is not only in terms of total amount of precipitation, but also in terms of its unpredicted distribution with regard to the crop growth cycle, mainly the critical growth periods. Dryland agriculture occurs in regions where evapotranspiration exceeds precipitations, and therefore is associated with arid, semiarid, and subhumid temperate regions. The limited rainfall in conjunction with its erratic distribution often induces drought periods that can vary in

285

terms of intensity and frequency and are becoming even more amplified and structural with climatic changes. These conditions call for particular soil management practices, especially for water and nutrients.

Drylands are defined based on the ratio of average annual precipitation (Pr) to potential evapotranspiration (PET) (UNEP-WCMC 2007). Lands where this ratio lies between 0.05 and 0.65 are considered drylands, while lands with a ratio less than 0.05 are not included in drylands but are considered as desert areas where no crop growth occurs without irrigation (Table 12.1, Figure 12.1). Aridity is also assessed on the basis of the number of days for which the soil water balance is favorable to plant growth (growing season). The delineation of drylands has been updated since 2007 to consider additional dryland areas of relevance to the Convention on Biological Diversity (CBD) with dry features despite their Pr/PET greater than 0.65 (Figure 12.1).

Unlike the common perception that drylands are limited to major parts of Africa, Latin America, and the Middle East, drylands are actually present in many regions around the world, such as Central Asia, Russia, the northwest of America (the United States and Canada), and

TABLE 12.1

Dryland Categories according to UNEP-WCMC (2007)

Classification	Aridity Index (Pr/PET)	Rainfall (mm)
Hyperarid	<0.05	<200
Arid	0.05 < Pr/PET < 0.20	<200 (winter) or <400 (summer)
Semiarid	0.20 < Pr/PET < 0.50	200–500 (winter) or 400–600 (summer)
Dry subhumid	0.50 < Pr/PET < 0.65	500–700 (winter) or 600–800 (summer)

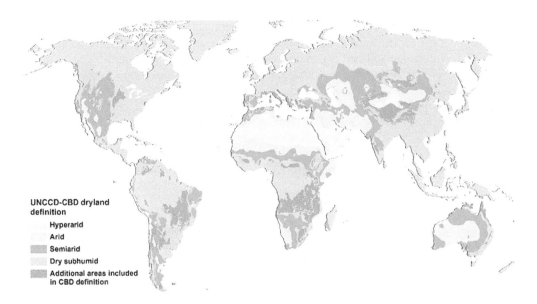

UNCCD-CBD dryland definition

Hyperarid
Arid
Semiarid
Dry subhumid
Additional areas included in CBD definition

FIGURE 12.1 Word distribution of drylands areas. (From UNEP-WCMC, A spatial analysis approach to the global delineation of dryland areas of relevance to the CBD Programme of Work on Dry and Subhumid Lands, dataset based on spatial analysis between WWF terrestrial ecoregions [WWF-US 2004] and aridity zones [CRU/UEA; UNEPGRID 1991], dataset checked and refined to remove many gaps, overlaps, and slivers [July 2014], 2007; drawn using GIS data available at www.unep-wcmc.org.)

Australia. Drylands occur in all continents and occupy about 6.31 Bha (47.2% of the global earth land area). They cover about 2.0 Bha in Africa, 2.0 Bha in Asia, 0.68 Bha in Australasia, 0.76 Bha in North America, 0.56 Bha in South America, and 0.3 Bha in Europe (UNEP 1992).

Dryland agriculture encompasses a diversity of cropping systems, varying from subsistence farming to intensive and high-cash crops, and is fundamental to the economy of the regions where it is prevailing. Although irrigation often occurs, rainfed agriculture generally predominates and represents the primary production sector. In the example of Morocco, a typical Mediterranean dryland country, where the economy is highly dependent on agriculture, the country's agriculture gross domestic product (AGDP) fluctuations follow the same trend as that of annual precipitations (Figure 12.2).

Enhancing dryland productivity is a lesser constraint in developed countries where agricultural practices benefit from advanced research and technologies. These difficulties are yet to be overcome in developing countries, where, despite valuable indigenous knowledge and satisfactory research, farmers are still struggling with basic needs to cope with subsistence farming and to bring their practices (especially nutrient inputs) to a level that guarantees the expression of the production potentials of their lands and crops. The challenges are multiple if drylands are to meet the needs of a vulnerable population, especially with the rising issues of climate change.

Many factors of soil degradation threaten the potential production capacity in these areas. Nutrient depletion, wind and water erosion, and salinity are very common issues. Although dryland farming has known significant changes in the past few decades in many parts of the world, mainly with a shift from traditional agriculture to more intensive cropping, nutrient inputs and management, especially nitrogen, remain the main concern for crop production. The use of fertilizers is at its lowest levels in many countries. In China, deficiency of nitrogen is reported to be almost everywhere, and that of phosphorus affects at least one-third of the arable lands (Li et al. 2009a). About 68 M ha of dryland soils in Pakistan with less than 300 mm have a negative nutrient balance (Shah and Arshad 2006). Nitrogen use efficiency (NUE) is considered more critical than that of other nutrients as it is highly affected by the amount and distribution of rain during crop season. Nitrogen is also related to organic matter dynamics, and the latter also affects water status necessary for nutrient bioavailability and uptake under water stress conditions. Therefore, this chapter will put more focus on the nitrogen dynamic as it is the major critical factor in dryland agriculture.

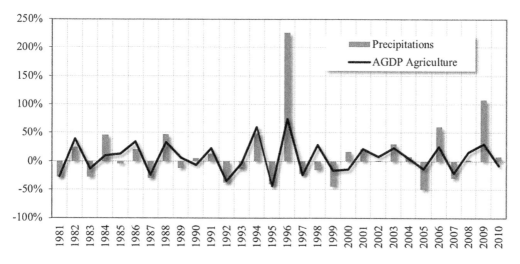

FIGURE 12.2 Variation of relative agriculture gross domestic product (AGDP) and annual precipitation (the value for each year is normalized with that of the previous year). (From Balaghi, R., Changement climatique, environnement et risques naturels au Maroc, Paper presented at: Atelier de formation et d'appui CEDRIG pour les projets ASAP-M, PAMPAT et PAMPAT Oriental, Rabat, 20–22 janvier 2015 [unpublished], 2015.)

12.2 MAJOR SOILS OF DRYLANDS

Although soil fertility and its evolution are significantly dependent on cropping systems and the degree of crop intensification, its overall status is also determined by soil type. Soil fertility cannot be dissociated from soil pedogenesis. As drylands are characterized in general by restraining soil-forming factors (mainly climate, vegetation, and living organisms), the soils are usually low to moderately differentiated. The rate of soil formation and the degree of differentiation are inversely related to aridity (Nettleton and Peterson 1983). In low rainfall areas, the variation in parent materials and topography are the main driving factors for differences of soil types and their properties. They are, in general, not very deep and have low organic matter and high base saturation due to low leaching. In many cases, they are affected by calcium carbonates inherited from calcareous parent materials or from secondary precipitation, which is responsible for high alkalinity. A variety of soils can be encountered in drylands depending on the other soil forming factors, mainly Aridisols (2.12 B ha), Entisols (2.33 B ha), Mollisols (0.8 B ha), Alfisols (0.38 B ha), Vertisols (0.21 B ha), and others (0.47 B ha) (Dregne 1976; Noin and Clark 1997). In Mediterranean-type climates, Red soils are common due to the differentiation of iron oxides, while Spodosols, Ultisols, and Oxisols are rarely encountered in these kinds of environments. Soils in drylands are not necessarily the result of pedogenesis of present climates, but may have developed under ancient climates. Soils of tropical climates found presently in many dryland areas are the result of paleoclimates. Deep and well-differentiated Alfisols and Oxisols in North Africa, India, and Australia are examples of such conditions.

Although vegetation and crop production in drylands are controlled by climate and water, the soil characteristics determine how much of that water will be stored and consequently be available for plants during their growth stages, mainly during the dry or drought periods. Soil moisture is a primary condition for organic matter and nutrient cycling and bioavailability for crops, and therefore becomes the sine qua non condition for production. Depth, texture, clay type, and organic matter content are important factors that determine the storage capacity and the retention forces of soil-available water. Organic matter, and therefore organic carbon cycles, are dramatically affected by dry conditions, as a result of natural and anthropic factors. The low amounts of residues returned and the high rates of decomposition are often not in favor of carbon sequestration.

12.3 NITROGEN DYNAMICS AND USE EFFICIENCY

Besides the shortage and sporadic distribution of water, the low-productivity potential of soils in drylands is commonly attributed to low and imbalanced use of fertilizers, which does not satisfy crop nutrient requirements. Some of the soil characteristics in drylands are also in favor of particular conditions affecting nutrient cycling and availability (high pH, calcium carbonates, swelling-shrinking clays, high base saturation). Drylands are in most areas fragile and vulnerable to degradation factors that affect the physical, chemical, and biological settings of the soils, which in turn affect the essential nutrients pool. Nutrient sources in poor drylands are from organic sources, and fertilizer use in developing countries is of appreciable use only in the more favorable rainfed areas.

Under conditions of low moisture, the majority of mineral nutrients (when present) are less mobile, often fixed or precipitated in the soil, and are therefore less available to plants. In India, it has been recognized that "dry soils are as hungry as they are thirsty" (Venkateswarlu 1987; Singh and Kumar 2009; Nicra 2012). Nitrogen and phosphorus availability are of universal concern in all dryland soils. Potassium is more variable as some parent materials may contain significant amounts of potassium-bearing minerals, but coarse-textured soils are usually potassium deficient (Sivanappan 1995; Ryan and Sommers 2010; Ryan et al. 2011). In rainfed agriculture, phosphorus and potassium are usually applied as deep fertilizers at the start of the season. As they are relatively of low mobility in the soil, they present fewer constraints in terms of management compared to nitrogen. However, since nitrogen is more mobile, its dynamics along the crop cycle are more intricate and more related to organic matter.

Fertilizer input in dryland agriculture varies with soil and climate, but is in general very low in rainfed farming in developing countries. For instance, in North Africa and the Middle East, fertilizer use is less than 20 kg ha^{-1}, all nutrients considered. Soil of drylands of Africa undergo important nutrient loss (Roy and Nabhan 1999) that may reach 30 kg ha^{-1}yr^{-1} of N-P-K (Henao and Baanante 2006). In Pakistan, dryland soils are reported to be almost entirely deficient in nitrogen (Shah and Arshad 2006; Irshad et al. 2007). Similar situations were stated for India (Singh and Venkateswarlu 1985) and Kenya (Murage et al. 2000). When nutrients are not replenished in the soil, a progressive mining takes place. The low nutrient status in drylands in many parts of the world is actually a vicious circle that involves low nutrient cycling, low productivity, low residues return, and low soil conditioning and protection, and therefore affects the sustainability of production systems as well as food security in the regions concerned.

12.3.1 NITROGEN AND ORGANIC MATTER

Organic matter is an important component of the soil and is an important source of nutrients. It plays a major role in supplying the soil with nitrogen and other nutrients as a result of microbial decomposition. It acts also as a sink when immobilization takes place. Organic matter is the main cycling component of soil fertility in low-input farming. Soils of drylands have in general low organic matter contents resulting from low biomass productivity and low returns (Ryan et al. 1997; Lal 2002, 2004). Nitrogen supply from organic matter depends on the importance of the two processes of organic matter decomposition, i.e., humification and mineralization. The former results in stable humus, and the latter releases CO_2 and mineral nutrients to the soils. While both processes in general co-occur, the dominance of one over the other depends on various abiotic and biotic factors, mainly soil temperature and moisture, the nature of organic matter, and the pool of microorganisms. The rates of these processes have been widely studied under controlled incubation conditions and under humid temperate field conditions. However, only limited knowledge is available for dry or rainfed conditions. Various models for organic matter humification and mineralization have been proposed (Hénin and Dupuis 1945; Kolenbrander 1969; Campbell et al. 1984; Jenkinson and Rayner 1977; Sauerbeck and Gonzalez 1977; Janssen 1984; Biederbeck et al. 1994; Mary and Guérif 1994). Most studies agree that organic matter mineralization depends on the C/N ratio. Low C/N ratio leads to higher mineralization of organic matter with high net N-mineralized, while high C/N leads to more humification and high N-immobilization. Vigil and Kissel (1991) reported that the break point between net N-mineralization and net-immobilization was at a C/N ratio of 40. Under warm dryland conditions, moderate temperatures favor rapid decomposition of OM by microbial activity during periods of adequate soil moisture.

Organic matter cycling and nitrogen release from its mineralization is an issue that can be looked at from different angles (Soudi 1988; Soudi et al., Chapter 5, present volume). While the accumulation of organic matter (i.e., humus) in the soil is highly recommended for soil conditioning and carbon sequestration, rapid mineralization on a season basis is of concern to low-input smallholder farmers as it is the main source of nutrients for their crops. In warm drylands or Mediterranean-type climates, annual organic matter depletion can be very important and may exceed 60% (Corbeels et al. 1998). Application of nitrogen fertilizers (even in small quantities), or the presence of residual nitrogen from previous legume crops, can lead to rapid decomposition of organic matter. Singh and Singh (1994) found that combined nitrogen fertilizer with straw resulted in a higher nitrogen mineralization compared to straw and fertilizer applied singly (Figure 12.3). The balance of mineral nitrogen can be appreciable for the season crop, but the accumulation of stable humus that plays other important roles (structure, moisture retention, aggregate protection, etc.), will be negatively affected.

Farming systems favoring residues return or the addition of farm manure with high C/N ratio may contribute to some buildup of stable humus in the soil. However, if low or no nitrogen fertilizer is used, a negative nitrogen balance will be carried over and yield potential will not be achieved.

The predilection of the work of humification and mineralization is a hidden issue in crop management systems in drylands and raises the question of which of the processes farming practices

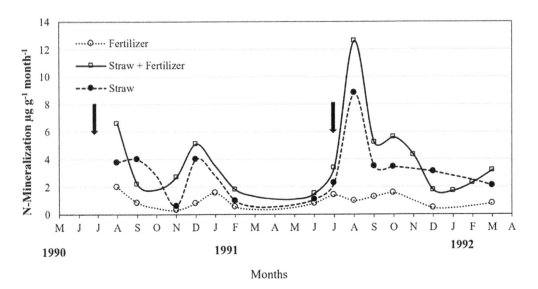

FIGURE 12.3 Effect of application of straw and fertilizer singly and in combination (arrows indicate time of application), on N-mineralization (μg g^{-1} month^{-1}). (Redrawn from Singh, H., and Singh, K.P., *Soil Biol. Biochem.*, 26, 695–702, 1994. With permission.)

should aim to use. Where soil nutrient status is very low, the management of organic matter is usually directed toward getting higher mineralization rates, as farmers rely on the nutrients pool resulting from this process. Despite the appreciable annual additions of organic matter of different forms (leftover residues, farm manure), we tend to see rapid depletion of organic matter. This is even more marked in systems adopting rotations with legume crops. The residual nitrogen from biological nitrogen fixation and the recycling of the crop residues with low C/N ratio lead to higher mineralization rates and therefore to rapid loss of soil organic matter (SOM). The lack of humic substances may also reduce the low chelation potential of the soil with respect to micronutrients, especially in coarse-textured or high pH calcareous soils. In farming systems favoring the return of residues of high C/N ratio, some storage of stable humus can occur, but would be too small to evolve to a significant buildup; on the contrary, the mineral nutrient pool, especially that of nitrogen, will not be able to guarantee good crop growth and yield potential will not be achieved.

Therefore, in drylands, especially those of warm climates, the turnover of SOM affects, and is affected by, nitrogen. Complex interactions drive the mineralization of organic matter and the bioavailability of mineral nitrogen, and therefore require good management practices to achieve the compromise goals of nitrogen supply while maintaining sufficient stable organic matter in the soil.

12.3.2 Nitrogen and Soil Water Status

Nitrogen is a major yield-limiting nutrient for crop production in drylands. In arid and semiarid rainfed regions, soil nitrogen content is closely related to soil moisture regime. Appreciable amounts of residual mineral nitrogen are usually found soon after the start of the rainy season (Chiang et al. 1983; Warren et al. 1997), but will change depending on the amount and type of organic matter available. When dry conditions persist, the topsoil may not provide enough nitrogen to the crop. If the root system is well developed, uptake may occur more in deeper soil layers (Campbell et al. 1977; Strong and Cooper 1980).

In dryland soils, nutrient use efficiency is highly related to water use efficiency (Shepherd et al. 1987; Palta and Fillery 1995). Nutrient dynamics and availability have great interactions with soil-available water (He and Dijkstra 2014). These interactions can affect crop growth and yield positively or negatively, depending on crop growth stages. Water limitations, low use of fertilizers, and low organic matter return to soils make nutrient management a challenge in drylands. Amounts, forms, timing, and methods of application of fertilizers need to be reasoned taking into consideration not only soil nutrient dynamics and supply capacity but also soil water status.

Important year-to-year rainfall variations significantly influence soil water and nutrient status, which in turn causes important repercussions on nutrient inputs. For the same initial soil nutrient status, the same nutrient input may lead to differences in yield when rainfall is different or when rainfall is irregular from one year to another. As the climatic conditions of the year progress, nutrient inputs need to be adjusted accordingly. A reduction or increase of the amount of nitrogen initially planned is necessary depending on whether the year is dry or relatively wet. Dry or wet conditions affect the soil pool as well, and the adoption of a different balance approach is necessary in order to predict crop needs with the changing season conditions. In the case of drought conditions, water stress not only affects nutrient mobility and availability, but also affects negatively the growth of the crop itself.

When soil moisture is adequate, moderate to high temperatures can promote nitrification of ammonium. Several studies reported that nitrification is high when soil water content is within an adequate range of water-holding capacity (40% to 70%), but become low or is even inhibited beyond this range (at extremes) as the soil becomes dry or waterlogged (Justice and Smith 1962; Malhi and McGill 1982; Flower and Challagha 1983; Klemedtsson et al. 1988; Shelton et al. 2000; Sahrawat 2008). The nitrification process can transform the entire $N\text{-}NH_4^+$ from organic matter mineralization or from ammonium-N based fertilizer within two to three days. This is also true for the transformation of urea. Since most crops absorb more nitrogen in the form of nitrate-N compared to ammonium-N, this transformation is advantageous if leaching risks are minimal. However, with high risks of leaching following storm events during the rainy season, nitrates may leach out of the rhizosphere. The loss of ammonium-N by volatilization in high pH conditions is lesser when nitrification is important. Since nitrate absorption by plants occurs mainly by passive mechanisms, mobility of nitrogen in the form of nitrate is important for plant uptake. Abdelmoumen et al. (2010) evaluated nitrogen loss using ^{15}N mass balance under two rainfall conditions (270 mm and 340 mmyr[-1]) in calcareous soils using different forms of urea. The results showed that N loss by volatilization was relatively low (11% and 18%) for both rainfall conditions. They also reported that urea hydrolysis is delayed by dry conditions and that nitrification is delayed by cold conditions following urea hydrolysis.

In semiarid areas, nitrogen requirements of crops are dictated by the amount of seasonal precipitation (Myers 1984). The input of mineral nitrogen for a crop my undergo important depletion in one year, while in another year large amounts of residual nitrogen can stay in the soil as a result of limited rainfall, low leaching, and poor crop growth (Noy-Meir and Harpaz, 1978; Seligman et al. 1986). In the latter situations, N recovery is expected to be important for the subsequent crop whether the same or the following year. Corbeels et al. (1998), studying N recovery by sunflower (*Helianthus annuus*) after wheat (*Triticum aestivum*) in a Typic Haploxerert in Morocco, found that the residual labeled ^{15}N fertilizer represented 3.6% of the fertilizer added to the previous winter wheat crop. This low plant recovery was attributed to microbial immobilization of N fertilizer and to the limited N uptake due to drought conditions during sunflower season. After sunflower harvest, 49.7% of added labeled fertilizer was left in the soil profile. Total recovery of the ^{15}N fertilizer over the two growing seasons was 83%. Seligman et al. (1986) reported low N recovery from residues of high C/N ratio, and Wood et al. (1996) reported low nitrogen recovery of various crops using ^{15}N labeled urea and ammonium sulfate under different rotations.

Generally, in dry years there are small or nonsignificant differences in terms of N uptake among unfertilized and fertilized crops. However, important differences in N uptake occur during rainy years. For the latter case, N mineralization can account for a large part of available N to the crop when sufficient organic matter is present. Figure 12.4 illustrates important differences in terms of N uptake and dry matter accumulation by wheat in three years with different rainfall conditions in northern Morocco (Corbeels 1997).

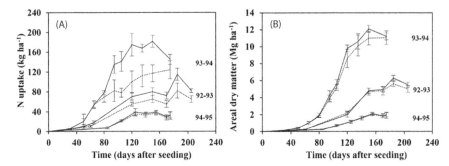

FIGURE 12.4 N uptake (A) and areal dry matter accumulation (B) of winter wheat during three rainfall contrasting years: 401, 332, and 264 mm for 1993/1994, 1992/1993, and 1994/1995, respectively (A: Anthesis; PM: physiological maturity). (Redrawn from Corbeels, M. et al., *Biol. Fertil. Soils*, 28, 321–328, 1999. With permission.)

Nitrogen use efficiency is not only affected by the total amount of rainfall, but also by rainfall patterns. The occurrence of rainfall during vegetative and reproductive stages is determinant for water use as well as for nutrient use by crops. The amount and distribution of rain in drylands affects the efficacy of applied nitrogen fertilizers. However, the unpredicted occurrence of rain makes it difficult to adopt a unique strategy for the choice of the form, the timing, and the method of application, especially when splitting is a common practice.

Crop nutrient uptake depends on the solubility of mineral nutrients and their movement in the soil. Under dry conditions, not only the form and the amount of mineral nutrients are affected, but also the mobility by the two processes of mass flow and/or diffusion. Under the action of water movement and roots suction, different forms of nitrogen behave differently as a result of complex interactions. It is generally accepted that nitrogen is more influenced by mass flow compared to phosphorus and potassium (Barber 1984). However, limited knowledge is available on the behavior of nitrogen mobility under dry soil conditions. Song and Li (2006) studied root function in the uptake of nitrogen of maize and the effect of soil water on the transfer and distribution of nitrates and ammonium under irrigated and nonirrigated regimes. Their findings showed that both root growth and water supply had a significant effect on nitrate-N transfer. Under irrigation, the difference of nitrate-N concentrations at different distance points from the maize plant were smaller, while clear differences of nitrate-N concentrations were observed under conditions of limited root growth space without water supply. They concluded that nitrate-N is transferred as solute to plant root systems with water uptake by plants. However, the transfer and distribution of ammonium-N transfer differed from that of nitrate-N and were not influenced by root growth or by soil water supply. This implies that, in dry conditions, nitrate-N placement (especially at early stages) is as critical as for other, less mobile nutrients. Localized fertilizer placement in the seed rows for cereals in drylands shows better nutrient use compared to broadcast, especially when rain occurs after sowing (McKenzie et al. 2001; Kelley and Sweeney 2005; Zhang et al. 2013). The effect of roots on nitrogen absorption will depend also on the root density and specific area. Any factors that favor root growth would favor root absorption as well.

The application of fertilizers in general, and nitrogen in particular, followed by dry conditions can result in increased soil solute concentrations in the topsoil. High amounts of fertilizers may even reduce the uptake. The interactions among fertilizer use and crop uptake involve a dynamic process that can have different effects depending on stages of crop growth. There is a strong need in rainfed dryland agriculture to have a good understanding of the interactions among soil water status, nutrient dynamics, and crop behavior in order to achieve the best nutrient use efficiency.

12.3.3 NITROGEN MINERALIZATION AND IMMOBILIZATION

Organic nitrogen mineralization is an important process that drives nitrogen changes in the soil. It represents a major process of nitrogen-supply potential, and an important pool of nitrogen that needs to be taken into consideration in N-balance and N-fertilization, especially in low input systems. N-mineralization is intimately related to organic matter decomposition, and their processes are consequently affected by the same factors. Soil moisture and temperature status are the main factors driving N-mineralization in general, and under dry conditions in particular. The populations of living organisms and microorganisms responsible for organic matter and nitrogen transformations decline with aridity due to the limited moisture needed for their survival.

N-mineralization results from the interaction between soil moisture and temperature. It often follows a linear process (Figure 12.5) with increasing soil moisture content at different temperatures (Stanford and Epstein 1974; Ju and Li 1998; Li et al. 2009b). Within a range of water and temperature conditions suitable for biological activity, N-mineralization occurs and varies with these two factors. However, under extreme moisture or temperature regimes, N-mineralization is very low or totally stopped. In most dryland soils, with the exception of those of North America (dry and cold), high temperatures and low moisture conditions are the main concerns.

N-mineralization is the highest during wetter years compared to dryer years, especially during periods of optimal moisture and temperature conditions for microbial activity (i.e., spring in Mediterranean climates), and varies with the type of cropping systems. N-mineralization was found to be highest under fallow soils compared to cropped soils. This trend is attributed to the fact that fallow practices are in favor of soil moisture conservation. Figure 12.6 shows significant differences in terms of mineralized N under different cropping for two years with contrasting rainfall in rainfed conditions of Morocco (Corbeels et al. 1999). There are major differences among the two years, but also important differences between fallow and cultivated plots within the "normal" year (401 mm) compared to the dry year (264 mm). In addition, the study underlines the constraints related to N fertilizer applications among the two years. The 401 mm rainfall allowed three splits of nitrogen, while the 264 mm rain imposed only one N application.

When compaction is not a concern, soils in drylands are rarely water saturated for long periods and most of the time are well aerated. These conditions support rapid organic N mineralization as well as nitrification of ammonium, either from fertilizers or from organic matter mineralization. Unless there is an important fixation by high-charge 2:1 clays, nitrogen is present in the soil mainly

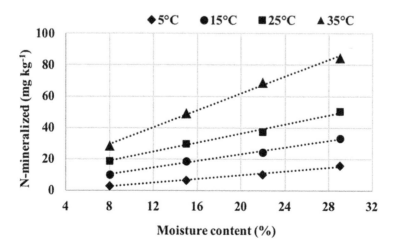

FIGURE 12.5 Effect of water and temperature on N-mineralization. (Drawn using data reported by Li, S.X. et al., *Adv. Agron.*, 102, 223–265, 2009b.)

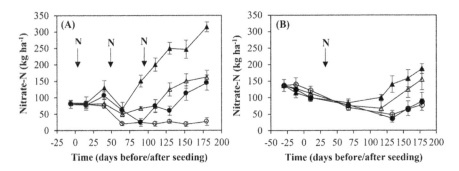

FIGURE 12.6 Change in soil nitrate-N from mineralization in total rooting depth: (o) unfertilized fallow plot; (●) fertilized fallow plot; (△) unfertilized wheat cultivated plot; (▲) fertilized wheat cultivated plot; (N) dates of N applications; (A) year with 401 mm; (B) year with 264 mm. (Redrawn from Corbeels, M. et al., *Biol. Fertil. Soils*, 28, 321–328, 1999. With permission.)

in the form of nitrates. In such cases, assuming no major rain events at the start of the season to cause nitrate leaching from the topsoil, nitrate-N is the dominant form of nitrogen potentially available for crops and can be used as a reliable quick soil test index reflecting soil nitrogen supplying capacity.

Various models were used to assess nitrogen mineralization by combining the potentially mineralizable nitrogen (N_o) with functions representing the effect of temperature and soil moisture on the mineralization rate constant (k) (Cabrera et al. 2005). Such models can be adopted to better predict nitrogen mineralization under limited water conditions, such as those of dryland soils. Campbell et al. (1988) compared two versions of such models using data obtained with different treatments: (1) summer fallow, and wheat following wheat grown on summer fallow on (2) dryland and with (3) irrigation. While the models showed close results for estimated and measured values under irrigation, they tended to underestimate nitrogen values under dryland conditions. Since the model was considered not dynamic, as it does not allow for N_o to be replenished continuously by nitrogen derived from decomposition of fresh residues and rhizosphere microbial biomass, net nitrogen mineralized from this source might explain the underestimation predicted under dry conditions, which was always obtained whenever the soil became very dry and was rewetted by rainfall. This phenomenon may be attributed to the possible flush of mineral nitrogen that can occur after rewetting, but is not taken into consideration in the model. Soudi et al. (1990a) and Elherradi et al. (2003) observed a similar flush of initial mineral nitrogen in the first few days of incubation, with amounts ranging from 26% to 40% of total mineralized N. This flush was attributed to the pretreatment effect of soil samples by drying and rewetting before incubation. This phenomenon is important to consider because it simulates the mineral-N increase observed under field conditions of rainfed drylands following drastic seasonal changes upon the first seasonal rain. The early mineralization flush can be attributed to the partial decay of the microbial biomass following summer desiccation. Dead microbial tissues have a low C/N ratio that makes them easily biodegradable by the surviving microbial biomass when soil moisture becomes favorable. High amounts of mineral-N (N-NO$_3$ + N-NH$_4$) varying from 58% to 73% of total mineralized-N supply were reported under field conditions (Soudi et al. 1990b) (Figure 12.7). This shows that early season N-flush can be beneficial for crops (such as cereals) if the rewetting of the soil after dry or drought periods does not cause significant leaching of the mineralized nitrogen, and needs to be taken into consideration in the estimation of crop nitrogen fertilization.

Nitrogen availability to crops, mainly at early stages, can be significantly affected by nitrogen immobilization before remineralization takes place. The incorporation of residues of high C/N from previous crops or from the addition of organic amendments induces a net N immobilization during initial decomposition. However, the amount and rate of immobilization and subsequent

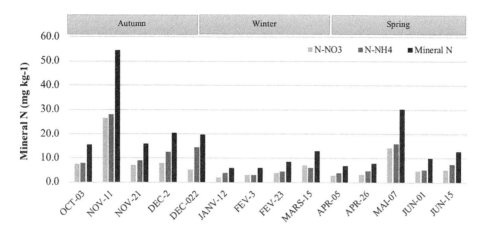

FIGURE 12.7 Seasonal variation of mineral-N in a Calcixeroll in Morocco. (Adapted from Soudi, B. et al., *Actes Inst. Agron. Vet.*, 10, 29–38, 1990b.)

remineralization vary largely depending on the nature of the residues as well as on several biotic and abiotic factors, mainly soil moisture conditions and N fertilizer use. Understanding the immobilization–remineralization trends upon the incorporation of residues in dryland soils is of great importance in relation to the synchronization of N supply and crop demand during critical growth stages, especially those occurring during early season decomposition (Vigil and Kissel 1991; Jensen 1997; Reinertsen et al. 1984; Recous et al. 1995; Vanlauwe et al. 1996; Corbeels et al. 2000; Li et al. 2009a).

Although the C and N dynamics during decomposition of plant residues in the soil are well understood, differences may be important among soils in rainfed drylands and those in humid regions. N immobilization by microbial activity can account for important impounding of N from the soil when residues are of high C/N ratio, which adds up to its reduced availability because of low soil moisture storage. Immobilization can happen at different periods of the growing cycle, but is more likely to occur in the early season when organic matter from previous crops has not yet undergone much decomposition. Immobilization can affect the overall available mineral N, as well as its use efficiency, and depends on the nature of the organic residues incorporated in the soil. In a study on N cycling in response to the decomposition of residues in a typical wheat–sunflower rotation in a Vertisol in Morocco, Coorbeels et al. (2000) reported that N availability may be severely restrained during the growing season when past wheat straw residues are incorporated into the soil. The amount of inorganic N immobilized due to such incorporation may compete with the plant demand, engendering low N efficiencies. Under dryland conditions, the rate of release of the immobilized N is usually very slow, and affects the overall recovery of nitrogen by the crop. Practices that increase organic matter of high C/N (wheat straw) may contribute to reducing organic matter mineralization, but can result in important microbial immobilization. However, the incorporation of organic matter of low C/N ratio (legumes residues) supports high N-mineralization, but would lead to rapid organic matter loss.

12.3.4 NITROGEN AND NUTRIENT BALANCE

Crop nutrition is driven by a balanced nutrient absorption. Nutrients need to be available and absorbed in adequate quantities required by the crop. When a nutrient is lacking in the soil at the required threshold, it limits the absorption and functioning of the others. Imbalanced nutrient use has been reported to be a major factor affecting nutrient use efficiency and nutrient depletion in low input regions in general, and in dryland regions in particular (Murage et al. 2000; Rashid and Ryan 2004; Li et al. 2009b, 2017; Ryan 2008b; Ryan et al. 2011 Nawaz and Farooq 2017). Fertilizer use is rarely based on soil testing. Furthermore, in many countries, several fertilizer formulae distributed in the market are not appropriate for the soil fertility status and for the requirements of the crops. This situation can lead to low response of the crops to the fertilizers used. In low input situations, this may contribute to the depletion and imbalance of one or more nutrients.

Combined use of N and phosphorus (P) has been reported to improve N use efficiency, reduce N loss, and improve crop yield (Li et al. 2009; Sharma et al. 2007; Ma et al. 2010). Phosphorus is critical to root growth, which is related to water and N use efficiency. Jin et al. (2006) found that the addition of P enhanced the concentration and accumulation of N and P in shoots and seeds of soybean cultivars. The addition of P alleviated the effect of drought stress on plant growth, P accumulation, and grain yield. They suggested that phosphorus fertilization could mitigate drought stress at the reproductive stage, resulting in fewer yield effects and improved grain quality. Studer et al. (2017) reported an interactive effect of drought and the application of nutrient N-P-K individually or in combination under drought stress for maize. Al-Karaki et al. (1996) found that high proline accumulation in leaves of water stressed crops at high P levels might be an adaptive response to drought for sorghum compared to bean. Several other studies in drylands corroborate the combined positive effect of N and P on N use efficiency and crop production (Sharma 2007; Ma et al. 2010).

Potassium (K) is an essential nutrient that affects many physiological processes involved in biotic and abiotic stresses resistance (Kant and Kafkafi 2002; Cakmak 2005; Wang et al. 2013). It is critical in plant tolerance to dry and drought conditions, as it is involved in stomatal regulation, osmotic adjustment, membrane stability, cell elongation, and photosynthesis (Mengel 2007; Benlloch-Gonzalez et al. 2010; Marschner 2012). Crops with adequate K fertilization show greater adaptation to water stress (Lindhauer 1985; Pervez et al. 2004; Premachandra et al. 2009). Asgharipour and Heidari (2011) reported that adequate K nutrition increased N uptake and improved drought resistance of sorghum. Zahoor et al. (2017) also found that K application regulates N metabolism and osmotic adjustment in cotton functional leaf under drought stress.

12.3.5 Biological Nitrogen Fixation and Water Stress

Biological nitrogen fixation (BNF) is an important process for N supply in low input cropping systems. BNF is even more critical for the sustainability of farming systems in dryland soils. Water stress, as a result of dry or drought conditions, is one of the most influencing factors that affects both legume plants as well as nodule formation and functioning. The decline of the BNF under dry conditions affects legume crop N uptake, plant growth, and subsequent N enrichment of the soil awaited from the rotation system. Although the sensitivity to water stress varies among legume crops, some crops (soybean [*Glycine max*], cowpea [*Vigna unguiculate*], and black gram [*Vigna mungo*]) have been reported to be more vulnerable to modest dry conditions (Sinclair and Serraj 1995b).

The inhibition of BNF under drought has been widely reported and is related to various factors such as reduced nitrogenase activity, reduced nodule permeability, reduced CO_2 flux, low O_2 conductance, inhibition of photosynthetate, and shoot nitrogen-feedback signaling (Fin and Burn 1980; Sinclair and Serraj 1995b; Duan et al. 2014; Gonzáles et al. 1998, 2001; Serraj et al. 1999, 2001; Galvez et al. 2005; Marino et al. 2007; Valentine et al. 2011). Growth and nitrogen fixation are depressed under water stress. The ability of plants to recover and the time for recovery are both related to the duration of the stress period. Weisz et al. (1985) found that drought resulted in decreased nodule conductance, which was detected as early as three days following water stress application. The effect on BNF is more important as drought is prolonged. The loss of nodule activity is also related to nodule water potential (Bennett and Albrecht 1884; Sprent 1971; Durand et al. 1987). Nodule respiration increases with an increase in soil water potential and declines upon water rehydration as a result of decaying (Nandwal et al. 1991). Under water stress, reduced BNF is attributed to depressed oxygen uptake and therefore to respiration, as a result of either a loss of oxygen conductance through the nodule or inhibition of oxygen-requiring reactions, or probably both (Pankhurst and Sprent 1975a, 1975b; Sinclair and Gourdiaan 1981; Serraj and Sinclair 1996a, 1996b, 1997).

Water stress sensitivity and tolerance are attributed to several molecules in the nodules, including ureides (Atkins et al. 1992; Sinclair and Serraj 1995b; Serraj et al. 1999, 2001; Vadez Sinclair et al. 2001; et al. 2000; King and Purcell 2005; Coleto et al. 2014), glutamine (Neo and Layzell 1997; Curioni et al. 1999), and asparagine (Bacanamwo and Harper 1997; Vadez et al. 2000; Lima and Sodek 2003). An extensive review by Arrese-Igor et al. (2011) addresses the various physiological processes involved in nodule BNF under drought. In general, BNF is a more drought-sensitive mechanism when ureides are the main nodule exporters, which is the case of soybean, and it is more drought tolerant when amides were the main nodule exporters, which is the case of pea (*Pisum sativum*), chickpea (*Cicer arietinum*) and faba bean (*Vicia faba*) (Sinclair and Serraj 1995a, 1995b, 1996; Vadez et al. 2000; Purcell et al. 2000; Vadez and Sinclair et al. 2001; Serraj 2003). Such traits can be used in the segregation of legume genotypes as to their BNF tolerance to water stress under dryland conditions. Despite the available research on the mechanisms involved in the sensitivity or tolerance of BNF under water-limiting conditions, soybean (a ureide type exporter) is the most studied crop (Arrese-Igor et al. 2011), and only limited research is available on other crops.

Although the sensitivity of BNF is important among legumes, the recovery capacity is an even more fundamental trait for crop adaptation and productivity after periods of water limitation. The rapidity of recovery of both transpiration and nodule nitrogen-fixation activity after soil water deficits is an important criterion for rainfed dryland conditions, especially at the establishment stage of the crop. Legume genotypes having such a capacity are of great value to sustain nitrogen supply and ensure good yields under water deficit conditions (Engin and Sprent 1973; Cerezini et al. 2016, 2017).

A metadata analysis study by Daryanto et al. (2015) using research on legume yield responses to drought under field conditions between 1980 and 2014 showed that the amount of water reduction was positively related with yield reduction, but the extent of the impact varied with legume species and the phenological stage during which drought occurred. Overall, lentil (*Lens culinaris*), groundnut (*Arachis hypogaea*), and pigeon pea (*Cajanus cajan*) were found to experience lower drought-induced yield reduction compared to other legumes such as cowpea and green gram (*Vigna radiate*). Yield reduction was generally more severe when legumes experienced drought during their reproductive stage. Higher yield reductions are also observed for legumes grown on soils with medium texture compared to those cultivated on soils of either coarse or fine texture. In contrast, legume yield reductions were not related to regions and their associated climatic factors.

12.3.6 ROOT SYSTEMS AND NUTRIENT USE AND EFFICIENCY

Crop roots are the hidden part of the plant. A prolific root system is essential to uptake of water and nutrients in drought-prone and nutrient-limiting situations to ensure good growth and productivity. The effect of water and nutrients on crop growth and productivity is often observed and assessed based on how the areal parts of the crop are progressing. It is only rarely that the rooting system is examined to see how it is behaving under dry conditions. Nutrient use efficiency, mainly N, is closely related to the root system density and depth (Zhang et al. 2013). It is even more important in drylands where topsoil can be subject to significant water deficits by evapotranspiration. Plants may adapt by extending their rooting system deep into the soil to reach for water and nutrients. In rainfed dryland farming, crop adaptation and yield are correlated to root mass (Passioura 1983; Palta et al. 2011). Nutrient use efficiency is improved by root system development. In North African countries where olive production is common, farmers consider from experience that when a young tree is subject to water stress, it develops a denser and deeper root system that allows it to become vigorous with age, tolerate dry conditions, and make better use of nutrients. All conditions that can improve root development will increase water and nutrient use efficiency. Good soil structure becomes a determining factor. Porous soils will allow root growth, but may not favor upward capillary water evaporation. Therefore, appropriate soil tillage and surface residues management are needed. Crops have varying rooting systems, and the essential nutrients have differing impacts on root growth and physiological functions. The addition of organic matter is a factor that can contribute to ensuring healthy structure. Nitrogen and phosphorus are the key elements for root development, and K is a key element for regulating water uptake and osmotic pressure of the roots. The complex interactions among soil conditions and plant behavior affect tremendously the faculty of the crop to express its potential under favorable conditions or to adapt and survive under stressful conditions.

12.4   NITROGEN MANAGEMENT AND AGRICULTURAL PRACTICES

Nutrient- and water-management practices are a major challenge for increasing crop production under rainfed dryland soils. The techniques include the use of organic and inorganic fertilizers, recycling crop residues, rotations including legumes, fallow, and soil and water conservation. The adaptation and the degree of technological development of any of these techniques depend on the local physical and socioeconomic context. Many practices have evolved from indigenous knowledge of the farmers, and others have been adopted based on successful experience in similar areas. Nutrient use efficiency is improved with the use of selected cultivars, planting date, weed control, and pest management.

12.4.1   COMBINED USE OF ORGANIC AND INORGANIC N SOURCES

Organic matter is considered a good soil conditioner, especially for its favorable impact on water status. Soil humus improves aggregate formation and soil structure. The increase of humus can increase significantly the soil available water capacity (AWC) (Bouyoucos 1939; Russel et al. 1952; Hollis et al. 1977). Hudson (1994) found good relationships between SOM and AWC. He reported that for the example of silt loam soils, the AWC for a soil with 4% OM (by weight) was more than twice that of a soil with 1% OM. The use of manure improves SOM content, which in turn improves soil AWC (Salter and Howarth 1961; Salter and Williams 1963). Bauer (1974) reported that a change of 1% of soil humus from the addition of manure resulted in a change of AWC in the range of 0.06 to 0.11 inch per foot in soils with coarse texture and in the range of 0.01 to 0.03 in soils with moderately fine and fine textures.

Combined use of inorganic and organic fertilizers is adopted to sustain soil fertility and improve crop productivity in drylands (Palm et al. 1997; Sharma et al. 2002; Place et al. 2003; FAO 2004; Hadda and Arora 2006; Bationo et al. 2007; Li et al. 2009; Kihara et al. 2011; Harraq et al. 2016; Calderon et al. 2017). In dryland farming, the combined use of inorganic and organic fertilizer is more beneficial than their separate use. Bouraima et al. (2015) reported that application of manure reduced surface runoff and soil erosion, and N and P loss were reduced by 41% and 33%, respectively, in the case of combined use of manure and chemical fertilizers.

The use of manure is indeed important as an organic amendment for soils in general, and for dryland soils in particular. However, its use can carry some problems depending on the origin of the organic materials (e.g., infested fields or barns). Insects, pathogens, and weed seeds can be propagated through the use of manure. When manure travels from one area to another, problems associated with the use of manure can be disseminated at different spatial scales (Petit et al. 2013). Therefore, proper management of manure is needed to avoid such problems.

12.4.2   MANAGEMENT OF FERTILIZER APPLICATIONS

One of the challenges for soil nutrient uptake by plants is soil moisture availability. When soil water is limited, not only is there a water stress for the plant, but also the nutrient movement and absorption by the crops are slowed down or even stopped. Fertilizers can only be applied during the periods of the season that have a good chance of coinciding with rain. This is particularly true for nitrogen, which is split in two or more dressing fractions. For instance, in most

Mediterranean areas, the practice for cereal fertilization is that P, K, and a fraction of N are applied as basal application at sowing. N is usually split in two or three dressing applications that target tillering and heading stages. The risk of a shortage of adequate moisture in midseason can jeopardize crop yields (Nageswara Rao et al. 1985; Matthews et al. 1990; Manyowa 1994; Ben Naceur et al. 1999; Fischer et al. 2003). In many situations, farmers are reluctant to apply nitrogen when there is no upcoming rain, fearing that nitrogen may be wasted and avoiding additional salinity stress to the crops.

IAEA (2005) reported on nutrient and water management project cases in rainfed dominated agriculture (<300 mm) in 11 countries with a wide range of cropping systems, characterized by low nutrient and organic matter status. These included wheat–maize systems on the Loess Plateau of China (Cai et al. 2005), sorghum–castor rotations in Andhra Pradesh of India (Ramana et al. 2005), maize-based systems in the Machakos District of Kenya (Sijali and Kamoni 2005), peanut production in Senegal (Sene and Badiane 2005), and wheat–vetch rotations in the Safi-Abda region of Morocco (El Mejahed and Aouragh 2005). The results from all the studies showed that irrespective of management practices, an important portion (20% to 60%) of applied fertilizer N was lost. These losses were attributed mainly to alkaline soil pH conditions, and occurred essentially at the beginning of the crop vegetative period. On the other hand, the residual value of applied N available to subsequent crops was very low, rarely exceeding 9% of the crop N requirement. Their findings underlined the importance of fertilizer-management practices to minimize losses, especially during the early part of the cropping season. Split application during the dry season allowed an important increase of wheat yields, but the amount of N to be applied at each split needs to be based on the soil N status and crop demand for N. The combined manure with mineral fertilizers in correct proportions could provide 10% to 15% of the crop-N requirement and contribute to increasing observed yields.

Another example of constraining fertilizer use is the case of rainfed fruit tree production. Olive orchards are one of the main crops in many dryland areas, especially the Mediterranean. The cycle of the olive tree starts with blooming, which usually occurs in mid or late spring (the start of the dry season). Most of the active growing cycle occurs in summer and early fall. If P and K can be applied at once early in the season, N splitting is a major problem. Most farmers can apply a first fraction with P and K, but the rest will depend on the rainfall conditions in late spring. The most important vegetative growth and fruit setting occur in summer when no water is available for N mobility in the soil. P and K mobility in this case is even worse.

Methods of application and placement of fertilizers are critical to achieving good distribution of nutrients in the rhizosphere and their uptake by the roots. They should be adapted depending on the planting density and root system (structure and depth). For instance, wheat's root system is shallow compared to that of maize. The latter is often row planted and tends to have a deep and pivoted root. When soil moisture is deficient, the more the nutrients are away from the root, the less chance there is for their uptake, regardless of the absorption mechanism. Interception is reduced due to limited root growth, and mass flow is hindered by the lack of water needed for the nutrients to be drawn by water movement exerted by roots in response to transpiration.

12.4.3 Tillage and Surface Residues Management

Tillage practices significantly affect soil properties that are in relation to soil moisture, organic matter, and nutrient dynamics in the soil. Under dryland conditions, conventional tillage can expose the soil to high evaporation, rapid loss of organic matter, and potential wind and water erosion. Water use efficiency is generally low with inappropriate tillage practices. Hatfield et al. (2001) reported that it is possible to improve water use efficiency by 25% to 40% through soil management practices that involve tillage. When adequate amounts of crop residues are present, conservation tillage is highly effective for conserving soil and water, achieving favorable crop yields, maintaining soil organic carbon contents, and enhancing soil and water quality (Unger et al. 1997).

Conservation tillage and reduced tillage have gained increasing attention in drylands (Unger 2002; Vere 2005; Cai et al. 2006; Wang 2006; Avci 2011; Mrabet et al. 2012). They are defined as any tillage and planting practices that can leave, respectively, 15% to 30% and 30% or more crop residues on the soils after planting (CTIC 2018). Among the benefits of such a management approach are the maintenance and accumulation of organic matter, as well as water and nutrient use efficiencies in the soil. Results of long-term experiments in drylands of China (Wang 2006) showed that the positive effect of reduced tillage on soil water availability, nutrient balance, and soil fertility indices were strongly improved by use of crop residue, either incorporated or applied as surface mulch. Simulations using the Century model revealed a positive effect on carbon sequestration under reduced tillage. Johnson et al. (2018) reported that conservation tillage systems are much better options for the cultivation of different drought-tolerant common bean varieties in semiarid areas of Kenya due to their soil moisture conservation ability as compared to conventional tillage systems, which are not a sustainable practice in such marginal areas. Sainju et al. (2012) and Sainju et al. (2016) studied N balance in dryland agroecosystems in response to tillage, crop rotations (involving cereals and legumes), and cultural practices (involving planting date, seeding rates and spacing, and application of N fertilization), over several years in the northern Great Plains of the United States. They reported that surface residue N was 30%–34% greater in no-till (NT) than in conventional tillage (CT). They also found that N sequestration rate at 0–20 cm from 2004 to 2011 varied from 29 to 89 kg N ha^{-1} $year^{-1}$ under CT and NT with spring wheat–pea rotations. Long-term experiments conducted using NT under Mediterranean conditions showed significant increase of SOM and N (Mrabet et al. 2001; Bessam and Mrabet 2003). Lal (2004) suggested that the adoption of conservation tillage practices is among the appropriate management practices that can increase soil carbon sequestration.

Surface residue management is recognized as a water conservation practice and has received important research attention in arid and semiarid regions (Duley and Russel 1939; Van Doren 1978; Unger 1984; Smika and Unger 1986; Bussière and Cellier 1994; Peng et al. 2014). Surface residues management and residues incorporation greatly affect nutrient cycling, especially that of N (Sauerbeck and Gonzalez 1977; Seligman et al. 1986; Schomberg et al. 1994; Vanlauwe 1996; Rescous 1995; Jensen 1997; Liwang et al. 1999; Schomberg and Steiner 1999a). Mulching, as a surface residue management, is of great importance in water conservation in drylands. Mulch farming, when it is adopted in dryland farming, contributes to modulate soil temperature, reduce water evaporation, protect soil from erosion, and contribute to the sequestration of C (FAO 2004). As stated earlier, any practice that contributes to improving soil moisture will contribute also to better conditions for nutrient bioavailability and uptake.

Several studies conducted in drylands of the Loess Plateau of China reported on the importance of mulching on water and N dynamics. Jin et al. (2008) showed that minimum tillage and mulching over a period of seven years consistently increased the yield of winter wheat, primarily by better water harvest. N uptake by grain and straw, N export, and residual soil N were the highest compared to conventional tillage with no mulching. Bayala et al. (2011) evaluated several studies relative to the effect of several conservation agricultures in drylands of Africa (Burkina Faso, Mali, Niger, and Senegal) for various crops. They also found that mulching was among the best practices for improving soil fertility and crop yields, especially when rainfall is less than 600 mm. Qin et al. (2015) reviewed the effects of mulching on wheat and maize, using data from 74 studies conducted in 19 countries. They indicated that mulching significantly increased yields, water use efficiency, and N use efficiency by up to 60%, compared with no-mulching. Effects were larger for maize than wheat, and larger for plastic mulching than for straw mulching. Plastic mulching performed better at relatively low temperatures, while straw mulching had the opposite trend. The effects of mulching tended to decrease with increasing water input. Mulching effects were not related to SOM content. They concluded that soil mulching can significantly increase maize and wheat yields and improve water and nitrogen use efficiencies, and thereby may contribute to closing the yield gap between attainable and actual yields, especially in dryland and low nutrient input agriculture. Wang and Xing (2016) reported that mulching improved significantly soil water content, soil nitrate-N content, and its vertical distribution in maize root zones. The results of a three-year field experiment by Gao et al. (2009) assessing the effect of mulching with different materials (straw and plastic) showed that N uptake, NUE, and yield of wheat were higher with mulching. In addition, after three years, residual nitrate-N in 0–200 cm soil averaged 170 kg ha^{-1}, which was equivalent to ~40% of the total N uptake by wheat in the three growing seasons.

Although plastic mulching has been evaluated in many studies and proved effective for water and N conservation, organic residues mulching is a much-preferred practice because it is the most appropriate for smallholder farmers and has limited or no environmental impact. Plastic mulching, used for a variety of crops including row vegetables and fruit crops, despite its multiple uses, is slowly degradable and represents a threat to the environment. Organic mulches, on the other hand, provide a source of organic matter and nutrients, and their cycling in the soil contributes to improving soil properties. When previous crop or fallow residues mulching is practiced with conservation tillage in dryland agriculture, the benefits are not only on soil water and nutrients but also on the soil health as a whole (Schomberg et al. 1994; Schomberg and Jones 1999; Fuentes et al. 2003).

12.4.4 LEGUME INCLUSION IN CROP ROTATIONS

Monoculture is a common practice in many dryland areas. After harvest, the soil is left bare for several months, and is often subject to grazing for the leftover residues. In summer, wind erosion can be very significant, while water erosion can happen during the fall before the next season's sowing. Grazing on crop residues causes depletion of organic matter, reduced topsoil protection, and nutrient imbalance. Such a practice can lead to nutrient mining when inputs are lower than outputs.

The N status of soils can be improved by the integration of legumes in crop rotations. This is especially important in drylands where N from previous legume crops is not subject to as much leaching as in humid conditions. Food legumes, such as chickpea, lentils, and faba bean, are commonly used in Mediterranean arid and semiarid regions as alternatives to fallow and continuous cropping (Ryan 2008a; Ryan and Sommers 2010). The adoption of legume crops in the rotations can contribute to increasing organic matter and therefore the potential release of N from mineralization (Ryan 1998; Ryan et al. 2008, 2010). When the residues left in the soil are of low C/N ratio, they are susceptible to higher rates of mineralization, and therefore contribute to releasing more N, P, and other nutrients in the soil during the cycle of the following crop. Depending on the crop, biological fixation by legume crops can provide variable amounts of mineral nitrogen during a growth cycle (Hardarson et al. 1987). Montañez (2000) reported net nitrogen supplies from legume crops varying from 50 to more than 300 kg N ha^{-1}yr^{-1}.

The contribution of N from BNF of legumes in a rotation depends on many factors, among which are the availability and compatibility of the rhizobium strain, soil water and temperature conditions, soil aeration, and the amount of initial nitrogen in the soil. In drylands, soil water status is often a determinant factor in the early stages of crop growth, and may also be influenced by the availability of starter nutrients. Molybdenum is a key element for rhizobia N fixation. This element is not very mobile in the soil if soil moisture is limited and can be affected by soil pH conditions as well. In calcareous soils, common in drylands, high pH is not a limiting factor to molybdenum compared to other micronutrients; however, this is not the case in acid soils.

Legume crops are also used in intercropping systems, either with field crops or fruit trees (Vandermeer 1989), mainly in smallholder farming. Intercropping is a common practice in many dryland areas (especially in India and China), and has been reported to improve crop yields and soil nutrient status compared to sole crops (Ramana 2015; Zhang et al. 2015; Rekha et al. 2017; Singh 2017). Legume crops have different BNF capacities that can be taken into consideration for efficient intercropping (Montañez 2000). Using ^{15}N techniques, Bationo et al. (2003) reported that cowpea fixed more than twice the amount of atmospheric N compared with groundnuts, and that the inclusion of cowpea in millet-based cropping systems improved nitrogen use efficiency by about 30%. Ramirez-Garcia (2015) found that barley intercropped with vetch had improved root growth and N uptake.

Song et al. (2007) investigated crop yield and various chemical and microbiological properties in the rhizosphere of wheat, maize, and faba bean grown in the field solely and intercropped (wheat/faba bean, wheat/maize, and maize/faba bean). They found that intercropping increased crop yield, changed N and P availability, and affected the microbiological properties in the rhizosphere of the three species compared to sole cropping. Bouhafa et al. (2015) and Daoui et al. (2012) reported that legumes improve soil N and P and contribute indirectly to improving soil nutrient status for olive (*Olea europea*) trees in Mediterranean rainfed olive orchards. The question remains whether intercropping can be adapted in all situations. It may be feasible in smallholder farming systems, but would be of controversial practicability in the more intensive large farms of favorable rainfed areas.

12.4.5 INTEGRATED SOIL FERTILITY MANAGEMENT

Integrated Soil Fertility Management (ISFM) is defined as the set of sound soil fertility management practices that include both the use of fertilizers as well as organic inputs in combination with the knowledge of how to adapt these practices to local conditions, in order to maximize nutrient use efficiency and optimize crop productivity (Vanlauwe et al. 2010, 2015). The approach of ISFM is of particular interest in farming systems that use low chemical fertilizers and rely more on organic sources of nutrients. As discussed earlier, nitrogen dynamics are closely related to organic matter, and therefore are of particular interest in the context of ISFM. Improving SOM content also aims at favoring other soil characteristics, such as water status, which in turn affects nitrogen bioavailability and use efficiency under dry conditions. ISFM is seen also as a soil conservation measure that can contribute to carbon sequestration.

The perception and adoption of soil fertility management in relation to agronomic efficiency need to be considered from multiple socioeconomic angles, with changes that can occur in steps and with increasing knowledge (Figure 12.8) (Vanlauwe et al. 2010, 2015). Innovations adapted to large market-oriented farms in developed countries may not be suitable to smallholders in developing countries where farmers are still struggling with subsistence agriculture. The move through the various steps of the ISFM depends also on the land's degree of responsiveness. The conditions might be the same, but the concerns, the practices, and the means to face these conditions are widely

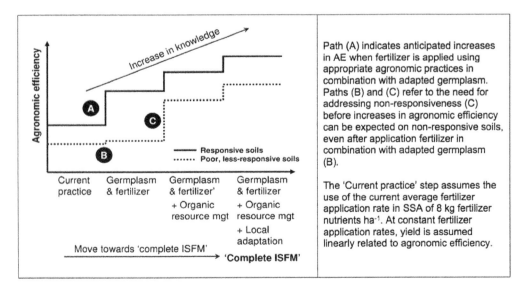

FIGURE 12.8 Conceptual relationship between the agronomic efficiency of fertilizers and organic resources and the implementation of various components of ISFM. (From Vanlauwe, B. et al., *Outlook Agric.*, 39, 17–24, 2010.)

different. For instance, land ownership status is a major factor toward investing in the long-term improvement of soil fertility. Itinerant farming as a result of drought and soil degradation causes continuous decline of soil fertility in many countries in the Sahel and sub-Saharan Africa (Reich et al. 2001; Darkoh 2003; FAO 2005a, 2005b).

12.4.6 THE "R" PRINCIPLES FOR NITROGEN MANAGEMENT IN DRYLANDS

FAO (2004) recommended that soil fertility management practices can be grouped according to the movement of nutrients into, within, and out of a system and that the practices can be sorted in four groups as suggested by Hilhorst and Muchena (2000):

- Adding nutrients to the soil
- Reducing losses of nutrients from the soil
- Recycling nutrients
- Maximizing the efficiency of nutrient uptake

A number of management practices were drawn from several case studies in Senegal and Sudan (FAO 2004). Any soil fertility management approach to be adopted or adapted to drylands needs to comply with the principles of one or more of these categories of practices.

N management needs also to obey to four stewardship principles, commonly referred to as the four Rs: "Right amount," "Right time," "Right place," and "Right form" (Ryan et al. 2011). In the context of drylands, since N availability to crops is highly dependent on several management approaches, especially those related to organic matter and water, we suggest considering an additional, fifth "R," which is the "Right approach," as explained below:

R1: Right amount
- The amount of N needs to be based on crop requirements, target optimum yields, and soil N balance. The knowledge of initial N stock and the potential of N mineralization from organic matter are key factors in dryland soils. The apparent N use efficiency indices for specific conditions are required to make reasonable estimations of the amount to apply to reach potential yields. Since leaching is generally in rainfed early season conditions, nitrate-N can be used as a good soil test.
- Since rainfall can change from year to year, N rates should be adjusted to the actual rainfall conditions that coincide with the critical stages of the target crop. A drier year will require reducing the N rates initially planned, while a rainier season would impose increasing the N amounts in order to achieve higher yields and to compensate for the loss by leaching.
- Volatilization is a common N loss process in dryland soils, especially in calcareous soils with high pH and lime content. Appropriate adjustments based on assessment studies are needed.
- In dryland agriculture, nitrogen is given more importance by farmers compared to other nutrients. Caution is needed to avoid excess N that may affect crop growth and production.
- The response of N depends on other nutrients and also influences the uptake of other nutrients. Understanding N interactions with other macro- and micronutrients, especially under water stress conditions, is essential.

R2: Right Time
- Due to limited rainfall, N applications need to be planned according to the probabilities of the occurrence of rain and should follow the critical growth stages of the crop.
- Basal N fertilization at the start of the season should depend on the stock of mineral N as well on SOM content. The quality of organic matter (C/N) will determine the importance of N-mineralization versus N-immobilization and the resulting N-balance to be considered for subsequent N-topdressings.

- N-topdressing is intimately linked to rain occurrence in order to increase the N use efficiency and avoid temporary stress to roots for water and nutrient uptake.
- Root development is important for water and nutrient uptake, and N is important for root development and for the synergies with other nutrients involved, such as phosphorus.
- Early available N to the crop for its root development is a prerequisite for water use efficiency in later stages.

R3: Right Place

- Geographically, N management depends on the degree of aridity from one area to another. Responses to N are higher in the more favorable areas that are less prone to sporadic rainfall occurrence. N rates and time of application will depend ultimately on the amounts and spatial variability of rainfall in a given region.
- At the parcel level, N is more efficient where roots are dominantly present. Where banding is possible, local applications of N give a better response. The incorporation of N with seeders at sowing is found to trigger better initial growth compared to broadcast application followed by tillage that usually returns N to depths beyond the zone of initial root development.

R4: Right Source

- The forms of N need to be applied according to the limited soil-water conditions, which determine N mobility, as well as to soil properties, which determine both the mobility as well as the chemistry of N.
- Different forms of N can be adopted at different growing stages of the crop.
- Ammonium-N can be subject to important losses under dry conditions, high pH, and high lime content. Nitrification of ammonium can contribute significantly to buffer soil pH.
- Urea is a dominant N fertilizer. Its use under cool and moist conditions can give good response. When soil moisture is favorable, urea can be transformed rapidly to ammonium, which in turn goes through rapid nitrification (within a few days). However, urea is prone to volatile loss under dry conditions, and can cause injuries to crop leaves when broadcasted as topdressing.
- Loss by denitrification is very rare in dryland soils.
- The use of organic matter is a good source of N and a good soil conditioner for N-cycling. The knowledge of N-mineralization rates needs to be taken into consideration in the N-balance.

R5: Right approach

- Nutrient balance: the right amounts of all essential elements is crucial to guarantee positive interactions for plant growth, including water and nutrient use and efficiency.
- Surface residues management and mulching play important roles in protecting the soil and improving its fertility, mainly N and C cycling.
- Integrated soil fertility management is particularly beneficial in dryland soils.
- A good knowledge of the trends of mineralization versus immobilization of OM residues and amendments (manure, compost, etc.) is necessary to make compromises between nutrient release (mainly N) and SOM buildup. Different strategies are required depending on the nature of the OM used/recycled (mainly the C/N ratio). Mineralization is often sought in low fertilizer-use systems, which is the dominant situation in drylands. However, the need for increasing SOM in such systems is well justified.
- Early inorganic N applications help prevent N-immobilization when the C/N ratio of SOM high.
- Crop rotation with N-fixing legume crops is an ancient practice to be sustained to improve soil conditions, N, and the status of other nutrients. Crop rotation is also a good alternative to fallow under favorable dryland conditions.

- Conservation agriculture practices including reduced tillage, no-till, or direct seeding can improve organic matter, water status, and N bioavailability.
- Weed control reduces competition for water and N, especially under constrained dryland agriculture conditions.

12.5 CONCLUSIONS

Dryland agriculture is constrained by many interacting factors, but the main one remains the limitation of water, which in turn affects soil nutrient bioavailability, organic matter dynamics, and overall productive capacity. Limitation of water is not only in terms of total amount of rainfall, but also in terms of its erratic distribution with regard to crop growth stages. Dryland agriculture encompasses a diversity of cropping systems, varying from subsistence farming to intensive agriculture. Many factors of soil degradation threaten the potential production capacity in these areas. Nutrient depletion, wind and water erosion, and salinity are very common issues. However, nutrient inputs and management, especially N, remain the main concern for crop production. Most drylands around the world suffer from low fertilizer use and inappropriate management practices. N use and use efficiency is considered more critical than that of other nutrients, as it is highly affected by the amount and distribution of rain during crop season. N and organic matter dynamics are intimately interrelated, and both are affected by soil water status, which is determinant for nutrient bioavailability and uptake in water stress conditions. Mineralization of organic matter is a process favored by warm conditions. It is regarded as a good source of N in low input farming systems. However, it tends to deplete SOM, which is needed in dryland soils for other positive traits. Biological nitrogen fixation represents a good source of N, especially in rotations and intercropping systems involving legume crops, but can be significantly affected by water stress in dry conditions. Combining organic and mineral fertilizer sources can enhance mineral N availability as well as crop growth better than either form alone. Synergies among nutrients are important under dryland soil conditions. N uptake and use by plants is highly affected by other nutrients such as potassium, a key element for water stress tolerance, and phosphorus, a crucial element for root development. Enhancing N use and use efficiency is more of a challenge under dryland conditions. It requires a good understanding of the processes governing N dynamics under water-limited conditions as well as the choice of appropriate management practices (fertilizer amounts and forms, crop rotations, reduced tillage, mulching, intercropping, etc.). These practices have interactive effects on soil properties, water status, N-cycling, organic matter dynamics, etc. They need to be adapted to the multitude of situations that can be encountered under dryland areas as a result of the interaction of variable climatic, physical, and socioeconomic factors, in order to guarantee effective N strategies that may also impact C strategies. It is recommended that N management needs to obey the "5Rs" stewardship principles – which consider an additional "R" for "Right approach" – a set of practices and methods that would ensure sustainable and optimal N-dynamics under dryland conditions.

LIST OF ABBREVIATIONS

AGDP	agriculture gross domestic product
AWC	available water capacity
Bha	billion hectares
BNF	biological nitrogen fixation
C	carbon
CBD	Convention on Biological Diversity
C/N	carbon/nitrogen ratio
CT	conventional tillage
CTIC	Conservation Technology Information Center

FAO	Food and Agriculture Organization
g	gram
GDP	gross domestic product
ha	hectare
ISFM	integrated soil fertility management
K	potassium
k	mineralization rate
kg	kilogram
M ha	million hectares
Mg	megagram
mg	milligram
μg	microgram
IAEA	International Atomic Energy Agency
N	nitrogen
N_o	potentially mineralizable nitrogen
NT	no-till
NUE	nitrogen use efficiency
OM	organic matter
P	phosphorus
PET	potential evapotranspiration
Pr	precipitation
SOM	soil organic matter
SOC	soil organic carbon
UNEP-WCMC	United Nations Environment Programme World Conservation Monitoring Centre

REFERENCES

Abdel Monem, M., W.L. Lindsay, R. Sommer, and J. Ryan. 2010. Loss of nitrogen from urea applied to rainfed wheat in varying rainfall zones in northern Syria. *Nutr. Cycl. Agroecosys.* 86(3): 357–366.

Al-Karaki, G., R.B. Al-Karaki, and C.Y. Al-Karaki. 1996. Phosphorus nutrition and water stress effects on proline accumulation in sorghum and bean. *J. Plant Physiol.* 148(6): 745–751.

Arrese-Igor, C., E.M. González, D. Marino, R. Ladrera, E. Larrainzar, and E. Gil-Quintana. 2011. Physiological response of legume nodules to drought. *Plant Stress* 5 (Special Issue 1): 24–31.

Asgharipour, M.R., and M. Heidari. 2011. Effect of potassium supply on drought resistance in sorghum: Plant growth and macronutrient content. *Pak. J. Agr. Sci.* 48(3): 197–204.

Atkins, C.A., M. Fernando, and S. Hunt. 1992. A metabolic connection between nitrogenase activity and the synthesis of ureides in nodulated soybean. *Physiol. Plant* 84: 441–447.

Avci, M. 2011. Conservation tillage in Turkish dryland research. *Agron. Sustain. Dev.* 32(2): 299–307.

Bacanamwo, M., and J.E. Harper. 1997. The feedback mechanism of nitrate inhibition of nitrogenase activity in soybean may involve asparagine and/or products of its metabolism. *Physiol. Plant.* 100: 371–377.

Balaghi, R. 2015. Changement climatique, environnement et risques naturels au Maroc. Paper presented at: Atelier de formation et d'appui CEDRIG pour les projets ASAP-M, PAMPAT et PAMPAT Oriental, Rabat, 20–22 Janvier 2015 (unpublished).

Barber, S.A. 1984. *Soil Nutrient Bioavailability: A Mechanistic Approach.* John Wiley & Sons, New York.

Bationo, A., J. Kihara, B. Vanlauwe, B. Waswa, and J. Kimetu. 2007. Soil organic carbon dynamics, functions and management in West African agro-ecosystems. *Agric. Syst.* 94: 13–25.

Bationo, A., U. Mokwunye, P.L.G. Vlek, S. Koala, B.I. Shapiro, and C. Yamoah. 2003. Soil fertility management for sustainable land use in the West African Sudano-Sahelian zone. In *Soil Fertility Management in Africa: A Regional Perspective*, edited by M.P. Gichuru, A. Bationo, M.A. Bekunda, P.C. Goma, P.L. Mafongoya, D.N. Mugendi, H.M. Murwira, S.M. Nandwa, P. Nyathi and M.J. Swift. TSBF-CIAT.

Bauer, A. 1974. Influence of soil organic matter on bulk density and available water capacity of soils North Dakota Agricultural Experimental Station. *Farm Res.* 31: 44–52.

Bayala, J., G.W. Sileshib, R. Coec, et al. 2012. Cereal yield response to conservation agriculture practices in drylands of West Africa: A quantitative synthesis. *J. Arid Environ.* 78: 13–25.

Bennett, J.M. and S.L. Albrecht. 1984. Drought and flooding effects on N fixation, water relations and diffusive resistance of soybean. *Agron. J.* 76: 735–740.

Benlloch-Gonzalez, M., J. Romera, S. Cristescu, F. Harren, J.M. Fournier, and M. Benlloch. 2010. K+ starvation inhibits water-stress-induced stomatal closure via ethylene synthesis in sunflower plants. *J. Exp. Bot.* 61: 1139–1145.

Ben Naceur, M., M. Naily, and M. Selmi. 1999. Effet d'un deficit hydrique, survenant a differents stades de developpement du ble, sur l'humidite du sol, la physiologie de la plante et sur les composantes du rendement. *New Medit.* 10(2): 53–60.

Bennett, J.M., and S.L. Albrecht. 1984. Drought and flooding effects on N_2 fixation, water relations, and diffusive resistance of soybean. *Agron J.* 76: 735–740.

Bessam, F., and R. Mrabet. 2003. Long-term changes in soil organic matter under conventional and no-tillage systems in semiarid Morocco. *Soil Use Manag.* 19: 139–143.

Biederbeck, V.O., H.H. Janzen, C.A. Campbell, and R.P. Zentner. 1994. Labile soil organic matter as influenced by cropping practices in an arid environment. *Soil Biol. Biochem.* 26(12): 1647–1656.

Bouhafa, K., L. Moughli, K. Daoui, A. Douaik, and Y. Taarabt. 2015. Soil properties at different distances of intercropping in three olive orchards in Morocco. *Int. J. Plant Soil Sci.* 7(4): 238–245.

Bouraima, A.K., B. He, and T. Tian. 2015. Runoff, nitrogen (N) and phosphorus (P) losses from purple slope cropland soil under rating fertilization in Three Gorges Region. *Environ. Sci. Pollut. Res.* 23(5): 4541–4550.

Bouyoucos, G.J. 1939. Effect of organic matter on the water-holding capacity and wilting point of mineral soils. *Soil Sci.* 47(5): 377–384.

Bussière, F., and P. Cellier. 1994. Modification of the soil temperature and water content regimes by a crop residue mulch – Experiment and modelling. *Agric. For. Meteorol.* 68(1–2): 1–28.

Cabrera, M.L., D.E. Kissel, and M.F. Vigil. 2005. Nitrogen mineralization from organic residues: Research opportunities. *J. Environ. Qual.* 34: 75–79.

Cai, D.X., J. Ke, X.B. Wang, W.B. Hoogmoed, O. Oenema, and U.D. Perdok. 2006. Conservation tillage for dryland farming in China. In: *Proceedings of the 17th ISTRO Conference*, 28 August–3 September 2006. International Soil Tillage Research Organization, Kiel, Germany, pp. 1627–1633.

Cai, G., T. Dang, S. Guo, and M. Hao. 2005. Effects of nitrogen and mulching management on rainfed wheat and maize in northwest China. In *Nutrient and Water Management Practices for Increasing Crop Production in Rainfed Arid/Semi-Arid Area. Proceedings of a Coordinated Research Project.* IAEA-TECDOC-1468. 209-2016. pp. 77–88.

Cakmak, I. 2005. The role of potassium in alleviating detrimental effects of abiotic stresses in plants. *J. Plant Nutr. Soil Sci.* 168: 521–530.

Calderón, F.J., M.F. Vigil, and J. Benjamin. 2017. Compost input effects on dryland wheat and forage yields and soil quality. *Pedosphere* 28(3): 451–462. doi:10.1016/S1002-0160(17)60368-0.

Campbell, C.A., D.R Cameron., W. Nicholaichuk, and H.R. Davidson. 1977. Effects of fertilizer N and soil moisture on growth, N content and moisture use by spring wheat. *Can. J. Soil Sci.* 57: 289–310.

Campbell, C.A., Y.D. Jame, and R. De Jong. 1988. Predicting net nitrogen mineralization over a growing season: Model verification. *Can. J. Soil Sci.* 68: 537–552.

Campbell, C.A., Y.W. Jame, and G.E. Winkleman. 1984. Mineralization rate constants and their use for estimating nitrogen mineralization in some Canadian prairie soils. *Can. J. Soil Sci.* 64: 333–343.

Cerezini, P., D. dos Santos, L. Fagotti, A.E. Pípolo, M. Hungria, and M.A. Nogueira. 2017. Water restriction and physiological traits in soybean genotypes contrasting for nitrogen fixation drought tolerance. *Sci. Agric.* 74(2): 110–117.

Cerezini, P., M.K. Riar, and T.R. Sinclair. 2016. Transpiration and nitrogen fixation recovery capacity in soybean following drought stress. *J. Crop Improv.* 30(5): 562–571.

Chiang, C., B. Soudi, and A. Moreno. 1983. Soil nitrogen mineralization and nitrification under Moroccan conditions. In *Nutrient Balances and the need for Fertilizers in Semi-Arid and Arid Regions.* 17th Colloquium of the International Potash Institute, Rabat and Marrakech. International Potash Institute, Bern Switzerland. pp. 129–139.

Coleto, I., M. Pineda, A.P. Rodino, A.M. De Ron, and J.M. Alamillo. 2014. Comparison of inhibition of N2 fixation and ureide accumulation under water deficit in four common bean genotypes of contrasting drought tolerance. *Annals Botany.* 113: 1071–1082.

Corbeels, M. 1997. Nitrogen availability and its effect on water limited wheat growth on Vertisols in Morocco. PhD thesis.

Corbeels, M., G. Hofman, and O. Van Cleemput. 1997. Response of rainfed bread wheat to nitrogen fertilizer on Moroccan Vertisols. 11th World Fertilizer Congress. Ghent, Belgium, 7–13 September, 1997.

Corbeels, M., G. Hofman, and O. van Cleemput. 1998. Residual effect of nitrogen fertilization in a wheat-sunflower cropping sequence on a Vertisol under semi-arid Mediterranean conditions. *Eur. J. Agron.* 9: 109–116.

Corbeels, M., G. Hofman, and O. van Cleemput. 1999. Soil mineral nitrogen dynamics under bare fallow and wheat in vertisols of semi-arid Mediterranean Morocco. *Biol. Fertil. Soils* 28: 321–328.

Corbeels, M., G. Hofman, and O. Van Cleemput. 2000. Nitrogen cycling associated with the decomposition of sunflower stalks and wheat straw in a Vertisol. *Plant Soil.* 218: 71–82.

CTIC (Conservation Technology Information Center). 2018. Tillage type definitions. http://www.ctic.purdue.edu/resourcedisplay/322/. Accessed July 2018.

Curioni, P.M.G., U.A. Hartwig, J. Nosberger, and K.A. Schuller. 1999. Glycolytic flux is adjusted to nitroge-nase activity in nodules of detopped and argon-treated alfalfa plants. *Plant Physiol.* 119: 445–453.

Daoui, K., Z. Fatemi, A. Bendidi, R. Razouk, A. Chergaoui, and A. Ramdani. 2012. Olive tree and annual crops association's productivities under Moroccan conditions. In: *First European Scientific Conference on Agriforestry, 9–10 October, Brussels:Book of Abstracts.* http://www.eurafagroforestry.eu/action/conferences/I_EURAFConference.

Darkoh, M.B.K. 2003. Desertification in the drylands: A review of the African situation. *Ann. Arid Zone* 42(3–4): 289–307.

Daryanto, S., L. Wang, and P.A. Jacinthe. 2015. Global synthesis of drought effects on food legume produc-tion. *PLoS One* 10(6): e0127401. doi:10.1371/journal.pone.0127401.

Dregne, H.E. 1976. *Soils of Arid Regions.* Elsevier, Amsterdam, the Netherlands.

Duan, Y., X. Shi, S. Li, X. Sun, and X. He. 2014. Nitrogen use efficiency as affected by phosphorus and potas-sium in long-term rice and wheat experiments. *J. Integrative Agric.* 13(3): 588–596.

Duley, F.L., and J.C. Russel. 1939. The use of crop residues for soil and moisture conservation. *J. Am. Soc. Agron.* 31: 703–709.

Durand, J.L., J.E. Sheehy, and F.R. Minchin. 1987. Nitrogenase activity, photosynthesis and nodule water potential in soya bean plants experiencing water deprivation. *J. Exp. Bot.* 38: 311–321.

El Mejahed, K. and L. Aouragh. 2005. Green manure and nitrogen fertilizer effects on soil quality and profit-ability of a wheat-based system in semi-arid Morocco. In *Nutrient and Water Management Practices for Increasing Crop Production in Rainfed Arid/Semi-Arid Area. Proceedings of a Coordinated Research Project.* IAEA-TECDOC-1468. 209-2016. pp. 89–106.

Elherradi, E., B. Soudi, and K. Elkacemi. 2003. Evaluation de la minéralisation de l'azote de deux sols amen-dés avec un compost d'ordures ménagères. *Étude et Gestion des Sols* 10(3): 139–154.

Engin, M., and J.I. Sprent. 1973. Effects of water stress on growth and nitrogen-fixing activity of Trifolium repens. *New Phytol.* 72: 117–126.

FAO (Food and Agriculture Organization). 2004. *Carbon Sequestration in Dryland Soils.* World Soils Resources Reports no. 102. FAO, Rome, Italy.

FAO (Food and Agriculture Organization). 2005a. The importance of soil organic matter: Key to drought-resistant soil and sustained food production, edited by A. Bot and J. Benites. FAO Soils Bulletin 80. FAO. 2005. Rome, Italy.

FAO (Food and Agriculture Organization). 2005b. *Land and Environmental Degradation and Desertification in Africa: Issues and Options for Sustainable Economic Development with Transformation*, edited by S.C. Nana-Sinkam. Monograph no 10. UN-ECA/FAO Agricultural Division, FAO, Rome, Italy.

Finn, G.A., and W.A. Brun. 1980. Water stress effects on CO_2 assimilation, photosynthate partitioning, stoma-tal resistance, and nodule activity in soybean. *Crop Sci.* 20: 431–434.

Fischer, K.S., R. Lafitte, S. Fukai, G. Atlin, and B. Hardy. 2003. *Breeding Rice for Drought-Prone Environments.* International Rice Research Institute, Los Baños, Philippines. http://books.irri.org/9712201899_content.pdf.

Flower, T.N., and J.R.Q. Challagha. 1983. Nitrification in soils incubated with pig slurry or ammonium sul-phate. *Soil Biol. Biochem.* 15(3): 337–342.

Fuentes, J.P., M. Flury, D.R. Huggins, and D.F. Bezdicek. 2003. Soil water and nitrogen dynamics in dryland cropping systems of Washington State. *Soil Till. Res.* 71: 33–47.

Galvez, L., E.M. Gonzalez, and C. Arrese-Igor. 2005. Evidence for carbon flux shortage and strong carbon/nitrogen interactions in pea nodules at early stages of water stress. *J. Exp. Bot.* 56: 2551–2561.

Gao, Y., Y. Li, J. Zhang, W. Liu, Z. Dang, W. Cao, and Q. Qiang. 2009. Effects of mulch, N fertilizer, and plant density on wheat yield, wheat nitrogen uptake, and residual soil nitrate in a dryland area of China. *Nutr. Cycl. Agroecosys.* 85(2): 109–121.

Gimenez, C., F. Orgaz, and E. Fereres. 1997. Productivity in water-limited environments: Dryland agricultural systems. In *Ecology in Agriculture*. L.E. Jakson (ed.). Academic Press.

González, E., L. Gálvez, M. Royuela, P. Aparicio-Tejo, and C. Arrese-Igor. 2001. Insights into the regulation of nitrogen fixation in pea nodules: Lessons from drought, abscisic acid and increased photoassimilate availability. *Agronomie, EDP Sciences* 21(6–7): 607–613.

González, E.M., P.M. Aparicio-Tejo, A.J. Gordon. F.R. Minchin, M. Royuela, and C. Arrese-Igor. 1998. Water-deficit stress effects on carbon and nitrogen metabolism of pea nodules. *J. Exp. Bot.* 49(327): 1705–1714.

Haas, H.J., C.E. Evans, and E.F. Miles. 1957. *Nitrogen and Carbon Changes in Great Plains Soils as Influenced by Cropping and Soil Treatments*. Technical Bulletin no. 1164. U.S. Department of Agriculture, Washington, DC.

Hadda, M.S., and S. Arora. 2006. Soil and nutrient management practices for sustaining crop yields under maize-wheat cropping sequence in sub-mountain Punjab, India. *Soil Environ.* 25(1): 1–5.

Hardarson, G., S.K.A. Danso, and F. Zapata. 1987. Biological nitrogen fixation in field crops. In: *Handbook of Plant Science in Agriculture*, edited by B.R. Christie. CRC Press, Boca Raton, FL, pp. 165–192.

Harraq, A., R. Bouabid, H. Bahri, and H. Boumchita. 2016. Dynamique de minéralisation de l'azote des fer-tilisants organiques et contribution des différentes formes d'azote à l'absorption de cet élément par la pomme de terre conduite en mode biologique. *Eur. J. Sci. Res.* 137(2): 196–213.

Hatfield, J. L., T.J. Sauer, and J.H. Prueger. 2001. Managing soils to achieve greater water use efficiency: A review. *Agron. J.* 93: 271–280.

He, M., and F.A. Dijkstra. 2014. Drought effect on plant nitrogen and phosphorus: A meta analysis. *New Physiol.* 204: 924–931.

Henao, J., and C. Baanante. 2006. *Agricultural Production and Nutrient Mining in Africa: Implications for Resource Conservation and Policy Development*. International Fertilizer Development Center (IFDC), Muscle Shoals, AL. www.ifdc.org.

Hénin, S., and M. Dupuis. 1945. Essai de bilan de la matière organique du sol. *Ann. Agron.* 11: 17–29.

Hilhorst, T., and F. Muchena. 2000. Managing soil fertility in Africa: Diverse settings and changing practice. In: *Nutrients on the Move: Soil Fertility Dynamics in African Farming Systems*, edited by T. Hilhorst and F. Muchena. IIED Drylands Programme, London, UK.

Hollis, J.M., R.J.A. Jones, and R. C. Palmer. 1977. The effects of organic matter and particle size on the water retention properties of some soils in the West Midlands of England. *Geodema* 17: 225–2310.

Hudson, B.D. 1994. Soil organic matter and available water capacity. *J. Soil Water Conserv.* 49(2): 189–194.

IAEA. 2005. Nutrient and water management practices for increasing crop production in rainfed arid/semi-arid area. *Proceedings of a Coordinated Research Project.* IAEA-TECDOC-1468.

IFPRI (International Food Policy Research Institute). 1996. *Feeding the World, Preventing Poverty and Protecting the Earth: A 2020 Vision*. International Food Policy Research Institute, Washington, DC. http://www.ifpri.org/publication/feeding-world-preventing-poverty-and-protecting-earth.

Irshad, M., T. Honna, S. Yamamoto, A.E. Eneji, and N. Yamasaki. 2005. Nitrogen mineralization under saline conditions. *Commun. Soil Sci. Plant.* 36(11–12): 1681–1689.

Irshad M., M. Inoue, M. Ashraf, H.K. Faridullah, M. Delower, K.M. Hossain, and A. Tsunekawa. 2007. Land desertification-an emerging threat to n. Environment and Food Security of Pakistan. *J. Appl. Sci.* 8: 1199–1205.

Janssen, B.H. 1984. A simple method for calculating decomposition and accumulation of 'young' soil organic matter. *Plant Soil.* 76: 297–304.

Jenkinson, D.S., and J.H. Rayner. 1977. The turnover of soil organic matter in some of the Rothamsted classi-cal experiments. *Soil Sci.* 123: 298–305.

Jensen, E.S. 1997. Nitrogen immobilization and mineralization during initial decomposition of ^{15}N labeled pea and barley residues. *Biol. Fert. Soils* 24: 39–44.

Jin, K., S. De Neve, B. Moeskops, J. Lu, J. Zhang, D. Gabriels, D. Cai, and J. Jin. 2008. Effects of different soil management practices on winter wheat yield and N losses on a dryland loess soil in China. *Aust. J. Soil Res.* 46(5): 455–463.

Johnson, Y.K., F.O. Ayukeb, J.M. Kinamac, and I.V. Sijali. 2018. Effects of tillage practices on water use efficiency and yield of different drought tolerant common bean varieties in Machakos County, Eastern Kenya. *Am. Sci. Res. J. Eng. Technol. Sci.* 40(1): 217–234.

Ju, X.T., and S.X. Li. 1998. The effect of temperature and moisture on nitrogen mineralization in soils. *Plant Nutr. Fertil. Sci.* 4(1): 37–42.

Justice, J.K., and R.L. Smith. 1962. Nitrification of ammonium sulfate in a calcareous soil as influenced by combinations of moisture, temperature, and levels of added nitrogen. *Soil Sci. Soc. Am. Proc.* 26: 246–250.

Kant, S., and U. Kafkafi. 2002. Potassium and abiotic stresses in plants. In: *Potassium for Sustainable Crop Production*, edited by N.S. Pasricha and S.K. Bansal. Potash Institute of India, Haryana, India, pp. 233–251.

Kelley, K.W., and D.W. Sweeney. 2005. Tillage and urea ammonium nitrate fertilizer rate and placement affects winter wheat following grain sorghum and soybean. *Agron. J.* 97: 690–697.

Kihara, J., A. Bationo, D.N. Mugendi, C. Martius, and P.L.G. Vlek. 2011. Conservation tillage, local organic resources and nitrogen fertilizer combinations affect maize productivity, soil structure and nutrient balances in semi-arid Kenya. *Nutr. Cycl. Agroecosys.* 90: 213–225.

King, C.A., and L.C. Purcell. 2005. Inhibition of N_2 fixation in soybean is associated with elevated ureides and amino acids. *Plant Physiol.* 137, 1389–1396.

Klemedtsson, L., B.H. Svensson, and T. Rosswall. 1988. Relationships between soil moisture content and nitrous oxide production during nitrification and denitrification. *Biol. Fert. Soils* 6(2): 106–111.

Kolenbrander, G.J. 1969. *De bepaling van de waarde van verschillende soorten organische stof ten aanzien van hun effect op her humusgehalte bij bouwland*. Inst. Bodemvruchtbaarheid, Haren, the Netherlands, C 6988.

Lal, R. 2002. Carbon sequestration in dryland ecosystems of West Asia and North Africa. *Land Degrad. Dev.* 13(1): 45–59.

Lal, R. 2004. Carbon sequestration in dryland ecosystems. *Environ. Manage.* 33(4): 528–544.

Li, P., H. Dong, A. Liu, J. Liu, M. Sun, Y. Li, S. Liu, X. Zhao, and S. Mao. 2017. Effects of nitrogen rate and split application ratio on nitrogen use and soil nitrogen balance in cotton field. *Pedosphere* 27(4): 769–777.

Li, S.X., Z.H. Wang, Y.J. Gao and B.A. Stewart. 2009a. Nitrogen in dryland soils of China and its management. *Adv. Agron.* 101: 123–181.

Li, S.X., Z.H. Wang, T.T. Hu, S.S. Malhi, S.Q. Li, Y.J. Gao, and X.H. Tian. 2009b. Nutrient and water management effects on crop production and nutrient and water use efficiency in dryland areas of China. *Adv. Agron.* 102: 223–265.

Lima, J.D., and L. Sodek. 2003. N-stress alters aspartate and asparagine levels of xylem sap in soybean. *Plant Sci.* 165: 649–656.

Lindhauer, M.G. 1985. Influence of K nutrition and drought on water relations and growth of sunflower (*Helianthus-annuus* L.). *J. Plant Nutr. Soil Sci.* 148: 654–669.

Liwang, M.A., G.A. Peterson, L.R. Ahuja, L. Sherrod, M.J. Shaffer, and K.W. Rojas. 1999. Decomposition of surface crop residues in long-term studies of dryland agroecosystems. *Agron. J.* 91: 401–409.

Ma, Q., W.T. Yu, S.M. Shen, H. Zhou, Z.S. Jiang, and Y.G. Xu. 2010. Effects of fertilization on nutrient budget and nitrogen use efficiency of farmland soil under different precipitations in Northeastern China. *Nutr. Cycl. Agroecosys.* 88: 315–327.

Malhi, S.S., and W.B. McGill. 1982. Nitrification in three Alberta soils: Effect of temperature, moisture and substrate concentration. *Soil Biol. Biochem.* 15: 397–399.

Manyowa, N.M. 1994. Maize production in Zimbabwe: Coping with drought stress in marginal agro-ecological zones. In: *Bilan hydrique agricole et sécheresse en Afrique tropicale: Vers une gestion*, edited by F.N. Reynier and L. Netoyo. Universités Francophones, John Libbey Eurotext, Paris, France.

Marino, D., P. Frendo, R. Ladrera, A. Zabalza, A. Puppo, C. Arrese-Igor, and E.M. Gonzalez. 2007. Nitrogen fixation control under drought stress. localized or systemic? *Plant Physiol.* 143: 1968–1974.

Marschner, P. 2012. *Marschner's Mineral Nutrition of Higher Plants*. 3rd ed. Academic Press, London, UK, pp. 178–189.

Mary, B., and J. Guérif. 1994. Intérêts et limites des modèles de prévision de l'évolution des matières organiques et de l'azote dans le sol. *Cahiers Agricultures*. 3: 247–257.

Matthews, R.B., D.M. Reddy, A.U. Rani, S.N. Azam-Ali, and J.M. Peacock. 1990. Response of four sorghum lines to mid-season drought. I. Growth, water use and yield. Field Crops Res. 25: 279–296.

McKenzie, R.H., A.B. Middleton, and M. Zhang. 2001. Optimal time and placement of nitrogen fertilizer with direct and conventionally seeded winter wheat. *Can. J. Soil Sci.* 81: 613–622.

Mengel, K. 2007. Potassium. In: *Handbook of Plant Nutrition*, edited by A.V. Baker and D.J. Pilbeam. CRC Press, Taylor & Francis Group, Boca Raton, FL.

Montañez, A. 2000. Overview and case studies on biological nitrogen fixation: Perspectives and limitations. *Sci. Agric.* 1–11.

Mrabet, R., K. Ibno-Namr, F. Bessam, and N. Saber. 2001. Soil chemical quality changes and implications for fertilizer management after 11 years of no-tillage wheat production systems in semiarid Morocco. *Land Degrad. Dev.* 21: 1–13.

Mrabet, R., R. Moussadek, A. Fadlaoui, and E. van Ranst. 2012. Conservation agriculture in dry areas of Morocco. *Field Crops Res.* 132: 84–94.

Murage, E.W., N.K. Karanja, P.C. Smithson, and P.L. Woomer. 2000. Diagnostic indicators of soil quality in productive and non-productive smallholders' fields of Kenya's Central Highlands. *Agr. Ecosyst. Environ.* 79: 1–8.

Myers, R.J.K. 1984. A simple model for estimating the nitrogen fertilizer requirement of a cereal crop. *Fert. Res.* 5, 95–108.

Myers, R.J.K., C.A. Campbell, and K.L. Weier. 1982. Quantitative relationship between net nitrogen mineralization and moisture content of soils. *Can. J. Soil Sci.* 62: 111–124.

Nageswara Rao, R.C., Sardar Singh, M.V.K. Sivakumar, K.L. Srivastava, and J.H. Williams. 1985. Effect of water deficit at different growth phases of peanut. I. Yield response. *Agronomy J.* 77: 782–786.

Nandwal, A.S., S. Bharti, I.S. Sheoran, and M.S. Kuhad. 1991. Drought effects on carbon exchange and nitrogen fixation in pigeonpea (*Cajanus cajan* L.). *J. Plant Physiol.* 38: 125–127.

Nawaz, A., and M. Farooq. 2016. Nutrient management in dryland agriculture systems. In *Innovations in Dryland Agriculture*, Farooq, M., and Siddique K.H.M. (eds.), Springer, Cham, pp. 115–142.

Neo, N.N., and D.B. Layzell. 1997. Phloem glutamine and the regulation of O_2 diffusion in legume nodules. *Plant Physiol.* 113: 259–267.

Nettleton, W.D., and F.F. Peterson. 1983. Aridisols. In: *Pedogenesis and Soil Taxonomy*, edited by L.P. Wilding, N.E. Smeck, and G.F. Hall. Elsevier, Amsterdam, the Netherlands, pp. 165–216.

NICRA (National Innovations on Climate Resilient Agriculture). 2012. All India Coordinated Research Project for Dryland Agriculture. Central Research Institute for Dryland Agriculture, Santoshnagar, India. Technical program of 2011–12. pp.132. http://www.nicra-icar.in/nicrarevised/images/Books/nicra-aicrpda.pdf.

Noin, D., and J.I. Clarke. 1997. Population and environment in arid regions of the world. In: *Population and Environment in Arid Regions*, edited by J. Clarke and D. Noin. MAB/UNESCO, vol. 19. Parthenon Publishing Group, New York. pp. 1–18.

Noy-Meir, I., Y. Harpaz. 1978. Agro-ecosystems in Israel. In *Cycling of Mineral Nutrients in Agricultural Ecosystems*, Frissel, M.J. (ed.), vol. 4, pp. 143–167. Agro-Ecosystems.

Palm, C.A., R.J.K. Myers, and S.M. Nandwa. 1997. Combined use of organic and inorganic nutrient sources for soil fertility maintenance and replenishment. In *Replenishing Soil Fertility in Africa*, Buresh R.J., Sanchez, P.A., and Calhoun, F. (eds.), pp. 193–217. Soil Science Society of America Madison, Wis.

Palta, J.A, and I.R.P. Fillery. 1995. N application enhances remobilization and reduces losses of pre-anthesis N in wheat grown on a duplex soil. *Aus. J. Agric. Res.* 46(3): 519–531.

Palta, J.A., X. Chen, S.P. Milroy, C.G.J. Rebetzke, M.F. Dreccer, and M. Watt. 2011. Large root systems: Are they useful in adapting wheat to dry environments? *Funct. Plant Biol.* 38: 347–354.

Pankhurst, C.E. and J.I. Sprent. 1975a. Effects of water stress on the respiratory and nitrogen-fixing activity of soybean root nodules. *J. Exp. Bot.* 26: 287–304.

Pankhurst, C.E., and J.I. Sprent. 1975b. Surface features of soybean root nodules. *Protoplasma* 85: 85–98.

Passioura, J.B. 1983. Roots and drought resistance. *Agric. Water Mgmt.* 7: 265–280.

Peng, Z., W. Ting, W. Haixia, W. Min, M. Xiangping, M. Siwei, Z. Rui, J. Zhikuan, and H. Qingfang. 2014. Effects of straw mulch on soil water and winter wheat production in dryland farming. *Sci. Rep.* 5: 10725.

Pervez, H., M. Ashraf, and M.I. Makhdum. 2004. Influence of potassium nutrition on gas exchange characteristics and water relations in cotton (*Gossypium hirsutum* L.). *Photosynthetica* 42: 251–255.

Petit, S., A. Alignier, N. Colbach, A. Joannon, and D. Cœur. 2013. Weed dispersal by farming at various spatial scales: A review. *Agron. Sustain. Dev.* 33(1): 205–217.

Place, F., C.B. Barret, H.A. Freeman, J.J. Ramisch, and B. Vanlauwe. 2003. Prospects for integrated soil fertility management using organic and inorganic inputs: Evidence from smallholder African agricultural systems. *Food Policy.* 28: 365–378.

Premachandra, G.S., H. Saneoka, and S. Ogata. 2009. Cell membrane stability and leaf water relations as affected by potassium nutrition of water-stressed maize. *J. Exp. Bot.* 42: 739–745. 33.

Purcell, L.C., C.A. King, and R.A. Ball. 2000. Soybean cultivar differences in ureides and the relationship to drought tolerant nitrogen fixation and manganese nutrition. *Crop Sci.* 40: 1062–1070.

Qin, W., C. Hu, and O. Oenema. 2015. Soil mulching significantly enhances yields and water and nitrogen use efficiencies of maize and wheat: A meta-analysis. *Sci. Rep.* 5: 16210.

Ramana, M.V., K.M. Khadke, T.J. Rego, J.V.D.K. Kumar Rao, R.J.K. Myers, G. Pardhasaradhi, and N. Venkata Ratnam. 2005. Management of nitrogen and evaluation of water-use efficiency in traditional and improved cropping systems of the southern Telangana region of Andhra Pradesh, India. In *Nutrient and Water Management Practices for Increasing Crop Production in Rainfed Arid/Semi-arid Area. Proceedings of a Coordinated Research Project.* IAEA-TECDOC-1468. pp. 139–154.

Ramirez-Garcia, J., H.J. Martens, M. Quemada, and K. Thorup-Kristensen. 2015. Intercropping effect on root growth and nitrogen uptake at different nitrogen levels. *J. Plant Ecol.* 8(4): 380–389.

Rashid, A. and J. Ryan. 2004. Micronutrient constraints to crop production in soils with Mediterranean-type characteristics: A review. *J. Plant Nutr.* 27: 959–975.

Recous, S, D. Robin, D. Darwis, and B. Mary. 1995. Soil inorganic N availability: Effect on maize residue decomposition. *Soil Biol. Biochem.* 27: 1529–1538.

Reich, P.F., S.T. Numbem, R.A. Almaraz, and H. Eswaran. 2001. Land resource stresses and desertification in Africa. In: *Responses to Land Degradation: Proceedings of the 2nd International Conference on Land Degradation and Desertification, Khon Kaen, Thailand,* edited by E.M. Bridges, I.D. Hannam, L.R. Oldeman, F.W.T. Pening de Vries, S.J. Scherr, and S. Sompatpanit. Oxford Press, New Delhi, India.

Reinertsen, S.A., L.F. Elliott, V.L. Cochran, and G.S. Campbell. 1984. Role of available carbon and nitrogen in determining the rate of wheat straw decomposition. *Soil Biol. Biochem.* 16: 459–464.

Rekha, R.G., B.K. Desai, R. Satyanarayan, M.R. Umesh, and S. Shubha. 2017. Effect of intercropping system and nitrogen management practices on nutrient uptake and post-harvest soil fertility. *Int. J. Sci. Nat.* 8(2): 236–239.

Roy, R.N., and H. Nabhan. 1999. Soil and nutrient management in sub-Saharan Africa in support of the soil fertility initiative. In: *Proceedings of the Expert Consultation Lusaka, Zambia, 6–9 December 1999.* FAO, Rome, Italy.

Russell, M.B., A. Klute, and W.C. Jacob. 1952. Further studies on the effect of long-time organic matter additions on the physical properties of Sassafras silt loam. *Soil Sci. Soc. Amer. Proc.* 16: 156.

Ryan, J. 1998. Changes in organic carbon in long-term rotation and tillage trials in northern Syria. In: *Management of Carbon Sequestration in Soil,* edited by R. Lal, J. Kimble, R. Follett, and B.A. Stewart. Advances in Soil Science. CRC Press, Boca Raton, FL, p. 285295.

Ryan, J. 2008a. Crop nutrients for sustainable agricultural production in the drought-stressed Mediterranean region. *J. Agric. Sci. Technol.* 10: 295–306.

Ryan, J. 2008b. A perspective on balanced fertilization in the Mediterranean region. *Turk. J. Agric. For.* 32: 79–89.

Ryan, J., S. Masri, S. Garabet, J. Diekmann, and H. Habib. 1997. *Soils of ICARDA's Agricultural Experiment Stations and Sites: Climate, Classification, Physical and Chemical Properties, and Land Use.* ICARDA Technical Bulletin. ICARDA, Aleppo, Syria.

Ryan, J., M. Pala, H. Harris, S. Masri, and M. Singh. 2010. Rainfed wheat-based rotations under Mediterranean-type climatic conditions: Crop sequences, N fertilization, and stubble grazing intensity in relation to cereal yields parameters. *J. Agric. Sci.* 48: 205–216.

Ryan, J., M. Singh, and M. Pala. 2008. Long-term cereal-based rotation trials in the Mediterranean region. Implications for cropping sustainability. *Adv. Agron.* 97: 276–324.

Ryan, J. and R. Sommers. 2010. Fertilizer best management practices in the dryland Mediterranean area – Concepts and perspectives. In: *19th World Congress of Soil Science: Soil Solutions for a Changing World, Brisbane, Australia, 1–6 August 2010: Proceedings.* World Congress of Soil Science, Brisbane, Queensland, Australia.

Ryan, J., R. Sommer, and H. Ibrikci. 2011. Fertilizer best management practices: A perspective from the dryland West Asia–North Africa Region. *J. Agron. Crop Sci.* 198: 57–67.

Sahrawat, K.L. 2008. Factors affecting nitrification in soils. *Commun. Soil Sci. Plant* 39(9–10): 1436–1446.

Sainju, M., A.W. Lenssen, T. Caesar-Ton, et al. 2012. Dryland soil nitrogen cycling influenced by tillage, crop rotation, and cultural practice. *Nutr. Cycl. Agroecosys.* 93(3): 309–322.

Sainju, U.M., A.W. Lenssen, B.L. Allen, W.B. Stevens, and J.D. Jabro. 2016. Nitrogen balance in response to dryland crop rotations and cultural practices. *Agric. Ecosys. Environ* .233: 25–32.

Salter, P.J., and F. Howarth. 1961. The available-water capacity of a sandy loam soil. II. The effects of farmyard manure and different primary cultivations. *J. Soil Sci.* 12: 335–342.

Salter, P.J., and J.B. Williams. 1963. The effect of farmyard manure on the moisture characteristics of a sandy loam soil. *J. Soil Sci.* 14: 73–81.

Sauerbeck, D.R., and M.A. Gonzalez. 1977. Field decomposition of carbon-14-labelled plant residues in various soils of the Federal Republic of Germany and Costa Rica. In: *Soil Organic Matter Studies.* International Atomic Energy Agency, Vienna, Austria, 1: 159–170.

Schomberg, H.H., and O.R. Jones. 1999. Carbon and nitrogen conservation in dryland tillage and cropping systems. *Soil Sci. Soc. Am. J.* 63: 1359–1366.

Schomberg, H.H., and J.L. Steiner. 1999. Nutrient dynamics of crop residues decomposing on a fallow no-till soil surface. *Soil Sci. Soc. Am. J.* 63: 607–613.

Schomberg, H.H., J.L. Steiner, and P.W. Unger. 1994. Decomposition and nitrogen dynamics of crop residues: Residue quality and water effects. *Soil Sci. Soc. Am. J.* 58: 372–381.

Seligman, N.G., S. Feigenbaum, D. Feinerman, and R.W. Benjamin. 1986. Uptake of nitrogen from high C-to-N ratio ^{15}N-labeled organic matter residues by spring wheat grown under semi-arid conditions. *Soil Biol. Biochem.* 18: 303–307.

Sene, M., and A.N. Badiane. 2005. Optimization of water and nutrient use by maize and peanut in rotation based on organic and rock phosphate soil amendments. In *Nutrient and Water Management Practices for Increasing Crop Production in Rainfed Arid/Semi-arid Area. Proceedings of a Coordinated Research Project.* IAEA-TECDOC-1468. pp. 197–208.

Serraj, R., V. Vadez, and T.R. Sinclair. 2001. Feedback regulation of symbiotic N_2 fixation under drought stress. *Agronomie* 21: 621–626.

Serraj, R., and T.R. Sinclair. 1996a. Inhibition of nitrogenase activity and nodule oxygen permeability by water deficit. *J. Exper. Botany.* 47: 1067–1073.

Serraj, R., and T.R. Sinclair. 1996b. Processes contributing to N2-fixation insensitivity to drought in the soybean cultivar Jackson. *Crop Sci.* 36: 961–968.

Serraj, R., and T.R. Sinclair. 1997. Variation among soybean cultivars in dinitrogen fixation response to drought. *Agron. J.* 89: 963–969.

Serraj, R., T.R. Sinclair, and L.C. Purcell. 1999. Symbiotic N_2 fixation response to drought. *J. Exp. Bot.* 50: 143–155.

Shah, Z.H., and M. Arshad. 2006. Land degradation in Pakistan: A serious threat to environments and economic stability. Eco Services International. https://www.eco-web.com/edi/060715.html. Accessed April 2018.

Sharma, K.L., K. Srinivas, S.K. Das, K.P.R Vittal, and G.J. Kusuma. 2002. Conjunctive use of inorganic and organic sources of nitrogen for higher yield of sorghum in dryland Alsol. *Indian J. Dryland Agric. Res. Dev.* 17: 79–88.

Sharma, K.L., K.P.R. Vittal, Y.S. Ramakrishna, K. Srinivas, B. Venkateswarlu, and J. Kusuma Grace. 2007. Fertilizer use constraints and management in rainfed areas with special emphasis on N use efficiency. In: *Agricultural Nitrogen Use & Its Environmental Implications*, edited by Y.P. Abrol, N. Raghuran, and M.S. Sachdev. I.K. International Publishing House, New Delhi, India.

Shelton, D.R., A. Sadeghi, and G.W. McCarty. 2000. Effect of soil water content on denitrification during cover crop decomposition. *Soil Sci.* 165: 365–371.

Shepherd, K.D., P.J.M. Cooper, A.Y. Allan, D.S.H. Drennan, J.D.H. Keatinge. 1987. Growth, water use and yield of barley in Mediterranean-type environments. *J. Agric. Sci.* 108: 365–378.

Sijali, I.V., and P.T. Kamoni. 2005. Optimization of water and nutrient use in rain-fed semi-arid farming through integrated soil-, water- and nutrient-management practices. In *Nutrient and Water Management Practices for Increasing Crop Production in Rainfed Arid/Semi-arid Area. Proceedings of a Coordinated Research Project.* IAEA-TECDOC-1468. 209–2016.

Sinclair, T.R., and J. Goudriaan. 1981. Physical and morphological constraints on transport in nodules. *Plant Physiol.* 67: 143–145.

Sinclair, T.R., L. Purcell, V. Vadez, and R. Serraj. 2001. Selection of soybean (Glycine max) lines for increased tolerance of N_2 fixation to drying soil. *Agronomie, EDP Sciences* 21(6–7): 653–657.

Sinclair, T.R., and R. Serraj. 1995a. Dinitrogen fixation sensitivity to drought among legume species. *Nature* 1995: 378–344.

Sinclair, T.R., and R. Serraj. 1995b. Legume nitrogen fixation and drought. *Nature* 378: 344–344.

Singh, A.K., and P. Kumar. 2009. Nutrient management in rainfed-dryland agro-ecosystems in the impending climate change scenario. *Agricultural Situation in India* 66(5): 265–269.

Singh, H., and K.P. Singh. 1994. Nitrogen and phosphorus availability and mineralization in dryland reduced tillage cultivation: Effects of residue placement and chemical fertilizer. *Soil Biol. Biochem.* 26(6): 695–702.

Singh, M. 2017. Production potential and profitability of pigeonpea [*Cajanus cajan* (L.) Millsp.] as influenced by intercropping with blackgram [*Vigna mungo* (L.) Hepper] and integrated nutrient management. PhD thesis. Faculty of Agriculture, Indira Gandhi Krishi Vishwavidyalaya, Raipur, India. http://krishikosh.egranth.ac.in/bitstream/1/5810038158/1/thesis%20Madhulika%20singh.pdf.

Singh, R.P., and P. Venkateswarlu. 1985. Role of all India coordinated research project for dryland agriculture in research development. *Fertilizer News* 30(4): 43–55.

Sivanappan, R.K. 1995. Soil and water management in the dry lands of India. *Land Use Policy* 12(2): 165–175.

Smika, D.E., and P.W. Unger. 1986. Effect of surface residues on soil water storage. In: *Advances in Soil Science*, edited by B.A. Stewart, 5. Springer, New York.

Smith, S. J., L.B. Young, and G.E Miller. 1977. Evaluation of soil nitrogen mineralization potentials under modified field conditions. *Soil Sci. Soc. Am. J.* 4l: 74–76.

Song, H. X., and S.X. Li. 2006. Root function in nutrient uptake and soil water effect on NO_3^--N and NH_4^--N migration. *Agric. Sci. China* 5(5): 377–383.

Song, H. X., and S.X. Li. 2006. Root function in nutrient uptake and soil water effect on NO_3^--N and NH_4^--N migration. *Agric. Sci. China* 5(5): 377–383.

Song, Y. N., F.S. Zhang, P. Marschner, F.L. Fan, H.M. Gao, X.G. Bao, J.H. Sun, and L. Li. 2007. Effect of inter-cropping on crop yield and chemical and microbiological properties in rhizosphere of wheat (*Triticum aestivum* L.), maize (*Zea mays* L.), and faba bean (*Vicia faba* L.). *Biol. Fertil. Soils* 43: 565–574.

Soudi, B. 1988. Etude de la dynamique de l'azote dans les sols marocains: Caractérisation et pouvoir minéralisateur. PhD thesis. es-Sciences Agronomiques, Institut Agronomique et Vétérinaire Hassan II, Rabat, Morocco.

Soudi, B., A. Sbai, and C.N. Chiang. 1990a. Nitrogen mineralization in semi-arid area of Morocco: Rate constant variation with depth. *Soil Sci. Soc. Am. J.* 54: 756–761.

Soudi, B., A. Sbai, and C.N. Chiang. 1990b. Variations saisonnières de l'azote minéral et effet combiné de la température et de l'humidité du sol sur la minéralisation. *Actes Inst. Agron. Vet.* 10(1): 29–38.

Sprent, J.I. 1971. The effects of water stress on nitrogen fixing root nodules. I. Effects on the physiology of detached soybean nodules. *New Physiol.* 70: 9–17

Stanford, G., and E. Epstein. 1974. Nitrogen mineralization and water relations in soils. *Soil Sci. Soc. Am. Proc.* 38: 103–106.

Strong, W.M., and J.E. Cooper. 1980. Recovery of nitrogen by wheat from various depths in a cracking clay soil. *Aust. J. Exp. Agric. Anim. Husb.* 20: 82–87.

Studer, C., Y. Hu, and U. Schmidhalter. 2017. Interactive effects of N-, P- and K-nutrition and drought stress on the development of maize seedlings. *Agriculture* 7(11): 90. doi:10.3390/agriculture7110090.

UNEP. 1992. *World Atlas of Desertification.* 1st ed. Edward Arnold, London, UK.

UNEP-WCMC. 2007. A spatial analysis approach to the global delineation of dryland areas of relevance to the CBD Programme of Work on Dry and Subhumid Lands. Dataset.

Unger, P.W. 1984. Tillage and residue effects on wheat, sorghum, and sunflower grown in rotation. *Soil Sci. Soc. Am. J.* 48: 885–891.

Unger, P.W. 2002. Conservation tillage for improving dryland crop yields. *Ciencia del Suelo* 20(1): 1–8.

Unger, P.W., H.H. Schomberg, T.H. Dao, and O.R. Jones. 1997. Tillage and crop residue management practices for sustainable dryland farming systems. *Ann. Arid Zone* 36(3): 209–232.

Vadez, V., T.R. Sinclair, and R. Serraj. 2000. Asparagine and ureide accumulation in nodules and shoots as feedback inhibitors of N_2 fixation in soybean. *Physiol. Plant* 110: 215–223.

Vadez, V., and T.R. Sinclair. 2000. Ureide degradation pathways in intact soybean leaves. *J. Exp. Bot.* 51: 1459–1465.

Valentine, A.J., V.A. Benedito, and Y. Kang. 2011. Legume nitrogen fixation and soil abiotic stress: From physiology to genomics and beyond. *Annu. Plant Rev.* 42: 207–248.

Van Doren, D.M., Jr., and R.R. Allmaras. 1978. Effect of residue management practices on the soil physical environment, microclimate, and plant growth. In: *Crop Residue Management Systems*, edited by W.R. Oschwald. Special Publication no. 31. American Society of Agronomy, Madison, WI, pp. 49–83.

Vandermeer, J. 1989. *The Ecology of Intercropping.* Cambridge University Press, Cambridge, UK.

Vanlauwe, B., A. Bationo, J. Chianu, K.E. Giller, R. Merckx, U. Mokwunye, O. Ohiokpehai, et al. 2010. Integrated soil fertility management: Operational definition and consequences for implementation and dissemination. *Outlook Agric.* 39: 17–24.

Vanlauwe, B., K. Descheemaeker, K.E. Giller, J. Huising, R. Merckx, G. Nziguheba, J. Wendt, and S. Zingore. 2015. Integrated soil fertility management in sub-Saharan Africa: Unravelling local adaptation. *Soil* 1: 491–508.

Vanlauwe, B., O.C. Nwoke, N. Sanginga, and R. Merckx. 1996. Impact of residue quality on the C and N mineralization of leaf and root residues of three agroforestry species. *Plant Soil* 183: 221–231.

Venkateswarlu, J. 1987. Efficient Resource Management Systems for Drylands of India. In *Advances in Soil Science. Advances in Soil Science*, Stewart, B.A. (ed.), vol 7. Springer, New York.

Vere, D. 2005. *Research into Conservation Tillage for Dryland Cropping in Australia and China.* Project Report no. 33. Impact Assessment Series. Australian Centre for International Agricultural Research, Canberra, Australia. www.aciar.gov.au.

Vigil, M.F., and D.E. Kissel. 1991. Equations for estimating the amount of nitrogen mineralized from crop residues. *Soil Sci. Soc. Am.* 5: 757–761.

Wang, M., Q. Zheng, Q. Shen, and S. Guo. 2013. The critical role of potassium in plant stress response. *Int. J. Mol. Sci.* 14.

Wang, X. 2006. Conservation tillage and nutrient management in dryland farming in China. PhD thesis. Wageningen University, Wageningen, the Netherlands.

Wang, X., and Y. Xing. 2016. Effects of mulching and nitrogen on soil nitrate-N distribution, leaching and nitrogen use efficiency of maize (*Zea mays* L.). *PLoS One* 11(8): e0161612. doi:10.1371/journal. pone.0161612.

Warren, G.P., S.S. Atwal, J.W. Irungu. 1997. Soil nitrate variations under grass, sorghum and bare fallow in semi-arid Kenya. *Exp. Agric.* 33: 321–333.

Weisz, P.R., R.F. Denison, and T.R. Sinclair. 1985. Response to drought stress of nitrogen fixation (acetylene reduction) rates by field-grown soybeans. *Plant Physiol.* 78(3): 525–530.

Wood, M., A.M. McNeill, C.J. Pilbeam, R.S. Swift, H.C. Harris, and P.G. Mugane. 1996. Sustainability of nitrogen use in two dryland farming systems. In: *Progress in Nitrogen Cycling Studies*, edited by O. Van Cleemput, G. Hofman and A. Vermoesen. Kluwer Academic Publishers, Berlin, Germany, pp. 303–306.

Zahoor, R., W. Zhao, H. Dong, J.L. Snider, M. Abid, B. Iqbal, and Z. Zhou. 2017. Potassium improves photosynthetic tolerance to and recovery from episodic drought stress in functional leaves of cotton (*Gossypium hirsutum* L.). *Plant Physiol. Biochem.* 119: 21–32.

Zhang X., G. Huang, X. Bian, and Q. Zhao. 2013. The effect of root interaction and nitrogen fertilization on the chlorophyll content, root activity, photosynthetic characteristics of intercropped soybean and microbial quantity in the rhizosphere. *Plant Soil Environ.* 50(2): 80–88.

Zhang, Y., J. Liu, J. Zhang, H. Liu, and S. Liu. 2015. Row ratios of intercropping maize and soybean can affect agronomic efficiency of the system and subsequent wheat. *PLoS One* 10(6): e0129245. doi:10.1371/journal.pone.0129245.

13 Tailored Fertilizers
The Need of the Times

Amit Roy

CONTENTS

13.1 INTRODUCTION

The quality of man's life is directly dependent upon that thin layer of the earth's crust we refer to as "soil." The soil supports our food and fiber production, and thus the welfare of the world's inhabitants. Our management of the soil determines its productivity over the long term. Management has many facets, including ensuring an adequate supply of plant nutrients for use by growing plants. What we know as plant nutrients are composed of 16 chemical elements that are generally found in the soil, but their supply can be depleted with year after year of growing crops. Also, some nutrients are lost by natural processes. Farmers, ranchers, and generally growers use chemical fertilizers to supplement depleted nutrients for optimum crop production.

Plant nutrients usually are grouped in three categories: primary, secondary, and micronutrients. The primary nutrients are nitrogen, phosphorus, and potassium. Secondary nutrients are calcium, magnesium, and sulfur. Micronutrients are boron, chlorine, copper, iron, manganese, molybdenum, and zinc; they are needed in much smaller amounts than other nutrients. Carbon, hydrogen, and oxygen round out the 16.

Agriculture has been constantly evolving since the initial agricultural revolution when domestication of plants and animals led to food production in excess of population growth. But along with this increase in food production came increases in human population growth (Figure 13.1), that put

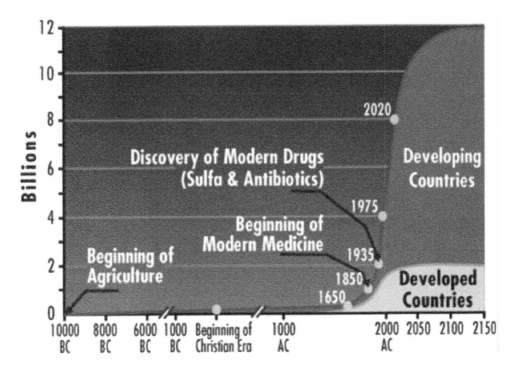

FIGURE 13.1 World population growth.

further pressure on food supplies. More land was required to produce more food, since the increase in crop production was limited by the relative scarcity of the inputs in the agricultural process (Hayami and Rutten 1971) with the resultant decline in soil fertility of existing croplands. The system relied heavily upon forage and green manure crops and higher levels of manure application, all coming exclusively from the farming sector. But this approach could not sustain soil productivity; thus the drive for manufactured fertilizers.

The reliance on manufactured (often referred to as chemical or mineral) fertilizers can be traced back to the nineteenth century when Justus von Liebig articulated the theoretical foundations of crop production and when John Bennett Lawes began producing fertilizers containing phosphorus (Smil 1997). However, only since the 1960s, when global starvation became a real possibility, have fertilizers assumed a predominant role in increasing agricultural productivity. Fertilizer was an integral part of the technological trinity – seed, water, and fertilizer – responsible for bringing about the "Green Revolution" that helped many densely populated countries, including India, China, and Indonesia, achieve food self-sufficiency in a short span of 20–25 years.

While recognizing the importance of maintaining soil fertility, advances in breeding and crop protection products, and the expansion of irrigation, there is evidence that 40%–60% of the production gains are directly attributable to fertilizer use, and this has kept food prices much lower than they otherwise would have been (Figure 13.2) (Stewart et al. 2005). Chemical fertilizers are available in solid, liquid, and gaseous forms. Solid and liquid fertilizers are available as single nutrient or multinutrient with varying nutrient release rates. The quantity, type, and form of fertilizers depend on several factors, including: (1) characteristics of the soil, (2) the crops grown, (3) the natural level of soil nutrients available for plant growth, and (4) desired economic returns. Globally, about 190 million metric tons of fertilizer nutrients were used for crop production, with China, India, Brazil, and the United States accounting for about 50% of the total amount of nutrients applied through fertilizers in the world. Fertilizer consumption and consequently food production in most countries has increased except in sub-Saharan Africa (SSA), where per capita food production has

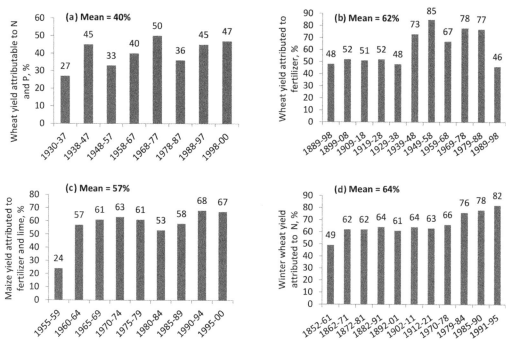

FIGURE 13.2 Yield attributed to fertilizer: (a) N and P from 1930 to 2000, in the Oklahoma State University Magruder plots; (b) N, P, and K from 1989 to 1998 in the University of Missouri Sanborn Field plots; (c) N, P, K, and lime from 1955 to 2000 in the University of Illinois Morrow plots; and (d) N with adequate P and K versus P and K alone from 1852 to 1995 (years between 1921 and 1969 excluded because part of the experiment was fallowed each year for weed control) in the Broadbalk Experiment at Rothhamsted, England. (From Stewart, W. et al., *Agron. J.*, 97, 2005.)

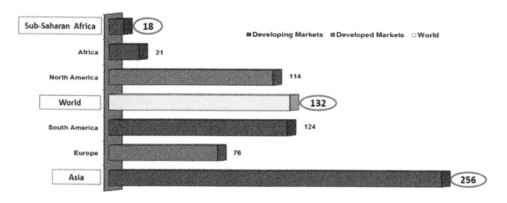

FIGURE 13.3 Per hectare fertilizer use by regions and markets, 2018 (kg/ha).

decreased since the 1970s. This decline can be attributed to several factors, including low soil fertility, agroclimatic conditions, and low fertilizer use. Compared with a world average of 132 kg of nutrients per hectare, SSA uses only 18 kg of nutrients per hectare, resulting in a significant mining of inherent nutrients from soils that are already low in nutrient status (Figure 13.3). Over 132 million tons of N, 15 million tons of P, and 90 million tons of K have been lost from cultivated land in 37 African countries in 30 years due to nutrient mining (Smaling 1993).

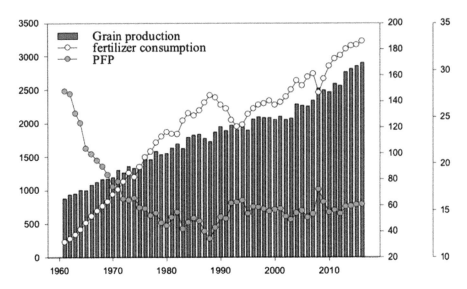

FIGURE 13.4 Global cereal production, fertilizer consumption, and partial factor productivity (PFP) since 1960. (Derived from Food and Agriculture Organization of the United Nations (FAO), FAOSTAT database http://faostat.fao.org, Rome, Italy, 2019.)

The global crop yields driven by developing countries continue to increase in direct proportion to fertilizer application; this has resulted in progressive decrease in partial factor productivity (PFP), which indicates kilogram of grain produced per kilogram of fertilizer used (Figure 13.4). This progressive decrease in PFP is partly due to fertilizers that do not match the need of the soil and plant, particularly the absence of secondary and micronutrients. Application timing and place are also contributing factors. For example, urea, a commonly used nitrogen fertilizer, breaks down in the soil and results in the release of reactive nitrogen in the air and nitrates that travel easily through the soil (Figure 13.5). Because it is water soluble and can remain in groundwater for decades, the addition of more nitrogen over the years has an accumulative effect. Groundwater contamination has been linked to some forms of cancer and birth defects. Similarly, excess

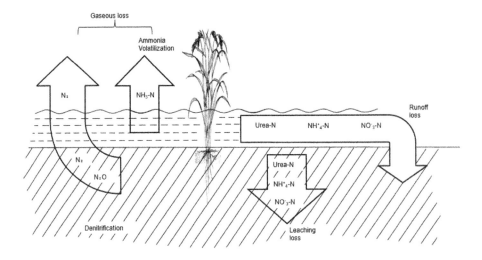

FIGURE 13.5 Pathways to N losses from urea fertilizer.

phosphates in rivers and tributaries cause algae growth, depleting water of oxygen and resulting in fish kill. Hence, the overarching issue is the drive to enhance nutrient use efficiency through a combination of new fertilizer formulation, including enhanced-efficiency fertilizers, and better management practices.

Since the 1960s, the population has more than doubled and is estimated to reach 9.8 billion by 2050 (UN 2019). While the population increase in developed regions is expected to remain virtually unchanged at 1.3 billion, developing regions will see an increase from 5.7 billion in 2010 to 8.3 billion by 2050. Given this population growth, coupled with increasing income levels and economic development, and changes in the food preferences of consumers, food demand is predicted to increase by nearly 70%. However, the absolute increase in food required by 2050 will be as large as the increase since the Green Revolution was launched in the 1960s – as available arable land and water become scarcer. Going forward we not only have to produce more food, but it must be sufficiently nutritious to overcome the malnutrition that continues to disproportionately affect children, particularly in SSA and Southeast Asia. As the need to produce more food and fiber accelerates, so too does the urgency to reduce air and water pollution from overuse and misuse of fertilizers.

This chapter reviews the effects of fertilizer management, fertilizer products, and government policies on food production and environmental issues from the use of fertilizers. Recent developments related to new fertilizers, new formulations, and management are reviewed for meeting the need of the crops in the face of climate changes, and also help produce more nutritious food.

13.2   SOILS

The environmental components of crop production systems are soil, water, air, and solar energy. The following is a definition for "soil" given by Buckman and Brady (1969):

> The soil may be defined as a natural body, synthesized in profile form from a variable mixture of broken and weathered minerals and a decaying organic matter, which covers the earth in a thin layer and which supplies, when containing the proper amounts of air and water, mechanical support and, in part, sustenance for plants.

Fertile and productive soils are vital for sustenance for plants, so maintaining soil fertility is essential for plant growth. Soil productivity is a measure of the ability of soil to produce a crop or sequence of crops under a specified management system. Soil productivity includes soil fertility plus factors such as soil moisture and temperature. Soil acidity and salinity and biotic stresses can reduce the productivity of even the most fertile soils. Farmers have no control over climate but can influence soil productivity through past and present activities that can maintain and/or enhance the organic carbon level to influence nutrient availability to the plant.

Soils differ widely in their ability to meet the nutrient requirements of plants since most have only natural fertility. To meet production targets, more nutrients are usually required than can be supplied by the soil. High crop yield results in removal of more nutrients unless additional nutrients are added to maintain its fertility. There are soils with high soil fertility, such as in Uganda, that can produce substantial crop yields without added fertilizer for a period of cropping cycle. However, under continuous cultivation, yields tend to decline, primarily due to the removal (mining) of nutrients by crops (Henao and Baanante 2006).

13.3   FERTILIZERS

Over the past eight decades, particularly after the creation of the Tennessee Valley Authority in 1934, many chemical fertilizers have been developed to supplement nutrients already available in the soil and to meet the high requirements of crops. The customary classification into single- or multinutrient fertilizers usually refers only to the three primary nutrients (nitrogen, phosphorous, and potassium). Fertilizer grade is used to classify different fertilizer materials based on the content of the three primary nutrients. The nutrient content, or grade, may refer either to the total or to the available nutrient content, and may be expressed traditionally for some nutrients in oxide form (P_2O_5, K_2O) or in elemental form (N, P, K). For example, a fertilizer grade of 7-28-14 is 7% N, 28% P_2O_5, and 14% K_2O.

13.3.1 NITROGEN FERTILIZERS

Nitrogen fertilizers are manufactured in a variety of formulations, each with different properties and uses for crop production systems. These fertilizers essentially begin with anhydrous ammonia, which is synthesized by reacting nitrogen and hydrogen under high temperature and pressure by the Haber–Bosch process through the chemical reaction [$3H_2 + N_2 \rightarrow 2NH_3$]. The nitrogen is obtained from air, while hydrogen is generally from natural gas. This process, developed in Germany just before World War I, is sometimes considered the most important technological development of the twentieth century. Nitrogen supply for ammonia synthesis is truly inexhaustible since the atmosphere contains 3.8 quadrillion tons of the element (Roy 2019). Various feedstocks can be used to obtain hydrogen, and during the last several decades the focus has been to improve the energy efficiency of ammonia synthesis. Natural gas is the preferred feedstock, and the best natural gas–based plants now use less than 30 GJ/t N.

The ammonia produced through the Haber–Bosch process accounts for more than 90% of the world's nitrogen fertilizers and without which we would produce 48% less food (Erisman et al. 2008). Besides direct application (as a fertilizer), ammonia is also used as raw material in the production of urea, ammonium nitrate, and other N fertilizers, as well as in the production of monoammonium phosphate (MAP), diammonium phosphate (DAP), and other multinutrient fertilizers (Figure 13.6). Fertilizers produced from ammonia provide farmers with a wide range of N source options for managing N to best meet their crop needs and logistical requirements.

13.3.2 PHOSPHATE FERTILIZERS

Phosphorous-containing fertilizers are produced from mined phosphate rock found in several locations around the world (UNIDO 1998). Phosphorus in fertilizer materials is usually expressed in the oxide form (P_2O_5). Although this form does not actually exist in fertilizer materials, it has been adopted as the standard form for comparison among different P fertilizers. Phosphate rock is a finite, nonrenewable resource. There is no substitute for phosphorus. Phosphate deposits are known in every continent of the world except Antarctica. Recently published total world phosphate rock reserves range from 60 to 65 billion tons (Van Kauwenbergh 2010) and are an indication of what can be mined at the current market price and technologies. The total world phosphate rock resources may range from 290 billion tons to 460 billion tons of phosphate rock at varying P_2O_5 grades (Herring and Fantel 1993). Both reserves and resources are a "snapshot at a point in time" and not an indication of "all there is." Phosphate deposits are not evenly distributed around the globe. The most abundant known reserves, in descending order of abundance, are in Morocco, China, the United States, Jordan, and Russia. At the current rates of extraction, these reserves can support the current rate of use for 300–400 years. Of course, in the future, based mainly on requirements due to population growth, increased demand for phosphorus-based products will require increased rates of phosphate rock production and the rate of depletion of known reserves may increase. This time horizon can be extended by exploiting known, incompletely explored, and yet-unknown phosphate deposits at higher costs. Conservation and efficient use of this valuable natural resource and recycling of P is receiving increased attention through sustainable P management initiatives (Scholz et al. 2014).

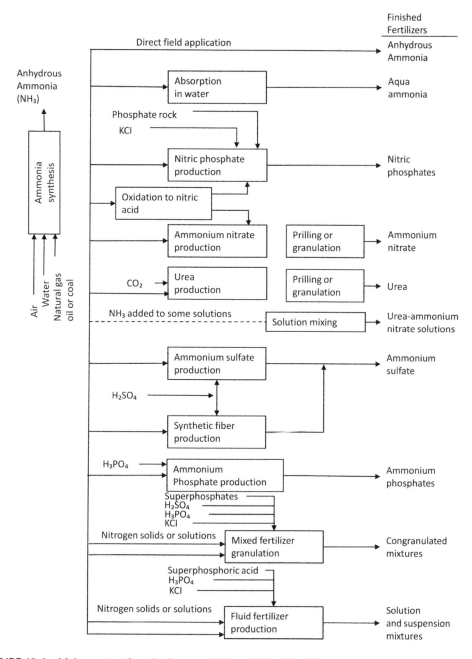

FIGURE 13.6 Major routes of synthetic ammonia into finished fertilizers.

Unprocessed phosphate rock may be applied as a source of P nutrition on acidic soils (soils below pH 5.5), but most (more than 90%) phosphate rock is processed for production of phosphate fertilizers by reacting it with either sulfuric or nitric acid. Phosphate products include single superphosphate (SSP), triple superphosphate (TSP), MAP, DAP, and polyphosphates (Figure 13.7).

13.3.3 POTASSIUM FERTILIZERS

Potassium fertilizers are mainly derived from geological saline deposits and as mined materials can be used directly. Most fertilizer products now in use are high-concentration materials that are water

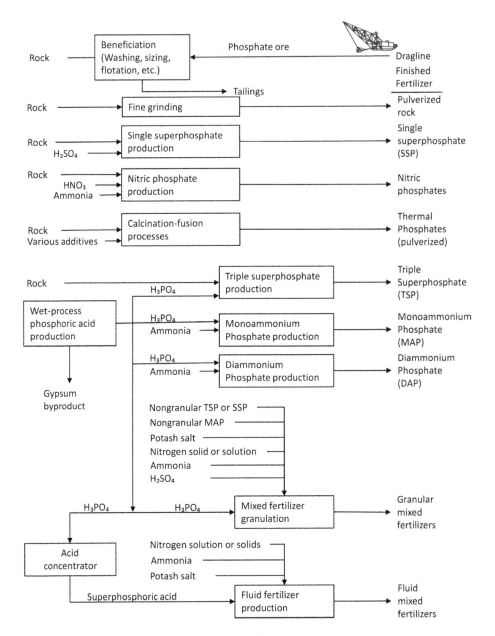

FIGURE 13.7 Major routes of phosphate rock into fertilizers.

soluble and quick acting. Potassium fertilizer is usually described in the oxide form (K_2O). As is the case with P, this form is a standard for comparison among K fertilizers, but it is not actually found in K-containing fertilizer materials.

Potassium is of least concern among the three primary nutrients. Not only is this element abundantly present in the earth's crust, but also it can be found in conveniently concentrated deposits in both deeply buried and near-surface sediments. Potassium deposits in descending order of known reserves are in North America (Canada and the United States), Germany, Russia, Belarus, Brazil, Israel, Ethiopia, and Jordan. Even the most conservative reserve base estimates indicate more than 400 years of reserve at the projected rate of use.

Potassium-containing fertilizers include potassium chloride (commercially known as potash), potassium sulfate, potassium magnesium sulfate (commercially known as sulfomag), and potassium nitrate.

13.3.4 SECONDARY AND MICRONUTRIENTS

Sulphur, calcium, and magnesium are considered secondary nutrients, because while these are essential to crop development, seasonal crop uptake is usually lower than for the primary nutrients (N, P, and K). The micronutrients, Zn, Fe, Mn, Cu, B, Mo, and Cl, are needed in very small amounts but are essential for plant growth. A deficiency of secondary and micronutrients usually becomes the limiting factor in crop production.

Secondary nutrients, such as sulfur, can be added to ammonium phosphates. Sulfur is also supplied through ammonium sulfate, calcium sulfate, and SSP. Calcium and, to a lesser extent, magnesium are present in many of the commonly used fertilizers. However, calcium and magnesium compounds are often applied separately for correction of soil acidity. The usual materials include agricultural limestone (principally calcium carbonate), dolomite (a double carbonate of calcium and magnesium), or dolomitic limestones, which are mixtures of the two minerals. Other sources, depending upon local availability and economic considerations, are hydrated limes, marl, oyster shells, sludges, or slags.

The principal sources of commonly used micronutrients are borax, sodium molybdate, and the soluble sulfate salts of the metallic elements. Other materials that are used include oxide and carbonate salts, chelates, and complexes of micronutrients and organic compounds.

13.4 SHORTCOMINGS OF TODAY'S FERTILIZERS

N, P, and K fertilizers are the most widely used fertilizers, primarily because virtually all crops require significant amounts of these nutrients to maximize production. Legume crops are the only exception because they have symbiotic bacterial colonies associated with their roots and therefore can meet plant N needs through biological N fixation. However, legumes do require large amounts of P and K. Of the primary nutrients, N and P fertilizers are subjected to more chemically and energy-intensive production processes, are available in several formulations and forms (solid blends and liquids), and often "carry" much smaller (often minute) amounts of secondary and micronutrients that are essential for maximum economic yields and improved nutritional value. Of all three nutrients, nitrogen requires the most energy to produce fertilizers from it (Figure 13.8). These fertilizers, however, are largely unchanged from the formulations and forms that were manufactured in the 1970s and 1980s and, in the absence of best management practices, result in significant loss of nutrients to the ecosystem. These losses produce negative economic and environmental impacts (VFRC 2012).

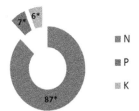

FIGURE 13.8 Energy requirement for fertilizer production. * % of total.

13.4.1 Economic and Environmental Sustainability of Fertilizers in Use

The availability of fertilizer N to plants is largely controlled by soil microbial processes. The N cycle in soils is complex, and under certain conditions large amounts of plant-available N can be lost from the soil to the atmosphere or in surface and subsurface water bodies. The N lost to the atmosphere is in various forms of nitrous oxide gases (collectively referred to as NO_x gases). In fact, fertilizer use contributes more greenhouse emission than production and transportation of fertilizers (Figure 13.9).

The N entering water bodies via runoff or leaching is in the form of nitrates. Most nitrates found in subsurface and surface waters result from crop production. The other primary nutrients (P and K) are not readily lost from soils, although runoffs containing P nutrients from crop production and animal waste can be significant pollutants in some bodies of water. Importantly, the amount of applied nutrient that is taken up by plants (referred to as "nutrient use efficiency," or NUE) is a function of the type of fertilizer(s) and how well a farmer can match the placement, timing, and quantity of applied nutrient to the plant's needs throughout its growing cycle. In 2010, the NUE of N (NUE-N) ranged between 25% and about 70%, depending on the region and country (Figure 13.10). For example,

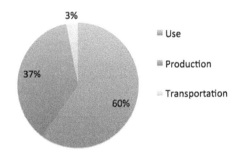

FIGURE 13.9 Breakdown of fertilizer greenhouse gas emissions from production, transportation, and use.

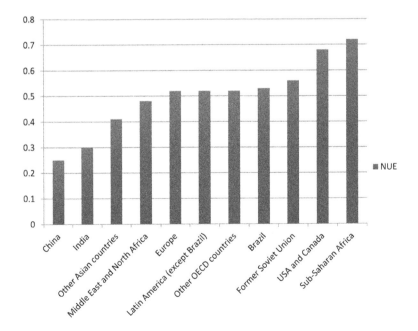

FIGURE 13.10 Nutrient use efficiency of N (NUE-N) in selected countries and regions. (Derived from Zhang, X. et al., Managing nitrogen for sustainable development, http://www.nature.com/doifinder/10.1038/nature 15743, 2015.)

in Canada and the United States, the NUE-N is about 65%, while China and India have rates of 25% and 30%, respectively. In contrast, the NUE-N in SSA is about 70%. There, despite a fertilizer application rate that is nearly an order of magnitude lower than the world average of 130 kg nutrients per hectare of arable land, the NUE-N remains high due to inherently low soil nutrient content and nutrient needs of the plant. Even though SSA's NUE-N is comparably high, SSA contributes to greenhouse gas (GHG) emissions through slash-and-burn practices.

The economic impact of low NUE-N is significant. For example, assuming an annual fertilizer-N consumption of 120 mmt and an NUE-N of 50%, up to $25 billion of the total price paid by farmers is lost annually. In addition to the economic impact of low NUE-N, air and water pollution from unused N is a major factor. Manure also contributes to air and water pollution. In 2014, manure-N on cropland and pastures equaled manufactured fertilizer-N application of around 100 million tons of N per year (FAO 2018). The production of current fertilizers involves several steps where losses occur resulting in less than 10% of mined phosphates is in food and feed (Scholz et al. 2014). Permanent losses from the field include leaching in the drainage, soluble P runoff, and erosion of soil particles containing P (Syers et al. 2008). Another component that is more difficult to account for are what are commonly referred to as fixation reactions, which are P precipitation and other sorption processes with calcium in calcareous soils, and iron and aluminum mostly in acid soils.

In addition to three primary nutrients, namely, N, P, K, there are 14 other elements identified by Justus von Liebig, a nineteenth-century German scientist, as essential for plant growth. In soil studies in India, there is evidence of the gradual reduction of essential secondary and micronutrients required by plants (Figure 13.11), suggesting that these essential elements may now be limiting yields and reducing efficiency of primary nutrients.

There are several reasons for these unintended nutrient deficiencies, but two critical factors appear to be (1) fertilizers not containing enough secondary and micronutrients, and (2) imbalanced fertilizer applications. The latter, in some instances, relates to policies that disproportionately subsidize one nutrient more than another. For example, in India, N, P, and K have been highly subsidized. To rein in the subsidy, the government implemented the nutrient-based subsidy scheme

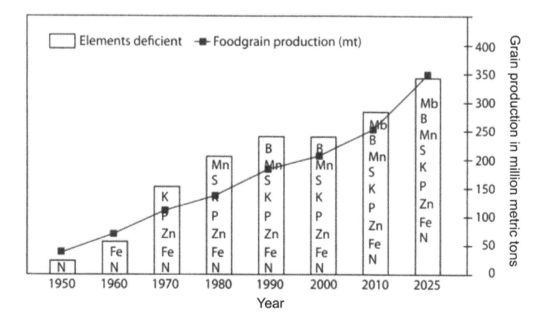

FIGURE 13.11 Emerging deficiencies of plant nutrients in relation to increased food grain production.

to progressively deregulate the prices of N, P, and K fertilizers. Unfortunately, the scheme was only implemented for P and K fertilizers, saving some subsidy but creating further imbalanced use of N, P, and K (Gulati 2014). A recent decision by the government of India to introduce specialty urea for agricultural application is designed to increase NUE-N, and thereby reduce N consumption and correct nutrient imbalance.

13.5 APPROACHES TO IMPROVING EFFICIENCY

In most countries, fertilizer application is based on optimum economic return anchored on crop-response data averaged over large areas. But studies have shown that farmers' fields show large variability in terms of nutrient-supplying capacity and crop response to nutrients. Thus, blanket fertilizer application recommendations may lead farmers to overfertilize in some areas and under-fertilize in others or to apply an improper balance of nutrients for their soil or crop. Thus, fertilizer applications based on blanket recommendations result in economic loss for the farmers and lead to air and water pollution. There are three approaches to improving the economics of fertilizer use and minimizing air and water pollution. These approaches – management, technology, and policy – are interconnected and must be implemented simultaneously.

13.5.1 MANAGEMENT

Site- and crop-specific production techniques and practices offer one approach to maximizing economic, social, and environmental benefits. This approach includes soil and crop management to fit the different conditions encountered in cropping areas. Sometimes this is also referred to as site-specific nutrient management (SSNM), which, among others, includes a balanced use of all essential crop nutrients. The SSNM approach has proven to be an efficient way to reduce NOx emissions to the atmosphere and leaching losses to surface and subsurface waters (Richards et al. 2015).

In Europe, a combination of balanced fertilizer application and management more than doubled cereal yield per unit fertilizer N. For example, at an application rate of 200 kg, fertilizer N/ha yield increased from 4t/ha to more than 8t/ha (Bindraban et al. 2008). In SSA, soil fertility is very low, and they are, besides NPK, highly deficient in secondary and micronutrients and organic matter. Addressing all soil fertility constraints, including macronutrients (NPK) as well as secondary and micronutrients and pH correction, is vital to achieving optimal fertilizer use efficiency and economic returns. Integrated soil fertility management (ISFM), which includes the use of organic matters, soil amendments, and balanced fertilizers, is an integral part of revitalizing SSA soils and building up soil organic carbon, which in turn improves NUE.

In Burundi, Mozambique, and Rwanda, average field trial results clearly show that balanced fertilizers (NPK + secondary + micronutrient) based on soil analyses performed better than control plots and those with application of fertilizers containing just NPK (Figure 13.12). In all cases, adding secondary and micronutrients (SMNs) significantly increased maize yield by more efficient uptake of nutrients, resulting in less loss to the environment. In Mozambique, SMNs were Mg, S, Zn, and B. In Rwanda, SMNs were S, Zn, and B. In Burundi, SMNs were Ca, Mg, S, Zn, and B (Wendt 2017).

Certainly, application of the right type of fertilizer in the right amount at the right time and in the right place ensures increased crop production, better utilization of nutrients, and reduced losses of nitrogen to the environment. This strategy, referred to as the 4Rs, has been adopted in several countries with excellent results (Bruulsema et al. 2012). An advancement of the 4R principles is precision farming (also known as precision agriculture), which employs detailed, site-specific information to precisely manage production inputs. The idea is to know the soil and crop characteristics unique to each part of the field, and to only apply inputs (seed, fertilizer, chemicals, etc.) within small portions of the field. Precision farming includes a set of technologies that combines sensors, information systems, advanced machinery, and information management to optimize production, allowing for better use of resources to maintain the quality of the environment while improving the sustainability of the food supply.

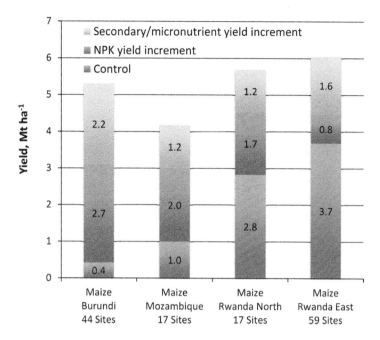

FIGURE 13.12 Effect of balanced fertilization on maize yields in Burundi, Mozambique, and Rwanda. (Wendt J, Senior Scientist, International Fertilizer Development Center [IFDC], 2017, personal communications, 2017.)

An adaptation of the 4Rs principle is subsurface application of urea, which is the prime nitrogen fertilizer for the cultivation of rice and accounts for about 85% of rice fertilizer N (Gregory et al. 2010), but it is prone to high losses (as much as 70%) under the present application practice of broadcasting, particularly in developing regions. However, research over several decades has shown that if urea, instead of being broadcast, is point-placed below the soil surface (subsurface) near the root zone of the rice plant, the N losses are lowered with simultaneous increase in yield. This application method involves producing larger urea granules (1–3 g) which are inserted below the soil surface (7–10 cm) approximately equidistant from four rice plants two weeks after transplanting of rice seedlings once during the growing season (Figure 13.13). The weight of urea granules depends

FIGURE 13.13 Commercial urea (prilled) and urea briquettes produced by compacting prilled urea.

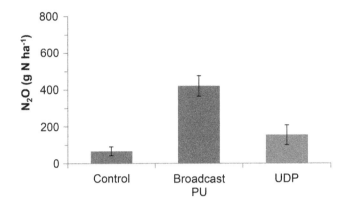

FIGURE 13.14 The deep placement of urea reduces N losses to the environment.

on the rate of application. This technology, known as urea deep placement (UDP), makes N available to the rice plant throughout its growth cycle, thereby reducing losses and increasing yields. Results in the greenhouse and in farmers' fields show a consistent increase in yield compared with broadcasting (Savant and Stangel 1990). The deep placement of urea also significantly reduces N concentration in floodwater and NO_x emissions under certain conditions (Figure 13.14).

13.5.2 TECHNOLOGY

Improving NUE in agriculture has been a concern for decades (Dobermann 2005), and the focus continues to be on technology to develop new products to achieve this. Increasing the efficiency of mineral nitrogen (N) fertilizer use is complicated by the fact that plants take up N normally as nitrate or ammonium ions through their roots from the soil solution. However, ammonium-N, unlike nitrate-N, can be retained in the soil, where soil and plants compete for available ammonium-N in the soil (Trenkel 1997). This competition for N is the main problem when it is added as mineral fertilizer to feed plants. Only a certain proportion of the N is taken up by the growing plants. The unused portion is lost to the environment mainly through ammonium volatilization, denitrification, and leaching. Therefore, researchers are pursuing several pathways to overcome such losses, including development of new fertilizers. Two important special types of fertilizers are (1) slow- and controlled-release fertilizers with the release of nutrients over several months, and (2) stabilized fertilizers (fertilizers associated with nitrification or urease inhibitors), delaying either the nitrification of ammonia or the ammonification of urea.

These special types of fertilizers are those that release, either by design or naturally, their nutrient content over an extended period to match the nutrient requirements of the crop (Roy 2018). Because of economics and environmental considerations, the slow- and controlled-release and stabilized N fertilizers have been much more important than phosphate (or potash), particularly under certain soil and climatic conditions. In most cases, unutilized phosphate and potash remain available for subsequent crops. Nevertheless, there are a few controlled-release water-soluble phosphate and potash fertilizers in the market for specialty application.

13.5.2.1 Slow-Release Nitrogen Fertilizers

Slow-release nitrogen fertilizers (SRNFs) release their nutrient at a slower rate than urea and ammonium nitrate, but the rate, pattern, and duration of release are controlled by soil and climatic conditions and cannot be predicted with any accuracy. Example of SRNFs are organic-N low-solubility compounds such as urea formaldehyde (UF) and isobutylidene-diurea (IBDU) (Roy 2018).

13.5.2.2 Controlled-Release Nitrogen Fertilizers

Controlled-release nitrogen fertilizers (CRNFs) are products whose rate, release pattern, and duration of release are predictable and controllable through a physical barrier applied onto granules. These coatings can be hydrophobic polymers or matrices in which the soluble active material restricts the dissolution of the fertilizer. The coated fertilizers can be further divided into fertilizers with organic polymer coatings – that are either thermoplastic or resins – and fertilizers coated with inorganic materials such as sulfur (S) or mineral-based coatings (Figure 13.15). Additional information regarding various coating materials is discussed by Shaviv (2005).

Sulfur-coated urea (SCU), one of the early CRNFs, was developed by the Tennessee Valley Authority (TVA) in the early 1960s and involved spraying molten sulfur onto urea granules in a rotating drum. The applied sulfur oxidized over time and was an important secondary nutrient, but the coating had many imperfections that hindered the prediction of the precise release rate, and additional paraffin coating was applied to cover the imperfections. The relatively high price of SCU compared with conventional fertilizers makes it uneconomical for use in crop production, but SCU has seen acceptance in nonagricultural markets such as golf courses and high-value specialty crops.

SCU provided the clue for the industry to develop improved coatings using cross-linked thermosetting and thermoplastic polymers, resulting in a polymer-coated commercial product known as Osmocote. Subsequently, recent technology enhancements have resulted in new slow-release products that are cost-effective for corn (*Zea mays*), wheat (*Triticum aestivum*), potatoes (*Solanum tuberosum*), rice (*Oryza sativa*), and others. For example, Nutrien (formerly Agrium Inc.) manufactures a product sold mainly in North America under the trade name ESN®, which is a polymer-coated urea-N product whose release mechanism is temperature and moisture controlled (Figure 13.16) (Blaylock 2010).

In Japan, rice is a staple crop, and it is intensively cultivated. In order to maximize yield and reduce N release to air and water, the farmers use a polymer-coated CRNF sold under the trade name Meister® and realize an NUE-N of nearly 80% compared with only 30% under broadcast

FIGURE 13.15 Uncoated commercial urea and polymer-coated urea.

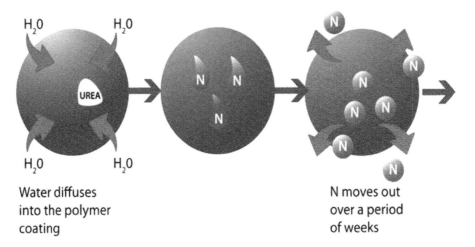

Water diffuses   N moves out
into the polymer   over a period
coating   of weeks

FIGURE 13.16 Release of N from polymer-coated urea.

TABLE 13.1

Effect of Controlled-Release Urea (Meister®) on Nutrient Use Efficiency-N for Rice Production in Japan

Application	N Rate, Kg/ha	N Uptake by Rice, Kg N/ha	NUE-N, %
Urea	100	30	30
Meister®	40	32	80

---

application (Shoji 2005) (Table 13.1). Although not currently practical for use in developing economies, the concept of polymer coating as used in ESN® and Meister® has spurred other companies to research alternative chemicals to produce an affordable slow-release product for use in developing economies.

13.5.2.3 Stabilized Nitrogen Fertilizers, Urease Inhibitors

These are compounds that prevent or suppress the transformation of amide-N in urea to ammonium hydroxide and ammonium through the action of the enzyme urease. By slowing down the rate at which urea is hydrolyzed in the soil, volatilization losses of ammonia to the air (as well as further leaching losses of nitrate) are either reduced or avoided. Thus, the efficiency of urea and of N fertilizers containing urea is increased, and any adverse environmental impact from their use is decreased. Among many inhibitors, N-(n-butyl) thiophosphoric triamide (NBPT), which is marketed under the trade name Agrotain®, has gained practical and commercial importance in agriculture (Roy 2018).

13.5.2.4 Stabilized Nitrogen Fertilizers, Nitrification Inhibitors

These are compounds that delay the oxidation of the ammonium ion (NH_4^+) by the *Nitrosomonas* bacteria in the soil. These bacteria transform ammonium ions into nitrite (NO_2^-), which is further transformed into NO_3^- by *Nitrobacter* and *Nitrosolobus* bacteria. Thus, the nitrification inhibitors control the loss of NO_3^- by leaching or the production of nitrous oxide (N_2O) by denitrification, resulting in increasing NUE-N. Nitrification inhibitors, including dicyandiamide (DCD), 3,4-dimethylpyrazole phosphate (DMPP), and 2-chloro-6-(trichloromethyl)pyridine (nitrapyrin), are commercially produced and used in certain markets, mainly in developed economies (Roy 2018).

The neem (*Azadipachta indica*) tree is indigenous to many tropical and semiarid countries. Meliacins, a component of the neem oil, has the characteristic of retarding the nitrification process (Mangat and Narang 2004). Neem cake has been used to coat urea to improve NUE-N. The efficacy of neem-coated urea (NCU) has been confirmed through several years of field trials, mainly in India. Compared with commercial urea, NCU results in 2%–10% higher rice yields. In some cases, the rice yields with 80% NCU are comparable to that by application of 100% commercial urea. In general, the farmers apply the same amount of NCU as uncoated urea but realize 5%–10% yield increases and subsequently lower N losses. The total global market for SRNF, CRNF, and stabilized N fertilizer is about 18% of the global nutrient consumption.

Unlike nitrogen fertilizers, there has been limited research with P fertilizers in altering and/or improving existing products in order to improve uptake by plants and reduce losses. However, the recent incidents of eutrophication of waterbodies that have decreased water quality both for aquatic life and for humans using these sources for drinking and recreation have created an urgency to develop new products and/or management techniques to preserve this finite resource.

The use of "as-mined" phosphate rock is obviously the most inexpensive option, but its effectiveness depends on its solubility in neutral ammonium citrate (NAC); the higher the solubility, the more reactive the rock and the more effective it is as a fertilizer (Figure 13.17). Grinding the rock to micron particle size (nanoparticles) increases its solubility in NAC and thus makes it more effective

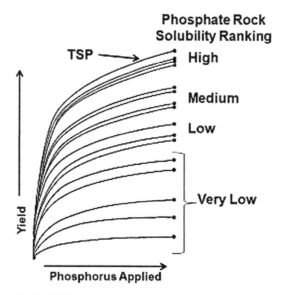

Soluble P_2O_5 in Neutral Ammonium Citrate	RAE	Solubility Ranking
(% P_2O_5)	(%)	
>5.9	>90	High
3.4–5.9	90–70	Medium
1.1–3.4	70–30	Low
<1.1	<30	Very low

$$RAE\ \% = \frac{\left(yield\ of\ ground\ PR\right) - \left(yield\ of\ check\right)}{\left(yield\ of\ TSP\right) - \left(yield\ of\ check\right)}\ x$$

FIGURE 13.17 Effectiveness of direct application rock as a function of its reactivity.

as a phosphate fertilizer. Partial acidulation of a phosphate rock with sulfuric or phosphoric acid is another option, particularly for medium-to-low-reactivity rocks. This product is known as partially acidulated phosphate rock (PAPR). Acidulation levels vary between 25% and 50%, and the product has a portion of its P_2O_5 in water-soluble form and the rest as phosphate rock. The water-soluble portion helps to establish the plant and develop its root system, which can then utilize the unacidulated portion of the rock (Shaviv 2005).

13.5.3 POLICY

New technologies and management practices are essential to improving productivity and NUE, increasing farmers' income, and reducing environmental degradation. But without conducive policies, neither technology nor management practices will be effective.

In Europe, NUE-N has progressively increased since the 1990s. In 2016, 30% more crop was produced with 20% less fertilizer-N application (Hoxa 2019). The European Union Common Agricultural Policy partly triggered this increased production at a lower application rate by reducing crop subsidies, and the European Union Nitrates Directive limited the manure application rates on agricultural land (Zhang et al. 2015).

China and India, which collectively account for more than 30% of global nitrogen consumption, both have a relatively low NUE-N (ranging between 25% and 30%). In China, an input subsidy reached $17 billion in 2014, and farmers' use of fertilizer-N exceeded the recommended amount by 20%–60% in cereals and even more in vegetables, whose NUE-N is around 15%. In China, farmers made a conscious decision to shift to fruit and vegetable production partly because the crop-to-fertilizer price ratio is more attractive than that of cereal production (particularly soybean, which is imported mainly from the United States and has a much higher NUE-N (70%–80%). The Chinese government, driven by low NUE-N and environmental challenges, has decided to cap fertilizer consumption at the 2020 level, which will see an increased use of more efficient products and better management practices (Reuters 2015).

In India, fertilizers are highly subsidized, and nitrogen more so than other nutrients, which results in the overuse of nitrogen-based fertilizer, which distorts nutrient (nitrogen-phosphorus-potassium) ratios, all of which results in a low NUE-N. To correct nutrient ratio distortion, the Indian government introduced a nutrient-based subsidy in 2010 (Chander 2016). The subsidy had only a minimal impact on improving NUE-N since urea was excluded from the scheme (Figure 13.18). To combat the

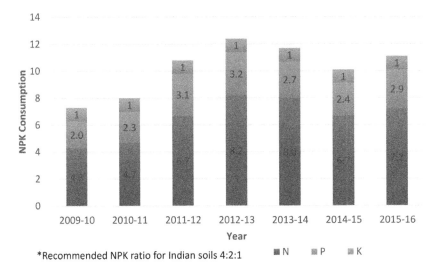

FIGURE 13.18 Imbalance in NPK use in India.

continued overuse of nitrogen, the government attempted to make fertilizer-N more efficient by requiring that all agricultural urea be coated with an extract from neem cake, a nitrification inhibitor. While early, the policy appears to yield the desired results of increasing NUE-N by 5% to 7%.

In general, Asian countries' fertilizer subsidies are universal, whereby all farmers are entitled to subsidized fertilizers. Consequently, as fertilizer consumption increases, so does the total subsidy bill. In some cases, one nutrient is subsidized more than other nutrients, which leads to unbalanced application and lowers NUE. In contrast, SSA countries have adopted the "smart subsidy" model (IFDC 2007), which involves providing subsidized fertilizer to farmers and/or crops. The total subsidy is dependent on the country's budget. The challenge for the "smart" subsidy is the selection of recipient farmers. The selection criteria differ by country, and political interference in the selection process is a challenge that several countries are now addressing.

13.6 NEXT FERTILIZERS – NUTRITIOUS FOOD AND HEALTHY SOILS

As mentioned earlier, most of today's fertilizers were developed several decades ago and for temperate agricultures, with plant-available nutrient contents (mainly NPK) as high as possible (high analyses) with the intent of delivering them at the lowest cost to farmers. These fertilizers dramatically increased the agricultural productivity of temperate-zone countries. These same fertilizers were a part of the package that triggered the "Green Revolution" in Asia in the 1970s. Farmers broadcasted fertilizers according to blanket fertilizer application recommendations. Production increased but not proportionally to the fertilizer application rate, resulting in a progressive decrease in NUE-N, resulting in NOx emissions to air, contributing to GHG (IPCC 2014). The soil characteristics, average soil temperature, and rainfall pattern in the tropics and subtropics are different from the temperate regions for which most of the current suit of fertilizers were developed. Hence there is a need for new fertilizers, new application techniques, and more conducive policies for the tropics and subtropics regions.

Reflecting on (1) the magnitude of the future food supply challenge in the face of increasing land and water scarcity and uncertain climatic conditions, (2) the increasing role that smallholder farmers will have to continue to play in the food supply chains in developing countries, and (3) the inherent flaws in current fertilizers, advanced technologies must be applied to introduce new fertilizers and improve production. But fertilizer alone is not the answer. It must be a part of a package that is anchored in soil fertility management practices, particularly nutrient management, conducive policies, extension, and information and knowledge management. With the vast amount of data that are generated through research and satellite systems, information must be converted into knowledge and packaged for easy access by smallholder farmers through new and emerging communication tools. Nevertheless, manufactured fertilizers will be an essential component for increasing food production.

The Haber–Bosch is a high-temperature (~500°C) and high-pressure process (~200 atm) to produce ammonia that is subsequently reacted with CO_2 to produce urea. Natural gas is the energy and feedstock of choice. Even though a new plant's energy consumption is close to theoretical, the energy requirement to produce nitrogen fertilizer is about 87% of the energy requirement for fertilizer production (Figure 13.8). Therefore, the current research focus should be to produce ammonia as close to the room temperature and atmospheric pressure as possible. If successful, lowering the operating temperature and pressure would reduce the energy requirement for nitrogen fertilizers and reduce GHG emissions, mainly CO_2.

But the low NUE-N efficiency is a bigger contributor to GHG than that from production and use (Figure 13.9). While product characteristics make them more susceptible to losses, policies and management are also contributing factors. For example, overapplication of nitrogen fertilizer due to heavy product subsidies has reduced NUE-N and increased losses to air and water. Nevertheless, new nitrogen products that are needed to improve efficiency are produced for selected markets. Additionally, there is a drive to produce fertilizers containing micronutrients to match the nutrient requirements of biofortified crops (Roy 2015).

As mentioned earlier, compared with efforts to improve the efficiency of nitrogen, improving the efficiency of phosphate fertilizers has not been a priority. But recently that has changed, and one approach being heavily explored to improve P use efficiency is to limit the physical and temporal association of the nutrient with the reactive component of the soil. The products that are being commercially produced or are under development can be grouped in the following categories: (1) coatings, (2) scaffolds, and (3) organic matrices and minerals of limited solubility. A detailed discussion of these approaches is in Weeks and Hettiarachchi (2019). At a more fundamental level, there is a need to evaluate the effect of phosphate-solubilizing bacteria in making the nutrient available more predictably.

13.7   SUMMARY

The global population is projected to reach 9.8 billion by 2050, with more than 90% of this increase in less-developed countries. Food production must increase by at least 70% to keep pace with the population growth. Fertilizers will be key to increasing agricultural productivity and keeping pace with population growth. But the type and application rates of these fertilizers need to be in sync with plants' nutrient requirements and need to be environmentally friendly and affordable. The raw materials needed to produce fertilizers' primary nutrients – nitrogen, phosphorus, and potassium – are adequate for the foreseeable future. However, phosphorus and potassium are finite resources and need to be sustainably managed to ensure availability for future generations. The deficiency of secondary and micronutrients is affecting crop yields, and their application separately or as a component of compound fertilizers is becoming essential. The current suite of fertilizers developed several decades ago for temperate regions has characteristics that make them prone to losses to air and water unless managed. To reduce this impact, adoption of technologies and management practices at the farm scale is needed. Some common principles include the "4Rs" approach of applying the right source, at the right rate, at the right time, in the right place. However, the technologies and management practices needed to achieve the 4Rs vary depending on the cropping systems, soil types, climate, and socioeconomic situations. Improvements in plant breeding, irrigation, and application of available 4R technologies have already made large gains in improving NUE, but new technological developments may be needed to achieve further gains, such as more affordable slow- and controlled-release fertilizers, and nitrification and urease inhibitors. Precision agriculture that uses a combination of information technology, remote sensing, and ground measurements is very promising, but for developing regions to effectively use this technology, information should be readily available, accessible, affordable, and site specific. However, the progress made so far is insufficient to achieve the projected 2050 goals of both food security and environmental stewardship. Hence the development of the next generation of fertilizers is vital to achieving the twin objectives of producing adequate nutritious food while protecting the environment by reducing losses of nutrients to the environment. There are several pathways to developing new fertilizers, but it would require a strong public–private partnership where the public sector invests in developing fundamental knowledge of nutrient movements in the soils and their uptakes by plants, while the private sector develops products that are affordable, eco-sensitive, and result in improving farmers' income. This approach also requires the fertilizer industry to invest more in research to develop new products.

REFERENCES

Bindraban PS, Loffler H, and Rabbinge R. (2008). How to close the ever-widening gap of Africa's agriculture. *International Journal of Technology and Globalisation* 4(3): 276–295.

Blaylock A. (2010). Enhanced efficiency fertilizers. Colorado State University Soil Fertility Lecture , Loveland, Colorado. October 20, 2010.

Bruulsema TW, Fixen PE, Sulewski GD (Eds.). (2012) *4R Plant Nutrition Manual: A Manual for Improving the Management of Plant Nutrition*. International Plant Nutrition Institute (IPNI), Norcross, GA.

Buckman HD and Brady NC. (1969). *The Nature and Properties of Soils*. Macmillan, New York.

Chander S, Director General, Fertilizer Association of India, New Delhi, India (2016), Personal Communication.

Dobermann A. (2005). Nitrogen use efficiency – State of the art. IFA International Workshop on Enhanced-Efficiency Fertilizers, Frankfurt. International Fertilizer Association, Paris, France.

Erisman JW, Sutton MA, Galloway J, Klimont Z, and Winiwarter W. (2008). How a century of ammonia synthesis changed the world. *Nature Geoscience* 1: 636–639.

FAO (Food and Agriculture Organization of the United Nations). (2018). Nitrogen inputs to agricultural soils from livestock manure. http://fao.org/3/18/18153EN/i8153en.pdf.

FAO (Food and Agriculture Organization of the United Nations). (2019). FAOSTAT Database. FAO, Rome, Italy. http://faostat.fao.org.

Gregory DI, Buresh RJ, Haefele SM, and Singh U. (2010). Fertilizer use, markets and management. In *Rice in the Global Economy: Strategic Research and Policy Issues for Food Security*, Pandey S, Byerlee D, Dawe D, Dobermann A, Mohanty S, Rozelle S and Hardy B (Eds.), pp. 231–263. International Rice Research Institute, Manila, Philippines.

Gulati A. (2014). Fertiliser pricing and subsidy policy in India: Exploring options for change. Paper Presented at the FAI Annual Seminar "Unshackling the Fertiliser Sector," Aerocity, New Delhi, 10–12 December 2014.

Hayami Y, and Ruttan VW. (1971). *Agricultural Development: An International Perspective*. Johns Hopkins Press, Baltimore, MD.

Henao J, and Baanante CA. (2006). *Agricultural Production and Soil Nutrient Mining in Africa*. International Fertilizer Development Center, Muscle Shoals, AL.

Herring JR, and Fantel JR. (1993). Phosphate rock demand into the next century: Impact on world food supply. *Nonrenewable Resources* 2(3): 226–241.

Hoxa A. (2019). Technical Director, Fertilizer Europe, Brussels, Belgium, Personal Communication.

IFDC (International Fertilizer Development Center). (2007). Africa Fertilizer Summit Proceedings, Special Publication-39, Muscle Shoals, Alabama, USA.

IPCC (Intergovernmental Panel on Climate Change). (2014). Summary for Policymakers. In *Climate Change 2014, Mitigation*, Edenhofer O, Pichs-Madruga R, Sokona Y, Farahani E, Kadner S, Seyboth K, Adler A, Baum I, Brunner S, Eickemeier P, Kriemann B, Savolainen J, Schlömer S, von Stechow C, Zwickel T, and Minx JC (Eds.). Contribution of Working Group III to the Fifth Assessment Report of the Intergovernmental Panel on Climate Change. Cambridge University Press, Cambridge, UK, and New York.

Mangat GS, and Narang JK. (2004). Agronomical trial for efficacy of NFL neem-coated urea. *Fertiliser Marketing News* 35(11): 1–3, 5, 7.

Reuters. (2015). China targets zero growth in chemical fertilizer use in 2020. http://www.producer.com/daily/china-targets-zero-growth-in-chemical-fertilizer-use-in-2020/.

Richards MB, Butterbach-Bahl K, Jat ML, Lipinski B, Ortiz-Monasterio I, and Sapkota T. (2015). Site-specific nutrient management: Implementation guidance for policymakers and investors. Global Alliance for Climate-Smart Agriculture.

Roy AH. (2015). Global fertilizer industry: Transitioning from volume to value. The 29th Francis New Memorial Lecture. Proceedings 769. International Fertilizer Society, London, UK.

Roy AH. (2018). Formulations of slow release fertilizers for enhancing N use efficiency. In *Soil Nitrogen Uses and Environmental Impacts*, Lal R, Stewart BA, and Babst BA (Eds.). CRC Press, Boca Raton, FL.

Roy AH. (2019). The fertilizer dilemma. In *Soil Degradation and Restoration in Africa*, Lal R, and Stewart BA (Eds.). Taylor & Francis Group, Boca Raton, FL.

Savant NK, and Stangel PJ. (1990). Deep placement of urea supergranules in transplanted rice: Principles and practices. *Fertilizer Research* 25, 1–83.

Scholz RW, Roy AH, Band FS, Hellums DT, and Ulrich AE (Eds.). (2014). *Sustainable Phosphorus Management – A Global Transdisciplinary Roadmap*. Springer, Dordrecht, the Netherlands.

Shaviv A. (2005). Controlled Release Fertilizers. *IFA International Workshop on Enhanced-Efficiency Fertilizers*, Frankfurt, Germany. International Fertilizer Association, Paris, France.

Shoji S. (2005). Innovative use of controlled availability fertilizers with high performance for intensive agriculture and environmental conservation. *Science in China Series C: Life Sciences* 48(2): 912–920.

Smaling EMA. (1993). *An Agro-ecological Framework of Integrated Nutrient Management with Special Reference to Kenya*. Wageningen Agricultural University, Wageningen, the Netherlands.

Smil V. (1997). *Cycles of Life*. Scientific American Library, New York.

Stewart W et al. (2005). The contribution of commercial fertilizer nutrients to food production. *Agronomy Journal* 97(1): 2–4.

Syers JK, Johnston AE, and Curtin D. (2008). Efficiency of soil and fertilizer phosphorus use. FAO fertilizer and plant nutrition bulletin 18. Food and Agriculture Organization of the United Nations, Rome, p. 108.

Trenkel ME. (1997). *Improving Fertilizer Use Efficiency: Controlled-Release and Stabilized Fertilizers in Agriculture*. International Fertilizer Association, Paris, France.

UN (United Nations). (2019). *World Population Prospects: The 2019 Revision*. United Nations, New York.

UNIDO (United Nations Industrial Development Organization) and IFDC (International Fertilizer Development Center). (1998). *Fertilizer Manual*. Kluwer Academic Publishers, Dordrecht, the Netherlands.

Van Kauwenbergh SJ. (2010). *World Phosphate Rock Reserves and Resources*. International Fertilizer Development Center Technical Bulletin T-75.

VFRC (Virtual Fertilizer Research Center). (2012). *A Blueprint for Global Food Security*. International Fertilizer Development Center, Muscle Shoals, AL.

Weeks JJ Jr., and Hettiarachchi M. (2019). A review of the latest in phosphorus fertilizer technology: Possibilities and pragmatism. *Journal of Environmental Quality* 48: 1300–1313.

Wendt J, Senior Scientist, International Fertilizer Development Center (IFDC), (2017). Personal Communications.

Zhang X, Davidson E, Mauzerall D, Searchinger T, Dumas P, and Shen Y. (2015). Managing nitrogen for sustainable development. http://www.nature.com/doifinder/10.1038/nature 15743

14 Managing Soils for Reducing Dependence on Chemicals and Import of Resources into Agroecosystems

Rattan Lal

CONTENTS

14.1 INTRODUCTION

14.1.1 THE SOIL–CLIMATE–VEGETATION NEXUS

The dynamic interaction between soil biota and the climate (Jenny 1941) is the basis of soil functions for humans and for nature (Figure 14.1). Drastic perturbations of the soil-biota-climate nexus, by natural phenomena or anthropogenic activities, can adversely impact the provisioning of essential ecosystem services and also lead to some disservices (e.g., accelerated erosion, disruption in elemental and water cycling, energy imbalance, and shifts in flora and fauna). Perturbation of the nexus in agroecosystems is also the cause of the high import of resources such as fertilizers, pesticides, tillage, irrigation, etc. Such ameliorative measures aggravate the environmental impact and footprint of agroecosystems. Among the symptoms of drastic perturbations are the drought–flood syndrome, eutrophication and contamination of waters (e.g., algal bloom, hypoxia/anoxia of coastal waters), desertification, heat waves, and extreme events. Anthropogenic activities are also impacting the carbon (C) cycle and the attendant global warming (Lal 2004, 2010; Jackson et al. 2017; Chapin III et al. 2009), the water cycle and the associated pedological and hydrological drought (Lal 2013), and hypoxia (Zillen et al. 2008). Therefore, a judicious and prudential management of the nexus (Figure 14.1) can minimize the need for the import of resources and enhance sustainability.

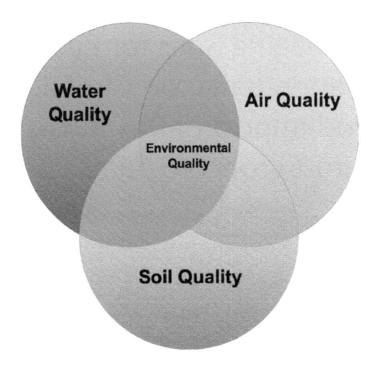

FIGURE 14.1 The soil–biota–climate nexus is the basis for provisioning of essential ecosystem services for nature including humans.

Global fertilizer use has increased drastically since the 1960s and is still increasing. Total annual average fertilizer use between 2015 and 2020 is about 200 million Mg and increasing (Table 14.1). The use of chemical fertilizer since 1960 is considered responsible for feeding a large proportion of the world population. The number of people (billions) fed by fertilizer is estimated at 0.4 in 1960, 0.9 in 1970, 1.34 in 1980, 2.13 in 1990, 2.70 in 2000, and 3.54 in 2015 (Table 14.2; Richie 2017). However, the environmental impact is very high. The global average ratio of $N:P_2O_5:K_2O$ is 3.3:1.3:1.0, and has a decreasing trend over time. However, the ratio is more skewed in favor of N in some emerging economies (i.e., India) because of high subsidies for N. The ratio should be

TABLE 14.1
Global Fertilizer Nutrient Demand

	Global Demand (10^6 Mg)				Ratio
Year	N	P_2O_5	K_2O	Total	N:P:K
2015	110.0	41.2	32.8	184.0	3.4:1.3:1.0
2016	111.6	41.9	33.1	186.7	3.4:1.3:1.0
2017	113.6	43.2	34.0	190.9	3.3:1.3:1.0
2018	115.4	44.1	34.9	194.4	3.3:1.3:1.0
2019	117.1	45.0	36.0	198.1	3.3:1.3:1.0
2020	118.8	45.9	37.0	201.7	3.2:1.2:1.0
Average	114.4	43.6	34.6	192.6	3.3:1.3:1.0

Source: FAO, World Fertilizer Trends and Outlook to 2020: Summary Report, FAO, Rome, Italy, 14 pp, 2017.

TABLE 14.2

World Population Supported with and without Fertilizer

Year	World Population (billion) Supported		
	Total	Without Fertilizer	With Fertilizer
1901	1.66	1.66	0.0
1960	3.04	2.64	0.4
1970	3.70	2.88	0.9
1980	4.46	3.12	1.34
1990	5.33	3.20	2.13
2000	6.15	3.44	2.70
2015	7.38	3.84	3.54

Source: Richie, H., How Many People Does Synthetic Fertilizer Feed?, https://ourworldindata.org/how-many-people-does-synthetic-fertilizer-feed, 2017.

about 4:2:1 for input of $N:P_2O_5:K_2O$. Whereas prudential and supplemental use of chemical fertilizers, supplemental irrigation, and tillage are justified as ameliorative strategies, indiscriminate and excessive/unbalanced use of inputs (i.e., pesticides, fertilizers, flood-based irrigation) can lead to disruption of the nexus and the attendant degradation of the environment. In this regard, management of the soil surface is critical to sustainable use of agroecosystems (Wassenaar et al. 2006; Amundson et al. 2015; Zhao et al. 2019). Management of soil C concentration and stock is essential to several of these services and benefits (Lal 2004, 2010, 2018b; Milne et al. 2015).

14.1.2 Soil Quality and Functionality

Soil quality is indicative of its capacity to sustain the nexus (Figure 14.1) and generate ecosystem services in perpetuity. There are some key soil properties (Zhao et al. 2019) with a strong control on soil quality and functionality (Figure 14.2). For example, concentration of soil organic carbon (SOC) affects soil bulk density (Ruehlman and Körshens 2009), soil erodibility (Wang et al. 2013; Wischmeier and Manneing 1989), plant-available water capacity (Hudson 1994), microbial processes (Wieder et al. 2013), and storage of atmospheric carbon dioxide (Lal 2018a; Lal et al. 2018; Wiesmeier et al. 2019). Identification and management of these key soil properties is critical not only to achieving sustainable agriculture but also to advancing sustainable development goals (SDGs) or the Agenda 2030 of the United Nations (Lal 2018b). Excessive, indiscriminate, and unbalanced use of reactive nitrogen (N, chemical fertilizer) has created environmental issues (Sebilo et al. 2013) and jeopardized the safe operating space (Rockström et al. 2009). The perturbation of the N cycle by the use of chemical fertilizers and leguminous crops including green manure (Galloway et al. 2003, 2004) has also perturbed the global C cycle because of their coupled cycling (Lal 2010) and possibility of increased mineralization of soil organic matter or SOM (Khan et al. 2007; Mulvaney et al. 2009).

14.1.3 Sustainability of Agroecosystems

Sustainability is a multidimensional attribute (Figure 14.3) (Lal et al. 2016b). It is composed of environmental, social, economic, and institutional components. Management of the soil-biota-climate

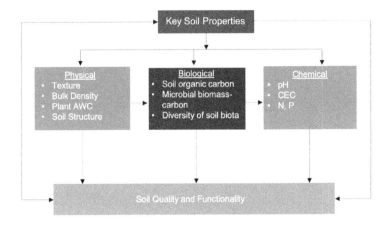

FIGURE 14.2 Ten key soil properties that impact soil quality and functionality.

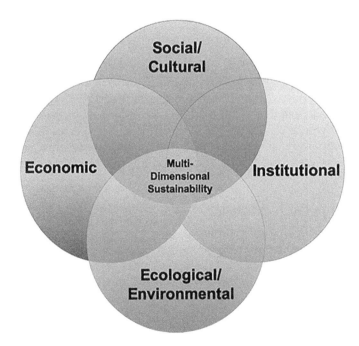

FIGURE 14.3 Components of the multidimensional sustainability.

nexus (Figure 14.1) is also critical to advancing multidimensional sustainability. All components interact with one another (Figure 14.3), and are integral to advancing multidimensional sustainability. Indeed, environmental/ecological sustainability, the key component of the whole (Figure 14.3), depends on other components (i.e., social/cultural, economic, and institutional). The economic aspect, often assessed on a short-term basis, needs an objective consideration. The long-term sustainability of agroecosystems depends on the economic profitability for nature rather than just for humans, and is a long-term issue over centennial and millennial scales rather than on an annual scale.

Therefore, the objective of this chapter is to synthesize the material presented in the previous 13 chapters and outline some research and development priorities so that dependence on chemical fertilizers and the environmental footprint of agroecosystems are minimized.

14.2 STRATEGIES TO REDUCE USE OF CHEMICAL FERTILIZERS AND MINIMIZE THE ENVIRONMENTAL IMPACT

Rather than increasing the rate of fertilizer input, it is incumbent upon soil/plant researchers and fertilizer manufacturers to identify and implement strategies that enhance the use efficiency of fertilizers and decrease losses and their transfer into the environment. Such strategies can be grouped into three categories (Figure 14.4), which depend on site-specific conditions including soil type, climate, farming system, and socioeconomic and cultural factors (Figure 14.5).

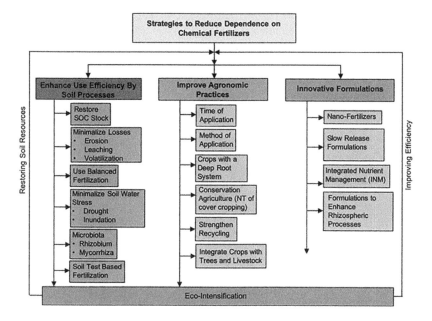

FIGURE 14.4   Strategies to reduce the input of chemical fertilizers and minimize the environmental footprint.

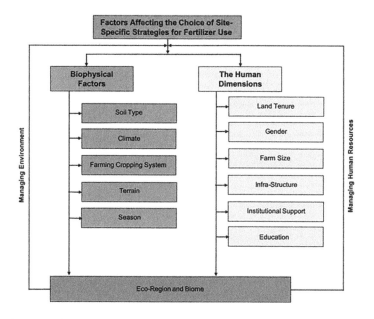

FIGURE 14.5   Controls of fertilizer management options.

14.3   STRATEGIES TO REDUCE DEPENDENCE ON CHEMICAL FERTILIZERS

It is important that nutrients harvested in farm produce are replaced in one form or another. Soils of agroecosystems must not have a negative nutrient budget. A large quantity of plant nutrients (both macro and micro) are removed in grains (Table 14.3), stover/straw (Table 14.4), and also in animal and forest products. The amount of nutrients removed is more in a high-productivity system compared with that of a low-productivity system. Since the 1960s, harvested nutrients have mostly been replaced by chemical/synthetic fertilizers such as highly reactive nitrogenous formulations. Leakage of reactive N along with that of phosphorus (P) into aquatic ecosystems has created hypoxia, algal bloom, and numerous other environmental hazards. Similarly, transfer of reactive N into the atmosphere as nitrous oxide (N_2O) has exacerbated global warming. Thus, the strategy is to identify alternatives to chemical/synthetic fertilizers for replacing the harvested nutrients.

Three determinants of alternate strategies of replacing the harvested nutrients include those that (1) enhance use efficiency, (2) improve agronomic performance, and (3) use innovative formulations of fertilizers (Figure 14.4). All three strategies depend on soil quality and functionality. Specifically, the use efficiency of inherent and imported plant nutrients is governed by a range of factors that impact soil quality and functionality, such as SOC concentration and stock in the rooting depth. Restoration of soil quality and functionality through increase in SOC concentration to the critical threshold level (~2%) can minimize losses by accelerated erosion, surface runoff and leaching, and volatilization. Further, nutrient use efficiency can also be enhanced by a balanced application of fertilizers along with that of biomass-C (i.e., CNPK rather than NPK). Yield-based and soil-test-based application of fertilizers is needed to develop a strategy for the balanced application of fertilizers. The use efficiency of nutrients also depends on microbial activity, especially that of mycorrhiza and

TABLE 14.3
Estimates of Macronutrients Harvested in Grains

	Grain	N	P_2O_5	K_2O	Ca	Mg	S	Total
		Nutrients Harvested (Kg/Mg)						
I	**Cereals**							
	Barley	18.2	7.8	5.2	0.5	1.0	1.6	34.3
	Corn	13.8	7.5	5.5	0.2	1.4	1.2	29.6
	Oats	19.5	7.8	5.9	0.8	1.2	2.0	37.2
	Rye	20.9	6.0	6.0	1.2	1.8	4.2	40.1
	Sorghum	14.9	7.4	4.5	1.2	1.5	1.5	31.0
	Wheat	20.8	10.4	6.3	0.2	2.5	1.3	41.5
	Average	18.0	7.8	5.6	0.7	1.6	2.0	35.6
II	**Pulses**							
	Chickpea	60.7	9.2	39.2	8.7	18.7	7.3	143.8
	Lentil	57.0	14.9	21.6	3.5	7.0	2.0	106.0
	Pigeon pea	70.8	15.3	16.0	7.5	19.2	12.5	141.3
	Soybeans	70.7	30.9	57.7	6.7	14.0	7.6	187.6
	Average	64.8	17.6	33.6	6.6	14.7	7.4	144.7

Source: FAO, Current World Fertilizer Trends and Outlook to 2016, FAO, Rome, Italy, 2012; Bender, R., *Nutrient Uptake and Partitioning in High-Yielding Corn*, University of Illinois at Urbana-Champaign, Urbana, IL, 2012.

TABLE 14.4
Nutrient Contents in Crop Residues

	Crop Residue	Grain: Straw Ratio	Concentration on Dry Mass Basis (%)		
			N	P_2O_5	K_2O
I	**Cereals**				
	Corn stover	1: 1.5	0.59	0.31	1.31
	Millet stock	1: 2.0	0.65	0.75	2.50
	Rice straw	1: 1.5	0.58	0.23	1.66
	Sorghum stalk	1: 2.0	0.40	0.23	2.17
	Wheat straw	1: 1.5	0.49	0.25	1.28
II	**Legumes**				
	Chickpea	1: 1.0	1.19	–	1.25
	Pigeon pea	1: 2.5	1.10	0.58	1.28
	Pulses (average)	1: 1.0	1.60	0.15	2.0
III	**Sugarcane**				
	Trash	1: 0.2	0.35	0.04	0.50

Source: FAO, Sources of Plant Nutrients and Soil Amendments, in *Plant Nutrition for Food Security: A Guide for Integrated Nutrient Management*, FAO, Rome, Italy, pp. 91–235, 2006.

rhizobium (see Section 14.5.2). Maintaining an optimal soil moisture regime by minimizing the drought–flood syndrome is also critical to achieving the dual goal of enhancing use efficiency of both water and nutrients (e.g., N, P).

Management of soil quality requires adoption of improved practices of soil, crop, animal, and tree management through a judicious integration of crops with trees, livestock, forages, and cover crops. The strategy is to strengthen cycling of C and plant nutrients so that these elements are neither transported out of the ecosystem (by erosion, leaching, or volatilization) nor released into the environment where they can be pollutants.

Then, there is a possibility of improved formulations of fertilizers so that use efficiency is maximized and losses into the environment (water, atmosphere) are minimized. Examples of some such formulations include nanofertilizer (Liu and Lal 2015, 2017; Liu et al. 2016) and slow-release formulations of highly soluble nitrogenous fertilizers (Staufenbeil 2018; Silva 2011).

14.3.1 FACTORS AFFECTING THE CHOICE OF SITE-SPECIFIC FERTILIZER OPTIONS

Indiscriminate use of chemical fertilizers can have severe environmental consequences in relation to eutrophication of water and pollution of air, among others. The dependence of fertilizers should be minimized, and these must be used as amendments following scientific guidelines (e.g., soil test, crop requirement). Factors affecting the choice of site-specific strategies for the use of fertilizers include biophysical, socioeconomic, cultural, and others related to the human dimensions. Important among biophysical parameters are soil type, climate, farming/cropping system, terrain, season, and weather patterns during the season. Pertinent issues of the human dimensions, especially in developing countries with predominantly small landholders, are infrastructure (e.g., roads, access to market, credit availability), institutional support (e.g., extension services, IT communication), and education and technical skills of the farming community (Figure 14.5). These factors vary among ecoregions and biome and have strong implications for the overall goal of judicious management of the environment and of human resources (Figure 14.5).

14.4 ADDRESSING THE FERTILIZER ADDICTION

14.4.1 GLOBAL FOOD DEMAND BY 2050

The projected increase in global agricultural production of 60% by 2050 compared with that of 2005–07 (FAO 2012) necessitates a closer look at the role of fertilizers in enhancing and sustaining food production. Excessive and indiscriminate use of chemical fertilizers can have severe environmental and sustainability concerns and can also adversely affect soil quality (Kotschi 2013; Mulvaney et al. 2009). Thus, site-specific options need to be considered to minimize dependence on the use of chemical fertilizers. Pedercini et al. (2014, 2015) suggested sustainable management of SOM as one of the options because increasing SOM can enhance soil and crop resilience.

14.4.2 MANAGING SOIL ORGANIC MATTER AND SOIL QUALITY

Restoration of SOM in depleted and degraded soils can improve soil quality and functionality and decrease the rate of fertilizer application. Some of the options for reducing dependence on fertilizers by managing SOM (Figure 14.6) include system-based conservation agriculture (CA) done in conjunction with no-till (NT) residue retention, cover cropping with appropriate legume and grass species or a mixture of species that enhance biodiversity, integrated systems of nutrient and pest management, and a close integration of crops in association with trees and livestock. In-field burning of crop residues must be avoided (Shyamsundar et al. 2019). Use of compost and increasing SOC contents can enhance the disease-suppressive attributes of a soil through

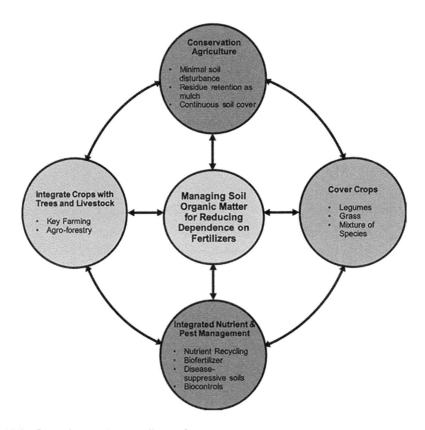

FIGURE 14.6 Strategies to enhance soil organic matter content.

improvements in soil health. Although there are some niches for the establishment of biofuel plantations on degraded and drastically disturbed lands, bioenergy is not always a feasible option (Stokstad 2019).

14.5 GENERAL CONCLUSIONS

14.5.1 MEETING GLOBAL FOOD DEMAND BY 2050

Meeting global food demand by 2050 and beyond does not necessarily imply increasing land area under agriculture, fertilizer and pesticide use, water use for irrigation, intensity and frequency of tillage, and other energy-based inputs. Resources used for agriculture are already more than what is needed, and the amount of food already produced is adequate to feed the world population of ~10 billion. Thus, the first priority is to reduce food waste, which may be as much as 30% of the total amount produced. The next step is to improve access and distribution by addressing poverty, civil strife, political unrest, and climate/soil refugees. Any justifiable allocation of additional resources (e.g., land water, fertilizers, pesticides, tillage) must be critically considered and objectively assessed. By abridging the yield gap and enhancing the use efficiency of fertilizers, water, and other inputs, some of the hitherto allocated land resources must be returned back to nature (Lal 2016, 2018b). Furthermore, priority must be given to plant-based diet for the growingly and increasingly affluent population (Lal 2017; IPCC 2019).

14.5.2 WAY FORWARD

The answer to the question "Are fertilizers needed?" is not in the form of either/or. With the world population of 7.7 billion in 2019 and expected to reach 8.5 billion by 2030 and 9.7 billion by 2050, all options are on the table. The strategy is for the judicious, discriminate, and targeted use of fertilizers as determined by the desired agronomic yield and the soil test for each specific soil/crop/ecosystem. Poisoning of soil by excessive, over, and indiscriminate use of chemicals is not acceptable and must be avoided. Further, plant nutrients contained in soil may not be adequate to obtain the desired agronomic yield. Thus, the inherent supply must be supplemented with external imports for obtaining the desired agronomic yield. The external input of plant nutrients for improving crop growth and enhancing nutritional quality can come from diverse sources: natural, recycled by-products, microbial inoculants, and synthetic or chemical formulations (solid or liquid). Most sources of plant nutrients can be grouped as biological, organic, and mineral. The strategy of enhancing soil fertility by the use of diverse sources of amendment is called "integrated nutrient management" (INM) or integrated soil fertility management (ISFM). The latter also includes recycling of plant nutrients contained in the remains of plants and animals and by-products of their processing, nitrogen-fixing biofertilizers, mineral materials such as rock phosphate, and Chilean saltpeter or guano.

There is also a wide range of nitrogen-fixing biofertilizers based on a wide range of organisms including bacteria (Rhizobian, Azotobacter, Azospirildum, Acetobacter), algae (Blue Green Algae composed of Nostoc Anabaena, Aulosira, Tolypothrix, and Calothrix used in rice paddies), and a fern (Azolla) (FAO 2006). There are also phosphate-solubilizing biofertilizers including bacteria (*Bacillus megatherium*), fungus (*Aspergillus* sp, *Penicillium* sp), yeast (*Saccharomyces* sp) and actinomycetes (*Streptomyces* sp). In addition, there are also nutrient-mobilizing biofertilizers composed of soil fungi mycorrhizal (ecto- and endomycorrhiza). Amendments are used to alleviate soil-related constraints such as low pH by liming, high pH by acidification ($CaSO_4 \, 2H_2O$ or sypsum), coarse texture (by adding clay), fine texture (by adding sand), and soil structure by introducing conditioners (e.g., organic or polymers and inorganic compounds) (FAO 2006). The strategy of INM and ISFM is to minimize dependence on the use of synthetic (mineral) formulations (i.e., ammonium sulfate, muriate of potash, ammonium nitrate, urea). The application of balanced plant nutrients is important because an excess or deficiency of one can impact the availability or efficient use of another.

REFERENCES

Amundson, R., A. Berhe, J. Hopmans, C. Olson, A. Ester Sztein, and D. L. Sparks. 2015. Soil Science: Soil and Human Security in the 21st Century. *Science* 348(6235): 647–655.

Bender, R. 2012. *Nutrient Uptake and Partitioning in High-Yielding Corn*. University of Illinois at Urbana-Champaign, Urbana, IL.

Chapin, S. III, F. J. McFarland, A. David McGuire, E. S. Euskirchen, R. W. Ruess, and K. Kielland. 2009. The Changing Global Carbon Cycle: Linking Plant–Soil Carbon Dynamics to Global Consequences. *Journal of Ecology* 97(5): 840–850.

FAO (Food and Agriculture Organization of the United Nations). 2006. Sources of Plant Nutrients and Soil Amendments. In *Plant Nutrition for Food Security: A Guide for Integrated Nutrient Management*. FAO, Rome, Italy, pp. 91–235.

FAO (Food and Agriculture Organization of the United Nations). 2012. *Current World Fertilizer Trends and Outlook to 2016*. FAO, Rome, Italy.

FAO (Food and Agriculture Organization of the United Nations). 2017. *World Fertilizer Trends and Outlook to 2020: Summary Report*. FAO, Rome, Italy.

Galloway, J., J. D. Aber, J. W. Erisman, S. Seitzinger, R. Howarth, E. B. Cowling, and B. Cosby Jr. 2003. The Nitrogen Cascade. *Bioscience* 53: 341.

Galloway, J. N., F. J. Dentener, D. G. Capone, et al. 2004. Nitrogen Cycles: Past, Present, and Future. *Biogeochemistry* 70(2): 153–226.

Hudson, B. D. 1994. Soil Organic Matter and Available Water Capacity. *Journal of Soil and Water Conservation* 49: 189–194.

IPCC (Intergovernmental Panel on Climate Change). 2019. *Climate Change and Land*. An IPCC special report on climate change, desertification, land degradation, sustainable land management, food security and greenhouse gas fluxes in terrestrial ecosystems. Intergovernmental Panel on Climate Change, WHO, Geneva, Switzerland.

Jackson, R. B., K. Lajtha, S. Crow, G. Hugelius, M. Kramer, and G. Piñeiro. 2017. The Ecology of Soil Carbon: Pools, Vulnerabilities, and Biotic and Abiotic Controls. *Annual Review of Ecology, Evolution, and Systematics* 48: 419–445.

Jenny, H. 1941. *Factors of Soil Formation: A System of Quantitative Pedology*. McGraw-Hill, New York.

Khan, S., R. L. Mulvaney, T. Ellsworth, and C. Boast. 2007. The Myth of Nitrogen Fertilization for Soil Carbon Sequestration. *Journal of Environmental Quality* 36: 1821–1832.

Kotschi, J. 2013. *A Soiled Reputation: Adverse Impacts of Mineral Fertilizers in Tropical Agriculture*. AGRECOL – Association for AgriCulture and Ecology, Heinrich Böll Foundation, Berlin, German, and WWF Germany, Berlin, Germany.

Lal, R. 2004. Soil Carbon Sequestration Impacts on Global Climate Change and Food Security. *Science* 304: 1623–1627. www.sciencemag.org/cgi/content/full/304/5677/1623/d.

Lal, R. 2010. Managing Soils and Ecosystems for Mitigating Anthropogenic Carbon Emissions and Advancing Global Food Security. *BioScience* 60(9): 708–721.

Lal, R. 2013. The Nexus Approach to Managing Water, Soil and Waste under Changing Climate and Growing Demands on Natural Resources. In Hülsmann, S., and R. Ardakanian (Eds.), *White Book: Advancing a Nexus Approach to the Sustainable Management of Water, Soil and Waste*. UNU-FLORES, Dresden, Germany, pp. 19–40.

Lal, R. 2016. Degradation: Quality. In Lal, R. (Ed.), *Encyclopedia of Soil Science*. 3rd ed. Taylor & Francis Group, Boca Raton, FL, pp. 602–607.

Lal, R. 2017. Improving soil health and human protein nutrition by pulse-based cropping system. *Advances in Agronomy* 145:167–204.

Lal, R. 2018a. Accelerated Soil Erosion as a Source of Atmospheric CO_2. *Soil & Tillage Research* 188: 35–40.

Lal, R. 2018b. Sustainable Development Goals and the IUSS. In Lal, R., R. Horn, and T. Kosaki (Eds.), *Soil and Sustainable Development Goals*. Catena Schweizerbart, Stuttgart, Germany, pp. 189–196.

Lal, R., D. Kraybill, D. O. Hansen, B. R. Singh, and L. O. Eik. 2016a. Research and Development Priorities. In Lal, R., D. Kraybill, D. O. Hansen, B. R. Singh, T. Mosogoya, and L. O. Eik (Eds.), *Climate Change and Multi-dimensional Sustainability in African Agriculture*. Springer, Cham, Switzerland, pp. 679–698.

Lal, R., D. Kraybill, D. O. Hansen, B. R. Singh, T. Mosogoya, and L. O. Eik. (Eds.). 2016b. *Climate Change and Multi-dimensional Sustainability in African Agriculture*. Springer, Cham, Switzerland.

Lal, R., P. Smith, H. F. Jungkunst, W. Mitsch et al. 2018. The Carbon Sequestration Potential of Terrestrial Ecosystems. *Journal of Soil and Water Conservation* 73: 145A–152A.

Liu, R., and R. Lal. 2015. Effects of Low Level Aqueous Hydrogen Sulfide and Other Sulfur Species on Lettuce (*Lactuca sativa*) Seed Germination. *Communications in Soil Science and Plant Analysis* 46(5): 576–587.

Liu, R., and R. Lal. 2017. Enhancing Efficiency of Phosphorus Fertilizers through Formula Modifications. In Lal, R., and B. A. Stewart (Eds.), *Soil Phosphorus*. Advances in Soil Science. Taylor & Francis Group, Boca Raton, FL, pp. 225–246.

Liu, R., H. Zhang, and R. Lal. 2016. Effects of Stabilized Nanoparticles of Copper, Zinc, Manganese, and Iron Oxides in Low Concentrations on Lettuce (*Lactuca sativa*) Seed Germination: Nanotoxicants or Nanonutrients? *Water, Air and Soil Pollution* 227: 42.

Milne, E., S. A. Banwart, E. Noellemeyer et al. 2015. Soil Carbon, Multiple Benefits. *Environmental Development* 13: 33–38.

Mulvaney, R. L., S. A. Khan, and T. R. Ellsworth. 2009. Synthetic Nitrogen Fertilizers Deplete Soil Nitrogen: A Global Dilemma for Sustainable Cereal Production. *Journal of Environmental Quality* 38: 2295–2314.

Pedercini, M., G. Zullich, and K. Dianati. 2014. Addicted to Fertilizer – How Soil Mismanagement Makes Our Soils Increasingly Vulnerable. In *Proceedings of the International System Dynamics Society*, Delft, the Netherlands.

Pedercini, M., G. Zullich, and K. Dianati. 2015. *Fertilizer Addiction: Implications for Sustainable Agriculture*. Global Sustainable Development Report 2015 Brief, pp. 1–4.

Richie, H. 2017. How Many People Does Synthetic Fertilizer Feed? https://ourworldindata.org/how-many-people-does-synthetic-fertilizer-feed.

Rockström, J., W. Steffen, K. Noone, et al. 2009. A Safe Operating Space for Humanity. *Nature* 461: 472–475.

Ruehlman, J., and M. Körshens. 2009. Calculating the Effect of Soil Organic Matter Concentration on Soil Bulk Density. *Soil Science Society of America* 73: 876–885.

Sebilo, M., B. Mayer, B. Nicolardot, G. Pinay, and A. Mariotti. 2013. Long-Term Fate of Nitrate Fertilizer in Agricultural Soils. *Proceedings of the National Academy of Sciences* 110(45): 18185–18189.

Shyamsundar, P., N. P. Springer, H. Tallis, et al. 2019. Fields on Fire: Alternatives to Crop Residue Burning in India. *Science* 365(6453): 536–538. https://science.sciencemag.org/content/365/6453/536.

Silva, G. 2011. *Slow Release N Fertilizers*. Michigan State University Extension, East Lansing, MI.

Staufenbeil, J. 2018. Slow-Release vs. Quick Release Fertilizers. https://www.milogranite.com/blog/Lawn/slow-release-quick-release-fertilizer.

Stokstad, E. 2019. Bioenergy Not a Climate Cure-All, Panel Warns. *Science* 365(6453): 527–528. https://science.sciencemag.org/content/365/6453/527.

Wang, B., F. Zheng, M. J. M. Römkens, and F. Darboux. 2013. Soil Erodibility for Water Erosion: A Perspective and Chinese Experiences. *Geomorphology* 187: 1–10.

Wassenaar, L., M. Hendry, and N. Harrington. 2006. Decadal Geochemical and Isotopic Trends for Nitrate in a Transboundary Aquifer and Implications for Agricultural Beneficial Management Practices. *Environmental Science & Technology* 40: 4626–4632.

Wieder, W. R., G. B. Bonan, and S. D. Allison. 2013. Global Soil Carbon Projections Are Improved by Modelling Microbial Processes. *Nature Climate Change* 3: 909–912.

Wiesmeier, M., L. Urbanski, E. Hobley, et al. 2019. Soil Organic Carbon Storage as a Key Function of Soils – A Review of Drivers and Indicators at Various Scales. *Geoderma* 333: 149–162.

Wischmeier, W. H., and J. V. Manneing. 1989. Relation of Soil Properties to Its Erodibility. *Soil Science Society of America Proceedings* 33: 131–137.

Zhao, X., Y. Yang, H. Shen, X. Geng, and J. Fang. 2019. Global Soil-Climate-Biome Diagram: Linking Surface Soil Properties to Climate and Biota. *Biogeosciences Discussions* 16: 2857–2871.

Zillen, L., D. J. Couley, T. Andren, and S. Bjorck. 2008. Past Occurrences of Hypoxia in the Baltic Sea and the Role of Climate Variability: Environment Change and Human Influence. *Earth Science Reviews* 91: 77–92.

Index

Note: Page numbers in italic and bold refer to figures and tables, respectively.

Printed and bound by CPI Group (UK) Ltd, Croydon, CR0 4YY

17/10/2024

01775663-0004